# Advanced Structured Materials

Volume 155

Common engineering materials reach in many applications their limits and new developments are required to fulfil increasing demands on engineering materials. The performance of materials can be increased by combining different materials to achieve better properties than a single constituent or by shaping the material or constituents in a specific structure. The interaction between material and structure may arise on different length scales, such as micro-, meso- or macroscale, and offers possible applications in quite diverse fields.

This book series addresses the fundamental relationship between materials and their structure on the overall properties (e.g. mechanical, thermal, chemical or magnetic etc.) and applications.

The topics of *Advanced Structured Materials* include but are not limited to

- classical fibre-reinforced composites (e.g. glass, carbon or Aramid reinforced plastics)
- metal matrix composites (MMCs)
- micro porous composites
- micro channel materials
- multilayered materials
- cellular materials (e.g., metallic or polymer foams, sponges, hollow sphere structures)
- porous materials
- truss structures
- nanocomposite materials
- biomaterials
- nanoporous metals
- concrete
- coated materials
- smart materials

Advanced Structured Materials is indexed in Google Scholar and Scopus.

More information about this series at https://link.springer.com/bookseries/8611

Holm Altenbach · Victor A. Eremeyev ·
Alexander Galybin · Andrey Vasiliev
Editors

# Advanced Materials Modelling for Mechanical, Medical and Biological Applications

Springer

*Editors*
Holm Altenbach
Institut für Mechanik, Fakultät für Maschinenbau
Otto-von-Guericke-Universität Magdeburg
Magdeburg, Germany

Alexander Galybin
Schmidt Institute of Physics of the Earth
Moscow, Russia

Victor A. Eremeyev
DICAAR
Università degli Studi di Cagliari
Cagliari, Italy

Andrey Vasiliev
Don State Technical University
Rostov-on-Don, Russia

ISSN 1869-8433 ISSN 1869-8441 (electronic)
Advanced Structured Materials
ISBN 978-3-030-81704-6 ISBN 978-3-030-81705-3 (eBook)
https://doi.org/10.1007/978-3-030-81705-3



This Springer imprint is published by the registered company Springer Nature Switzerland AG
The registered company address is: Gewerbestrasse 11, 6330 Cham, Switzerland

# Preface

**Sergey Aizikovich** was born in a family of a navy officer on the 2nd of August, 1951, in Petropavlovsk–Kamchatskiy, the biggest city in Kamchatka, a land of volcanoes in the Far East of Russia. Being in a school, he met his future colleagues and friends who directed his future carrier. He graduated from Rostov State University in 1973. The diploma was written under supervision of Iosif Vorovich and Leonid Lebedev. From 1973, he has worked in the Scientific Institute of Mechanics and Applied Mathematics of Rostov State University. Under supervision of Victor Aleksandrov in 1980, he defended his Ph.D. thesis (candidate of physical–mathematical sciences) on "Asymptotic methods of solution of contact problems for inhomogeneous in depth elastic half-space and half-plane" and then in 2003 his habilitation (doctoral dissertation) entitled "Contact problems of elasticity for non-homogeneous media".

Sergey Aizikovich: Tilicho Lake near Annapurna, Nepal, 7.04.2013

a) b)

Sergey Aizikovich: (a) in Sun City, South Africa, 10.01.2009; (b) near Dombay, Caucasus, 13.06.2015

From 2009, Sergey Aizikovich is working in the Don State Technical University, Rostov, on Don. Nowadays, he is the head of the laboratory of functionally graded and composite materials of Research and Education Center "Materials." His scientific interests relate to the following topics:

- Mechanics of contact interactions;
- Functionally graded materials;
- Mechanics of fracture;
- Integral equations of mathematical physics;
- Inverse problems of the theory of elasticity;
- Applications of elasticity to biological and medical problems of mechanics of materials.

Sergey Aizikovich, Verona, Italy, 18.10.2015

He has published more than hundred papers in peer-reviewed journals and a few monographs, patents, and registered computer programs. During his scientific carrier, he got awards from the Government of Soviet Union (1983), the Government of the Russian Federation (2015), the Government of the Rostov region (2011, 2020), and the Government of National Committee of Republic of Kazakhstan (2017). He got grants from INTAS, Russian Foundation for Basic Research, Russian Science Foundation, and from the Ministry of Science and Education of Russia. He is a member of the Russian National Committee on Theoretical and Applied Mechanics.

From his student time, he was interested in alpinism. In 1970, he got a recognition "Alpinist of Soviet Union." Nowadays, he is an instructor of second category. He climbed over Mount Elbrus (Western part of the Caucasus) more than 15 times. He climbed over Lenin Peak (Pamir Mountains), over Khan Tengri (Tian Shan mountains), and over Nevado Sajama (extinct stratovolcano and the highest peak in Bolivia). He was nearby Annapurna, Himalayas (2013). In 2018, he climbed over volcano Klyuchevskaya Sopka, Kamchatka, and in 2021, he climbed over Mönkh Saridag (Munku-Sardyk) that is the highest mountain in the Sayan Mountains.

This volume of the Advanced Structured Materials Series is dedicated to the seventieth birthday of our friend and colleague Sergey Aizikovich and contains a selection of scientific papers prepared by scientists from different countries. It is devoted to actual research in mechanics of materials with applications in engineering, biology, and medicine.

Magdeburg, Germany — Holm Altenbach
Gdańsk, Poland/Cagliari, Italy — Victor A. Eremeyev
Moscow, Russia — Alexander Galybin
Rostov-on-Don, Russia — Andrey Vasiliev
June, 2021

# Contents

# Contributors

**Adair R. Aguiar** Departamento de Engenharia de Estruturas, Escola de Engenharia de São Carlos, Universidade de São Paulo, São Carlos, SP, Brazil

**Sergei Alexandrov** Ishlinsky Institute for Problems in Mechanics RAS, 101-1 Prospekt Vernadskogo, Moscow, Russia

**Harutyun Amirjanyan** Institute of Mechanics, National Academy of Sciences of Armenia, Yerevan, Armenia

**Aleksandr Anoshkin** Department of Mechanics of Composite Materials and Structures, Perm National Research Polytechnic University, Perm, Russia

**Vladimir Balandin** Research Institute for Mechanics, National Research Lobachevsky State University of Nizhny Novgorod, Nizhny Novgorod, Russia

**Vladimir Balandin Jr.** Research Institute for Mechanics, National Research Lobachevsky State University of Nizhny Novgorod, Nizhny Novgorod, Russia

**Evgeny Barkanov** Institute of Materials and Structures, Riga Technical University, Riga, Latvia

**Artem V. Basalin** Research Institute for Mechanics, National Research Lobachevsky State University of Nizhny Novgorod, Nizhny Novgorod, Russia

**Svetlana M. Bauer** St. Petersburg State University, St. Petersburg, Russian Federation;
Research and Education Center "Materials", Don State Technical University, Rostov-on-Don, Russian Federation

**Aleksandr A. Belov** National Research Lobachevsky State University of Nizhny Novgorod, Nizhny Novgorod, Russian Federation

**Aleksandr V. Boev** Research Institute for Mechanics, National Research Lobachevsky State University of Nizhny Novgorod, Nizhny Novgorod, Russian Federation

**Anatoly M. Bragov** Research Institute for Mechanics, National Research Lobachevsky State University of Nizhny Novgorod, Nizhny Novgorod, Russia

**Sergei A. Chizhik** Nanoprocesses and Technology Laboratory, A.V. Luikov Heat and Mass Transfer Institute of the National Academy of Science of Belarus, Minsk, Belarus

**Paolo D'Avanzo** Dipartimento di Scienze Fisiche, Università di Napoli "Federico II", Napoli, Italy

**Victor A. Eremeyev** Department of Mechanics of Materials and Structures, Faculty of Civil and Environmental Engineering, Gdansk University of Technology, Gdansk, Poland;
Don State Technical University, Rostov on Don, Russia;
DICAAR, Università degli Studi di Cagliari, Cagliari, Italy

**Andrey R. Filippov** National Research Lobachevsky State University of Nizhny Novgorod, Nizhny Novgorod, Russian Federation

**Sergey Fomenko** Institute for Mathematics, Mechanics and Informatics, Kuban State University, Krasnodar, Russia

**Alexander N. Galybin** The Schmidt Institute of Physics of the Earth (IPE RAS), Moscow, Russia;
Don State Technical University, Rostov-on-Don, Russia

**Adam Gilewicz** Faculty of Mechanical Engineering, Koszalin University of Technology, Koszalin, Poland

**Evgeny Glushkov** Institute for Mathematics, Mechanics and Informatics, Kuban State University, Krasnodar, Russia

**Natalia Glushkova** Institute for Mathematics, Mechanics and Informatics, Kuban State University, Krasnodar, Russia

**Mikhail E. Gonov** National Research Lobachevsky State University of Nizhny Novgorod, Nizhny Novgorod, Russian Federation

**Mikhail V. Grigoryev** Research Institute for Mechanics, National Research Lobachevsky State University of Nizhny Novgorod, Nizhny Novgorod, Russia

**Mikhail Guzev** Institute for Applied Mathematics, Far Eastern Branch of the Russian Academy of Sciences, Vladivostok, Russia;
Department of Oil and Gas Technologies, Perm National Research Polytechnic University, Perm, Russia

**Vahram Hakobyan** Institute of Mechanics, National Academy of Sciences of Armenia, Yerevan, Armenia

**Leonid A. Igumnov** Research Institute for Mechanics, National Research Lobachevsky State University of Nizhny Novgorod, Nizhny Novgorod, Russia

**Andrey B. Kachanov** North-Western State Medical University Named After I.I. Mechnikov, St. Petersburg, Russian Federation;
St. Petersburg Branch of the S.N. Fyodorov Eye Microsurgery Federal State Institution, St. Petersburg, Russian Federation

**Tsai-Wen Ko** National Cheng Kung University, Tainan, Taiwan (R.O.C.)

**Alexey M. Kolesnikov** Southern Federal University, Rostov-na-Donu, Russian Federation;
Don State Technical University, Rostov-na-Donu, Russian Federation

**Aleksandr Yu. Konstantinov** National Research Lobachevsky State University of Nizhny Novgorod, Nizhny Novgorod, Russian Federation

**Alexandr S. Kornievsky** Institute of Mathematics, Mechanics and Computer Science, Southern Federal University, Rostov-on-Don, Russia

**Andrejs Kovalovs** Institute of Materials and Structures, Riga Technical University, Kalku St. 1, Riga, Latvia

**Tatyana A. Kuznetsova** Nanoprocesses and Technology Laboratory, A.V. Luikov Heat and Mass Transfer Institute of the National Academy of Science of Belarus, Minsk, Belarus

**Polina A. Lapina** Research and Education Center "Materials", Don State Technical University, Rostov-on-Don, Russia

**Vasilina A. Lapitskaya** Nanoprocesses and Technology Laboratory, A.V. Luikov Heat and Mass Transfer Institute of the National Academy of Science of Belarus, Minsk, Belarus

**Daria A. Letunova** Southern Federal University, Rostov-na-Donu, Russian Federation

**Andrey K. Lomunov** National Research Lobachevsky State University of Nizhny Novgorod, Nizhny Novgorod, Russian Federation

**Elena Lyamina** Ishlinsky Institute for Problems in Mechanics RAS, 101-1 Prospekt Vernadskogo, Moscow, Russia

**Mohammad Malikan** Department of Mechanics of Materials and Structures, Faculty of Civil and Environmental Engineering, Gdansk University of Technology, Gdansk, Poland

**Ivan P. Markov** Research Institute for Mechanics, National Research Lobachevsky State University of Nizhny Novgorod, Nizhny Novgorod, Russian Federation

**Andrey V. Nasedkin** Institute of Mathematics, Mechanics and Computer Science, Southern Federal University, Rostov-on-Don, Russia

**Rostislav D. Nedin** Institute of Mathematics, Mechanics and Computer Sciences named after I.I. Vorovich, Southern Federal University, Rostov-on-Don, Russia;

Vladikavkaz Scientific Center of Russian Academy of Sciences, Southern Mathematical Institute, Vladikavkaz, Russia

**Vladimir Parkhachev** National Research Lobachevsky State University of Nizhny Novgorod, Nizhny Novgorod, Russia

**Andrey N. Petrov** Research Institute for Mechanics, National Research Lobachevsky State University of Nizhny Novgorod, Nizhny Novgorod, Russian Federation

**Alexander Pirumov** Department of Management Problems, Russian Technological University—MIREA, Moscow, Russia

**Pavel Pisarev** Department of Mechanics of Composite Materials and Structures, Perm National Research Polytechnic University, Perm, Russia

**Edmar B. T. Prado** Departamento de Engenharia de Estruturas, Escola de Engenharia de São Carlos, Universidade de São Paulo, São Carlos, SP, Brazil

**Alessio Ciro Rapisarda** Dipartimento di Scienze Fisiche, Università di Napoli "Federico II", Napoli, Italy

**Evgeniy V. Sadyrin** Laboratory for Mechanics of Biomaterials, Research and Education Center Materials, Don State Technical University, Rostov-on-Don, Russia

**Avetik Sahakyan** Institute of Mechanics, National Academy of Sciences of Armenia, Yerevan, Armenia

**Igor Sevostianov** Department of Mechanical and Aerospace Engineering, New Mexico State University, Las Cruces, NM, USA

**Salvatore Samuele Sirletti** Dipartimento di Scienze Fisiche, Università di Napoli "Federico II", Napoli, Italy;
Institut für Theoretische Physik, Johann Wolfgang Goethe-Universität Frankfurt am Main, Frankfurt am Main, Germany

**Kai-Wen Tan** National Cheng Kung University, Tainan, Taiwan (R.O.C.)

**Alexander O. Vatulyan** Institute of Mathematics, Mechanics and Computer Sciences named after I.I. Vorovich, Southern Federal University, Rostov-on-Don, Russia;
Vladikavkaz Scientific Center of Russian Academy of Sciences, Southern Mathematical Institute, Vladikavkaz, Russia

**Liudmila A. Venatovskaya** St. Petersburg State University, St. Petersburg, Russian Federation

**Eva B. Voronkova** St. Petersburg State University, St. Petersburg, Russian Federation;
Research and Education Center "Materials", Don State Technical University, Rostov-on-Don, Russian Federation

**Yun-Che Wang** National Cheng Kung University, Tainan, Taiwan (R.O.C.)

**Bogdan Warcholinski** Faculty of Mechanical Engineering, Koszalin University of Technology, Koszalin, Poland

**Tomasz Wiczenbach** Department of Mechanics of Materials and Structures, Faculty of Civil and Environmental Engineering, Gdansk University of Technology, Gdansk, Poland

**Tatiana Yuzhnina** National Research Lobachevsky State University of Nizhny Novgorod, Nizhny Novgorod, Russia

**Anna D. Zagrebneva** Research and Education Center "Materials", Don State Technical University, Rostov-on-Don, Russia

**Vladimir B. Zelentsov** Research and Education Center "Materials", Don State Technical University, Rostov-on-Don, Russia

# Chapter 1
# Estimate of Elastic Properties of Biological Tissues Using a Finite Element Methodology

**Adair R. Aguiar and Edmar B.T. Prado**

**Abstract** Experimental observations reveal that abnormal biological tissues, such as tumors, found in the breast and prostate, tend to be stiffer than healthy biological tissues. Classical linear elasticity is then used to model the responses of these tissues under small strains. In particular, soft tissues are modeled as linearly elastic, isotropic, and incompressible materials. For a particular class of plane problems, it has been shown that if the shear elastic modulus, $\mu$, is known at four different points in a sample, then it can be uniquely determined from the knowledge of two displacement fields obtained from two distinct experiments performed on the same sample. We use this result to propose a non-iterative numerical procedure to determine $\mu$ in a sample of soft tissue, which is subjected to two quasi-static experiments that are possible to reproduce in laboratory and are simulated numerically using the finite element method. No a priori knowledge of the shear elastic modulus is required, but it is assumed that resultant forces are known on complementary parts of the boundary of the sample. Results for the distribution of $\mu$ in a long cylinder of rectangular cross section containing an eccentric circular inclusion and an inclusion with a complex geometry are presented. The methodology yields numerical results that are both in very good agreement with analytical results and more accurate than numerical results obtained from another methodology presented in the literature. This work is of great interest in the detection of cancerous tumors and in the differential diagnosis of biological tissues.

**Keywords** Elastography · Hyperbolic equation · Inverse problem · Finite element method

---

A. R. Aguiar (✉) · E. B.T. Prado
Departamento de Engenharia de Estruturas, Escola de Engenharia de São Carlos, Universidade de São Paulo, Av. Trabalhador Sãocarlense, 400, Cx. P. 359, 13560-590 São Carlos, SP, Brazil
e-mail: aguiarar@sc.usp.br

E. B.T. Prado
e-mail: edmarbt@gmail.com

 
H. Altenbach et al. (eds.), *Advanced Materials Modelling for Mechanical, Medical and Biological Applications*, Advanced Structured Materials 155,
https://doi.org/10.1007/978-3-030-81705-3_1

## 1.1 Introduction

According to Fung (2004); Ophir et al. (1991) and Liu et al. (2003), experimental observations reveal that abnormal biological tissues respond differently to mechanical loads than normal biological tissues. In particular, Sarvazyan (1993) and Krouskop et al. (1998) observe that the abnormal tissues are stiffer than the normal tissues. This different mechanical response is widely used in the diagnosis of breast cancer. Health professionals touch the patient's breast to determine whether a lesion is present in the breast. Similar procedure is also used to diagnose prostate cancer. In several cases, however, the lesion may not be detected due to its reduced size or due to its localization deep inside the body.

A technique in medical physics to detect the presence of tumors consists of pressing lightly an ultrasonic device on the external surface of the human body, which in turn deforms the part of the body located near this surface. Mridha and Ödman (1986) mention that the deformation must be small, less than 2.5 %, in order to ignore the viscous behavior of the biological tissue and to treat the tissue as a linear elastic solid. According to Ophir et al. (2001), here too, the presence of lesions in the interior of the body may not be detected with conventional ultrasound exams.

The determination of elastic properties of biological tissues from experimental data has been the subject of intense investigation. Park and Maniatty (2006) present a direct inversion approach for reconstructing the shear elastic modulus $\mu$ in a nearly incompressible, isotropic, and linearly elastic solid from dynamic measurements of the interior displacement field during time-harmonic excitation. The authors employ a finite element methodology to find approximate solutions to the governing equations, which does not use boundary conditions. The authors use simulated data with and without noise on a two-dimensional domain to validate the resulting algorithm. This algorithm is then used to reconstruct $\mu$ with magnetic resonance measured experimental data involving a tissue-mimicking phantom.

Searching for ways to systematize the manual breast exam procedure, Olson and Throne (2010) have proposed an inverse technique to infer the elastic modulus throughout the breast tissue based on measured displacements on the external surface of a body. These measured displacements were obtained from the numerical solution of a forward two-dimensional problem using the finite element method (FEM). For this, the authors have used known loads on the body surface and a distribution of tumors inside the body. Numerical noise was added to this solution to simulate measurement errors. The associated inverse problem was solved using a genetic algorithm. The authors were able to detect the presence of tumors inside the body for different noise levels and mesh refinements.

Later, Olson and Throne (2013) have reformulated the earlier two-dimensional technique to make it more practical and have extended the approach to three dimensions. The authors have employed static indentations of the tissue and measurements of surface displacements with FEM to create detailed maps of the elastic modulus of the interior of the breast tissue. Recently, Olson et al. (2019) have developed an electromechanical device to automate and refine the manual breast exam process and have used inverse techniques to generate a tissue stiffness map of the breast tissue. The authors have validated experimentally the techniques by testing tissue phantoms with and without simulated tumors.

Barbone and Gokhale (2004) have considered plane problems in linear elastostatics and have shown that, except for four arbitrary constants, the general expression for the shear elastic modulus $\mu$ of an isotropic and incompressible material can be determined from two compatible and linearly independent displacement fields, which can be obtained from two distinct experiments. The knowledge of $\mu$ in four distinct points inside the body is sufficient for the determination of the arbitrary constants. The authors have assumed that both displacement fields are differentiable. In this work, we consider plane strain problems in classical linear elastostatics and propose a methodology to determine $\mu$ from two linearly independent displacement fields with less restrictive assumptions. Our methodology requires neither a priori knowledge of $\mu$ inside the body nor differentiability of the displacement fields.

To obtain the two linearly independent fields, in Sect. 1.2, we formulate the *direct problem* of finding the displacement field of an isotropic solid, which is nearly incompressible, subject to appropriate loading conditions. This field is then used to formulate the *inverse problem* of finding both the shear elastic modulus $\mu$ and the constraint reaction pressure $\pi$ in an incompressible solid. In view of the results of Barbone and Gokhale (2004) mentioned above, we introduce a weak formulation of the inverse problem, which consists of finding $\mu$ and $\pi$ in the incompressible solid deformed by the two linearly independent displacement fields and subjected to known resultant forces applied on complementary parts of the boundary of the solid.

In Sect. 1.3, we use a standard finite element methodology to construct discrete formulations for the weak version of the direct problem. The experiments are then simulated numerically by solving two discrete direct problems, which yield two linearly independent displacement fields. These solutions also yield resultant forces on complementary parts of the boundary for both experiments. Using both the displacement fields and the resultant forces, we construct discrete formulations for the weak version of the inverse problem stated above. The solution of this discrete problem allows the approximate reconstruction of $\mu$ in the solid.

In Sect. 1.4, we introduce the experiments that yield the linearly independent displacement fields and present numerical results obtained from the solutions of the discrete inverse problem. We use color maps to show numerical results that are in very good agreement with the corresponding analytical results. We then compare results obtained in this work with results obtained by Park and Maniatty (2006). In Sect. 1.5, we discuss future directions of this work and present some concluding remarks.

## 1.2 The Direct and Inverse Problems

In Sect. 1.2.1, we present the general statements of both the direct problem, which is used to simulate experiments that can be performed in the laboratory, and the inverse problem, which uses displacement fields obtained from the simulated experiments for the determination of the shear elastic modulus $\mu$. In Sect. 1.2.2, we review some results presented by Barbone and Gokhale (2004) with an example and propose a numerical procedure to compute $\mu$. In Sects. 1.2.3 and 1.2.4, we present the weak formulations of, respectively, the direct and the inverse problems.

### 1.2.1 Problem Formulation

Consider a long cylinder in a state of plane strain parallel to its bases and let $\mathcal{B} \subset \mathbb{R}^2$ be the undistorted natural reference configuration of a body. Points $\boldsymbol{x} \in \mathcal{B}$ are mapped to points $\hat{\boldsymbol{x}} \equiv \boldsymbol{x} + \boldsymbol{u}(\boldsymbol{x}) \in \mathbb{R}^2$, where $\boldsymbol{u}(\boldsymbol{x})$ is the displacement of $\boldsymbol{x}$. The boundary $\partial\mathcal{B}$ of $\mathcal{B}$ is composed of two non-intersecting parts, $\partial_d \mathcal{B}$ and $\partial_t \mathcal{B}$, $\partial_d \mathcal{B} \cup \partial_t \mathcal{B} = \partial\mathcal{B}$, $\partial_d \mathcal{B} \cap \partial_t \mathcal{B} = \emptyset$, such that $\boldsymbol{u}(\boldsymbol{x}) = \bar{\boldsymbol{u}}(\boldsymbol{x})$ for $\boldsymbol{x} \in \partial_d \mathcal{B}$ and a dead load traction field $\bar{\boldsymbol{t}}(\boldsymbol{x})$ is prescribed for $\boldsymbol{x} \in \partial_t \mathcal{B}$. Park and Maniatty (2006) present a brief discussion about different experimental techniques that can be used to measure the displacement of a point inside the body.

The cylinder undergoes a small strain and is in equilibrium in the absence of body force, so that

$$\operatorname{div} \boldsymbol{T}(\boldsymbol{x}) = \boldsymbol{0}, \quad \forall \boldsymbol{x} \in \mathcal{B}, \tag{1.1}$$

where $\boldsymbol{T}$ is the stress tensor, which must satisfy the loading condition

$$\boldsymbol{T}\boldsymbol{n} = \bar{\boldsymbol{t}} \quad \text{on } \partial_t \mathcal{B}, \tag{1.2}$$

with $\boldsymbol{n}$ being a unit normal to $\partial_t \mathcal{B}$.

We assume that the body is incompressible, isotropic, and linearly elastic, so that

$$\boldsymbol{T} = -\pi\,\boldsymbol{1} + 2\mu\boldsymbol{E}, \tag{1.3}$$

where $\pi$ is a constraint reaction field, $\mu$ is the shear elastic modulus, which may depend on $\boldsymbol{x} \in \mathcal{B}$, and

$$\boldsymbol{E} = \nabla_s \boldsymbol{u} \equiv \frac{1}{2}\left[\nabla\boldsymbol{u} + (\nabla\boldsymbol{u})^{\mathrm{T}}\right] \tag{1.4}$$

is the infinitesimal strain tensor. Observe from (1.3) that $\boldsymbol{T}$ is determined from both the strain tensor $\boldsymbol{E}$ and an arbitrary tensor $-\pi\,\boldsymbol{1}$, which represents the reaction of the body to local changes of volume. For incompressible materials, any infinitesimal deformation of $\mathcal{B}$ must satisfy

$$\operatorname{tr} \boldsymbol{E} = \operatorname{div} \boldsymbol{u} = 0. \tag{1.5}$$

Substituting (1.3) into both (1.1) and (1.2), we obtain

$$-\nabla \pi(\boldsymbol{x}) + 2\,\mathrm{div}(\mu\,\nabla_s \boldsymbol{u})(\boldsymbol{x}) = \boldsymbol{0}\,, \qquad \forall \boldsymbol{x} \in \mathcal{B}\,, \tag{1.6}$$

$$(-\pi + 2\,\mu\,\nabla_s \boldsymbol{u})\,\boldsymbol{n}\,(\boldsymbol{x}) = \bar{\boldsymbol{t}}\,, \qquad \forall \boldsymbol{x} \in \partial_t \mathcal{B}\,, \tag{1.7}$$

respectively.

The *direct problem* consists of finding both the displacement field $\boldsymbol{u} : \mathcal{B} \to \mathbb{R}^2$ and the pressure field $\pi : \mathcal{B} \to \mathbb{R}$ that satisfy the divergence-free condition (1.5) and the equilibrium equation (1.6) together with the traction condition (1.7) and the imposed displacement $\boldsymbol{u} = \bar{\boldsymbol{u}}$ on $\partial_d \mathcal{B}$. If $\partial_d \mathcal{B} = \emptyset$ then the given traction field $\bar{\boldsymbol{t}}$ must satisfy the usual global equilibrium condition that is essential for the existence of a solution.

The *inverse problem* consists of finding both the shear elastic modulus $\mu : \mathcal{B} \to \mathbb{R}$ and the pressure field $\pi : \mathcal{B} \to \mathbb{R}$ that satisfy the differential equation (1.6), where the displacement field $\boldsymbol{u} : \mathcal{B} \to \mathbb{R}^2$ is known everywhere in $\overline{\mathcal{B}} \equiv \mathcal{B} \bigcup \partial\mathcal{B}$ and satisfies the kinematical constraint (1.5).

Barbone and Gokhale (2004) and McLaughlin and Yoon (2004) show that, in general, the inverse problem stated above does not have a unique solution. Barbone and Gokhale (2004) show, however, that by knowing two compatible, linearly independent, and differentiable displacement fields from two experiments performed on the same body in a state of plane strain, the general expression of $\mu$ contains at most four arbitrary constants. We review these results below with an example and propose a procedure to compute the arbitrary constants, which can be extended to the more general case of discontinuous $\mu$.

### 1.2.2 Uniqueness

Let $(\boldsymbol{e}_1, \boldsymbol{e}_2, \boldsymbol{e}_3)$ be an orthonormal basis in $\mathbb{R}^3$ associated to a system of rectangular cartesian coordinates with the origin in $O$. The vectors $\boldsymbol{e}_1, \boldsymbol{e}_2$ are parallel to the plane that contains $\mathcal{B}$ and $\boldsymbol{e}_3$ is parallel to the axis of the straight cylinder. In this coordinate system, a point is represented by $\boldsymbol{x} + \xi_3 \boldsymbol{e}_3$, where $\boldsymbol{x} = \xi_1 \boldsymbol{e}_1 + \xi_2 \boldsymbol{e}_2 \in \mathcal{B}$ and $\xi_i \in \mathbb{R}$, $i = 1,\ 2,\ 3$. In addition, $\boldsymbol{u} = v_1 \boldsymbol{e}_1 + v_2 \boldsymbol{e}_2$, where $v_i \in \mathbb{R}\,,\ i = 1, 2$. It follows from (1.4) that $\boldsymbol{E} = \sum_{i,j=1}^{2} \varepsilon_{ij}\, \boldsymbol{e}_i \otimes \boldsymbol{e}_j$ , where $\boldsymbol{e}_i \otimes \boldsymbol{e}_j$ is the tensorial product between $\boldsymbol{e}_1$ and $\boldsymbol{e}_2$, which is defined by $\left(\boldsymbol{e}_i \otimes \boldsymbol{e}_j\right) \boldsymbol{e}_k = \delta_{j\,k} \boldsymbol{e}_i$, and $\varepsilon_{ij} \equiv \frac{1}{2}\left(\frac{\partial\, v_i}{\partial \xi_j} + \frac{\partial\, v_j}{\partial \xi_i}\right)$. Since the cylinder is under plane strain, it follows from (1.5) that $\varepsilon_{22} = -\varepsilon_{11}$.

Taking the curl of (1.6) and assuming that $\boldsymbol{u}$ is known everywhere in $\mathcal{B}$, we eliminate $\pi$ and obtain the second-order differential equation

$$L\,[\mu] \equiv \left(\frac{\partial^2}{\partial \xi_1^2} - \frac{\partial^2}{\partial \xi_2^2}\right)(\mu\;\varepsilon_{12}) - 2\frac{\partial^2}{\partial \xi_1 \partial \xi_2}\,(\mu\;\varepsilon_{11}) = 0 \text{ in } \mathcal{B}, \tag{1.8}$$

where $L[\mu]$ is a linear operator on $\mu$. Since $(\varepsilon_{11})^2 + (\varepsilon_{12})^2 > 0$, it is a standard procedure to show that $L$ is hyperbolic (Weinberger 1965) and that the characteristics of $L$ are parallel and perpendicular to the principal directions of $\boldsymbol{E}$.

To illustrate the fact that $\mu$ can not be determined from the differential equation (1.8) alone and to motivate the introduction of a weak formulation of the inverse problem stated in Sect. 1.2.1 for the determination of $\mu$, consider an isotropic, incompressible, and linearly elastic cylinder with square cross section in a state of plane strain perpendicular to its axis. The length of a side of the square section is $\bar{\xi}$. The cylinder is subjected to two experiments that yield the homogeneous deformation fields

$$\varepsilon^1_{11}(\xi_1, \xi_2) = -\varepsilon^1_{22}(\xi_1, \xi_2) = \varepsilon_1, \ \ \varepsilon^1_{12}(\xi_1, \xi_2) = 0, \quad \text{Bi-axial experiment}, \tag{1.9}$$

$$\varepsilon^2_{11}(\xi_1, \xi_2) = \varepsilon^2_{22}(\xi_1, \xi_2) = 0, \ \varepsilon^2_{12}(\xi_1, \xi_2) = \varepsilon_2, \quad \text{Shearing experiment}. \tag{1.10}$$

In the bi-axial experiment, the cylinder is being pulled in one direction while being compressed in the other direction. The resultant forces on both sides have normal components of intensity $\sigma_1$. In the shearing experiment, the same cylinder is subjected to tangential forces of intensity $\sigma_2$ on its lateral sides. Clearly, the deformation fields (1.9) and (1.10) satisfy the constraint (1.5) identically and, if $\varepsilon_1 = \varepsilon_2$, refer to the same state of simple shear. In this case, the principal directions, and hence the characteristics of $L$ in (1.8), are inclined with respect to each other by an angle of 45°.

Substituting the deformation fields (1.9) into (1.8) and solving for $\mu$, we obtain

$$\mu(\xi_1, \xi_2) = \phi_1(\xi_1) + \phi_2(\xi_2), \tag{1.11}$$

where $\phi_i$, $i = 1, 2$, are arbitrary functions of its arguments, which can not be determined from the bi-axial experiment alone. To determine these functions, we substitute (1.6) together with the deformation field (1.10) into (1.8). We then get

$$\mu(\xi_1, \xi_2) = \mu_0 + \mu_1\,\xi_1 + \mu_2\,\xi_2 + \mu_3\,\left(\xi_1^2 + \xi_2^2\right), \tag{1.12}$$

where $\mu_i \in \mathbb{R}$, $i = 0, ..., 3$, are arbitrary constants. The expression (1.12) is presented by Barbone and Gokhale (2004) to illustrate the fact that, except for four arbitrary constants, $\mu$ can be determined from two compatible and linearly independent displacement fields, which yield the deformation fields given by (1.9) and (1.10).

Corresponding to the deformation fields in both (1.9) and (1.10), there exist pressure fields $\pi^1$ and $\pi^2$ that can be determined from the substitution of (1.12) together with either (1.9) or (1.10) into (1.6), yielding

$$\pi^1(\xi_1, \xi_2) = 2\,\varepsilon_1\,\left[\mu_1\,\xi_1 - \mu_2\,\xi_2 + \mu_3\,\left(\xi_1^2 - \xi_2^2\right) + \hat{\pi}^1\right], \tag{1.13}$$

$$\pi^2(\xi_1, \xi_2) = 2\,\varepsilon_2\,\left[\mu_2\,\xi_1 + \mu_1\,\xi_2 + 2\,\mu_3\xi_1\,\xi_2 + \hat{\pi}^2\right], \tag{1.14}$$

where $\hat{\pi}^i \in \mathbb{R}, i = 1, 2,$ are arbitrary constants of integration.

To determine the constants $\mu_i \in \mathbb{R}, i = 0, ..., 3,$ in (1.12) and $\hat{\pi}^i \in \mathbb{R}, i = 1, 2,$ in (1.13) and (1.14), we assume that the resultant forces $\sigma_i$, $i = 1, 2,$ are known from experimental measurements. Since the traction field on the boundary is given by $\boldsymbol{t}(\boldsymbol{x}, \boldsymbol{n}) = \boldsymbol{T}(\boldsymbol{x})\,\boldsymbol{n}$, where $\boldsymbol{n}$ is the outward normal to the boundary at $\boldsymbol{x} \in \partial\mathcal{B}$ and $\boldsymbol{T}$ is given by (1.3), we obtain a system of equations for the determination of the constants above. Even though this system is overdetermined, we obtain the solution

$$2\,\mu_0\,\bar{\xi} = \frac{\sigma_2}{\varepsilon_2} = \frac{\sigma_1}{\varepsilon_1}, \quad \mu_1 = \mu_2 = \mu_3 = 0, \quad \hat{\pi}^1 = \hat{\pi}^2 = 0. \tag{1.15}$$

It then follows from (1.12) – (1.14) that $\mu$ is constant, which means that the cylinder is homogeneous and that $\pi^i = 0, i = 1, 2.$

The procedure outlined above for the determination of $\mu$ from both known deformation fields (1.9) and (1.10) and known resultant forces acting on complementary parts of the boundary of the cylinder can be extended to the general case of a non-homogeneous cylinder for which $\mu$ may not even be continuous. In this case, the determination of the fields $\mu$ e $\pi^i, i = 1, 2$ is not trivial and requires a numerical method for the construction of approximations to these fields. Below, we present a weak formulation of the inverse problem for the determination of $\mu$ e $\pi^i, i = 1, 2,$ which together with a finite element method allow this construction.

### 1.2.3 Weak Formulation of the Direct Problem

Let $\mathcal{L}^2(\mathcal{B})$ be the set of all square-integrable functions given by

$$\mathcal{L}^2(\mathcal{B}) = \{\varphi : \mathcal{B} \to \mathbb{R};\ |\varphi|_0 < \infty\}, \qquad |\varphi|_0 \equiv \left(\int_{\mathcal{B}} |\varphi|^2\, dA\right)^{1/2}, \tag{1.16}$$

and let $\left(\mathcal{H}^1(\mathcal{B})\right)^2$ be a Hilbert space defined by

$$\left(\mathcal{H}^1(\mathcal{B})\right)^2 = \left\{\boldsymbol{v} : \boldsymbol{B} \to \mathbb{R}^2;\ \|\boldsymbol{v}\|_1 < \infty\right\}, \tag{1.17}$$

with

$$\|\boldsymbol{v}\|_1 \equiv \left(\int_{\mathcal{B}} (\boldsymbol{v}\cdot\boldsymbol{v} + \nabla\boldsymbol{v}\cdot\nabla\boldsymbol{v})\, dA\right)^{1/2}.$$

In addition, we introduce

$$\mathcal{S} = \{\boldsymbol{v} : \mathcal{B} \to \mathbb{R}^2 | \boldsymbol{v} \in \left(\mathcal{H}^1(\mathcal{B})\right)^2, \boldsymbol{v} = \bar{\boldsymbol{u}} \text{ a.e. on } \partial_d \mathcal{B}\} \tag{1.18}$$

as the set of all kinematically admissible displacements and

$$\mathcal{V} = \{\boldsymbol{v} : \mathcal{B} \to \mathbb{R}^2 | \boldsymbol{v} \in \left(\mathcal{H}^1(\mathcal{B})\right)^2, \boldsymbol{v} = \boldsymbol{0} \text{ a.e. on } \partial_d \mathcal{B}\} \tag{1.19}$$

as the set of all admissible variations.

Then, a weak formulation of the direct problem stated in Sect. 1.2.1 consists of determining both $\boldsymbol{u} \in \mathcal{S}$ and $\pi \in \mathcal{L}^2(\mathcal{B})$ that satisfy

$$\begin{aligned} a_1(\boldsymbol{u}, \boldsymbol{v}) - a_2(\pi, \boldsymbol{v}) &= f(\boldsymbol{v}), \qquad && \forall \boldsymbol{v} \in \mathcal{V}, \\ a_2(p, \boldsymbol{u}) &= 0, && \forall p \in \mathcal{L}^2(\mathcal{B}), \end{aligned} \tag{1.20}$$

where

$$a_1(\boldsymbol{u}, \boldsymbol{v}) = 2 \int_{\mathcal{B}} \mu \nabla_s \boldsymbol{u} \cdot \nabla_s \boldsymbol{v} \, d\boldsymbol{x}, \qquad a_2(\pi, \boldsymbol{v}) = \int_{\mathcal{B}} \pi \operatorname{div} \boldsymbol{v} \, d\boldsymbol{x}, \tag{1.21}$$

$$f(\boldsymbol{v}) = \int_{\partial_t \mathcal{B}} \bar{\boldsymbol{t}} \cdot \boldsymbol{v} \, d\boldsymbol{x}. \tag{1.22}$$

Next, we introduce a penalty formulation of (1.20), which yields a simple and efficient numerical procedure to find the solution $(\boldsymbol{u}, \pi)$ of the direct problem. This formulation is analogous to a penalty formulation used by Aguiar and Fosdick (2000) in the numerical investigation of a class of plane prolems in nonlinear elasticity. In general terms, the *penalty formulation* is based upon relaxing the constraint equation (1.20b) and introducing an alternative sequence of problems corresponding to a one-parameter family of *compressible* elastic materials. When the parameter is zero, the constraint equation (1.20b) is recovered, and when the parameter is small, the solution to the corresponding alternative problem is supposed to be an approximation to the solution $(\boldsymbol{u}, \pi)$ of (1.20).

To implement the penalty formulation, let

$$b(\pi, p) = \frac{1}{2} \int_{\mathcal{B}} \pi \, p \, d\boldsymbol{x}. \tag{1.23}$$

For $\varepsilon > 0$, consider the problem of finding $\boldsymbol{u}_\varepsilon \in \mathcal{S}$ and $\pi_\varepsilon \in \mathcal{L}^2(\mathcal{B})$ that satisfy

$$\begin{aligned} a_1(\boldsymbol{u}_\varepsilon, \boldsymbol{v}) - a_2(\pi_\varepsilon, \boldsymbol{v}) &= f(\boldsymbol{v}), \qquad && \forall \boldsymbol{v} \in \mathcal{V}, \\ \varepsilon \, b(\pi_\varepsilon, p) + a_2(p, \boldsymbol{u}_\varepsilon) &= 0, && \forall p \in \mathcal{L}^2(\mathcal{B}), \end{aligned} \tag{1.24}$$

where the penalty parameter $\varepsilon > 0$ is sufficiently small. By using the the expression (1.23) and eliminating $\pi_\varepsilon(\cdot)$ from the system of equations (1.24), the alternative problem reduces to determining $\boldsymbol{u}_\varepsilon \in \mathcal{S}$ such that

$$a_1(\boldsymbol{u}_\varepsilon, \boldsymbol{v}) + c(\boldsymbol{u}_\varepsilon, \boldsymbol{v}) = f(\boldsymbol{v}), \qquad \forall \boldsymbol{v} \in \mathcal{V}, \tag{1.25}$$

where

$$c(\boldsymbol{u}_\varepsilon, \boldsymbol{v}) = \frac{2}{\varepsilon} \int_{\mathcal{B}} (\operatorname{div} \boldsymbol{u}_\varepsilon)(\operatorname{div} \boldsymbol{v})\, \mathrm{d}\boldsymbol{x}\,. \tag{1.26}$$

We then want to find $\boldsymbol{u}_\varepsilon \in \mathcal{S}$ that satisfies the equation (1.25), where $a_1(\cdot,\cdot)$, $f(\cdot)$, and $c(\cdot,\cdot)$ are given by, respectively, (1.21a), (1.22), and (1.26). It is well known that for $\varepsilon > 0$ and $\mu > 0$, this problem has a unique solution $(\boldsymbol{u}_\varepsilon, \pi_\varepsilon) \in \mathcal{S} \times \mathcal{L}^2(\mathcal{B})$ that converges to the solution $(\boldsymbol{u}, \pi)$ of (1.20) as $\varepsilon \to 0$.

The penalty formulation for the direct problem stated above is equivalent to the weak formulation of the boundary value problem for a compressible, isotropic, and linearly elastic solid. To see this, introduce

$$\boldsymbol{T} = \lambda \operatorname{div} \boldsymbol{u}\, \mathbf{1} + 2\,\mu\, \nabla_s \boldsymbol{u} \tag{1.27}$$

into both (1.1) and (1.2), where $\lambda$ is a Lamé modulus, and consider the weak formulation of the corresponding direct problem, which consists of determining $\boldsymbol{u} \in \mathcal{S}$ that satisfies

$$a(\boldsymbol{u}, \boldsymbol{v}) - f(\boldsymbol{v}) = 0, \qquad \forall \boldsymbol{v} \in \mathcal{V}, \tag{1.28}$$

where $f(\boldsymbol{v})$ is given by (1.22) and

$$a(\boldsymbol{v}, \boldsymbol{v}) = \int_{\mathcal{B}} (\lambda \operatorname{div} \boldsymbol{u} \operatorname{div} \boldsymbol{v} + 2\,\mu\, \nabla_s \boldsymbol{u} \cdot \nabla_s \boldsymbol{v})\, \mathrm{d}\boldsymbol{x}\,. \tag{1.29}$$

If we consider that $\lambda$ in (1.29) is constant and equal to $2/\varepsilon$ in (1.26), then both formulations are equivalent.

### 1.2.4 Weak Formulation of the Inverse Problem

Let $\boldsymbol{u}^1$ and $\boldsymbol{u}^2$ be two compatible and linearly independent displacement fields (Barbone and Gokhale 2004), which are not necessarily differentiable and are obtained from two distinct experiments performed on the same body, and let

$$\boldsymbol{R}_i^j = \int_{\partial_i \mathcal{B}} \left(-\pi^j \mathbf{1} + 2\,\mu\, \nabla_s \boldsymbol{u}^j\right) \boldsymbol{n}_i \; \mathrm{d}\,L \tag{1.30}$$

be resultant forces that are known on r complementary parts $\partial_i \mathcal{B},\ i = 1, ..., r,$ of $\partial\mathcal{B}$, so that $\partial\mathcal{B} \equiv \bigcup_{i=1}^{r} \partial_i \mathcal{B}$ and $\partial_i \mathcal{B} \bigcap \partial_j \mathcal{B} \neq \emptyset, i \neq j$. In (1.30), $\pi^j$, j = 1, 2, is the pressure field associated with the displacement field $\boldsymbol{u}^j$ and $\boldsymbol{n}_i$ is a unit normal to $\partial_i \mathcal{B}$. To satisfy the global equilibrium condition, we must have $\sum_{i=1}^{r} \boldsymbol{R}_i^j = 0$, j = 1, 2.

We want to determine the shear elastic modulus $\mu{:}\mathcal{B} \to \mathbb{R}^2$ and the pressure field $\pi^j{:}\mathcal{B} \to \mathbb{R}^2$, j = 1, 2, that satisfy both

$$- \nabla\pi^j\,(\boldsymbol{x}) + 2\,\mathrm{div}\left(\mu\ \nabla_s \boldsymbol{u}^j\right)(\boldsymbol{x}) \ = \ 0, \quad \forall \boldsymbol{x}\ \in \mathcal{B}, \tag{1.31}$$

and the expression (1.30) for known resultants $\boldsymbol{R}_i^j$, where the displacement field $\boldsymbol{u}^j{:}\mathcal{B} \to \mathbb{R}^2$, j = 1, 2, is known everywhere in $\overline{\mathcal{B}}$ and satisfy the kinematical constraint div $\boldsymbol{u}^j = 0$.

The weak form of the inverse problem that we shall consider in this work consists of finding $\mu \in \mathcal{L}^2(\mathcal{B})$ and $\pi^j \in \mathcal{L}^2(\mathcal{B})$, j = 1, 2, that satisfy

$$- \int_{\mathcal{B}} \pi^j\ tr\ \nabla_s \boldsymbol{v}\ dA + 2 \int_{\mathcal{B}} \mu\ \nabla_s \boldsymbol{u}^j\ \cdot \nabla_s \boldsymbol{v}\ dA = 0,\ \ \boldsymbol{u}^j\ \in \mathcal{S},\ \ \mathrm{j} = 1, 2,\ \ \ \forall \boldsymbol{v}\ \in \mathcal{V}, \tag{1.32}$$

together with (1.30) for $i = 1, ..., r$, where we recall from above that, here, $\partial_t\,\mathcal{B} = \emptyset$.

The weak form stated above together with a finite element methodology discussed in the next section allows an approximate reconstruction of $\mu$ from the given data $\boldsymbol{u}^j$ and $\boldsymbol{R}_i^j,\ j = 1,\ 2, i = 1, ..., r$. Observe from (1.32) that this formulation does not require differentiability of $\boldsymbol{u}^j$.

## 1.3 Discrete Formulation

We want to construct approximations to the solution $(\boldsymbol{u}_\varepsilon, \pi_\varepsilon) \in \mathcal{S} \times \mathcal{L}^2(\mathcal{B})$ of the direct problem, given by (1.25) together with (1.21a), (1.22), and (1.26), for a given $\varepsilon > 0$ and to the solution $(\mu, \pi^1, \pi^2) \in \left(\mathcal{L}^2(\mathcal{B})\right)^3$ of the inverse problem, given by (1.32) together with (1.30) and (1.31). For this, we consider a finite element formulation based on the introduction of discrete problems over finite-dimensional subsets of $\mathcal{S}$, given by (1.18), $\mathcal{V}$, given by (1.19), and $\mathcal{L}^2(\mathcal{B})$. Recall from Sect. 1.2.4 that $\partial_t\,\mathcal{B} = \emptyset$ in the case of the inverse problem.

We begin by assuming that $\mathcal{B} \subset \mathbb{R}^2$ is a polygonal domain composed of $m$ non-overlapping quadrilaterals $\mathcal{K}_i \in \mathbb{R}^2$, so that

$$\mathcal{B} = \bigcup_{i=1}^{m} \mathcal{K}_i\,, \tag{1.33}$$

and the intersection of any two of these quadrilaterals is either empty, a point or a straight line. For each $\mathcal{K}_i \subset \mathcal{B}$, the interior of the quadrilateral $\mathcal{K}_i$ is non-empty.

Let $\mathcal{K}_i, i = 1, \ldots, m$, be endowed with the set of nodes $\{\boldsymbol{x}_{ik} \in \mathcal{K}_i,\ k = 1, \ldots, 4\}$, which are the vertices of $\mathcal{K}_i$. We shall consider *Lagrange finite elements* $(\mathcal{K}_i, \mathcal{P}_i, \Sigma_i)$, $i = 1, \ldots, m$, where $\mathcal{P}_i$ is a set of smooth functions $\varphi : \mathcal{K}_i \to \mathbb{R}$, and $\Sigma_i$ is a set of degrees of freedom corresponding to the coefficients $\varphi(\boldsymbol{x}_{ik})$. The functions $\varphi$ are linear combinations of normalized basis functions $\varphi_{ik},\ k = 1, \ldots, 4$, so that $\varphi_{ik}(\boldsymbol{x}_{il}) = \delta_{kl},\ k,\ l = 1,\ \ldots,\ 4$. The functions $\varphi_{ik}$ are continuous over $\mathcal{K}_i$. In this way, a *finite element mesh* is the union of all the finite elements $(\mathcal{K}_i, \mathcal{P}_i, \Sigma_i)$.

Let us denote by $\mathcal{N} = \{\boldsymbol{x}_1, \ldots, \boldsymbol{x}_n\} \subset \bar{\mathcal{B}}$ the complete set of $n$ nodes in $\bar{\mathcal{B}}$. Then, for each $i = 1, \ldots, m$ and each $k = 1, \ldots, 4, \boldsymbol{x}_{ik} \in \mathcal{K}_i$ corresponds to a unique element from $\mathcal{N}$. We define the set of functions $\{\varphi_j : \mathcal{B} \to \mathbb{R}\,,\ j = 1, \ldots, n\}$, such that $\varphi_j(\boldsymbol{x}) = \varphi_{ik}(\boldsymbol{x})$ for $\boldsymbol{x} \in \mathcal{K}_i,\ j = 1, \ldots, n,\ k = 1, \ldots, 4$. Thus, $\varphi_j$ is continuous on $\mathcal{B}$ and satisifies $\varphi_j(\boldsymbol{x}_i) = \delta_{ji}$. In fact, the set of functions $\varphi_j$ is a finite-dimensional basis for the set of continuous functions defined by

$$\mathcal{P} = \left\{ \varphi : \mathcal{B} \to \mathbb{R} | \varphi(\boldsymbol{x}) = \sum_{j=1}^{n} \alpha_j\, \varphi_j(\boldsymbol{x})\,, (\alpha_1, \ldots, \alpha_n) \in \mathbb{R}^n \right\}. \tag{1.34}$$

## 1.3.1 Discrete Direct Problem

We now define the finite-dimensional space

$$\mathcal{V}_h = \{\boldsymbol{v}_h : \mathcal{B} \to \mathbb{R}^2 \big| \boldsymbol{v}_h \in \mathcal{P} \times \mathcal{P}\,,\ \boldsymbol{v}_h = 0 \text{ on } \partial\mathcal{B}\}\,, \tag{1.35}$$

where $h$ stands for the characteristic length of the finite element mesh. Since (1.33) holds by assumption, we note that $\mathcal{V}_h \subset \mathcal{V}$. A function $\boldsymbol{v} \in \mathcal{V}_h$ has the representation

$$\boldsymbol{v}_h(\boldsymbol{x}) = \sum_{i=1}^{2n} \vartheta_i \boldsymbol{w}_i(\boldsymbol{x}), \qquad \boldsymbol{x} \in \bar{\mathcal{B}} \equiv \mathcal{B} \cup \partial\mathcal{B}\,, \tag{1.36}$$

where $\vartheta_i \in \mathbb{R}$, and $\boldsymbol{w}_i$ is a vector in $\mathbb{R}^2$ of the form

$$\boldsymbol{w}_{2j-1} = (\varphi_j, 0)\,, \qquad \text{and} \qquad \boldsymbol{w}_{2j} = (0, \varphi_j)\,, \qquad j = 1, \ldots, n\,, \tag{1.37}$$

relative to a fixed orthonormal base, with $\varphi_j$ being a scalar basis function defined for the $j^{\text{th}}$ node in the set $\mathcal{N}$ defined above.

In particular, notice that $\boldsymbol{v}_h(\boldsymbol{x}_i) = (\vartheta_{2i-1}, \vartheta_{2i})$ for $i = 1, \ldots, n$. Each $\vartheta_i,\ i = 1, \ldots, 2n$, is a *degree of freedom associated with* $\mathcal{V}_h$. Thus, considering the fact that each node in $\mathcal{N}$ has two degrees of freedom and that $\boldsymbol{v} \in \mathcal{V}_h$, it is convenient to decompose the complete set of $2n$ degrees of freedom into two complementary integer sets $\mathcal{Z}^*$ and $\mathcal{Z}$ such that $\vartheta_i = 0\,,\ \forall\, i \in \mathcal{Z}^*$ and $\mathcal{Z} \equiv \{1, 2, \ldots, 2n\} \backslash \mathcal{Z}^*$.

Similar to the set $\mathcal{V}_h$, we also introduce the set

$$\mathcal{S}_h = \{\boldsymbol{u}_h : \mathcal{B} \to \mathbb{R}^2 \big| \boldsymbol{u}_h \in \mathcal{P} \times \mathcal{P}\,, \boldsymbol{u}_h = \bar{\boldsymbol{u}} \text{ on } \partial_d \mathcal{B}\}\,. \tag{1.38}$$

Observe that a displacement $\boldsymbol{u}_h \in \mathcal{S}_h$ has the representation

$$\boldsymbol{u}_h(\boldsymbol{x}) = \sum_{i=1}^{2n} \upsilon_i \boldsymbol{w}_i(\boldsymbol{x})\,, \tag{1.39}$$

where for each index $i \in \mathcal{Z}^*$, the corresponding value of $\upsilon_i$ is prescribed, $\upsilon_i = \bar{\upsilon}_i$.

Let us also define the space

$$\mathcal{L}_h^2 = \{p_h : \mathcal{B} \to \mathbb{R} \big|\ p_h \text{ is piecewise continuous on } \mathcal{B}\}\,, \tag{1.40}$$

An element $p_h \in \mathcal{L}_h^2$ has the representation

$$p_h(\boldsymbol{x}) = \sum_{i=1}^{m} p_i \tau_i(\boldsymbol{x})\,, \qquad \boldsymbol{x} \in \bar{\mathcal{B}}, \tag{1.41}$$

where $p_i \in \mathbb{R}$ and $\tau_i$ is a scalar basis function, which is piecewise constant, has support in $\mathcal{K}_i$, and is normalized so that $\tau_i(\boldsymbol{x}) = \delta_{ij}$ for $\boldsymbol{x} \in \mathcal{K}_j$, $i, j = 1, \ldots, m$.

Now, for a given penalty parameter $\varepsilon > 0$, we substitute both $\boldsymbol{u}_\varepsilon$, of the form in (1.39), and $\boldsymbol{v}$, given by (1.36), into the expressions (1.21.a), (1.22), and (1.26). Since the coefficients $\vartheta_i$ , $i \in \mathcal{Z}$, are arbitrary, we can rewrite the (1.25) in the form

$$\sum_{j=1}^{n} \left(\kappa_{i\,j} + \hat{\kappa}_{i\,j}\right) \upsilon_i = \gamma_i\,, \qquad i \in \mathcal{Z}\,, \tag{1.42}$$

where

$$\kappa_{i\,j} = a_1(\boldsymbol{w}_i, \boldsymbol{w}_j)\,, \qquad \hat{\kappa}_{i\,j} = c(\boldsymbol{w}_i, \boldsymbol{w}_j)\,, \tag{1.43}$$

$$\gamma_i = f(\boldsymbol{w}_i)\,, \tag{1.44}$$

The *discrete direct problem* associated to the space $\mathcal{S}_h$ and obtained from the penalty formulation in (1.25) consists of finding the coefficients $\upsilon_i$ , $i \in \mathcal{Z}$, in $\boldsymbol{u}_{\varepsilon h} \equiv \sum_{i=1}^{2n} \upsilon_i \boldsymbol{w}_i(\boldsymbol{x}) \in \mathcal{S}_h$, that satisfy the linear system of equations (1.42) together with (1.43) and (1.44). Recall from Sect. 1.2.3 that, for the direct problem, both $\mu$ in (1.21) and $\varepsilon$ in (1.26) are known and $\partial_t \mathcal{B}$ may not be empty.

Holding $h$ fixed and decreasing $\epsilon$, we generate a sequence of solutions parameterized by $\epsilon$ for the discrete problem that converges to a limit function $\boldsymbol{u}_h$. We then refine the finite element mesh by decreasing $h$ and repeat the process above. In so doing, we generate a sequence of solutions $\boldsymbol{u}_h$ parameterized by $h$ which converges to the solution $\boldsymbol{u}$ of the original problem given by (1.20) together with (1.21) and (1.22).

### 1.3.2 Discrete Inverse Problem

We want to construct finite element approximations for the solution $(\mu, \pi^1, \pi^2) \in (\mathcal{L}^2(\mathcal{B}))^3$ of the inverse problem given by (1.32) together with (1.30) for $i = 1, ..., r$. For this, we consider a finite element formulation based on the introduction of discrete problems over finite-dimensional subsets of $\mathcal{L}^2(\mathcal{B})$ and $\mathcal{V}$. The corresponding discrete problems can be solved using direct solvers, as opposed to iterative solvers used in the literature, such as in Park and Maniatty (2006).

We assume that both fields $\boldsymbol{u}^1$ and $\boldsymbol{u}^2$ are known. We then use the representation given by (1.36) to approximate $\mu$ and $\pi^j$, $j = 1, 2$, in the form

$$\mu_h(\boldsymbol{x}) = \sum_{k=1}^{m} \mu_k \tau_k \quad \text{and} \quad \pi_h^i(\boldsymbol{x}) = \sum_{k=1}^{m} \pi_k^i \tau_k, \quad \boldsymbol{x} \in \overline{\mathcal{B}}, \tag{1.45}$$

where $\mu_k \in \mathbb{R}$ and $\pi_k^i \in \mathbb{R}$, i = 1,2, $k = 1, 2, ..., m$. We also assume that $\partial_j\mathcal{B} \subset \partial\mathcal{B}$ is given by

$$\partial_j\mathcal{B} = \bigcup_{p \in \mathcal{Z}_j} \mathcal{D}_p, \tag{1.46}$$

where $\mathcal{D}_p$ is a side of $\mathcal{K}_p$ that also belongs to $\partial\mathcal{B}$ and $\mathcal{Z}_j$ is the set of integer numbers that identify the finite elements with sides contained in $\partial_j\mathcal{B}$.

Substituting the expressions in (1.45) and $\boldsymbol{v}_h$, given (1.36), into both (1.30) and (1.32), and using the fact that the coefficients $\vartheta_i$, $i \in \mathcal{Z}$, are arbitrary, these equations can be rewritten as

$$\sum_{q \in Z_i} \beta_{kq}^j \, \omega_q^j = \boldsymbol{e}_k \cdot \boldsymbol{R}_i^j, \qquad i = 1, 2, ..., r, \; k = 1, 2, \tag{1.47}$$

$$\sum_{q=1}^{2m} \alpha_{pq}^j \, \omega_q^j = 0, \qquad p \in \mathcal{Z}, \tag{1.48}$$

respectively, where j = 1, 2, and

$$\omega_{2q-1}^j = \mu_q, \qquad \omega_{2q}^j = \pi_q^j, \tag{1.49}$$

$$\alpha_{p\,(2q-1)}^j = 2 \int_{\mathcal{K}_q} \nabla_s \boldsymbol{u}^j \cdot \nabla_s \boldsymbol{w}_p \, dA, \qquad \alpha_{p\,(2q)}^j = - \int_{\mathcal{K}_q} tr\, \nabla_s \boldsymbol{w}_p \, dA, \tag{1.50}$$

$$\beta_{k\,(2q-1)}^j = 2\boldsymbol{e}_k \cdot \int_{\mathcal{D}_q} (\nabla_s \boldsymbol{u}^j)\, \boldsymbol{n}_q \, dA, \qquad \beta_{k\,(2q)}^j = -\boldsymbol{e}_k \cdot \int_{\mathcal{D}_q} \boldsymbol{n}_q \, dA. \tag{1.51}$$

Then, the discrete inverse problem associated to the problem given by (1.30) and (1.32) consists of finding the coefficients $\omega_q^j$, j = 1, 2, q = 1, …, m, defined by (1.49), that satisfy the linear system of equations given by (1.47)–(1.51), which, in general, is overdetermined. To solve the linear system, we use the singular value decomposition algorithm presented by Golub and van Loan (1996). The main input data are the coefficients on the right-hand side of both (1.47) and (1.48), the matrix $\boldsymbol{W}$ formed by the coefficients that multiply $\omega_q^j$ on the left-hand side of these equations, the dimensions of $\boldsymbol{W}$, and a tolerance that yields the largest nonsingular, square matrix of $\boldsymbol{W}$. Here, the tolerance is a non-negative number below which a singular value of $\boldsymbol{W}$ is considered to be zero. A preliminary study of the influence of the tolerance on the values of the coefficients $\mu_k$, k = 1, …, m, in (1.45a) shows that, for tolerances below $10^{-8}$, all the values obtained for these coefficients were physically plausible and differed very little from each other.

## 1.4 Numerical Results

We use the discrete formulations introduced in Sect. 1.3 to obtain the shear elastic modulus $\mu$ of a long cylinder that is in a state of plane strain parallel to its bases. Recall from Sect. 1.2.2 that we need two displacement fields obtained from two distinct experiments performed on this cylinder. We show in Fig. 1.1a schematic representation of two experiments that yield both the resultants $\boldsymbol{R}_i^j$ in (1.47) and the displacements $\boldsymbol{u}^j$ in both (1.50) and (1.51).

For this, we use the finite element package *Ansys 5.5*[1] and assumed that the shear elastic moduli of both the matrix, which is the cylinder without the inclusion, and the inclusion is given by, respectively,

$$\mu_M = 36\,\text{kPa} \qquad \text{and} \qquad \mu_I = C_R\,\mu_M, \tag{1.52}$$

where $C_R \geq 0$ is the shear modulus contrast ratio. Observe that $C_R = 0$ corresponds to an empty hole, $C_R = 1$ corresponds to a homogeneous cylinder, and $C_R > 1$ corresponds to an inclusion that is $C_R$ times stiffer than the matrix. Both the matrix and the inclusion are homogeneous and nearly incompressible with Poisson's ration $\nu_M = \nu_I = 0.499999$. The length $\bar{\xi}$ in Fig. 1.1 is $\bar{\xi} = 50$ mm and the radius of the circular section is $r = 6$ mm. The conditions on the boundary of the cylinder are the same ones considered for the experiments described in Sect. 1.2.2. Here, however, the resulting deformation fields are not homogeneous.

In Fig. 1.2, we show a mesh of finite elements used in the computation of both $\boldsymbol{u}^j$ and $\boldsymbol{R}_i^j$, where the inclusion is shown on the upper right side. In these simulations, we have also obtained approximations for the pressure fields $\pi^j$, $j = 1, 2$.

With the vectors $\boldsymbol{u}^j$ and $\boldsymbol{R}_i^j$ computed from the numerical simulations above, we have solved the linear system given by (1.47)–(1.51) using the non-uniform mesh of

[1] Ansys 5.5 is proprietary software of Ansys Inc., USA.

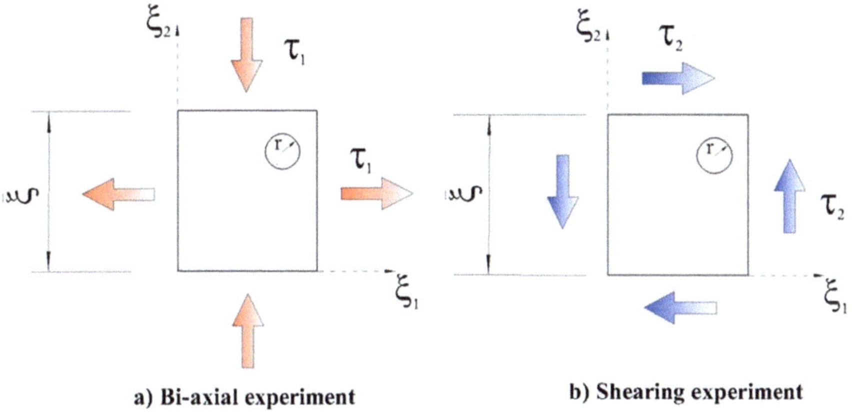


**Fig. 1.1** Bi-axial and shearing experiments on a cylinder containing an inclusion

**Fig. 1.2** Non-uniform mesh of finite elements to dicretize the square cross section of a cylinder containing an eccentric inclusion

**Table 1.1** Finite element meshes for the cylinder with an eccentric inclusion

| Mesh | Number of nodes | Number of elements |
|---|---|---|
| 1 | 733 | 668 |
| 2 | 931 | 866 |
| 3 | 1077 | 1012 |
| 4 | 1218 | 1153 |
| 5 | 1535 | 1470 |

Fig. 1.2 and computed approximations for both $\mu$ and $\pi^j$, $j = 1, 2$, from (1.45). We have verified that the approximations of $\pi^j$, $j = 1, 2$, are in very good agreement with the corresponding approximations obtained from the numerical simulations above.

In the next two figures, we show sequences of frames that use color maps to represent the distribution of the shear elastic modulus inside the cylinder of square cross section. Each frame has its own color map, which corresponds to a range of values for $\mu$. The smallest and largest values of $\mu$ correspond to the colors at, respectively, the bottom and the top of the color map. The frames on the upper left corner in both figures contain exact values of $\mu$ and are used as reference frames to be compared with the other five frames. These five frames are obtained from different finite element meshes, which are represented in Table 1.1 by increasing numbers of nodes and elements. In particular, Fig. 1.2 corresponds to Mesh 5 in Table 1.1.

In Figs. 1.3 and 1.4, we consider the cases $C_R = 0$[2] and $C_R = 6$, respectively. Comparing the reference frame (a) in each figure with the other five frames of the same figure and observing the values in the corresponding legend, we note that $\mu_h$, defined by (1.45a), converges to $\mu$ as the mesh is refined. In both cases, $\mu_h \to 36$ kPa everywhere inside the matrix. Inside the inclusions, we see from Fig. 1.3 that $\mu_h \to 0$ kPa and from Fig. 1.4 that $\mu_h \to 216$ kPa.

Next, in Fig. 1.5, we consider an inclusion with a cross section having a complex geometry. The color maps of both Frame (a) and Frame (b) represent, respectively, the exact and the calculated values of $\mu$ inside the inclusion and the matrix. By comparing both frames, we see that our numerical approach yields accurate values for $\mu$ and is capable of reconstructing the geometry of the inclusion.

Finally, we compare results obtained from our methodology of reconstruction of $\mu$ against results obtained by Park and Maniatty (2006) for the case of a cylinder with square cross section containing a centered circular inclusion. Here, too, the cylinder is subjected to a state of plane strain parallel to its bases. These authors present a direct inversion approach for reconstructing $\mu$ in a nearly incompressible, isotropic, and linearly elastic solid from dynamic measurements of the interior displacement field during time-harmonic excitation. They use a finite element methodology to

[2] Because of numerical difficulties experienced during the numerical simulations using ANSYS 5.5 , we used $C_R = 10^{-25}$ and Poisson's ratio $\nu = 0.3$ for the inclusion, instead of $C_R = 0$ and $\nu = 0.5$.

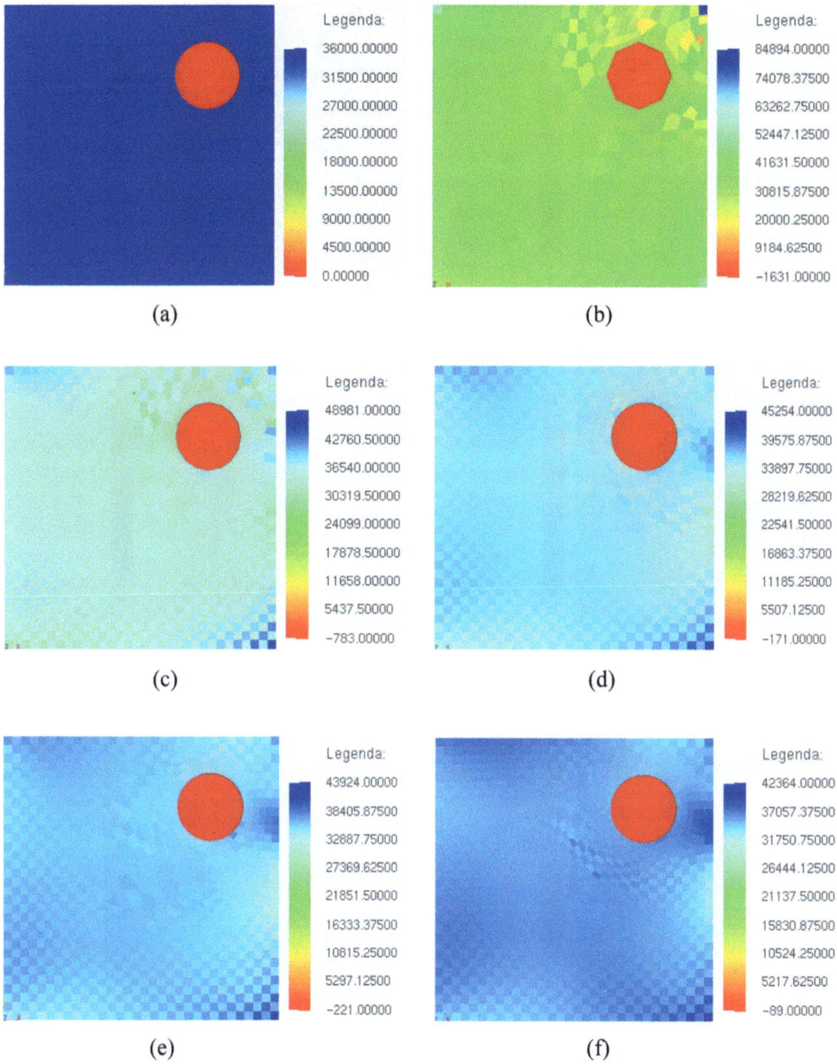


**Fig. 1.3** Reconstruction of $\mu$ in a cylinder containing an eccentric circular hole, ($C_R = 0$)

find approximate solutions to the governing equations, which does not use boundary conditions.

To obtain the displacement field, they solve a direct problem by considering that one of the horizontal lateral sides of the cylinder is kept fixed while the opposite side is submitted to a horizontal excitation of 150 MHz with an amplitude of 1 mm. The other two vertical lateral sides are traction free. The matrix has the same shear elastic modulus as ours, given by (1.52a), and a slightly smaller Poisson's ratio, given by $\nu = 0.499992$, which is the same one for the inclusion.

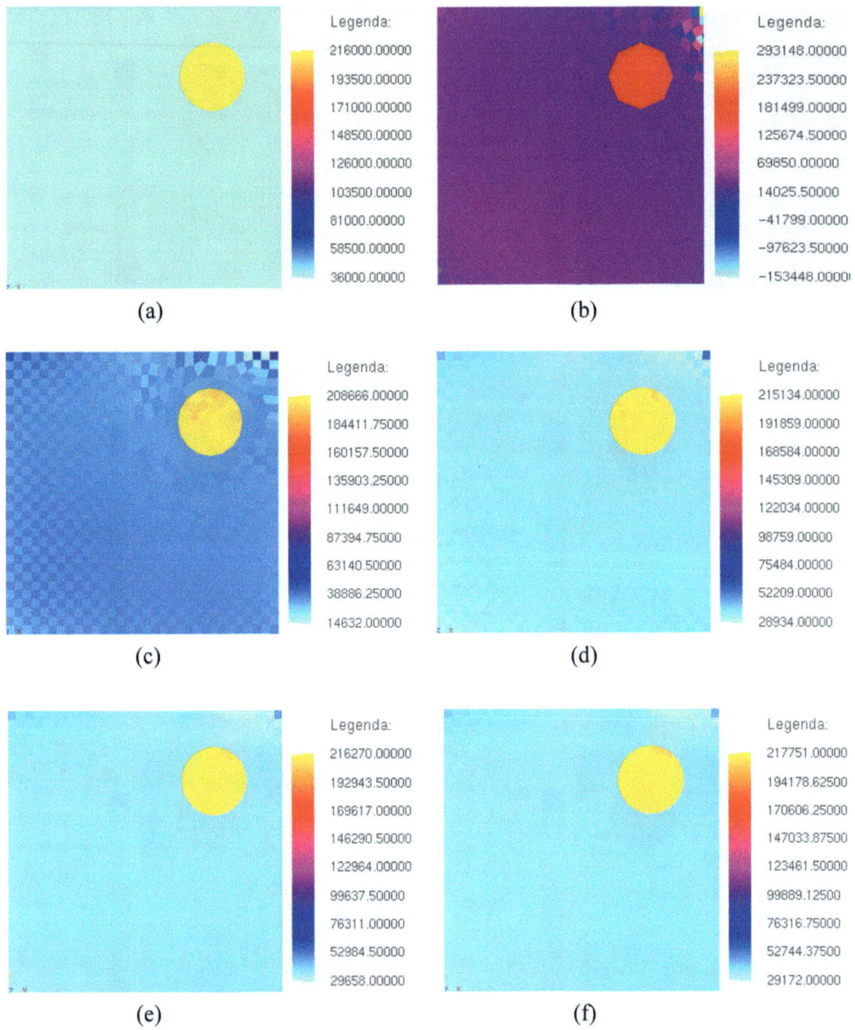


**Fig. 1.4** Reconstruction of $\mu$ in a cylinder containing an eccentric circular inclusion with $C_R = 6$

Park and Maniatty (2006) present values for the percent average error of the reconstructed shear modulus, which is defined by

$$E_\mu = \frac{100\,\Delta\mu}{\bar{\mu}}, \qquad \Delta\mu = \sqrt{\frac{\int_{\mathcal{B}} (\mu_h - \mu)^2\, d\boldsymbol{x}}{A_{\mathcal{B}}}}, \qquad \bar{\mu} = \sqrt{\frac{\int_{\mathcal{B}} \mu_h^2\, d\boldsymbol{x}}{A_{\mathcal{B}}}},$$

where $A_{\mathcal{B}} \equiv \bar{\xi}^2$ is the area of the square cross section and $\mu_h$ is their finite element approximation for the known distribution $\mu$. They also present values for the average

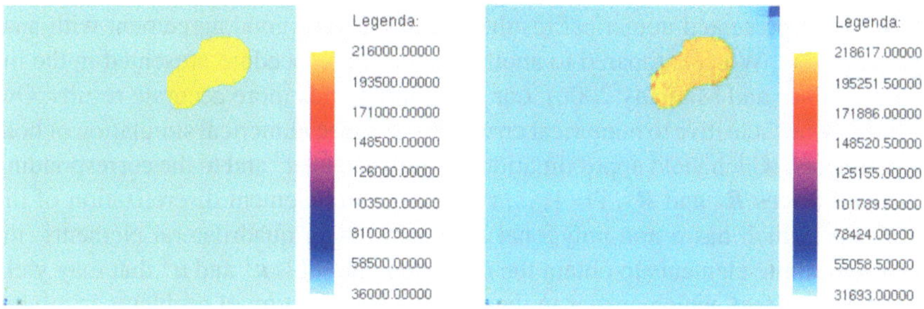


(a) (b)

**Fig. 1.5** Reconstruction of $\mu$ in a cylinder containing an inclusion with a complex geometry

reconstructed shear modulus of the inclusion, which we infer[3] from their work that it is given by $\bar{\mu}_I = \int_{\mathcal{B}_I} \mu_h \, d\boldsymbol{x} \big/ A_{\mathcal{B}_I}$, where $A_{\mathcal{B}_I} = \pi\, r^2$ is the area of the circular cross section. For the case of a circular inclusion of radius $r = 6$ mm and $C_R = 6$, these values are $E_\mu = 27.53\%$ and $\bar{\mu}_I = 191.2$ kPa.

Next, we adapt the authors dynamic experiment to a static experiment, so that we can use it in our simulations. For this, we consider that the cylinder is in equilibrium, fixed in one horizontal lateral side, and subjected to a horizontal displacement of 1 mm on the opposite lateral side. The conditions on the other vertical lateral sides are the same as in Park and Maniatty (2006).

The additional experiment needed in our simulations consists of compressing two lateral sides of the cylinder through rigid frictionless platens while keeping the other two lateral sides traction free.

The finite element mesh used in our simulation contains 1544 nodes and 1479 elements. We have verified that $\max_{\boldsymbol{x}_i \in \mathcal{B}_I} |\mu_i - \mu(\boldsymbol{x}_i)| \cong 2.1$ kPa, where we recall from Sect. 1.3 that $\mu_i$ is the approximate value of $\mu$ at the node $\boldsymbol{x}_i$. This difference corresponds to $\mu_i \cong 213.9$ kPa for a node $\boldsymbol{x}_i$ located inside the inclusion. Comparing this value with $\bar{\mu}_I = 191.2$ kPa of Park and Maniatty (2006) and the value $\mu_I = 216$ kPa that is obtained from (1.52.b) with $C_R = 6$, we see that our methodology yields a better approximation of $\mu$ than their methodology does.

## 1.5 Discussion and Conclusion

We have presented a numerical procedure based on the finite element method for the determination of the shear elastic modulus $\mu$ of a linearly elastic, isotropic, and incompressible solid in a state of plane strain. It is assumed that two linearly independent displacement fields are known inside the body from two distinct experiments and that resultant forces are known on complementary parts of the boundary of the solid.

---

[3] We use the expression @@(20) in Park and Maniatty (2006).

We have presented numerical results that are in very good agreement with analytical results. When compared to another numerical procedure presented in the literature, (Park and Maniatty 2006), our procedure yields more accurate results. Our results are not sensitive to numerical errors due to (i) the numerical simulation of both experiments, which yield approximations to both $\boldsymbol{u}^1$ and $\boldsymbol{u}^2$ and to the corresponding resultant forces $\boldsymbol{R}_i^1$ and $\boldsymbol{R}_i^2$, i = 1, ..., r; (ii) the finite element discretization of the inclusion, which has a non-polygonal geometry, using quadrilateral elements; iii) the use of finite elements to obtain the displacement fields $\boldsymbol{u}^1$ and $\boldsymbol{u}^2$ that may yield the locking effect, which appear in the numerical simulation of problems involving incompressible materials. See a discussion about this effect in Hughes (1987). In this work, we did not perform a sensitivity analysis of the experimental error in the values of the displacement fields, and this constitutes a limitation that will be addressed in future investigation. In spite of this limitation, the displacement fields are numerical approximations of the exact displacement fields. The differences between the exact and the approximated fields could be interpreted as perturbation errors. In this case, we see that our results are not very sensitive to these errors.

We are now investigating the application of the numerical approach (i) using regular meshes, instead of the non-uniform meshes used in this work, (ii) in the reconstruction of $\mu$ in problems that are more general than plane problems, (iii) using constitutive relations that take into account viscous and nonlinear behavior of living tissues.

Our numerical procedure has great potential of application in medical physics; especially, in the detection of tumors inside healthy biological tissues and in the identification of different tissues in parts of the human body. It also has the potential of being applied in all branches of engineering requiring non-destructive evaluation of the interior of a solid body in a state of plane strain.

**Acknowledgements** Financial support is gratefully acknowledged by the first author to National Council for Scientific and Technological Development (CNPq), grants n° 420099/2018-2 and n° 304732/2019-2, and by the second author to Coordination for the Improvement of Higher Education Personnel (CAPES)—Finance Code 001.

## References

Aguiar AR, Fosdick RL (2000) A singular problem in incompressible nonlinear elastostatics. Math Models Methods Appl Sci 10:1181–1207

Barbone PE, Gokhale NH (2004) Elastic modulus imaging: on the uniqueness and nonuniqueness of the elastography inverse problem in two dimensions. Inverse Probl 20:283–296

Fung YC (2004) Biomechanics: mechanical properties of living tissues, 2nd edn. Springer, New York

Golub GH, van Loan CF (1996) Matrix computations, 3rd edn. The John Hopkins University Press, Baltimore

Hughes T (1987) The finite element method—linear static and dynamic analysis. Prentice-Hall, Inc

Krouskop TA, Wheeler TM, Kallel F, Garra BS, Hall T (1998) Elastic moduli of breast and prostate tissues under compression. Ultrason Imaging 20:260–274

Liu HT, Sun LZ, Wang G, Vannier MW (2003) Analytic modeling of breast elastography. Med Phys 30(9):2340–2349
McLaughlin Yoon JR (2004) Unique identifability of elastic parameters from time-dependent interior displacement measurement. Inverse Probl 20:25–45
Mridha M, Ödman S (1986) Noninvasive method for the assessment of subcutaneous edema. Med Biol Eng Comput 24(4):393–398
Olson LG, Throne RD (2010) Numerical simulation of an inverse method for tumour size and location estimation. Inverse Probl Sci Eng 18(6):813–834
Olson LG, Throne RD (2013) An inverse problem approach to stiffness mapping for early detection of breast cancer. Inverse Probl Sci Eng 21(2):314–338
Olson LG, Throne RD, Nolte AJ, Crump A, Griffin K, Han T, Iovanac N, Janssen T, Jones M, Ling X, Samp M (2019) An inverse problem approach to stiffness mapping for early detection of breast cancer: tissue phantom experiments. Inverse Probl Sci Eng 27(7):1006–1037
Ophir J, Céspedes I, Ponnekanti H, Yadi Y, Li X (1991) Elastography: a quantitative method for imaging the elasticity of biological tissues. Ultrason Imaging 13(2):111–134
Ophir J, Kallel F, Varghese T, Konofagou E, Alam SK, Krouskop T, Garra B, Righetti R (2001) Optical and acoustical imaging of biological media: Elastography. Comptes Rendus de l'Acadéémie des Sciences - Series IV - Phys-Astrophys 2(8):1193–1212
Park E, Maniatty AM (2006) Shear modulus reconstruction in dynamic elastography: time harmonic case. Phys Med Biol 51:3697–3721
Sarvazyan A (1993) Shear acoustic properties of soft biological tissues in medical diagnostics. J Acoust Soc Am ASA 93(4):2329–2330
Weinberger H (1965) Partial differential equations with complex variables and transform methods. Wiley, New York

# Chapter 2
# An Efficient Method for Describing Plane Strain Bending of Viscoplastic Sheets at Large Strains

**Sergei Alexandrov, Elena Lyamina, and Alexander Pirumov**

**Abstract** An efficient method for describing plane strain bending of wide rigid viscoplastic sheets at large strains is developed. Both pure bending and bending under tension are considered. The approach is based on the transformation equations between Eulerian and Lagrangian coordinates. In addition, it is shown that it is advantageous to use the equivalent strain rate as an independent variable instead of the space coordinate. Due to this change in the independent variable, a uniform treatment of an arbitrary dependence of the yield stress on the equivalent strain rate is possible. The solution reduces to an ordinary differential equation that should be solved numerically. However, one term of this equation reduces to the expression $0/0$ at the initial instant. Therefore, analytic treatment of the differential equation is required to describe an initial stage of the process. An illustrative example is provided. This example illustrates the effect of the parameter involved in the Bingham model on the through-thickness distribution of stresses and the bending moment. It is shown that the thickness of the sheet decreases as the deformation proceeds.

**Keywords** Bending under tension · Viscoplasticity · Large strains · Analytic solution

S. Alexandrov · E. Lyamina (✉)
Ishlinsky Institute for Problems in Mechanics RAS,
101-1 Prospekt Vernadskogo,119526 Moscow, Russia
e-mail: lyamina@inbox.ru

S. Alexandrov
e-mail: sergei_alexandrov@spartak.ru

A. Pirumov
Department of Management Problems, Russian Technological University—MIREA,
78 Vernadsky Avenue, 119454 Moscow, Russia
e-mail: alpirumov@mail.ru

 
H. Altenbach et al. (eds.), *Advanced Materials Modelling for Mechanical, Medical and Biological Applications*, Advanced Structured Materials 155,
https://doi.org/10.1007/978-3-030-81705-3_2

## 2.1 Introduction

Bending is an important deformation mode in all sheet metal forming processes. A brief overview of this deformation mode in such processes has been provided in Ikumapayi et al. (2020). The present paper concerns the theoretical description of plane strain pure bending and bending under tension by analytic and semi-analytic methods. Such solutions are useful for the preliminary design of processes. They can also combine with methods used for identifying material properties (e.g., Kato et al. 2014; Laws 1981; Maeda et al. 2018; Oldroyd 1971; Urriolagoitia-Sosa et al. 2006).

Hill (1950) has provided the exact rigid perfectly plastic solution for both pure plane strain bending and plane strain bending under tension processes. The basic assumptions made in this work have also been adopted in all subsequent solutions of this class. Tan et al. (1995) have provided a semi-analytic solution based on anisotropic plasticity. Hill's quadratic yield criterion (Hill 1950) has been adopted in this work. The solution has been reduced to an ordinary differential equation, which is solved by the Runge–Kutta method. Tan et al. (1995) have also attempted to account for isotropic and kinematic hardening. However, the total strain found using the initial and final configurations has been used instead of the equivalent plastic strain, which should be calculated by integrating over the actual strain path, involved in the hardening laws. A method for calculating the correct value of the equivalent plastic strain in the bending process has been developed in Verguts and Sowerby (1975). An iterative procedure in conjunction with a numerical method for solving an ordinary differential equation is required for getting quantitative results. A simplified method to deal with rigid/plastic, hardening materials is based on an artificial strain distribution in the region between the current neutral line and the original center fiber (Dadras and Majless 1982).

The solutions reviewed above are for rigid plastic models. Elastic/plastic solutions are also available (e.g., Bruhns et al. 2003). However, it has been shown in Alexandrov et al. (2021) that the effect of elasticity is negligible in bending at large strains. It is even possible to accurately predict residual stresses using the method proposed in Lee and Dawson (1989). Therefore, the present paper focuses on a rigid viscoplastic model. The approach proposed in Alexandrov et al. (2006) for pure bending and then extended in Alexandrov et al. (2011) for bending under tension is employed.

## 2.2 Statement of the Problem

A wide sheet of initial thickness $H$ and initial width $2L$ is simultaneously bent by two couples $M$ and stretched by two forces $F$ (per unit length). The initial cross section is rectangular (Fig. 2.1a). The intermediate and final cross sections are ring sectors (Fig. 2.1b). The state of strain is plane. The outer and inner radii of the sheet are denoted as $r_{AB}$ and $r_{CD}$, respectively. The thickness of the sheet after any amount

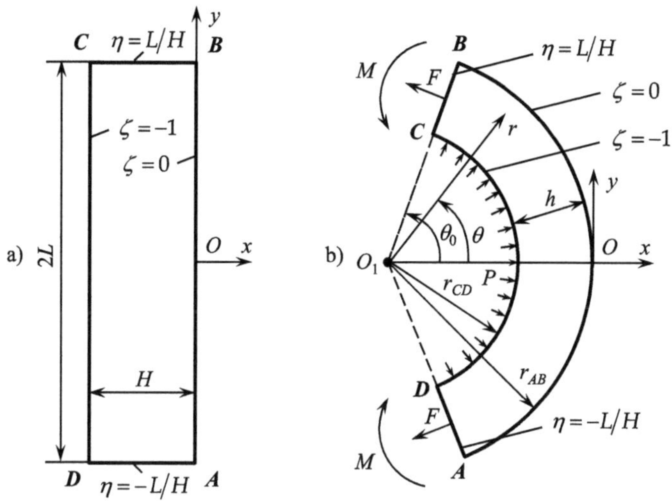


**Fig. 2.1** Geometry of the problem and coordinate systems

of deformation is $h$. It is assumed that pressure $P$ is uniformly distributed over the inner radius of the sheet. Then, the equilibrium equations require that

$$F = r_{CD} P. \tag{2.1}$$

The outer radius of the sheet is traction free. End effects are neglected. The elastic portion of the strain tensor is also neglected.

The constitutive equations comprise the pressure-independent yield criterion and its associated flow rule. Therefore, the material is incompressible, and the plane strain yield criterion takes the form:

$$|\sigma_1 - \sigma_2| = \frac{2}{\sqrt{3}}\sigma_Y. \tag{2.2}$$

Here, $\sigma_1$ and $\sigma_2$ are the principal stresses in the planes of flow, and $\sigma_Y$ is the tensile yield stress. The associated flow rule is equivalent to the equation of incompressibility, the condition that the principal stress and principal strain rate axes coincide, and the condition that the plastic work rate is positive. The incompressibility equation can be written as

$$\xi_1 + \xi_2 = 0. \tag{2.3}$$

Here, $\xi_1$ and $\xi_2$ are the principal strain rates in the planes of flow. The equivalent strain rate is defined as

$$\xi_{eq} = \sqrt{\frac{2}{3}}\sqrt{\xi_1^2 + \xi_2^2}. \tag{2.4}$$

It is assumed that the tensile yield stress is an arbitrary function of the equivalent strain rate. Then, Eq. (2.2) can be rewritten as

$$\frac{|\sigma_1 - \sigma_2|}{\sigma_0} = \frac{2}{\sqrt{3}}\Phi\left(\frac{\xi_{eq}}{\xi_0}\right). \tag{2.5}$$

Here, $\sigma_0$ is the static tensile yield stress and $\xi_0$ is a characteristic strain rate of the material.

## 2.3 General Solution

The general solution valid for any isotropic incompressible material model has been proposed in Alexandrov et al. (2006) for pure bending. This solution has been generalized on bending under tension in Alexandrov et al. (2011). These solutions are briefly overviewed in this section to provide the basis for developing an efficient method for solving the equations derived in Alexandrov et al. (2006, 2011).

The solutions above are based on the following transformation between the Eulerian Cartesian coordinate system $(x,\ y)$ and the Lagrangian coordinate system $(\zeta,\ \eta)$:

$$\begin{aligned} \frac{x}{H} &= \sqrt{\frac{\zeta}{a} + \frac{s}{a^2}}\cos(2a\eta) - \frac{\sqrt{s}}{a} \quad \text{and} \\ \frac{y}{H} &= \sqrt{\frac{\zeta}{a} + \frac{s}{a^2}}\sin(2a\eta). \end{aligned} \tag{2.6}$$

Here, $a$ is a time-like variable such that $a = 0$ at the initial instant and $s$ is a function of $a$. This function should be found from the solution. At the initial instant, $\zeta = x/H$ and $\eta = y/H$ if

$$s = \frac{1}{4} \tag{2.7}$$

at $a = 0$. The origin of the Cartesian coordinates coincides with the intersection of the outer radius of the sheet and the axis of symmetry of the process throughout the process of deformation. Then,

$$\zeta = 0 \quad \text{and} \quad \zeta = -1 \tag{2.8}$$

on lines *AB* and *CD*, respectively (Fig. 2.1). These equations are valid throughout the process of deformation. In addition to the coordinate systems above, one can introduce the following plane polar coordinate system $(r,\ \theta)$:

$$\frac{r}{H} = \frac{\sqrt{\zeta a + s}}{a} \quad \text{and} \quad \theta = 2a\eta. \tag{2.9}$$

It follows from (2.8) and (2.9) that

$$\frac{r_{AB}}{H} = \frac{\sqrt{s}}{a} \quad \text{and} \quad \frac{r_{CD}}{H} = \frac{\sqrt{s-a}}{a}. \tag{2.10}$$

Then, the current thickness of the sheet is

$$\frac{h}{H} = \frac{r_{AB}}{H} - \frac{r_{CD}}{H} = \frac{\sqrt{s}}{a} - \frac{\sqrt{s-a}}{a}. \tag{2.11}$$

One can determine the principal strain rates from (2.6). Then, the equivalent strain rate is found from (2.4) as

$$\xi_{\mathrm{eq}} = \frac{|\zeta + \mathrm{d}s/\mathrm{d}a\ |}{\sqrt{3}\,(\zeta a + s)}\frac{\mathrm{d}a}{\mathrm{d}t}. \tag{2.12}$$

Here, $t$ is the time. It is seen from this equation that the $\zeta-$coordinate of the neutral line is

$$\zeta = \zeta_n = -\frac{\mathrm{d}s}{\mathrm{d}a}. \tag{2.13}$$

Equation (2.12) can be rewritten as

$$\xi_{\mathrm{eq}} = \frac{(\zeta + \mathrm{d}s/\mathrm{d}a)}{\sqrt{3}\,(\zeta a + s)}\frac{\mathrm{d}a}{\mathrm{d}t} \tag{2.14}$$

in the range $0 \ge \zeta \ge \zeta_n$ and

$$\xi_{eq} = -\frac{(\zeta + \mathrm{d}s/\mathrm{d}a)}{\sqrt{3}\,(\zeta a + s)}\frac{\mathrm{d}a}{\mathrm{d}t} \tag{2.15}$$

in the range $\zeta_n \ge \zeta \ge -1$.

The transformation equations in (2.6) satisfy the equation of incompressibility. These equations also show that the principal strain rate trajectories coincide with the coordinate lines of $(\zeta, \eta)$—and $(r, \theta)$—coordinate systems. Therefore, the assumption that the principal stress trajectories coincide with the same coordinate lines allows one to satisfy the associated flow rule. Besides, in this case, the shear stress vanishes on lines *AB*, *BC*, *CD*, and *DA* (Fig. 2.1), as required by the stress boundary conditions. The only equilibrium equation that is not satisfied automatically is

$$\frac{\partial \sigma_r}{\partial r} + \frac{\sigma_r - \sigma_\theta}{r} = 0. \tag{2.16}$$

Here, $\sigma_r$ and $\sigma_\theta$ are the normal stresses referred to the polar coordinate system. The yield criterion (2.5) can be rewritten as

$$\frac{|\sigma_r - \sigma_\theta|}{\sigma_0} = \frac{2}{\sqrt{3}}\Phi\left(\frac{\xi_{eq}}{\xi_0}\right). \tag{2.17}$$

Since $\xi_\theta > \xi_r$ in the range $0 \geq \zeta \geq \zeta_n$ and $\xi_\theta < \xi_r$ in the range $\zeta_n \geq \zeta \geq -1$, equation (2.17) becomes

$$\frac{\sigma_r - \sigma_\theta}{\sigma_0} = -\frac{2}{\sqrt{3}}\Phi\left(\frac{\xi_{eq}}{\xi_0}\right) \tag{2.18}$$

in the range $0 \geq \zeta \geq \zeta_n$ and

$$\frac{\sigma_r - \sigma_\theta}{\sigma_0} = \frac{2}{\sqrt{3}}\Phi\left(\frac{\xi_{eq}}{\xi_0}\right) \tag{2.19}$$

in the range $\zeta_n \geq \zeta \geq -1$.

## 2.4 Method of Solution

Using (2.9), one can replace differentiation with respect to $r$ with differentiation with respect to $\zeta$ in (2.16). As a result,

$$\frac{\partial \sigma_r}{\partial \zeta} = \frac{a\left(\sigma_\theta - \sigma_r\right)}{2\left(\zeta a + s\right)}. \tag{2.20}$$

Consider the region $\zeta_n \leq \zeta \leq 0$. Equations (2.14) and (2.18) can be rewritten as

$$\varepsilon = \frac{\xi_{eq}}{\xi_0} = \frac{\mu\left(\zeta + \mathrm{d}s/\mathrm{d}a\right)}{\sqrt{3}\left(\zeta a + s\right)} \quad \text{and} \quad \frac{\sigma_r - \sigma_\theta}{\sigma_0} = -\frac{2}{\sqrt{3}}\Phi\left(\varepsilon\right). \tag{2.21}$$

Here, $\mu = \xi_0^{-1}\mathrm{d}a/\mathrm{d}t$. The general structure of equations (2.20) and (2.21) suggests using $\varepsilon$ as a new independent variable instead of $\zeta$. It follows from the first equation in (2.21) that

$$\frac{\partial \varepsilon}{\partial \zeta} = \frac{\mu}{\sqrt{3}}\frac{\left(s - a\mathrm{d}s/\mathrm{d}a\right)}{\left(\zeta a + s\right)^2}. \tag{2.22}$$

The chain rule yields

$$\frac{\partial \sigma_r}{\sigma_0 \partial \zeta} = \frac{\partial \sigma_r}{\sigma_0 \partial \varepsilon}\frac{\partial \varepsilon}{\partial \zeta}. \tag{2.23}$$

Substituting the second equation in (2.21) into (2.20) and, then, the resulting equation and (2.22) into (2.23) yields

$$\frac{\partial \sigma_r}{\sigma_0 \partial \varepsilon} = \frac{a\left(\zeta a + s\right)\Phi\left(\varepsilon\right)}{\mu\left(s - a\mathrm{d}s/\mathrm{d}a\right)}. \tag{2.24}$$

One can use the first equation in (2.21) to eliminate $\zeta$ in (2.24). As a result,

$$\frac{\partial \sigma_r}{\sigma_0 \partial \varepsilon} = \frac{a\Phi(\varepsilon)}{\mu - \sqrt{3}a\varepsilon}. \tag{2.25}$$

Let $\varepsilon_a$ be the value of $\varepsilon$ at $\zeta = 0$ (i.e., on *AB* in Fig. 2.1). Then, it follows from the first equation in (2.21) that

$$\varepsilon_a = \frac{\mu \mathrm{d}s/\mathrm{d}a}{\sqrt{3}s}. \tag{2.26}$$

The radial stress vanishes on *AB*. The solution of equation (2.25) satisfying this boundary condition can be written as

$$\frac{\sigma_r}{\sigma_0} = a\int_{\varepsilon_a}^{\varepsilon} \frac{\Phi(\beta)}{\left(\mu - \sqrt{3}a\beta\right)} \mathrm{d}\beta. \tag{2.27}$$

Here, $\beta$ is a dummy variable of integration. Let $\sigma_n$ be the value of the radial stress at the neutral line. The equivalent strain rate vanishes at this line. Therefore, it follows from (2.27) that

$$\frac{\sigma_n}{\sigma_0} = a\int_{\varepsilon_a}^{0} \frac{\Phi(\varepsilon)}{\left(\mu - \sqrt{3}a\varepsilon\right)} \mathrm{d}\varepsilon. \tag{2.28}$$

Consider the region $-1 \le \zeta \le \zeta_n$. Equations (2.15) and (2.19) can be rewritten as

$$\varepsilon = -\frac{\mu(\zeta + ds/da)}{\sqrt{3}(\zeta a + s)} \quad \text{and} \quad \frac{\sigma_r - \sigma_\theta}{\sigma_0} = \frac{2}{\sqrt{3}}\Phi(\varepsilon). \tag{2.29}$$

It follows from the first equation that

$$\frac{\partial \varepsilon}{\partial \zeta} = \frac{\mu}{\sqrt{3}} \frac{(a\mathrm{d}s/\mathrm{d}a - s)}{(\zeta a + s)^2}. \tag{2.30}$$

Substituting the second equation in (2.29) into (2.20) and, then, the resulting equation and (2.30) into (2.23) yields (2.24). One can use the first equation in (2.29) to eliminate $\zeta$ in (2.24). As a result,

$$\frac{\partial \sigma_r}{\sigma_0 \partial \varepsilon} = \frac{a\Phi(\varepsilon)}{\mu + \sqrt{3}a\varepsilon}. \tag{2.31}$$

Let $\varepsilon_b$ be the value of $\varepsilon$ at $\zeta = -1$ (i.e., on *CD* in Fig. 2.1). Then, it follows from the first equation in (2.29) that

$$\varepsilon_b = \frac{\mu\,(1 - \mathrm{d}s/\mathrm{d}a\,)}{\sqrt{3}\,(s - a)}. \tag{2.32}$$

The radial stress is equal to $-P$ on $CD$. The solution of Eq. (2.31) satisfying this boundary condition can be written as

$$\frac{\sigma_r}{\sigma_0} = a\int_{\varepsilon_b}^{\varepsilon} \frac{\Phi\,(\beta)}{\left(\mu + \sqrt{3}a\beta\right)}\mathrm{d}\beta - \frac{P}{\sigma_0}. \tag{2.33}$$

The radial stress must be continuous across the neutral line. Therefore, it follows from (2.33) that

$$\frac{\sigma_n}{\sigma_0} = a\int_{\varepsilon_b}^{0} \frac{\Phi\,(\varepsilon)}{\left(\mu + \sqrt{3}a\varepsilon\right)}\mathrm{d}\varepsilon - \frac{P}{\sigma_0}. \tag{2.34}$$

Equations (2.28) and (2.34) combine to give

$$a\int_{\varepsilon_b}^{0} \frac{\Phi\,(\varepsilon)}{\left(\mu + \sqrt{3}a\varepsilon\right)}\mathrm{d}\varepsilon - \frac{P}{\sigma_0} = a\int_{\varepsilon_a}^{0} \frac{\Phi\,(\varepsilon)}{\left(\mu - \sqrt{3}a\varepsilon\right)}\mathrm{d}\varepsilon. \tag{2.35}$$

In the case of pure bending, this equation reduces to

$$\int_{\varepsilon_b}^{0} \frac{\Phi\,(\varepsilon)}{\left(\mu + \sqrt{3}a\varepsilon\right)}\mathrm{d}\varepsilon = \int_{\varepsilon_a}^{0} \frac{\Phi\,(\varepsilon)}{\left(\mu - \sqrt{3}a\varepsilon\right)}\mathrm{d}\varepsilon. \tag{2.36}$$

One can eliminate $\varepsilon_a$ and $\varepsilon_b$ in (2.35) and (2.36) using (2.26) and (2.32). In the case of bending under tension, it is also necessary to use (2.1) and (2.10) to eliminate $P$. In either case, the resulting equation involves $a$, $s$, and $\mathrm{d}s/\mathrm{d}a$. This equation can be regarded as an ordinary differential equation for $s$ as a function of $a$, though a non-standard numerical technique is required to solve it. Equation (2.7) supplies the initial condition.

Once Eq. (2.35) or (2.36) has been solved, the distribution of the radial stress is immediate from (2.27) and (2.33). The distribution of the circumferential stress is then determined from (2.21) and (2.29). The bending moment per unit length, $M$, is given by (Alexandrov et al. 2011)

$$m = \frac{2\sqrt{3}M}{\sigma_0 H^2} = \frac{\sqrt{3}}{a}\int_{-1}^{0} \left(\frac{\sigma_\theta}{\sigma_0} - \frac{F}{\sigma_0 h}\right)\mathrm{d}\zeta. \tag{2.37}$$

Here, $m$ is the dimensionless bending moment and $h$ should be eliminated using (2.11). Note that $m = 1$ in the case of pure bending ($F = 0$) of rigid perfectly plastic material ($\Phi = 1$). One can use equations (2.22) and (2.30) to replace integration with respect to $\zeta$ with integration with respect to $\varepsilon$ in (2.37). As a result, this equation becomes

$$m = \frac{3\left(ads/da - s\right)}{a}\int_{\varepsilon_b}^{0}\left(\frac{\sigma_\theta}{\sigma_0} - \frac{F}{\sigma_0 h}\right)\frac{\mathrm{d}\varepsilon}{\left(\sqrt{3}a\varepsilon + 1\right)^2} - \frac{3\left(a\mathrm{d}s/\mathrm{d}a - s\right)}{a}\int_{0}^{\varepsilon_a}\left(\frac{\sigma_\theta}{\sigma_0} - \frac{F}{\sigma_0 h}\right)\frac{\mathrm{d}\varepsilon}{\left(\sqrt{3}a\varepsilon - 1\right)^2}. \tag{2.38}$$

## 2.5 Illustrative Example

In the case of the Bingham solid, the function $\Phi$ involved in (2.5) is

$$\Phi\left(\varepsilon\right) = 1 + A\varepsilon \tag{2.39}$$

where $A$ is constant. Consider the pure bending process. Substituting (2.39) into (2.36) and integrating yields

$$\frac{2A\mu}{\sqrt{3}a}\ln\mu + A\left(\varepsilon_b - \varepsilon_a\right) + \ln\left(\frac{\mu + \sqrt{3}a\varepsilon_b}{\mu - \sqrt{3}a\varepsilon_a}\right) - \frac{A\mu}{\sqrt{3}a}\ln\left[\left(\mu + \sqrt{3}a\varepsilon_b\right)\left(\mu - \sqrt{3}a\varepsilon_a\right)\right] = 0. \tag{2.40}$$

This equation suggests introducing the new variable $\lambda = \varepsilon/\mu$ and the new parameter $B = \mu A$. Then, Eq. (2.40) becomes

$$B\left(\lambda_b - \lambda_a\right) + \ln\left(\frac{1 + \sqrt{3}a\lambda_b}{1 - \sqrt{3}a\lambda_a}\right) - \frac{B}{\sqrt{3}a}\ln\left[\left(1 + \sqrt{3}a\lambda_b\right)\left(1 - \sqrt{3}a\lambda_a\right)\right] = 0. \tag{2.41}$$

It follows from (2.26) and (2.32) that

$$\lambda_a = \frac{\mathrm{d}s/\mathrm{d}a}{\sqrt{3}s} \quad \text{and} \quad \lambda_b = \frac{\left(1 - \mathrm{d}s/\mathrm{d}a\right)}{\sqrt{3}\left(s - a\right)}. \tag{2.42}$$

After eliminating $\lambda_a$ and $\lambda_b$ using (2.42), Eq. (2.41) can be solved numerically to find $s$ as a function of $a$. However, the last term in (2.41) reduces to the expression $0/0$ at $a = 0$. Therefore, the solution for small $a$ should be investigated analytically.

Due to symmetry, the neutral line is given by the equation $\zeta_n = -1/2$ at the initial instant. Therefore, it follows from (2.13) that $ds/da = 1/2$ at $a = 0$. Then, the function $s\,(a)$ satisfying this condition and (2.7) can be represented as

$$s = \frac{1}{4} + \frac{a}{2} + ka^2 + o\left(a^2\right) \tag{2.43}$$

as $a \to 0$. Substituting (2.43) into (2.41) and (2.42) shows that

$$k = \frac{3\sqrt{3} + 8B}{12\left(\sqrt{3} + 2B\right)}. \tag{2.44}$$

Equations (2.43) and (2.44) supply $s\,(a)$ and $\mathrm{d}s/\mathrm{d}a$ for small $a$.

Figure 2.2 depicts the dependence of $\zeta_n$ on $H/r_{CD}$ at several values of $B$ of the interval $1 \le B \le 5$. It is seen from this figure that the effect of $B$ on the position of the neutral line is not significant, and the difference between the curves is almost invisible. The lowest curve corresponds to $B = 1$. The through-thickness distribution of the radial and circumferential stresses at several stages of the process is shown in Figs. 2.3 and 2.4, respectively. These distributions correspond to $B = 3$. The dimensionless coordinate $X$ is

$$X = \frac{r}{H} - \frac{r_{CD}}{H}. \tag{2.45}$$

Thus $X = 0$ corresponds to the inner surface of the sheet. One can express the right-hand side of (2.45) as a function of $a$, $s$, and $\zeta$ using (2.9) and (2.10). Having found the through-thickness distribution of the circumferential stress, the dimensionless bending moment is readily calculated from (2.38) in which one should put $F = 0$. The dependence of $m$ on $H/r_{CD}$ at several values of $B$ is shown in Fig. 2.5.

## 2.6 Conclusions

A general solution has been derived to describe the plane strain bending process for rigid viscoplastic material. The solution is valid for any dependence of the yield stress on the equivalent strain rate. The transformation equations between Eulerian and Lagrangian coordinates proposed in Alexandrov et al. (2006) have been employed. However, it has been found that it is advantageous to replace one of the Lagrangian coordinates with the equivalent strain rate. This change in the independent variable allows for a unified treatment for the arbitrary dependence of the yield stress on

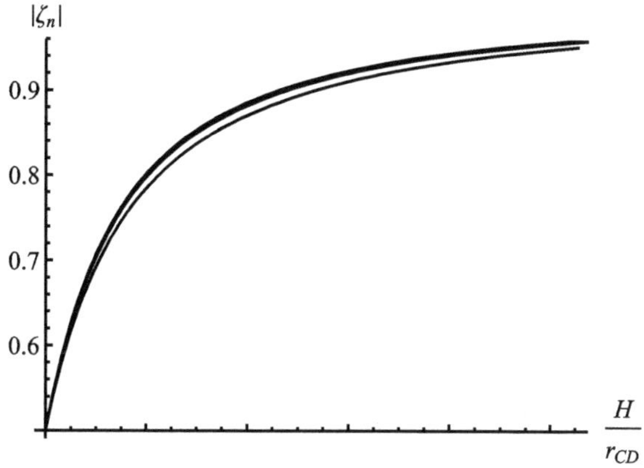


**Fig. 2.2** Variation of the $\zeta$−coordinate of the neutral line with the inner radius of the sheet for several $B$-values of the range $1 \leq B \leq 5$

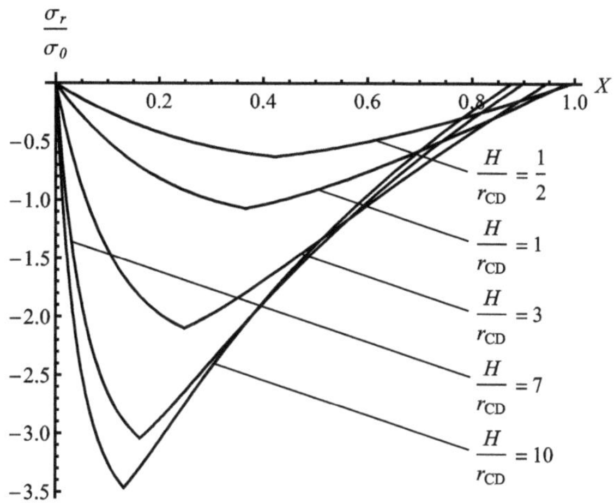


**Fig. 2.3** Through-thickness distribution of the radial stress at $B = 3$ and several values of the sheet's inner radius

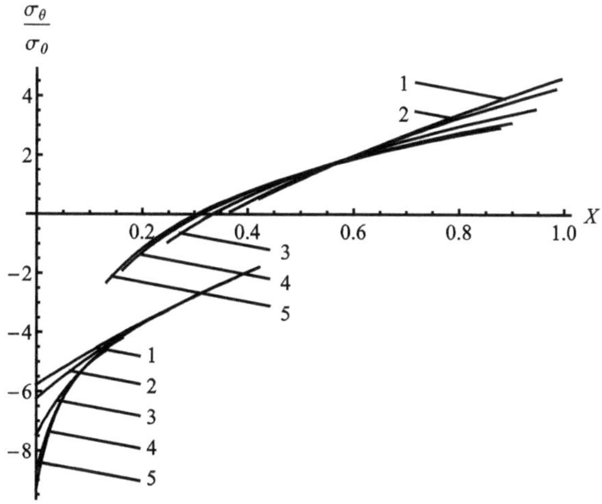


**Fig. 2.4** Through-thickness distribution of the circumferential stress at $B = 3$ and several values of the sheet's inner radius (curve 1 corresponds to $H/r_{CD} = 1/2$, curve 2 to $H/r_{CD} = 1$, curve 3 to $H/r_{CD} = 3$, curve 4 to $H/r_{CD} = 7$, and curve 5 to $H/r_{CD} = 10$)

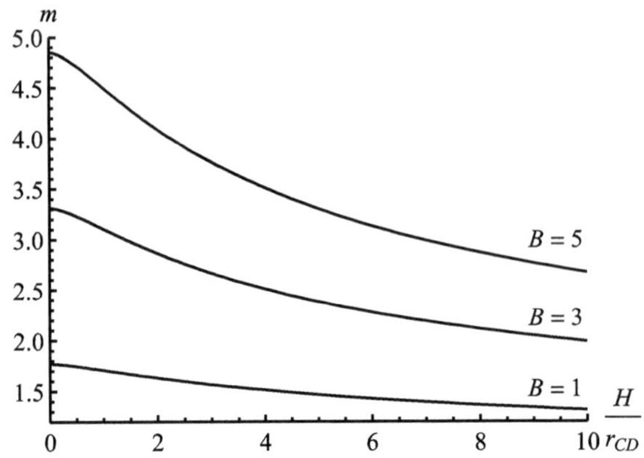


**Fig. 2.5** Variation of the dimensionless bending moment with the inner radius of the sheet for several $B$-values

the equivalent strain rate. The solution has been reduced to solving an ordinary differential equation.

The illustrative example has shown some interesting features of the bending of viscoplastic sheets. In particular, the position of the neutral line is not sensitive to the parameter that is involved in the Bingham model (Fig. 2.2). On the other hand, as expected, the bending moment is very sensitive to this parameter (Fig. 2.5). It is seen from Figs. 2.3 and 2.4 that the thickness of the sheet decreases as the deformation proceeds. This feature of the solution is very different from the rigid perfectly plastic solution, which predicts no change in the thickness in the pure bending process.

**Acknowledgements** This research was made possible by the grants RFBR-19-51-52003 (Russia) and AAAA-A20- 120011690136-2 (Russia).

## References


Alexandrov S, Kim J-H, Chung K, Kang T-J (2006) An alternative approach to analysis of plane-strain pure bending at large strains. J Strain Anal Eng Des. https://doi.org/10.1243/03093247JSA154

Alexandrov S, Manabe K, Furushima T (2011) A general general analytic solution for plane strain bending under tension for strain-hardening material at large strains. Arch Appl Mech. https://doi.org/10.1007/s00419-011-0529-9

Alexandrov S, Lyamina E, Hwang Y-M (2021) Finite pure plane strain bending of inhomogeneous anisotropic sheets. Symmetry. https://doi.org/10.3390/sym13010145

Bruhns OT, Gupta NK, Meyers ATM, Xiao H (2003) Bending of an elastoplastic strip with isotropic and kinematic hardening. Arch Appl Mech. https://doi.org/10.1007/s00419-002-0273-2

Dadras P, Majless SA (1982) Plastic bending of work hardening materials. Trans ASME J Eng Ind. https://doi.org/10.1115/1.3185823

Hill R (1950) The mathematical theory of plasticity. Clarendon Press, Oxford

Ikumapayi OM, Akinlabi ET, Madushele N, Fatoba SO (2020) A brief overview of bending operation in sheet metal forming. In: Emamian S, Awang M, Yusof F (eds) Advances in manufacturing engineering; Lecture notes in mechanical engineering. Springer, Singapore, pp. 149–159. https://doi.org/10.1007/978-981-15-5753-8_14

Kato H, Tottori Y, Sasaki K (2014) Four-point bending test of determining stress-strain curves asymmetric between tension and compression. Exp Mech. https://doi.org/10.1007/s11340-013-9791-9

Laws V (1981) Derivation of the tensile stress-strain curve from bending data. J Mater Sci. https://doi.org/10.1007/BF01033845

Lee Y, Dawson PR (1989) Obtaining residual stresses in metal forming after neglecting elasticity on loading. ASME J Appl Mech. https://doi.org/10.1115/1.3176086

Maeda T, Noma N, Kuwabara T, Barlat F, Korkolis YP (2018) Measurement of the strength differential effect of DP980 steel sheet and experimental validation using pure bending test. J Mater Process Technol. https://doi.org/10.1016/j.jmatprotec.2018.02.009

Oldroyd PWJ (1971) Tension-compression stress-strain curves from bending tests. J Strain Anal. https://doi.org/10.1243/03093247V064286

Tan Z, Persson B, Magnusson C (1995) Plastic bending of anisotropic sheet metals. Int J Mech Sci. https://doi.org/10.1016/0020-7403(94)00069-V

Urriolagoitia-Sosa G, Durodola JF, Lopez-Castro A, Fellows NA (2006) A method for the simultaneous derivation of tensile and compressive behaviour of materials under Bauschinger effect using bend tests. Proc Inst Mech Eng Part C. https://doi.org/10.1243/09544062JMES180

Verguts H, Sowerby R (1975) The pure plastic bending of laminated sheet metals. Int J Mech Sci. https://doi.org/10.1016/0020-7403(75)90061-2

# Chapter 3
# Using a Radio Interferometer for Measurement of the Dynamic Poisson's Ratio of Wood

**Vladimir Balandin, Vladimir Balandin Jr., Andrey K. Lomunov, Vladimir Parkhachev, and Tatiana Yuzhnina**

**Abstract** We proposed a new method for measuring the dynamic Poisson's ratio of both isotropic and anisotropic materials. The specimen was loaded with a compressive pulsed load generated by gas gun and a split Hopkinson bar. The longitudinal compression deformation of the specimen was determined from the signals recorded with the use of small-base strain gauges glued to the measuring pressure bars. To measure the temporal progress of the radial components of the specimen deformation, a millimeter-wave radio interferometer was used. To assess the possible asymmetry of the radial expansion, measurements were carried out using two independent channels irradiating diametrically opposite zones of the lateral surface of the specimen. The tests were carried out on a pine specimen with air humidity in the form of a cylinder 54 mm in diameter and 30 mm in height. A compressive pulsed load was applied along the fibers by using the split Hopkinson pressure bar (SHPB). The separate movement of the lateral surfaces of the sample was recorded both along and across the annual layers by using the two-channel radio interferometer. It was determined that the opposite displacements of the regions of the lateral surface of the specimen during its expansion along the annual layers are quite close; while during expansion across the annual layers, they are very different. The relative transverse deformation of the sample in both cases was determined as the sum of lateral opposite displacements relative to the specimen's diameter. As a result, two components of

V. Balandin · V. Balandin Jr. · A. K. Lomunov (✉)
Research Institute for Mechanics, National Research Lobachevsky State University of Nizhny Novgorod, Nizhny Novgorod, Russia
e-mail: lomunov@mech.unn.ru

V. Balandin
e-mail: balandin@mech.unn.ru

V. Balandin Jr.
e-mail: rustydog2007@yandex.ru

V. Parkhachev · T. Yuzhnina
National Research Lobachevsky State University of Nizhny Novgorod, Nizhny Novgorod, Russia
e-mail: wp@rf.unn.ru

T. Yuzhnina
e-mail: yuzhina_tatiana@mech.unn.ru

 
H. Altenbach et al. (eds.), *Advanced Materials Modelling for Mechanical, Medical and Biological Applications*, Advanced Structured Materials 155,
https://doi.org/10.1007/978-3-030-81705-3_3

the dynamic Poisson's ratio were obtained, which amounted to ~0.2 (in the direction along the annual layers) and ~0.24 (in the direction across the annual layers).

**Keywords** Dynamic Poisson's ratio · Kolsky method · Radio interferometer · Orthotropic medium · Wood

## 3.1 Introduction

Currently, computer-aided design (CAD) techniques such as finite element modeling (FEM) are used to design a variety of critical structures. Reliable FEM is based on trustworthy data on material properties such as elastic constants, including Poisson's ratio. For anisotropic materials such as wood, the Poisson's ratio in tension and compression for each orthogonal direction is the ratio of the lateral strain to the longitudinal strain. These parameters have mainly been studied in the last 4–5 decades using simple mechanical or electrical measuring systems (e.g., strain gauges, mechanical extensometer systems, inductive devices for strain measuring) due to the high cost and lack of availability of optical non-contact measuring systems (Davis 2004).

Poisson's ratio can be obtained using both static and dynamic test methods. Static testing is most often used to determine Poisson's ratio, however, when operating structures under impulse loads, for their reliable design and calculation, numerical models must include either the value of the dynamic Poisson's ratio (for isotropic bodies), or a set of Poisson's ratios in characteristic directions (for anisotropic bodies). The dynamic Poisson's ratio can be determined using various impact methods and analysis of the response to applied low amplitude impulse loads. Ahmed (2018) paid attention to some of the dynamic characteristics of concrete using shock resonance (ASTM C215-14, 2014), when a suspended sample is impulsively loaded with a small impact device. Then the sample's response is measured by accelerometers on the sample, which makes it possible to record the acceleration of particles in the form of the main longitudinal, torsional, and transverse resonant frequencies of concrete, respectively. Fundamental frequencies for three different vibration modes are obtained by choosing the right impact point and accelerometer mounting locations.

The first studies of Poisson's ratio for spruce wood were carried out by Carrington (1921,1922a; b). For anisotropic materials such as wood, Poisson's ratio for one orthogonal direction is the ratio of lateral compression (lateral strain) to axial tension (axial strain). Carrington derived Poisson's ratio from beam bending experiments by measuring with a telescope the curvature in the transverse direction (transverse strain ($\varepsilon q$)) and longitudinal direction (axial strain ($\varepsilon l$)). Hörig (1931) revised this data and adopted Voigt's (1882, 1887, 1966) ideas on the orthotropic behavior of wood materials. The Hörig (1935) model is the basis for the orthotropic description of wood, which is still used today. Further significant studies of spruce wood were carried out by Wommelsdorff (1966) and Neuhaus (1981). Through tensile and bending experiments, they determined six orthotropic Poisson's ratios using for strain measuring inductive devises as well as strain gauges. In addition, Niemz and Caduff (2008) investigated Poisson's ratio for spruce wood. A dynamic test method was proposed to determine the Poisson's ratio of sawn timber (Wang et al 2017), based on the curve of first-order bending vibrations of a cantilever slab. The free vibration of the cantilever specimen was excited by tapping. Using strain gauge rosettes, Poisson's ratios were measured along and across the fibers on radial sections and along the fibers on transverse sections of Sitka spruce samples. In Swamy (1971), Poisson's ratio for concrete was determined on the basis of an analysis of the resonant frequencies of longitudinal and torsional vibrations of a beam, as well as the speed of a pulse excited by a low-power generator of transverse vibrations. In Boiko et al. (2008), a method was proposed for determining the dynamic Poisson's ratio, based on measuring the dynamic axial and lateral deformations of two samples made of the same material. The method is based on the analysis of forced vibrations of a cylindrical specimen loaded with an inertial mass. The top and bottom surfaces of the sample's viscoelastic material were attached to a load disk and a rigid base plate, respectively. The rigid base plate was subjected to forced vibrations from a vibration exciter. Two accelerometers were used to track the movements of the base plate and the loading disk. The ratio of magnitudes and the phase difference between the oscillation of the load disk and the vibration of the base plate represents the axial dynamic deformation of the specimen. Gao et al. (2016) also proposed principles and methods for determining the dynamic Poisson's ratio of wood based on the analysis of bending vibrations of a cantilever slab using tensometry. Before gluing the strain gauge rosette to the surface of the board, a regression analysis was used to select a location where the shear stress inside the wood board was zero during first-order bending vibration. The ratio between the linear amplitudes of the transverse and longitudinal deformation at a first-order bending frequency was taken as the measured value of the Poisson's ratio of the material. In Cao et al. (2019), a method of dynamic tests for Poisson's ratio of sawn timber was proposed, based on the curve of first-order bending vibrations of a cantilever slab. Using rosette strain gauges, the longitudinal and transverse deformation of the cantilever plate was measured in which vibrations were excited by the impact mechanism with a rubber tip.

Methods based on ultrasonic non-destructive testing are quite common methods for indirect measurement of the dynamic Poisson's ratio. The ultrasonic pulse velocity (UPV) test method (Pal 2019) two transducers are used to generate and receive an ultrasonic pulse (electrical signal). In this case, the receiving transducer is located exactly on the opposite side of the test sample. The speed of ultrasonic pulses of cubic concrete specimens was determined from the time of transmission of longitudinal and transverse waves through the body of a specimen of known geometry. The recorded parameters were used to determine the dynamic values of the Poisson's parameter and the modulus of elasticity of the concrete. In Wang et al. (2015), using the concrete as an example, an energy approach was used to obtain a correlation between the shear modulus and the slab torsion frequencies based on the form of first-order torsional vibrations. Also, a relationship was obtained between the dynamic modulus of elasticity of concrete and the first-order bending of a free slab. Based on the analysis of stresses and strains in the first-order bending mode of a concrete cantilever slab and the physical relationship between stress and strain in a plane stress state, a dynamic method for determining Poisson's ratio for concrete was proposed.

More sophisticated measurement methods (electronic interferometry, speckle structures and a combination of laser and video extensometry) were used in Kumpenza et al. (2018) in experiments on uniaxial tension of spruce wood, which made it possible to determine all six basic Poisson's ratios. Keunecke et al. (2008) used digital image correlation (DIC) as a research method to measure strain distribution. DIC is a non-contact optical measurement of the deformation of the sample surface under dynamic loading (Valla et al. 2011). In Hoshino et al. (2018), a procedure was used to measure two independent properties of a viscoelastic material using the digital image correlation during dynamic viscoelasticity testing. The efficiency of the proposed method is demonstrated by measuring the properties of the soft epoxy resin material. The measurement example shows that with the proposed procedure, reference curves of the time-dependent dynamic Poisson's ratio as well as other material properties can be obtained over a wide range of times. The DIC technique is quite informative, but technically very difficult to implement; it requires a high-speed camera with a high resolution and licensed software for mathematical correlation analysis of sequential images of the working area of the sample recorded by a digital camera during its deformation. In addition, for this method to be successful, the specimen must be specially prepared by applying a random bit pattern (speckle) to its surface.

In this paper, we propose a new method for determining the dynamic Poisson's ratio of both isotropic and anisotropic materials. The main measuring instrument in the work is a millimeter-wave interferometer (radio interferometer).

## 3.2 Research Method

The specimen is loaded with a pulsed compressive load by an installation, which implements the Kolsky (1949) method using a split Hopkinson bar (SHB). SHB consists of two long pressure bars equipped with strain gauges for measuring elastic strain pulses (incident $\varepsilon^I(t)$, reflected $\varepsilon^R(t)$, and transmitted $\varepsilon^T(t)$) and a short specimen made of the material under study, placed between their ends (Fig. 3.1). The main relations of the Kolsky method are as follows:

$$\sigma_x(t) = E\frac{A}{A_0}\varepsilon^T(t), \quad \varepsilon_x(t) = -\frac{2C}{L_0}\int\limits_0^t \varepsilon^R(t)\mathrm{d}t,$$

$$\dot{\varepsilon}_x(t) = -\frac{2C}{L_0}\varepsilon^R(t), \tag{3.1}$$

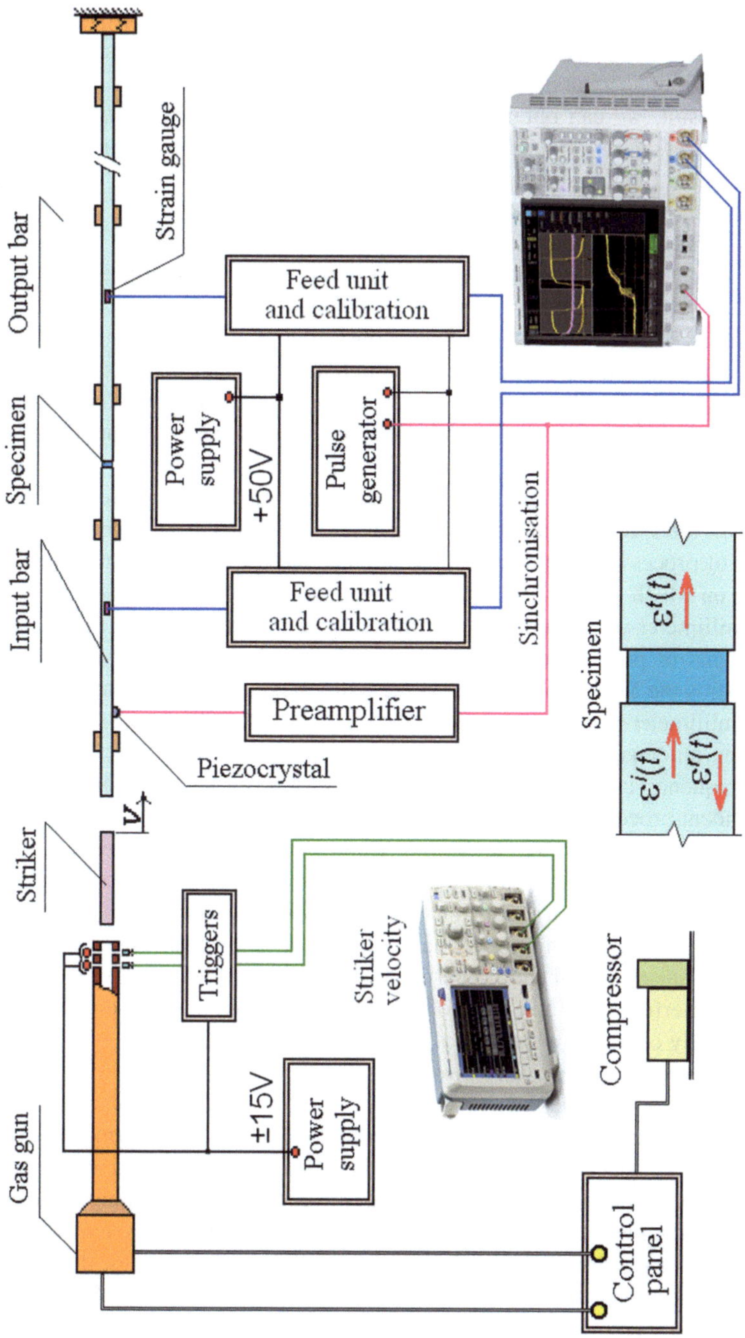


**Fig. 3.1** Compression test setup scheme

here $E$, $A$, and $C$ are the elasticity modulus of pressure bars, their cross section area, and elastic wave velocity, respectively, $A_0$ and $L_0$ are initial cross section area and initial length of specimen, respectively.

As a result, we can obtain the time dependences of the development of axial values of stress $\sigma_x(t)$, strain $\varepsilon_x(t)$, and strain rate $\acute{\varepsilon}_x(t)$ of the specimen and then, after excluding the time parameter, construct a stress–strain curve $\sigma_x(\varepsilon_x)$ and the curve $\acute{\varepsilon}_x(\varepsilon_x)$ (Bragov and Lomunov 1995).

A 57 mm gas gun was used to load the SHB with a pulsed load. Measuring bars with a diameter of 60 mm and a length of 1.5 m are made of high strength aluminum alloy.

A three-channel radio interferometer operating at a wavelength of 3.2 mm was used to measure the development of the radial component of the specimen deformation in time $\varepsilon_r(t)$. The undoubted advantages of the radio interferometer, when testing structurally inhomogeneous bodies, are a sufficiently large area of illumination, which makes it possible to neutralize the effect of structural inhomogeneity of the surface and to obtain integral displacement on a selected area of the lateral surface of the specimen. In addition, the radio interferometer is undemanding to the quality of processing of the reflecting surface, and it can successfully work with a surface on which there are inhomogeneities of 0.1–0.2 mm.

A millimeter-wave radio interferometer is a device designed to measure the movements of physical objects that can reflect radio waves. This interferometer implements the Michelson scheme in the radio range. The device includes a highly stable millimeter-wave generator of electromagnetic oscillations, transmitting and receiving antennas, a receiving device, and a data recording unit (oscilloscope). High-frequency circuit elements are connected by a waveguide path. Up to three independent processes can be registered at the same time.

The measurements were carried out using two independent channels irradiating diametrically opposite zones of the lateral surface of the specimen to exclude distortion of the recorded radial expansion of the specimen due to its possible displacement as a whole during loading, or during nonaxisymmetric expansion of the specimen due to the anisotropy of the properties of the material under study. The layout of the radio interferometer antennas for separate registration of radial displacements of diametrically opposite deformation zones is shown in Fig. 3.2.

In the experiments, we used two groups of receiving and transmitting antennas located on opposite sides of the specimen. The transmitting (1) and receiving (2) antennas were located in blocks of foamed polyethylene in the form of rectangular parallelepipeds, placed on the frame of the installation. The antennas were placed in a horizontal plane passing through the axis of the gun barrel at the same distance from the axis of the measuring bars and the specimen. The antenna system of the radio interferometer represents the "open ends" of dielectric waveguides, through which the probe radiation is removed from the interferometer, and the signal reflected from the specimen is returned back to the interferometer. Dielectric waveguides are placed in a flexible cylindrical shell made of foamed polyethylene along the entire length of the interferometer. The transmitting (1) and receiving (2) antennas were placed in a

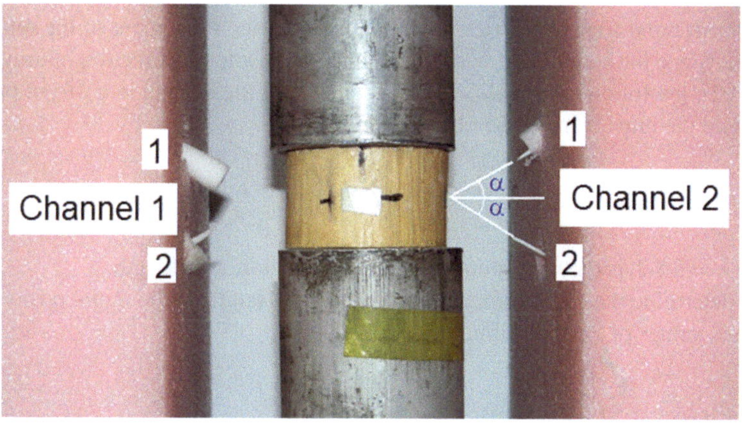


**Fig. 3.2** Scheme of recording the radial component of deformation with a radio interferometer

block of foamed polyethylene at a fixed angle α to the normal to the lateral surface of the specimen at the point of irradiation (Fig. 3.2).

The interferometer works as follows. Continuous electromagnetic oscillations of constant power at a fixed frequency created by the generator are fed through the elements of the waveguide path to the transmitting antenna and are radiated into the environment. The reflected signal from the moving object is received by the receiving antenna and through the waveguides is fed to the input of the receiving device. A reference oscillation with a frequency equal to the frequency of the radiated electromagnetic wave is supplied to the other input of the receiving device from the generator. The connection between the parameters of the interferogram and the parameters of the motion of the object under study is due to the Doppler effect. The current phase and instantaneous frequency of the interferogram, respectively, carry information about the displacement and instantaneous velocity of the object under study, from which the radio wave emitted by the interferometer has been reflected.

The probing signals have a wavelength $\lambda = 3.2$ mm in vacuum (frequency 93.75 GHz). The radio signals reflected from the surface have a frequency that differs by the magnitude of the Doppler shift. In the radio interferometer, the received signals containing the Doppler frequency shift are transferred by the local oscillator to an intermediate frequency of 2 GHz, after which they are recorded by a Tektronix DPO 7254C oscilloscope with a sampling rate of 10 GHz. At the same time, a reference signal (also transferred to the intermediate frequency) from the interferometer generator at an unbiased frequency is recorded. Further, the processing of the received signals (reference and reflected, containing information about the Doppler frequency shift) is performed in software. This made it possible to avoid the use of hardware signal detection and additional errors introduced by it.

The determination of the value of the phase difference between the reference and reflected signals for each recording point was carried out by the method of numerical optimization. Let us denote the phase difference for the $n$-th point by $\varphi_n$. The change in $\varphi_n$ from point to point carries information about the movement of the reflecting surface. Since the sounding is carried out at an angle α to the direction of movement (Fig. 3.2), the following equality is true, connecting the change in $\varphi_n$ from point to point and the movement of the object $\Delta x_n$:

$$\varphi_n - \varphi_{n-1} = 2k\Delta x_n \cos\alpha$$

where $k = 2\pi/\lambda$ is the wave number, λ is the radiation wavelength.

To determine the displacement of the lateral surface from the initial position, the obtained values of $\Delta x_n$ are summed

$$X_n = \sum \Delta x_n.$$

## 3.3 Test Results

The tests were carried out on a specimen of pine with air humidity in the form of a cylinder 54 mm in diameter and 30 mm in height (Fig. 3.3). For better reflection of the probe radiation from the lateral surface of the specimen, small (5 × 10 mm) pieces of aluminum foil were glued onto diametrically opposite zones of the lateral surface of the sample in the *T* and *R* directions (Fig. 3.3).

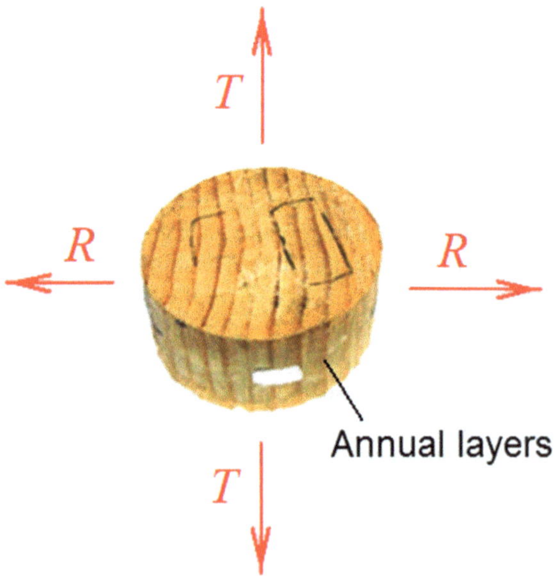


**Fig. 3.3** Specimen tested with the registration of transverse expansion along the annual growth rings (*T*-direction) and radial expansion across the growth rings (*R*-direction)

The specimens were loaded along the grains by compression pulses generated by loading the SHB with strikers made of high strength aluminum alloy with a diameter of 57 mm, accelerated in the barrel of a gas gun to speeds of 12–16 m/s. At these impact velocities, the specimen does not obtain irreversible deformations, and so, it can be tested many times. Compression pulses in the measuring bars were recorded simultaneously with the signals of the interferometer.

Since the radio interferometer has only three independent measuring channels, it is impossible to simultaneously register the radial expansion of the specimen in two mutually perpendicular directions (*T* and *R*). Therefore, in this work, the expansion of the specimen in the *T*-direction and in the *R*-direction was separately recorded in two independent experiments.

In the first experiment, two channels of the radio interferometer were used to separately register the displacements of the opposite lateral surfaces of the sample along the tree rings (direction *T*). The processes of the development of these transverse displacements obtained separately are shown in Fig. 3.4.

Figure 3.4 clearly shows that the displacements of the areas of the lateral surface of the specimen located on diametrically opposite sides of it are quite close, but have some differences associated with possible movement and rotation of the specimen as a whole. In this case, the transverse deformation of the specimen was determined as the sum of the lateral displacements is related to the diameter of the specimen. The process of development of longitudinal deformation of the specimen $\varepsilon_x(t)$ was determined from the signals from strain gauges on a loading pressure bar (1). The

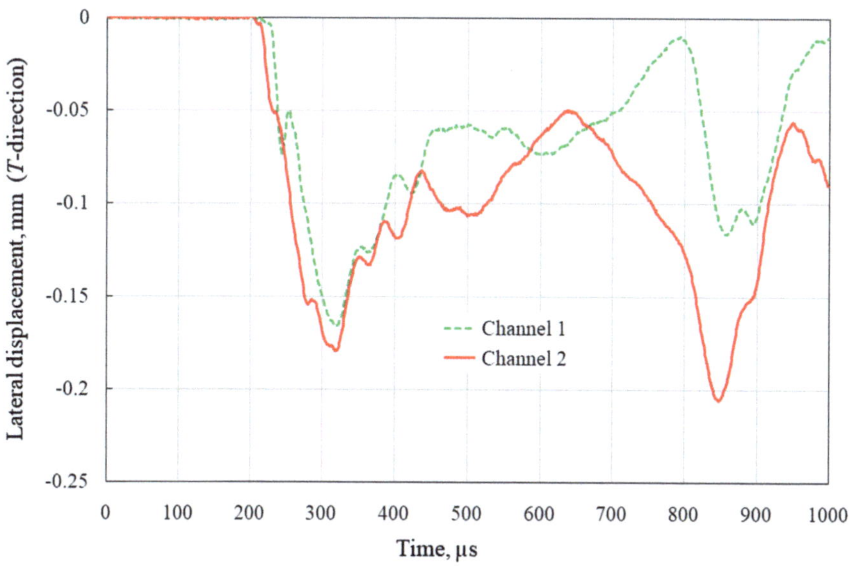


**Fig. 3.4** Displacement of the lateral surface of the sample in the *T*-direction (along the growth rings)

obtained dependences of the transverse and longitudinal deformations of the specimen were synchronized with each other in time (Fig. 3.5). Then the dependence of the transverse strain on the longitudinal strain was plotted (Fig. 3.6). This curve is well approximated by a linear dependence with a slope of ~0.2. This will be the required *TX*-component of Poisson's ratio in the direction *T* (along the tree rings).

Since the specimen did not receive residual deformations during the above test, the next experiment was carried out with the same specimen, but it was turned on 90°, and the transverse deformation was recorded across the wood layers (*R*-direction in Fig. 3.3). The speed of the striker (and, accordingly, the amplitude of the longitudinal deformation of the specimen) in this experiment was somewhat different

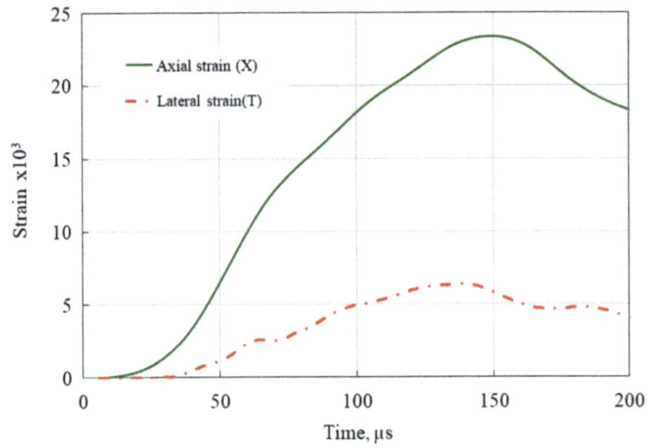


**Fig. 3.5** Dependences of the longitudinal and transverse deformations (direction *T*) of the specimen on time

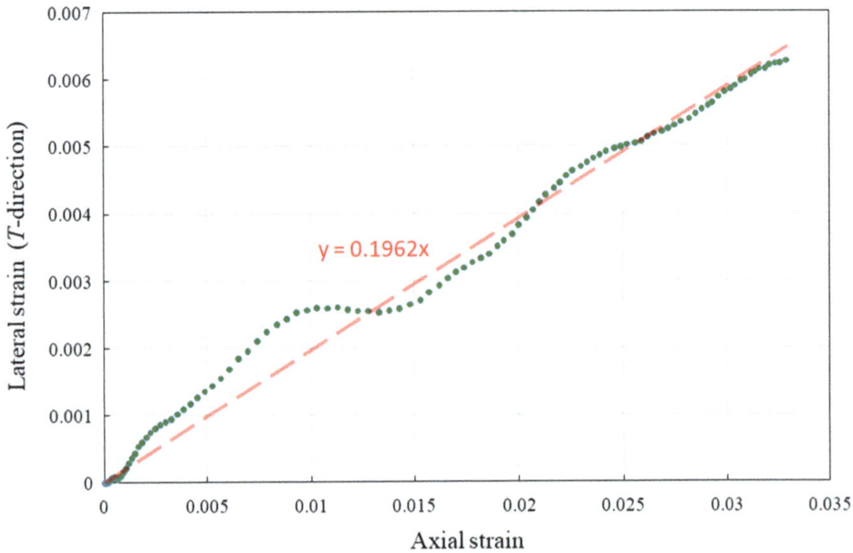


**Fig. 3.6** Dependence of the transverse deformation in the *T*-direction on the longitudinal

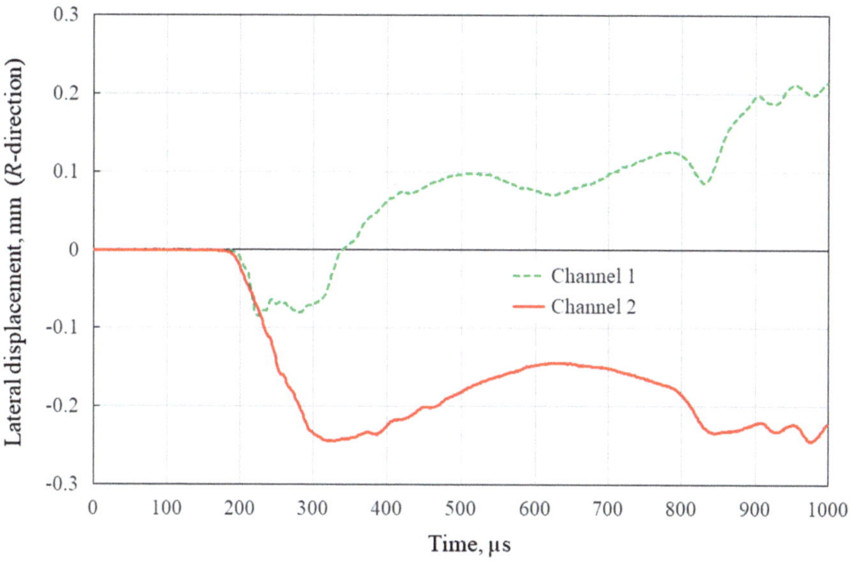


**Fig. 3.7** Displacement of the lateral surface of the sample in the *R*-direction (across the growth rings)

from the previous experiment. Figure 3.7 shows the time dependences of the lateral displacements obtained as a result of processing the interferometer signals.

It is clearly seen that due to the significant anisotropy of wood properties, the radial displacements from different sides of the specimen coincide only in the initial 50 μs of the process. With increasing time, they become significantly different, which confirms the need to use the selected scheme for measuring the transverse deformation of an anisotropic sample by two groups of antennas on opposite sides of the specimen. The relative transverse deformation of the sample, as in the first case, was determined as the sum of lateral displacements divided to the specimen diameter.

The process of development of longitudinal deformation of the sample $\varepsilon_x(t)$, as before, was determined from the readings of strain gauges on an input pressure bar (1). The obtained dependences of the transverse and longitudinal deformations of the specimen were synchronized with each other in time (Fig. 3.8), and then the dependence of the transverse deformation on the longitudinal deformation was plotted (Fig. 3.9). The obtained value of the tangent of the slope of the approximating straight line corresponding to the *RX*-component of Poisson's ratio in the direction *R* (across annual rings) is ~0.24.

Thus, by using a new technique for the separate measurement of two components of the transverse expansion of an anisotropic material under axial compression, it was possible to determine the values of the components of the dynamic Poisson's ratio *TX* and *RX*, which can be used for determining the parameters of the mathematical model of wood required for reliable calculations of dynamically loaded structures.

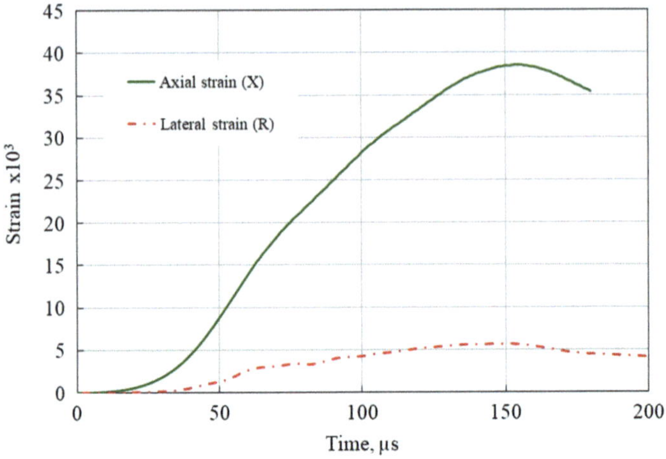


**Fig. 3.8** Dependences of the longitudinal and transverse deformations (direction $R$) of the specimen on time

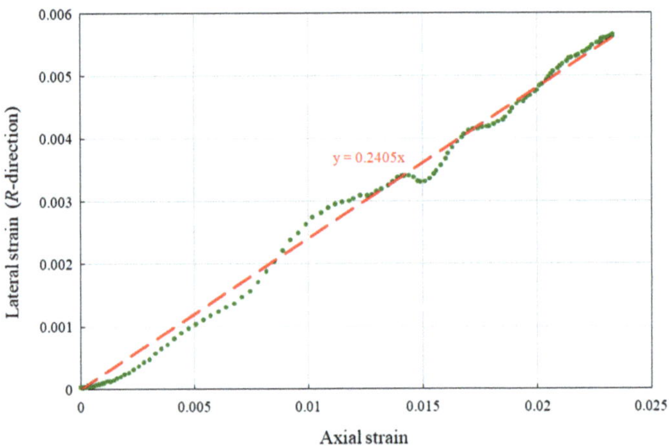


**Fig. 3.9** Dependence of the transverse deformation in the $R$-direction on the longitudinal

## 3.4 Conclusions

A technique is proposed for measuring the development of two orthogonal components of the radial deformation of a wood specimen under axisymmetric compression using a two-channel Michelson interferometer of the millimeter range. The measurements were carried out using two independent channels irradiating diametrically opposite zones of the lateral surface of the sample. It was found that due to the strong anisotropy of wood properties during axially symmetric compression of the specimen, its transverse expansion along the annual rings ($T$-direction) is close to symmetric, while the expansion across the annual rings (*R*-direction) has a significant difference from the symmetric one. The relative transverse deformation of the sample in both cases was determined as the sum of lateral displacements divided to the specimen diameter. As a result, two components of the dynamic Poisson's ratio were obtained, which amounted to ~0.2 (in the direction along the annual rings) and ~0.24 (in the direction across the annual rings), which can be used to identify the wood model required for reliable calculations of dynamically loaded designs.

**Acknowledgements** The experimental investigation of wood by the Kolsky method was carried out with the financial support from the RFBR (grant 19-38-90226). The development of a program for the numerical optimization of interferograms was financially supported by the Ministry of Science and Higher Education of the Russian Federation (task 0729-2020-0054).

## References

Ahmed L (2018) Dynamic measurements for determining Poisson's ratio of young concrete. Nordic Concrete Res Publ No NCR 58(1–6):95–106. https://www.researchgate.net/publication/329418208_Dynamic_Measurements_for_Determining_Poisson%27s_Ratio_of_Young_Concrete

ASTM C215-14 (2014) Standard test method for fundamental transverse, longitudinal, and torsional frequencies of concrete specimens. ASTM International; West Conshohocken, PA, USA, Dec 2014, 7p. https://www.astm.org/Standards/C215

Boiko AV, Kulik V, Seoudi B, Chun HH, Lee I (2008) Measurement of dynamic Poisson's ratio and form factor for a cylindrical sample. In: Advances in heterogeneous material mechanics (Chapter 22): experimental methods related to modeling, https://www.researchgate.net/publication/236589216_Measurement_of_dynamic_Poisson%27s_ratio_and_form_factor_for_a_cylindrical_sample

Bragov AM, Lomunov AK (1995) Methodological aspects of studying dynamic material properties using the Kolsky method. Int J Imp Eng 16(2) 321–330. https://www.sciencedirect.com/science/article/abs/pii/0734743X9593939G

Cao Y, Li M, Wang Z, Wang Y, Gao Z (2019) Dynamic testing and analysis of Poisson's ratio of lumbers based on the cantilever-plate bending mode shape method. J Test Eval. https://doi.org/10.1520/JTE20160521

Carrington H (1921) The determination of values of Young's modulus and Poisson's ratio by the method of flexures. Lond Edinb Dublin Philos Mag J Sci 41:206–210. https://doi.org/10.1080/14786442108636212

Carrington H (1922a) The elastic constants of spruce as influenced by moisture. Aeronaut J (Lond Engl 1897) 26:462–471. https://doi.org/10.1017/S2398187300139465

Carrington H (1922b) Young's modulus and Poisson's ratio for spruce. Lond Edinb Dublin Philos Mag J Sci 43:871–878. https://doi.org/10.1080/14786442208633943

Davis JR (2004) Tensile testing, 2nd edn. ASM International

Gao Z, Zhang X, Wang Y, Yang R, Wang G, Wang Z (2016) Measurement of the Poisson's ratio of materials based on the bending mode of the cantilever plate. BioResources 11(3):5703–5721. https://doi.org/10.15376/BIORES.11.3.5703-5721

Hörig H (1931) Zur Elastizität des Fichtenholzes. I. Folgerungen aus Messungen von H. Carrington an Spruce (To the elasticity of spruce wood. I. Consequences of the measurements conducted by H. Carrington on Spruce). Zeitschrift für technologische Physik 12:369–379 (in Germ.). https://www.scienceopen.com/document?vid=b97d5031-646a-41c2-b76a-58f166ef1717

Hörig H (1935) Anwendung der Elastizitätstheorie anisotroper Körper auf Messungen an Holz (Application of the elasticity theory of anisotropic bodies to wood measurements). Arch Appl Mech 6:8–14 (in German). https://link.springer.com/article/10.1007%2FBF02086407

Hoshino Y, Tamai K, Zhang Y, Yoneyama S (2018) Direct measurement and master curve construction of viscoelastic Poisson's ratio with digital image correlation. Strain 54(6). https://doi.org/10.1111/str.12294

Keunecke D, Hering S, Niemz P (2008) Three-dimensional elastic behaviour of common yew and Norway spruce. Wood Sci Technol 42:633–647. https://doi.org/10.1007/s00226-008-0192-7. https://link.springer.com/article/10.1007/s00226-008-0192-7

Kolsky H (1949) An investigation of the mechanical properties of material at very high rates of loading. Proc Phys Soc (London) 62B:676–700. https://iopscience.iop.org/article/10.1088/0370-1301/62/11/302

Kumpenza C, Matz P, Halbauer P, Grabner M, Steiner G, Feist F, Müller U (2018) Measuring Poisson's ratio: mechanical characterization of spruce wood by means of non-contact optical gauging techniques. Wood Sci Technol 52:1451–1471. https://link.springer.com/article/10.1007/s00226-018-1045-7

Neuhaus FH (1981) Elastizitätszahlen von Fichtenholz in Abhängigkeit von der Holzfeuchtigkeit (Elasticity constants of spruce wood in relation to the wood moisture content). Dissertation, Ruhr-Universität Bochum (in German). https://books.google.ru/books/about/Elastizit%C3%A4tszahlen_von_Fichtenholz_in_A.html?id=nRv0SgAACAAJ&redir_esc=y

Niemz P, Caduff D (2008) Untersuchungen zur Bestimmung der Poissonschen Konstanten an Fichtenholz (Investigations to determine the Poisson's ratio of spruce wood). Holz Roh Werkst 66:1–4. https://doi.org/10.1007/s00107-007-0188-2(inGerman)

Pal P (2019) Dynamic Poisson's ratio and modulus of elasticity of Pozzolana Portland Cement concrete. Int J Eng Technol Innov 9(2):131–144. https://core.ac.uk/download/pdf/228833802.pdf

Swamy RN (1971) Dynamic Poisson's ratio of Portland cement paste, mortar and concrete. Cement Concrete Res 1(5):559–583. https://www.sciencedirect.com/science/article/abs/pii/0008884671900603

Valla A, Konnerth J, Keunecke D et al (2011) Comparison of two optical methods for contactless, full field and highly sensitive in-plane deformation measurements using the example of plywood. Wood Sci Technol 45:755–765. https://doi.org/10.1007/s00226-010-0394-7

Voigt W (1882) Allgemeine Formeln für die Bestimmung der Elastizitätskonstanten von Kristallen durch Beobachtung der Biegung und Drillung von Prismen (General formulas to determine the elastic constants of crystals by observing the bending and twisting of prisms). Ann Phys 252:273–321 (in German). https://doi.org/10.1002/andp.18822520607

Voigt W (1887) Theoretische Studien über die Elastizitätsverhältnisse der Kristalle (Theoretical studies on the elasticity of crystals). Königliche Gesellschaft der Wissenschaften zu Güttingen (in German). https://www.worldcat.org/title/theoretische-studien-uber-die-elasticitatsverhaltnisse-der-krystalle/oclc/634403068

Voigt W (1966) Lehrbuch der Kristallphysik (Textbook of crystal physics). Vieweg + Teubner Verlag, Wiesbaden (in German). https://link.springer.com/book/10.1007/978-3-663-15884-4

Wang ZH, Gao ZZ, Wang YL, Cao Y, Wang GG, Liu B, Wang Z (2015) A new dynamic testing method for elastic, shear modulus and Poisson's ratio of concrete. Constr Build Mater 100:129–135. https://doi.org/10.1016/j.conbuildmat.2015.09.060

Wang G-G, Zhang X, Gao Z-Z, Wang Y-L, Yu C, Wang Z (2017) Dynamic testing and analysis of Poisson's ratio constants of timber. In: Mechanics and architectural design, pp 9–18. https://doi.org/10.1142/9789813149021_0002

Wommelsdorff O (1966) Dehnungs-und Querdehnungszahlen von Hölzern (Elongation and transverse strain constants of wood). Dissertation, Leibniz Universität Hannover (in German). https://www.tib.eu/de/suchen/id/TIBKAT%3A046346732/Dehnungs-und-Querdehnungszahlen-von-H%C3%B6lzern/

# Chapter 4
# An Application of Thermal Analogy in Active Control Problems

**Evgeny Barkanov, Andrejs Kovalovs, Aleksandr Anoshkin, and Pavel Pisarev**

**Abstract** An application of thermal analogy for the solution of active control problems is demonstrated for quasi-static beam testing and investigation of the active twist behaviour of smart helicopter rotor blade. To obtain accurate finite element solutions, rigorous convergence study using different finite element types and modelling approaches is carried out. The methodology for optimal placement of MFC actuators on lightweight structures is developed to control the structural vibrations in the low-frequency range. The results of numerical study are successfully confirmed by the results of physical experiments.

**Keywords** PZT and MFC patches · Active control · Finite element method · Thermal analogy · Experimental verification.

## 4.1 Introduction

Nowadays, lightweight structures with integrated piezoelectric materials Lead Zirconate Titanate (PZT) and Macro Fibre Composite (MFC) used as actuators or sensors have obtained considerable spreading in transport and aerospace engineering for the purpose of geometry control (Grohmann et al. 2006; Henry et al. 2019; Zhang et al. 2019), vibration and noise reduction (Kovalovs et al. 2007a, 2009; Nasser et al. 2012), as well as for overall performance (Barkanov et al. 2008; Basheer 2020; Kovalovs et al. 2007b) and safety improvement (Giurgiutiu 2015; Tatarinov et al.

E. Barkanov (✉) · A. Kovalovs
Institute of Materials and Structures, Riga Technical University, Kalku St. 1, 1658 Riga, Latvia
e-mail: barkanov@latnet.lv

A. Kovalovs
e-mail: andrejs.kovalovs@rtu.lv

A. Anoshkin · P. Pisarev
Department of Mechanics of Composite Materials and Structures, Perm National Research Polytechnic University, Komsomolski Ave. 29, 614990 Perm, Russia
e-mail: pisarev85@live.ru

 
H. Altenbach et al. (eds.), *Advanced Materials Modelling for Mechanical, Medical and Biological Applications*, Advanced Structured Materials 155,
https://doi.org/10.1007/978-3-030-81705-3_4

2018; Tuloup et al. 2020). To design such constructions, different numerical analyses have been developed in the last three decades. The piezoelectric response could be modelled with 2D and 3D coupled-field solid elements requiring for a description of electromechanical problem elastic, piezoelectric and dielectric constants (NN 2015). As the main disadvantage of this approach are an increase in the problem size, inefficient matrix reformulation and large storage requirements. For this reason, in the case of complex structures with large dimension of the numerical problem to be solved, the thermal analogy could be examined as an effective methodology. In this case, piezoelectric coefficients characterizing PZT or MFC patches are introduced into analysis as thermal expansion coefficients.

In the present study, an application of thermal analogy for the solution of active control problems is successfully demonstrated for static and dynamic analyses of simple beam and plate structures, and active twist analysis of smart helicopter rotor blades. The numerical results are validated by the results of physical experiments studying quasi-static behaviour of aluminium beam with glued MFC and active twist of helicopter rotor blade with MFC actuators implemented from both sides of the blade and cured together with glass fibre reinforced polymer (GFRP) skin.

## 4.2 Thermal Analogy

Thermal analogy between piezoelectric strains and thermally induced strains could be applied successfully for an active control of complex structures changing the piezoelectric analysis with corresponding thermo-elastic analysis. In this case, piezoelectric coefficients characterizing PZT or MFC patches are introduced into the finite element model as thermal expansion coefficients.

To find their relationship, two constitutive equations should be examined. For the converse effect, the equation of elasticity coupled with the charge equation of electrostatics by means of piezoelectric constants could be written in the following form:

$$\{\sigma\} = [D]\{\varepsilon\} - [D][d]\{E\} \tag{4.1}$$

where $\{\sigma\}$ is the stress vector, $[D]$ is the elasticity matrix evaluated at constant electric field, $\{\varepsilon\}$ is the elastic strain vector, $[d]$ is the piezoelectric strain matrix, and $\{E\}$ is the electric field intensity vector. Another equation to be examined is coupled thermo-elastic equation:

$$\{\sigma\} = [D]\{\varepsilon\} - [D]\{\alpha\}\Delta T \tag{4.2}$$

where $\{\alpha\}$ is the vector of thermal expansion coefficients and $\Delta T = T - T_{ref}$ is the temperature change relative to the reference temperature. The relationship between piezoelectric and thermal strains are established comparing both constitutive equations

$$[d]\{E\} = \{\alpha\}\Delta T \tag{4.3}$$

Further, the electric field intensity for an arbitrary MFC and PZT actuators could be written as

$$E = \frac{\Delta V}{\Delta_{es}} \tag{4.4}$$

where $\Delta V$ is the voltage difference between electrodes and $\Delta_{es}$ is the effective electrode spacing. Finally, the relationship between piezoelectric and thermal expansion coefficients characterising actuators could be determined as follows assuming that $\Delta V \cong \Delta T$ and substituting Eq. (4.4) into Eq. (4.3)

$$\alpha_{ij} = \frac{d_{ij}}{\Delta_{es}} \quad i, j = 1, 2, 3 \tag{4.5}$$

where $d_{ij}$ are the effective piezoelectric constants.

## 4.3 Methodology Validation

Validation of the presented methodology is carried out examining an active quasi-static behavior of aluminum beam with adhesively bonded MFC actuator (M8528-P1). The beam is clamped from one side and deflections at free end are measured by PANASONIC laser sensor (HL-G108-A-C5). The geometrical parameters of investigated object are presented in Fig. 4.1, and material properties are given in Table 4.1. It is necessary to note that the material properties of aluminum beam have been obtained from vibration tests solving an inverse problem (Barkanov et al. 2015), and the material properties of MFC have been taken from the brochure of smart material corporation.[1]

Rigorous convergence study using different finite element types and modelling approaches has been executed before the comparison of numerical and experimental results. So, in the first approach (Fig. 4.2a) the multilayered shell finite elements SHELL 181 have been used, in the second approach (Fig. 4.2b) aluminium—MFC construction has been modelled with two shell finite elements SHELL 181 through the thickness and in the third approach (Fig. 4.2b) aluminum – MFC construction has been described with two solid finite elements SOLID 185 (8 nodes) or SOLID 186 (20 nodes).

The transverse deflection of the beam end under 1000 [V] applied to the MFC actuator for different finite element meshes and modelling approaches is presented

[1] https://www.smart-material.com/media/Datasheets/MFC_V2.3-Web-full-brochure.pdf.

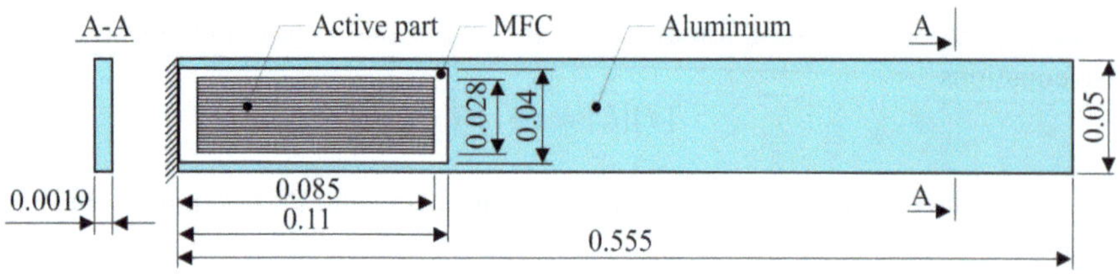


**Fig. 4.1** Aluminum beam with MFC (all dimensions in m)

**Table 4.1** Material properties for quasi-static testing

| Property | Aluminum | MFC | PZT |
|---|---|---|---|
| Elastic modulus, $E_1$ [GPa] | 69.47 | 30.336 | 48.30 |
| Elastic modulus, $E_2 = E_3$ [GPa] | 69.47 | 15.857 | 54.05 |
| Shear modulus, $G_{12} = G_{13}$ [GPa] | 27.32 | 5.515 | 29.41 |
| Poison's ratio, $\nu_{21}$ | 0.3 | 0.16 | 0.38 |
| Poison's ratio, $\nu_{12} = \nu_{13}$ | 0.3 | 0.31 | 0.41 |
| Density, $\rho$ [kg/m$^3$] | 2732 | 4700 | 7500 |
| Piezoelectric constant, $d_{33}$ [m/V] | – | $4.6 \times 10^{-10}$ | $4.6 \times 10^{-10}$ |
| Piezoelectric constant, $d_{31} = d_{32}$ [m/V] | – | $-2.1 \times 10^{-10}$ | $-2.1 \times 10^{-10}$ |

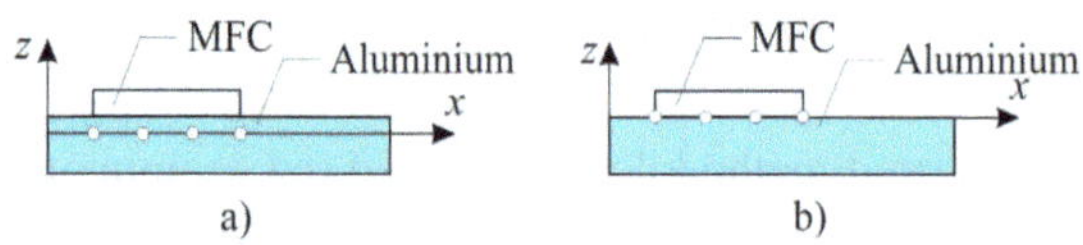


**Fig. 4.2** Finite element modelling approaches: **a** Multilayered SHell (SH181), **b** Two SHell (2SH181) or Two SoliD (2SD185 or 2SD186)

in Table 4.2 and corresponding dimensions of the finite element mesh used for a convergence study—in Fig. 4.3.

Table 4.2 shows that using multilayered and single-layer shell elements as well as higher order SOLID186 elements, the finite element mesh could be quite rough but at least two finite elements through the thickness of aluminium beam should be used utilizing SOLID185 finite elements as it is seen in Fig. 4.4. It is necessary to note that MFC volume have not been subdivided with solid finite elements through the thickness since it has been used as the heating source with an applied temperature in the thermo-elastic problem. For the next calculations, the finite element models with the dimension of finite elements marked with the bold font in Table 4.2 have been applied.

Results of quasi-static experimental and finite element solution applying thermal analogy in the form of beam end's deflection under different voltage applied is presented in Fig. 4.5. It is observed a quite large difference between experimental and numerical results. To improve an accuracy of the finite element model and taking

**Table 4.2** Solution convergence: beam end's deflection [mm]

| $l_{FE}$ [mm] | SH181 | 2SH181 | 2SD185 | | | 2SD186 |
|---|---|---|---|---|---|---|
| | | | $t_{Al}/1$ | $t_{Al}/2$ | $t_{Al}/3$ | $t_{Al}/1$, $t_{Al}/2$, $t_{Al}/3$ |
| 4 | 4.5 | 4.5 | 2.4 | 2.0 | 1.9 | 4.6 |
| 3 | 4.5 | 4.5 | 3.5 | 2.7 | 2.6 | 4.6 |
| 2 | 4.5 | 4.5 | 5.2 | 3.6 | 3.4 | 4.6 |
| 1 | 4.5 | 4.5 | 7.5 | 4.5 | 4.3 | 4.6 |
| 0.5 | 4.5 | 4.5 | 8.6 | 4.8 | 4.6 | 4.6 |
| 0.25 | 4.5 | 4.5 | 8.9 | 4.9 | 4.7 | 4.6 |

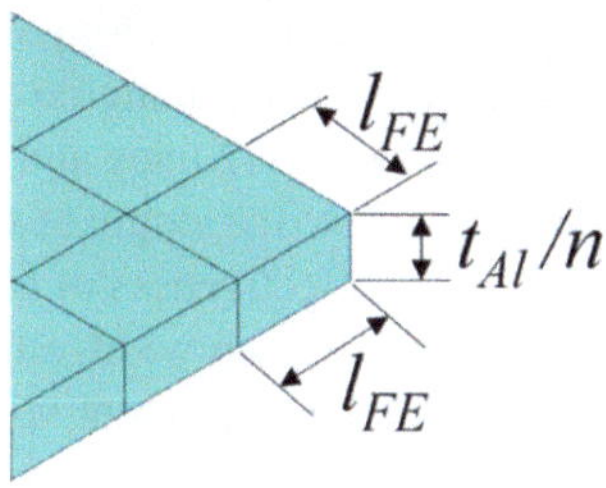


**Fig. 4.3** Finite element mesh

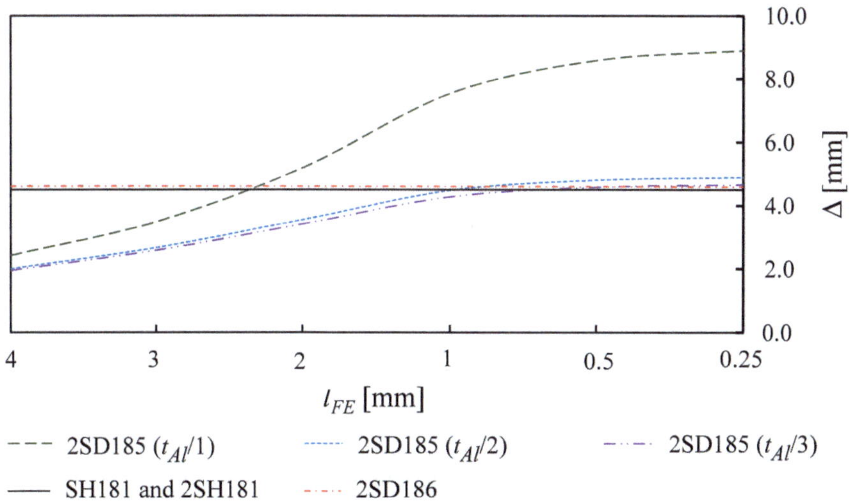


**Fig. 4.4** Finite element solution convergence

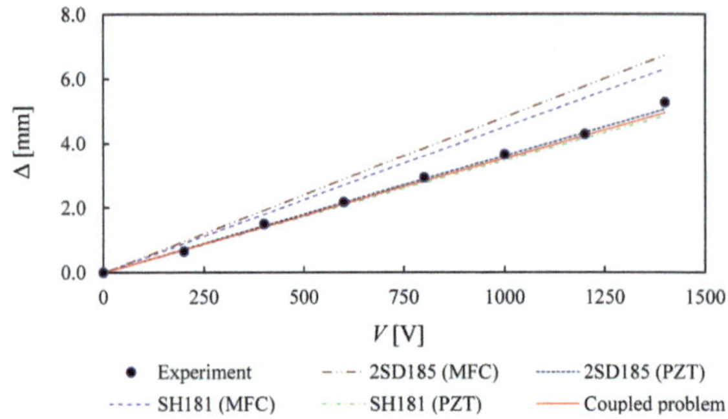


**Fig. 4.5** Beam end's deflection

into account, the rules of utilization of thermal analogy when the temperature field is applied only for the piezoelectric fibres, MFC material properties are changed with corresponding PZT fibre material properties (Kuang and Zhu 2019) and their volume in MFC. The length of PZT patch is equal to the length of an active part of MFC (length of piezoelectric fibres). In this case an effectiveness of MFC application remains the same for the bended beam. The ratio between width and thickness of PZT patch is not so important value like length, but an equivalent volume of piezoelectric fibres should be preserved. The numerical results obtained after replacing of MFC with PZT patch are given in the same Fig. 4.5 where good coincidence with experimental data is observed. Additionally, the thermal analogy is successfully validated by the solution of coupled electromechanical problem (Anoshkin et al. 2020).

## 4.4 Vibrations Control

To control the structural vibrations in the low-frequency range, the problem of optimal location of MFC actuators has been solved for a variety of isotropic and orthotropic panels. An effectiveness of application of MFC actuators for the vibration damping has been estimated with the factor describing the amplitude reduction related to the applied energy (voltage). In this case, the reduction coefficient is written as following:

$$k_R = \frac{R}{V} = \frac{A_O - A}{VA_O} 100\% \tag{4.6}$$

where $R$ is the amplitude reduction, $V$ is the voltage applied, $A_o$ is the amplitude of frequency response at resonance peak with $V = 0$ and $A$ is the corresponding amplitude with $V \neq 0$.

In the first example, simply supported aluminium beam is covered from the top side with MFC layer consisting of separate MFC patches. This is done to exclude a shift of system eigenfrequencies in the consequent numerical analysis. Each MFC patch is actuated separately according its location on the beam (Fig. 4.6). At the beginning beam with attached MFC patches but without voltage applied is loaded with a har-

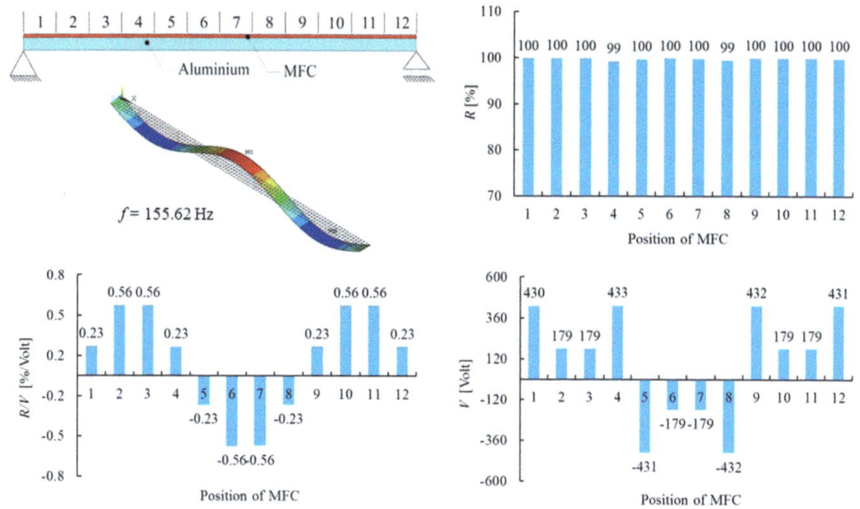


**Fig. 4.6** Optimal location of MFC patch on the beam for a damping of its third bending mode

monic force to find the frequency spectra of excitation. Then for each active MFC patch, the same analysis is carried out. Herewith such voltage (Fig. 4.6b) is seeking out for a separate eigenfrequency when the maximal reduction of the displacement amplitude (Fig. 4.6a) is obtained. Without limitation on energy applied very high and practically the same amplitude reduction $R$ could be obtained (Fig. 4.6a). In reality, the energy applied is limited by the maximum operational positive/negative voltage and not the same values of the amplitude reduction are obtained but this fact is not important now in the numerical analysis. Finally, the reduction coefficient in dependence on MFC position is presented in Fig. 4.6c where it is seen that the maximal amplitude reduction could be obtained with location of MFC patch in the places of mode's peak or in the places with maximal curvature. The same phenomenon is observed for other modes and beams made of another materials including composites as well as for the plates.

Moreover, an angle of piezoelectric fibres in MFC patch located on the mode's peak is also subjected to the same phenomenon. To study this effect, aluminium and laminated composite panels with active MFC square patch and under harmonic load applied have been investigated (Fig. 4.7). Different boundary conditions applied to the plate structure have been examined. The maximal reduction coefficients have been obtained for the systems when direction of piezoelectric fibres in MFC actuator coincides with the maximal curvature of corresponding mode's peak. So for the first mode of aluminium panel presented in Fig. 4.8a the reduction coefficient $k_R = 0.52$ $[\%/V]$ has maximum value for the angle of piezoelectric fibres $\alpha$=90° but for the fourth mode of the same plate the reduction coefficient $k_R = 0.25$ $[\%/V]$ has maximum value for the angle of piezoelectric fibres $\alpha = 0°$ (Fig. 4.8b). For more general case when mode peaks are rotated in the plane, the optimal angles of piezoelectric fibres

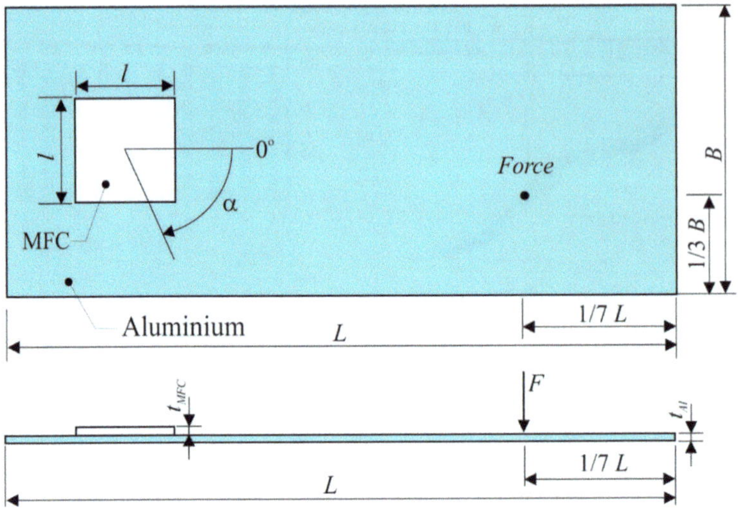


**Fig. 4.7** Plate with attached MFC actuator

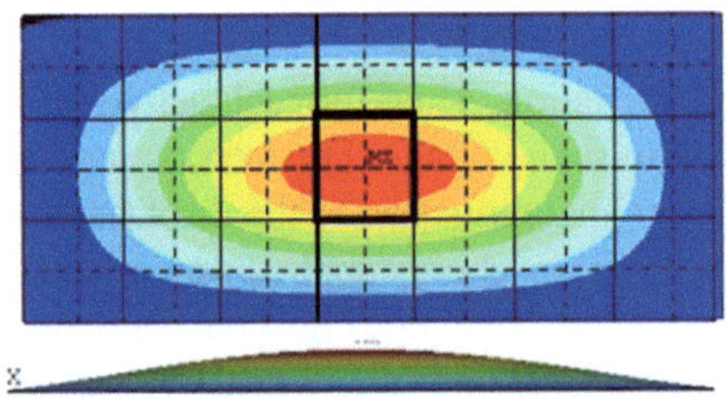

| $f_1 = 94.64$ Hz | | | |
|---|---|---|---|
| $\alpha$ [°] | $0^0$ | $45^0$ | $90^0$ |
| $R$ [%] | 99.0 | 99.4 | 99.5 |
| $V$ [V] | 540 | 282 | 192 |
| $k_R$ [%/V] | 0.18 | 0.35 | 0.52 |

a)

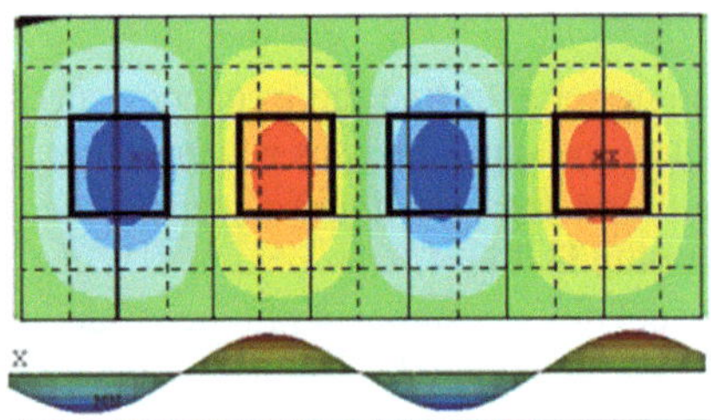

| $f_1 = 212.07$ Hz | | | |
|---|---|---|---|
| $\alpha$ [°] | $0^0$ | $45^0$ | $90^0$ |
| $R$ [%] | 99.6 | 99.7 | 99.8 |
| $V$ [V] | -397 | -484 | -615 |
| $k_R$ [%/V] | 0.25 | 0.21 | 0.16 |

b)

**Fig. 4.8** Optimal location of piezoelectric fibres in MFC patch located on the aluminium panel for a damping of its: **a** first mode, **b** fourth mode

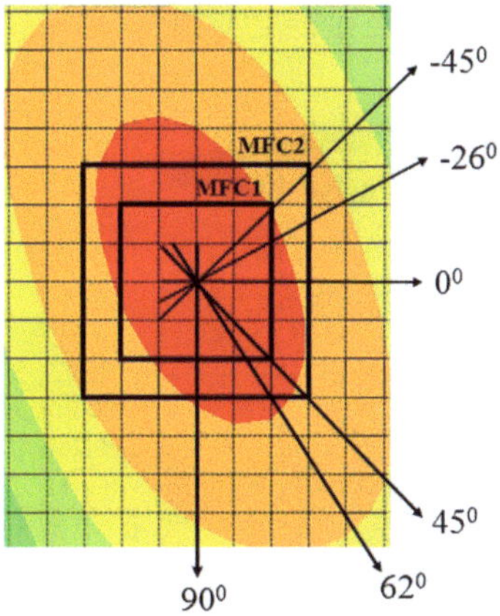


| MFC1 40x40 [mm] | | | | | | |
|---|---|---|---|---|---|---|
| $\alpha$ [⁰] | -45⁰ | -26⁰ | 0⁰ | 45⁰ | 62⁰ | 90⁰ |
| $R$ [%] | 99.5 | 99.5 | 99.5 | 99.3 | 99.2 | 99.3 |
| $V$ [V] | 1510 | 1340 | 1380 | 2220 | 2713 | 2670 |
| $k_R$ [%/V] | 0.07 | 0.07 | 0.07 | 0.04 | 0.04 | 0.04 |

| MFC2 50x50 [mm] | | | | | | |
|---|---|---|---|---|---|---|
| $\alpha$ [⁰] | -45⁰ | -26⁰ | 0⁰ | 45⁰ | 62⁰ | 90⁰ |
| $R$ [%] | 99.7 | 99.7 | 99.7 | 99.6 | 99.6 | 99.7 |
| $V$ [V] | 688 | 610 | 624 | 995 | 1210 | 1205 |
| $k_R$ [%/V] | 0.14 | 0.16 | 0.16 | 0.10 | 0.08 | 0.08 |

**Fig. 4.9** Optimal location of piezoelectric fibres in MFC patches located on the laminated composite panel for a damping of its mode with a peak rotated in the plane

in MFC patches attached to the laminated composite panel are presented in Fig. 4.9 for two different patch dimensions.

Numerous parametric studies have shown that the optimal location can be determined knowing the form of a separate mode which could be find easily by different experimental and numerical analyses. MFC patches should be located on the mode peaks with orientation of piezoelectric fibres along its maximal curvature.

## 4.5 Active Twist Analysis

To demonstrate an application of thermal analogy for the solution of more complicated problems, the design of an active twist of smart helicopter rotor blade is examined. For this purpose, well-known BO 105 model rotor blade (Fig. 4.10) consisting of unidirectional GFRP C-spar, laminated composite skin and foam core has been chosen. Since, it is planned to produce a smart rotor for quasi-static testing, its profile has been slightly modified into a symmetrical NACA 0012, due to manufacturing reasons. The material properties of smart rotor blade components are given in Table 4.3. To twist the rotor blade, 12 MFC actuators with operating voltage of −500 ... +1500 V have been implemented from both sides of the blade and then cured together with GFRP.

3D finite element model (Fig. 4.11) has been developed in NN (2015), where the rotor blade skin with MFC active layer and spar “moustaches” are modelled by the linear layered structural shell elements SHELL99, but the spar, foam and balance weight by 3D 20 node structural solid elements SOLID186. To preserve the rotor blade profile, the node offset option is applied for the joint MFC-skin-spar “moustaches” structure. In this case, the finite element nodes are located at the top

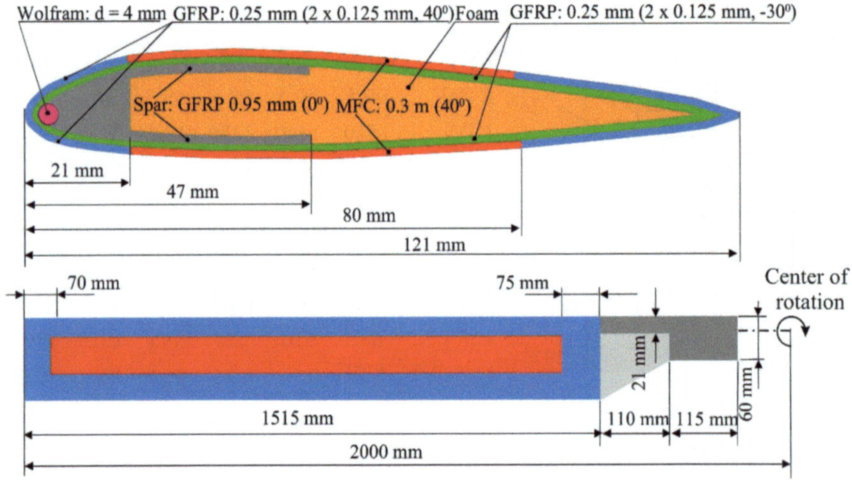


**Fig. 4.10** Smart helicopter rotor blade

**Table 4.3** Material properties of smart helicopter rotor blade components

| Property | MFC | GFRP | Rohacell 51FX | Wolfram |
|---|---|---|---|---|
| Elastic modulus, $E_x$ [GPa] | 30.336 | 45.166 | 0.035 | 400 |
| Elastic modulus, $E_y = E_z$ [GPa] | 15.5 | 11.981 | 0.035 | 400 |
| Shear modulus, $G_{yz}$ [GPa] | 5.7 | 1.289 | 0.014 | 160 |
| Shear modulus, $G_{xy} = G_{xz}$ [GPa] | 10.7 | 4.583 | 0.014 | 160 |
| Poison's ratio, $\nu_{yz}$ | 0.35 | 0.1325 | 0.25 | 0.28 |
| Poison's ratio, $\nu_{xy} = \nu_{xz}$ | 0.4 | 0.238 | 0.25 | 0.28 |
| Density, $\rho$ [kg/m$^3$] | 4700 | 2008 | 52 | 19400 |
| Piezoelectric constant, $d_{33}$ [m/V] | $4.18 \times 10^{-10}$ | – | – | – |
| Piezoelectric constant, $d_{31} = d_{32}$ [m/V] | $-1.98 \times 10^{-10}$ | – | – | – |

surface of laminate. The rotor blade is clamped from one side and is free from another to demonstrate an active twist.

The results of quasi-static test are given in Table 4.4.

They show that the difference between active twist angles obtained experimentally and numerically is negligible that confirms high accuracy of the developed 3D finite element model and applied thermal analogy.

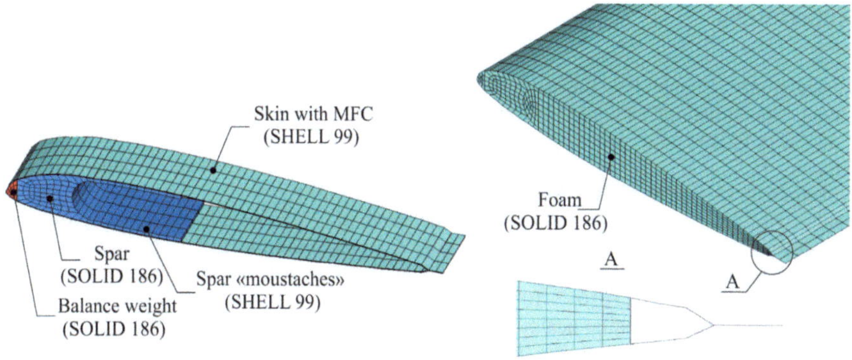


**Fig. 4.11** Finite element model of smart helicopter rotor blade

**Table 4.4** Results of quasi-static test

| Active twist angle, $\varphi$ [°] | | Relative error, $\Delta$ [%] |
|---|---|---|
| Experiment | FEM | |
| 3.93 | 4.06 | 3.3 |

## 4.6 Summary

The thermal analogy has been successfully applied for the solution of different active control problems with a confirmation of the obtained numerical results by the results of physical experiments. Rigorous convergence study using different finite element types and modelling approaches has allowed to obtain high accuracy of the finite element solutions.

To control the structural vibrations in the low-frequency range, the methodology for the optimal location of MFC actuators has been developed. It has been demonstrated that the optimal location can be determined if the form of a separate mode is known. In this case, MFC patches should be located on the mode peaks with orientation of piezoelectric fibres along its maximal curvature.

It is necessary to note that the thermal analogy has allowed considerably reduce calculation time and costs since it operates with the finite element model of less dimension. Moreover, it gives the possibility to use various modelling features of the finite element package which are absent in the coupled piezo-electric analysis.

**Acknowledgements** This work has been supported by the Russian Science Foundation (project No. 18-19-00722) and the European Regional Development Fund, measure 1.1.1.5 "Support to international cooperation projects in research and innovation of RTU" (project No. 1.1.1.5/18/I/008).

## References

Anoshkin AN, Pisarev PV, Bayandin SR, Kungurtseva EG (2020) Numerical calculation of composite structures equipped with flexible piezoactuators stress-strain state. AIP Conf Proc 2216(1):040013

Barkanov E, Gluhih S, Kovalov A (2008) Optimal design of the active twist for helicopter rotor blades with C-spar. Mech Adv Mater Struct 15(3–4):325–334

Barkanov E, Wesolowski M, Hufenbach W, Dannemann M (2015) An effectiveness improvement of the inverse technique based on vibration tests. Comput Struct 146:152–162

Basheer AA (2020) Advances in the smart materials applications in the aerospace industries. Aircraft Eng Aerosp Technol 92(7):1027–1035

Giurgiutiu V (2015) Structural health monitoring (SHM) of aerospace composites. In: Soutis C, Irving P (eds) Polymer composites in the aerospace industry. Woodhead Publishing, pp 449–507

Grohmann B, Maucher C, Jänker P (2006) Actuation concepts for morphing helicopter rotor blades. In: Grant I (ed) CAS 2006 CD-ROM proceedings : 25th International Congress of aeronautical sciences, 3–8 Sept 2006, Hamburg, Germany, International Council of the Aeronautical Sciences, ICAS 2006–10.6.4

Henry AC, Molinari G, Rivas-Padilla JR, Arrieta AF (2019) Smart morphing wing: optimization of distributed piezoelectric actuation. AIAA J 57(6):2384–2393

Kovalovs A, Barkanov E, Gluhihs S (2007a) Active control of structures using macro-fiber composite (MFC). J Phys Conf Ser 93:012034

Kovalovs A, Barkanov E, Gluhihs S (2007b) Active twist of model rotor blades with D spar design. Transport 22(1):38–44

Kovalovs A, Wesolowski M, Barkanov E, Gluhihs S (2009) Application of macro-fibre composite (MFC) as piezoelectric actuator. J Vibroeng 11(1):105–112

Kuang Y, Zhu M (2019) Evaluation and validation of equivalent properties of macro fibre composites for piezoelectric transducer modelling. Compos Part B Eng 158:189–197

Nasser H, Kiefer-Kamal EH, Hu H, Belouettar S, Barkanov E (2012) Active vibration damping of composite structures using a nonlinear fuzzy controller. Compos Struct 94(4):1385–1390

NN (2015) ANSYS mechanical user's guide. Release 16.2. ANSYS Inc., Canonsburg, PA

Tatarinov A, Barkanov EN, Davydov E, Mihovski M (2018) T- and L-types of long range guided waves for defect detection. In: Dumitrescu A, Parinov IA, Barkanov EN (eds) Non-destructive testing and repair of pipelines. Springer International Publishing, pp 15–29

Tuloup C, Harizi W, Aboura Z, Meyer Y (2020) Integration of piezoelectric transducers (pzt and pvdf) within polymer-matrix composites for structural health monitoring applications: new success and challenges. Int J Smart Nano Mater 11(4):343–369

Zhang Z, Li Y, Yu X, Li X, Wu H, Wu H, Jiang S, Chai G (2019) Bistable morphing composite structures: a review. Thin-Walled Struct 142:74–97

# Chapter 5
# Pneumo-dynamic Experimental Setup for Studying the Behaviour of Structural Materials at Strain Rates of the Order of 100 1/s

**Artem V. Basalin, Anatoly M. Bragov, and Aleksandr Yu. Konstantinov**

**Abstract** Investigation of the behaviour of structural materials in a wide range of the strain rates is an urgent task. The sources of dynamic loads originate from explosion, shock, and earthquake vibration. The loading rate is an important index corresponding to different velocities of impacts for construction materials. The strain rate of structures under impact loading is in the range of 0.1–200 $s^{-1}$, but it is more than 200 $s^{-1}$ under blast loading. The rock engineering often involves the dynamic loading scenarios, such as excavation engineering, civil engineering, blasting engineering, projectile impact, seismic events, and rock collapse. There is still limited data in strain rate regimes of relevance, specifically for drop shock applications. This paper describes a pneumo-dynamic experimental setup created in the dynamic testing laboratory of the Research Institute for Mechanics of Nizhny Novgorod State University, to study the dynamic behaviour of structural materials at strain rates of the order of 100 $s^{-1}$. The results of approbation are given on the example of dynamic tests of plate specimens made of steel HX220BD.

**Keywords** Strain rate · Experimental investigation · Plastic deformation · Yield stress · Numerical simulation · Experiment · Strain rate

## 5.1 Introduction

Interest in the study of dynamic deformation and fracture of materials is because a large number of important problems in assessing the strength of new critical

A. V. Basalin · A. M. Bragov · A. Yu Konstantinov (✉)
Research Institute for Mechanics, National Research Lobachevsky State University of Nizhny Novgorod, Gagarin ave. 23, Nizhny Novgorod, Russia 603950
e-mail: konstantinov@mech.unn.ru

A. V. Basalin
e-mail: artem.basalin@mech.unn.ru

A. M. Bragov
e-mail: bragov@mech.unn.ru



H. Altenbach et al. (eds.), *Advanced Materials Modelling for Mechanical, Medical and Biological Applications*, Advanced Structured Materials 155,
https://doi.org/10.1007/978-3-030-81705-3_5

structures require specifying the strength and deformation characteristics of materials as functions of the strain rate. Such dependences are extremely important for obtaining reliable estimates of the stress–strain state and strength of structures experiencing dynamic loads because of technological accidents, terrorist attacks, falls, etc. Dynamic problems of strength assessment are associated with air and railroad, road transportation of containers with spent nuclear fuel, dismantling of nuclear charges in explosion-proof containers, on-site operations at nuclear power plants (loading, transportation of fuel elements, assemblies, transport containers, equipment falling). In the aerospace industry, similar problems arise due to the breakage of the fan blades of an aircraft engine due to material fatigue, birds, ice floes, hail, stones hit into the engine and possible collision of space debris with orbital stations. In the automotive industry, dynamic simulation is required for crash test analysis. In metalworking such problems arise during development of technological operations for forging, stamping, rolling. The rock engineering often involves the dynamic loading scenarios, such as excavation engineering, civil engineering, blasting engineering, projectile impact, seismic events and rock collapse.

Currently, the quasi-static range of strain rates (from $10^{-3}$ to $10^{-1}$ $s^{-1}$) is well studied and worked out from the methodological point of view. In addition, the dynamic range for high strain rates above 1000 $s^{-1}$ is well studied. This area is typical for high-velocity impact and explosive loading problems. In the first case, standard experimental equipment is used for research, in the latter—the Split Hopkinson Pressure Bar method and its modifications (ASM 2000; Bragov and Lomunov 1984; Caverzan et al. 2012; Davies 1948; Eskandari and Nemes 2000; Gama et al. 2004; Jiang and Zhang 2006; Kolsky 1949; Lomunov 1987; Bragov et al. 2020).

The range of strain rates from $10^0$ $s^{-1}$ to $10^2$ $s^{-1}$ is currently poorly studied; however, this range of strain rates is critically important for assessing the strength of structures in a fall and low-speed impact. For example, in the problem of calculating the functional characteristics of transport packaging containers for nuclear fuel, high strain rates occur only in the contact zone for a short time, while the processes occurring at strain rates from $10^0$ to $10^2$ $s^{-1}$ are important for assessing the preservation of container tightness. A similar situation takes place when designing elements of aviation technology. For example, assessment of damage, when exposed to a low-speed, impacts during technological operations. The strain rates typical for the tasks of developing elements of passive safety of cars (crash tests) also lie in the range of low and medium strain rates.

In addition, dynamic deformation of structures, as a rule, consists of two stages. At the first stage of intense dynamic loading, the structure or its element is deformed locally at a high rate. As a rule, this stage is investigated. At the second stage—the aftereffect period—the deformation of the structure occurs on a large scale and much more slowly under the action of inertial loads. The second stage in a number of cases can determine the fracture of the structure; therefore, it is important that the model describes the material behaviour reliably over the entire range of strain rates.

It should be noted that the generally accepted practice is to interpolate the strain rate dependences of the characteristics of materials based on available experimental information. Moreover, the most common approach is that the characteristic is considered linearly dependent on the logarithm of the strain rate.

So, it should be noted, there is still limited data in strain rate regimes of relevance, specifically for drop shock applications.

There is a number of works dedicated to investigation of the behaviour of structural materials at medium strain range.

In Singh et al. (2019), orthogonal metal cutting as a novel method to characterize lead-free solder alloys was presented. Experiments are carried out using a wedge like tool that cuts through a work piece at a fixed depth and rake angle while maintaining a constant cutting velocity. These experiments are conducted at room temperature on Sn1.0Ag0.5Cu bulk test specimens with strain rates varying from 0.32 to 48 $s^{-1}$. The range of strain rates is only limited by the ball screw driven slide allowing higher strain rates if needed. The strains and strain rates are captured through particle image velocimetry (PIV) using sequential images taken from a high-speed camera just ahead of the cutting tool. Results for the stress–strain response obtained through the experiments are compared to prior work for validation. Orthogonal metal cutting is shown to be a potentially attractive method for characterization of solder at higher strain rates.

In Cadoni et al. (2016), the effects of strain rate on the mechanical properties in tension of a commercial aluminium alloy AA7081 are analysed. The tests have been carried out by means of three different set-ups for quasi-statics, medium and high strain rates. The medium strain rate tests have been carried out on a hydro-pneumatic machine (HPM) (Cadoni et al. 2012). This HPM has a cylindrical tank, which is divided into two chambers by a sealed piston. At the beginning of the test, one chamber, which is nearer to the specimen, is filled with gas (generally air) at high pressure (e.g. 150 bars) and the other one is filled with water. Equal pressure is established in both the chambers so that the forces acting on the two piston faces are in equilibrium. The end of the piston shaft is connected to one end of the specimen, and another end of the specimen is connected with an elastic bar, which is rigidly fixed to the machine. This elastic bar is instrumented with a strain gauge whose function is to measure the load resisted by the specimen during the test. The way the test is performed has been extensively reported in Cadoni et al. (2011a; b), Asprone et al. (2009). The strain rates obtained by using this setup were about 30 1/s.

To study the strength and deformation characteristics of sandstone subjected to low-medium speed impact loading, the complete stress–strain relationships of uniaxial compression at strain rates of $10^{-2}$–55 $s^{-1}$ were obtained utilizing MTS and drop weight impact test devices in Shi et al. (2020). The quasi-static uniaxial compression test was carried out with the MTS647.250 servo-hydraulic material testing device. The INSTRON (CEAST 9350) drop weight impact apparatus was used for dynamic compression test. The initial height of the hammer can be set between 0.3 and 1.1 m to achieve an impact speed range between 0.77 and 24 m/s. In the impact loading test, two different hammers of 16.175 kg and 21.175 kg were selected with impact velocities of 2.5 m/s, 3 m/s, 3.5 m/s and 4 m/s, respectively. In the centre of the sample, the resistance strain gauges connected to the dynamic strain instrument were pasted along the axial and circular direction. The stress–strain curve of the specimen under dynamic compression can be derived by combining the stress time-history curve collected by the force sensor on the hammer, and the strain time-history curve is collected by the dynamic strain instrument. Meanwhile, a high-speed camera was used to record the macroscopic crack development process of samples under impact loads.

Grote et al. (2001), Zhang and Zhao (2014), Ma et al. (2019, 2020) performed low-speed and high-speed impact tests on rocks in the strain rate ranges of $10^{-5}$–$10^{-2}$ $s^{-1}$ and $10^1$–$10^2$ $s^{-1}$ with the hydraulic testing machine and the large-diameter SHPB test apparatus, respectively. One of the reasons that make it difficult to use traditional circuits based on SHPB at moderate strain rates is the short duration of the load pulse. Thus, for a limited time of loading at a low rate, it is not possible to achieve significant deformations of the sample. This problem is partially solved in schemes with a pre-stressed loading elastic bar. The duration of the load in this case turns out to be significant.

In Solomos et al. (2001) using such setup, it was investigated the strain rate, temperature and size effects of three nuclear steels. The materials are ferritic steel 20MnMoNi55 (vessel head), austenitic steel X6CrNiNb1810 (Upper Internal Structure) and ferritic steel 26NiCr Mo146 (bolting). Smooth cylindrical tensile specimens of three sizes have been tested at strain rates from 0.001/s up to 300/s, at room and elevated temperatures (400–600 °C).

The uniaxial tests at high strain rate were performed on a modified Hopkinson bar device Albertini and Montagnani (1977). It consists of two half-bars, the incident and transmitter bar, respectively, with the specimen introduced in between. Elastic energy is stored in a pre-stressed bar, which is the solid continuation of the incident bar of the machine. By releasing this energy (rupturing the blocking brittle intermediate piece), a tension wave with small rise-time is generated and transmitted along the incident bar loading the specimen to fracture. This tension wave fulfils the requirements for being a uniaxial elastic plane stress wave because the pulse wavelength is long compared to the bar diameter, and the pulse amplitude does not exceed the yield strength of the bar.

In Nilson (2003), the study of the behaviour of steels Armox 500 T and Armox 600 T was carried out in the range of strain rates 100–1000 $s^{-1}$ s. A single bar tensile Hopkinson apparatus was used. The tensile force was generated with pendulum. Zabotkin et al. (2003) showed the design concept of an instrumented falling weight impact (IFWI) tester and tested 316L stainless steel, titanium alloy and alloy C22 under medium strain rates.

In (Kong et al., 2010), an investigation into the effect that medium strain rates have on the mechanical properties of cold drawn mild steel and 304L stainless steel was carried out using two 6 mm diameter by 14 mm long specimens machined from each grade of steel. The tests were performed using an Instron test machine under a quasi-static strain rate, and a high capacity instrumented falling weight axial impact (IFWI) test rig developed at the University of Wollongong was used to conduct the medium strain rate tests. The axial load time histories were measured by an Interface 1210 Precision model load cell with maximum capacity of 50 kN connected to each specimen and the deformation time history by a strain gauge attached to each specimen. The drop hammer weighs 590 kg. The height of the drop hammer can be raised up to 3 m high to achieve the desired impact energy and velocity. During the test, the drop hammer is released from a predetermined height to impact on top of the steel frame. The steel frame then accelerates downwards, pulling the specimen until it fractures. Obviously, varying the drop height means that the time needed to fracture the specimen varies, which means that different strain rates can be obtained.

At hammer height about 100 mm, it was obtained strain rate order of 80 $s^{-1}$. Maximum obtained strain rate equalled ~500 $s^{-1}$.

In Cui et al. (2019), high-speed tensile tests were performed on a Zwick/Roell HTM5020 high-speed testing system (Ulm, Germany) with a maximum test velocity of 20 m/s, as well as a maximum load of 50 kN. The mechanical properties and failure behaviour of long glass fibre reinforced thermoplastic composites were investigated. The dynamic tensile tests were carried out at the strain rates of 1, 10, 100, 200 and 400 $s^{-1}$.

In order to obtain full-field strain contours, strain evaluation was performed for all groups of tensile tests by the aid of the DIC system. DIC was an optical method to track changes in a series of deformed images. It was used to measure displacements and strains on the surface of a deforming material in a non-contact way (Mehdikhani et al. 2016). This method had been proved to be an efficient strain recording method for composites (Jones and Iadicola 2018; Li et al. 2019; Codolini et al. 2018; Röhrig et al. 2017; McCormick and Lord 2010). The post-processing analysis of all the measurement results was conducted with the DIC professional software (VIC 2D 6, Correlated Solutions Inc., Irmo, SC, USA).

This paper describes a pneumo-dynamic experimental setup created in the dynamic testing laboratory of the Research Institute for Mechanics of Nizhny Novgorod State University, to study the dynamic behaviour of structural materials at strain rates of the order of 100 $s^{-1}$.

## 5.2 Experimental Setup Description

The general scheme of the experimental setup is shown in Fig. 5..1. The numbers indicate: 1—compressor, 2—pressure chamber, *3*—pressure gauge, *4*—ball valves, *5*—connecting hoses, *6*—solenoid valves, *7*—pneumatic cylinder, *8*—lower movable bar, *9*—specimen, *10*—grips, *11*—laser distance sensor and *12*—piston.

Sample dimensions are shown in Fig. 5.2. The upper blade part of the sample (right part on the figure) is longer than the lower one for installation of the strain gauges.

Lower grip movements measured with a high-speed laser triangulation sensor RIFTEK RF603HS. The strain in the working part of the sample is determined from the data of the sensor using the following formula:

$$\varepsilon(t) = \frac{U(t)}{L_0}$$

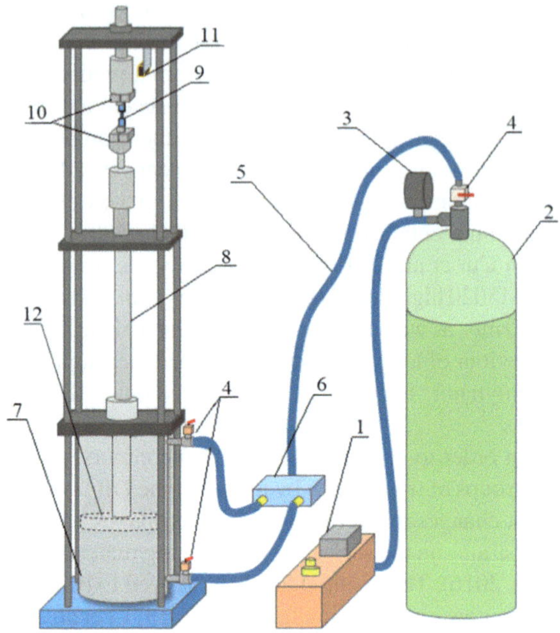


**Fig. 5.1** Experimental setup scheme

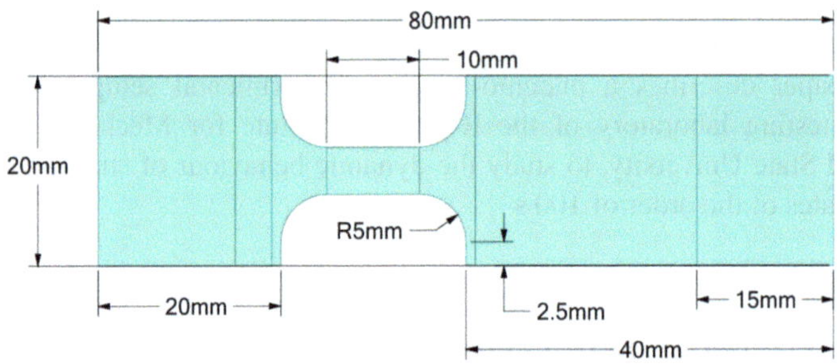


**Fig. 5.2** Sample dimensions

here $U(t)$—time history of the lower grip displacement, $L_0$—length oft the sample working part.

The strain rate realized in the experiment is determined by differentiating the time dependence $\varepsilon(t)$:

$$\dot{\varepsilon}(t) = \frac{\mathrm{d}}{\mathrm{d}t}\varepsilon(t)$$

To obtain the stress in the working part of the sample, strain gauges are installed on the upper blade part of the sample. Assuming uniform elastic deformation of the blade part of the sample, the time dependence of the stress can be determined as follows:

$$\sigma(t) = \frac{E \cdot S_1 \cdot \varepsilon_s(t)}{S_2}$$

here $E$—Young's modulus of the specimen, $S_1$—cross-section area of the blade part of the specimen, $\varepsilon_s(t)$—strain history measured by the strain gauges, $S_2$—cross-section area of the working part of the specimen.

Numerical analysis of the stress–strain state of the specimen, verification of the fulfilment of the uniformity and elasticity of specimen's blade part deformation assumptions are given in Sect. 5.3. Eliminating the time from dependencies $\varepsilon(t)$ and $\sigma(t)$, we obtain the deformation curve $\sigma(\varepsilon)$ with the known law of change in the strain rate $\dot{\varepsilon}(\varepsilon)$ during loading of the specimen.

We use a potentiometric circuit for connecting strain gauges. The measuring channel is calibrated using an electrical method. The oscilloscope records electrical voltages in the power supply circuit of the gauges. Deformation of a strain gauge is associated with a change of its resistance with the general strain gauge law:

$$k \cdot \varepsilon = \frac{\Delta R}{R},$$

here $k$—strain sensitivity factor, $\varepsilon$—deformation or the strain gauge, $\Delta R$—change of the strain gauge resistance and $R$—nominal strain gauge resistance.

A calibration procedure is used to determine the conversion factor of the voltage taken from the strain gauges circuit into their deformation. A resistor with a known resistance $R_{\text{calibr}}$ is periodically inserted into the power circuit with the gauges using a relay. Thus, the change in the resistance of the strain gauges by the value of $R_{\text{calibr}}$ is simulated. The oscilloscope records the corresponding changes in the electrical voltage in the power supply circuit of the gauges (Fig. 5.3).

Since the voltage rise is proportional to the change in circuit resistance, the following proportion can be made:

$$\frac{\Delta U^{\text{calibr}}}{R^{\text{calibr}}} = \frac{\Delta U}{\Delta R},$$

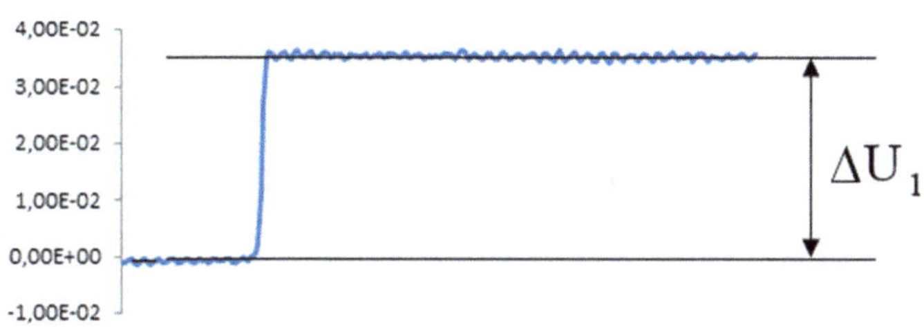


**Fig. 5.3** Voltage rise during electric calibration procedure

here $\Delta U^{\text{calibr}}$—voltage rise during electric calibration procedure, $R^{calibr}$—the nominal of the calibration resistor, $\Delta U$—change in voltage during experiment and $\Delta R$—change of strain gauges resistance due to their deformation.

From the above formulas, it follows:

$$\varepsilon = \frac{R^{\text{calibr}}}{k \cdot \Delta U^{\text{calibr}} \cdot R} \cdot \Delta U$$

Thus, the conversion factor (calibration factor) is:

$$K = \frac{R^{\text{calibr}}}{k \cdot \Delta U^{\text{calibr}} \cdot R}$$

To carry out an experiment on a pneumo-dynamic setup, compressed air is pumped by a compressor (1) into a high-pressure chamber (2). Pressure in chamber is controlled using a pressure gauge (3). After pumping, we open the ball valve (4) and the compressed air through the pneumatic hose (5) enters the electro-pneumatic distributor (6). Further, upper ball valve is opened, and the pressure is applied to the pneumatic cylinder (7), thereby bringing the pneumatic cylinder rod (8) to the desired height for fixing the experimental specimen (9) into special grips (10), which are attached to the pneumatic cylinder rod by threaded joins. To determine the deformation in the specimen, a high-speed laser triangulation position sensor (11) is used. Before the experiment, the pressure above and below the piston (12) of the pneumatic cylinder is the same. In order to produce dynamic tension of the specimen, it is necessary to open the lower ball valve of the pneumatic cylinder. The rod with some force pulls the specimen due to the difference of pressure above and below the piston. The maximum force reached by the pneumatic cylinder is 31kN. The value of applied force is controlled by the pressure in the cylinder chambers.

## 5.3 Numerical Analysis

To assess the stress–strain state of the working part of the specimen, as well as to analyse the possibility of measuring the force acting on the specimen by registering the deformation in the blade part of the sample, a numerical simulation of the process of high-speed deformation of the specimen was performed. The simulation was carried out using the LS DYNA programme code. The problem was solved using an explicit time integration scheme. The geometric formulation of the problem is shown in Fig. 5.4. The sample is a 0.65 mm thick plate. The sample has a working section and extended blades for fixing the sample in the experimental setup. One of the blade parts has a large length for installing strain gauges for the measurement of the force acting on the specimen during loading. The length of the working part is 10 mm, and the width is 5 mm. The length of the smaller blade is 20 mm, and the larger one is 40 mm. The blade width is 20 mm. A 15 mm portion of a blade is used

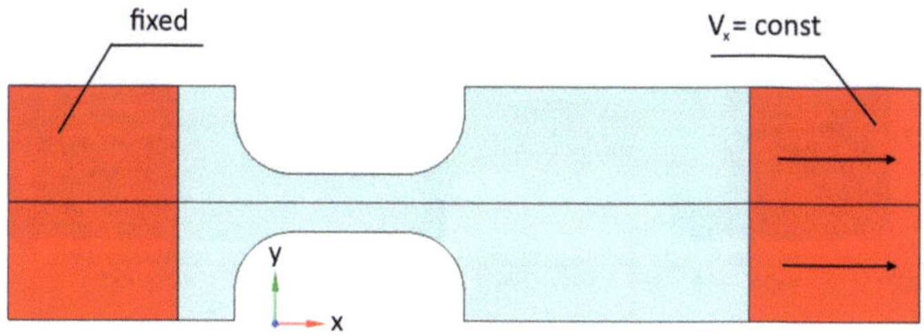


**Fig. 5.4** Geometry of the task and boundary conditions

to anchor the specimen in the experimental setup grip. One end of the specimen is fixed and remains motionless. On the other, a constant velocity equalled 1 m/s is set. This velocity provides an average strain rate of the sample of about 100 $s^{-1}$. Due to the presence of symmetry, half of the model was considered. Figure 5.5 shows a finite element model of a half part of the specimen.

Figure 5.6 illustrates the plastic strain field in the specimen at the moment when the value of strain in the working part reaches ~10%. It can be seen that plastic strain is concentrated and uniformly distributed in the working part of the sample.

Figure 5.7 illustrates the axial stress fields in the specimen (stresses in the direction of the $x$-axis). Figure 7a shows the entire range of stress values. Figure 7b shows the stress range from 30 to 200 MPa, for a clear visual representation of the stresses in the zone of the specimen's blades. This makes it possible to estimate the uniformity of the indicated fields in the zone of the strain gauges location. In addition, stress distributions were plotted over the width of the blade in sections at different distances

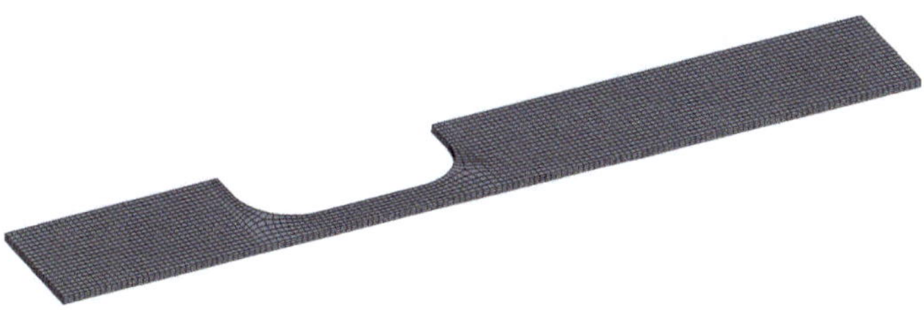

**Fig. 5.5** Finite element model

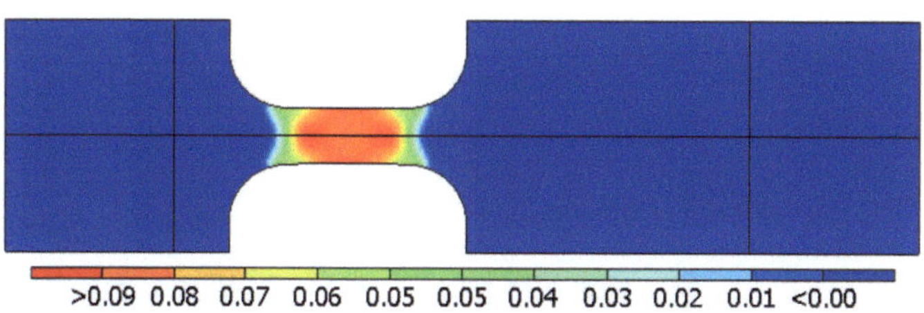


**Fig. 5.6** Plastic strains field in specimen

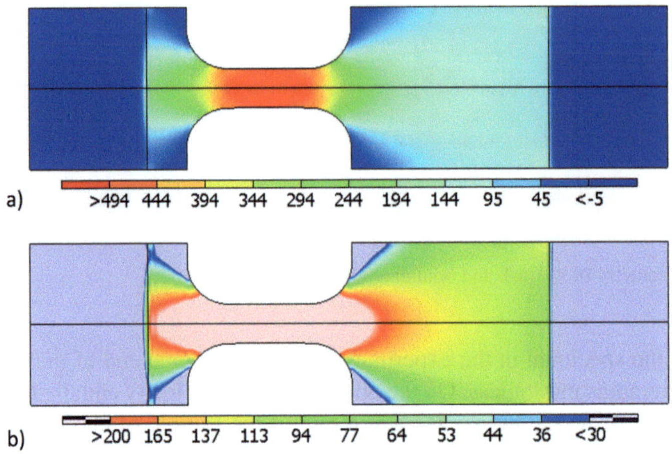


**Fig. 5.7** Axial stress in specimen, MPa

from the beginning of the blades (Fig. 5.8). The distribution of stresses along the width of the specimen at distances of 5, 10, 15 and 20 mm is shown in Fig. 5.9. It can be

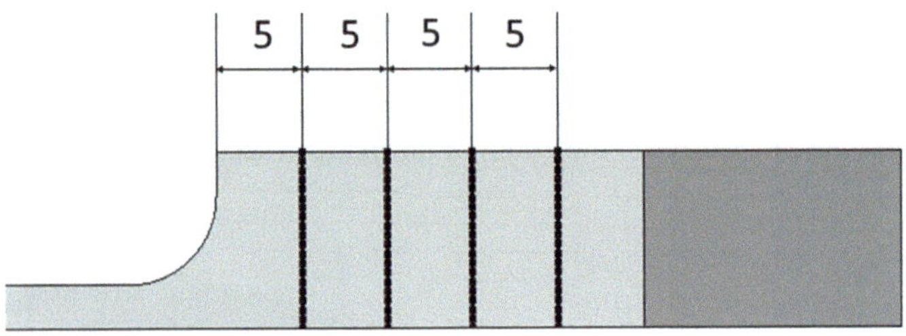


**Fig. 5.8** Cross sections definition

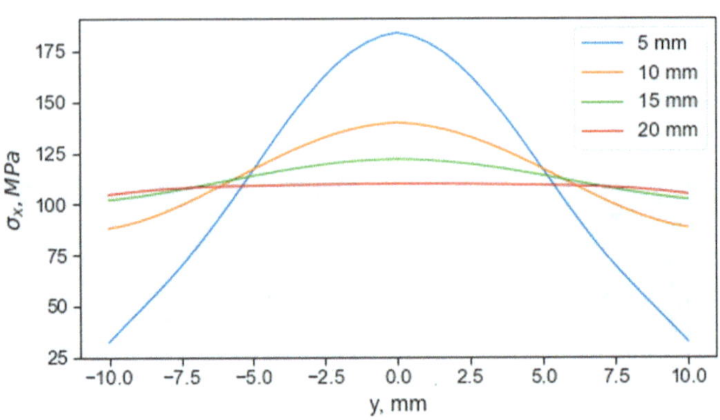


**Fig. 5.9** Stress distribution in different section

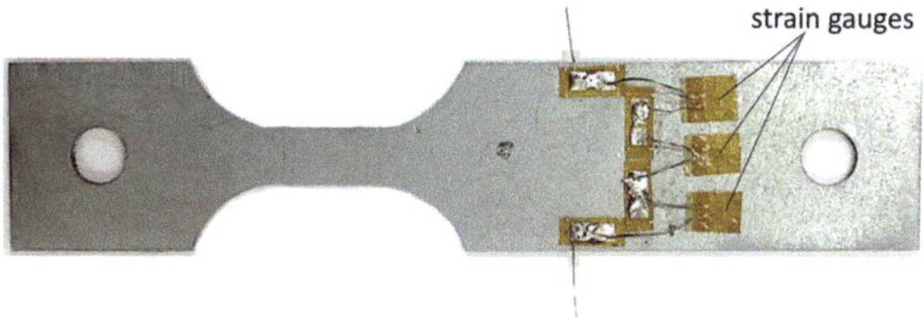


**Fig. 5.10** Strain gauges position

noted that at a distance of 20 mm from the beginning of the blade, the axial stresses are uniformly distributed.

Based on the performed numerical analysis, two conclusions can be drawn:

1. The blade portion of the sample is deformed in the elastic region.
2. Axial stresses at a distance of 20 mm are uniformly distributed over the blade width.

Thus, the position of the strain gauges was chosen (Fig. 5.10) to determine the force acting on the specimen during loading.

## 5.4 Experimental Results

This section presents the results of tests carried out on the developed experimental setup. The tests were carried out on HX220BD sheet steel specimens. Figure 5.11 shows the time histories of stress (black line), strain (red dashed line) and strain rate (blue dotted line). It can be noted that the average strain rate is 100 $s^{-1}$. The sample is ruptured at the time about 5 microseconds after the start of the loading. Figure 5.12 illustrates a deformation diagram. The initial yield stress of the investigated material is about 300 MPa. The rupture of the sample occurred at a relative elongation of the order of 50%. In general, it can be noted that the goal of creating an experimental setup has been achieved.

## 5.5 Conclusions

The experimental setup described in the work makes it possible to study the processes of deformation and fracture of plate specimens at moderate strain rates. In order to increase the accuracy of measuring the forces and expanding the types of tested specimens (including axisymmetric ones), it is planned to use a dynamic capacitive force sensor in the future. To improve the accuracy of measuring the deformation of the working part of the sample, high-speed video registration will be used in combination with the DIC method.

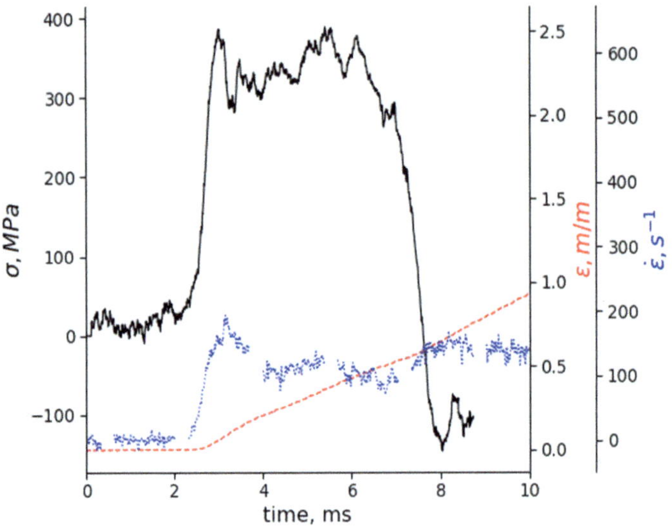


**Fig. 5.11** Time histories of the stress, strain and strain rate of the specimen

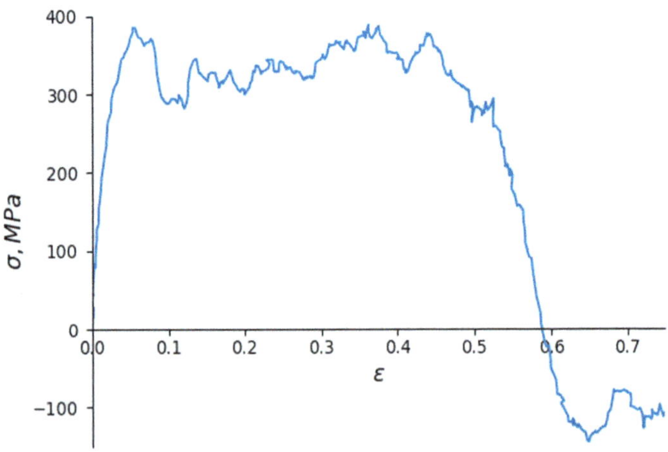


**Fig. 5.12** Stress–strain curve

**Acknowledgements** The theoretical study was done with financial support from the Ministry of Science and Higher Education of the Russian Federation (Project 0729-2020-0054). The experimental investigations were conducted with financial support from RFFI (Project 19-38-90217).

## References

Albertini C, Montagnani M (1977) Dynamic material properties of several steels for fast breeder reactor safety analysis. EUR 5787 EN, Ispra

ASM Int (2000) ASM Handbook vol 8, Mechanical Testing and Evaluation, ASM Int, Materials Park OH, pp 462–476

Asprone D, Cadoni E, Prota A, Manfredi G (2009) Strain-rate sensitivity of a pultruded e-glass/polyester composite. ASCE-J Compos Constr 13:558–564

Bragov AM, Lomunov AK (1984) Features of the construction of strain diagrams by the Kolsky method. Appl Problems Strength Plasticity 28:125–137 (in Russ)

Bragov AM, Igumnov LA, Konstantinov AY, Lomunov AK, Rusin EE, Eremeyev VA (2020) Experimental analysis of wear resistance of compacts of fine-dispersed iron powder and tungsten monocarbide nanopowder produced by impulse pressing. Wear 456–457. https://doi.org/10.1016/j.wear.2020.203358

Cadoni E, Dotta M, Forni D, Spaetig P (2011a) Strain-rate behavior in tension of the tempered martensitic reduced activation steel Eurofer97. J Nucl Mater 414:360–366

Cadoni E, Dotta M, Forni D, Tesio N (2011b) Dynamic behaviour of reinforcing steel bars in tension. Appl Mech Mater 82:86–91

Cadoni E, Dotta M, Forni D, Bianchi S, Kaufmann H (2012) Strain rate effects on mechanical properties in tension of aluminium alloys used in armour applications. EPJ Web Conf 26:05004

Cadoni E, Dotta M, Forni D, Kaufmann H (2016) Effects of strain rate on mechanical properties in tension of a commercial aluminium alloy used in armour applications. Procedia Struct Integr 2:986–993

Caverzan A, Cadoni E, di Prisco M (2012) Tensile behaviour of high performance fibre-reinforced cementitious composites at high strain rates. Int J Impact Eng 45:28–38

Codolini A, Li QM, Wilkinson A (2018) Mechanical characterization of thin injection-moulded polypropylene specimens under large in-plane shear deformations. Polym Testing 69:485–489

Cui J, Wang S, Wang S, Li G, Wang P, Liang C (2019) The effects of strain rates on mechanical properties and failure behavior of long glass fiber reinforced thermoplastic composites. Polymers (Basel) 11(12). https://doi.org/10.3390/polym11122019

Davies RM (1948) A critical study of the Hopkinson pressure bar. Philos Trans R Soc London, Ser A 240–821:375–457

Eskandari H, Nemes JA (2000) Dynamic testing of composite laminates with a tensile split Hopkinson bar. J Compos Mater 34(4):260–273

Gama BA, Lopatnikov SL, Gillespie JW Jr (2004) Hopkinson bar experimental technique: a critical review. Appl Mech Rev 57(4):223–250

Grote DL, Park SW, Zhou M (2001) Dynamic behavior of concrete at high strain rates and pressures: I. Experimental characterization. Int J Impact Eng 25(9):869–886

Jiang B, Zhang R (2006) Tensile properties in the through-thickness direction for a carbon fiber woven reinforced composite at impact loading rate. J Phys IV France 134:1071–1075

Jones I, Iadicola ME (2018) A good practices guide for digital image correlation. Int Digit Image Correl Soc

Kolsky H (1949) An investigation of the mechanical properties of materials at very high rates of loading. Proc Phys Soc London Sect B 62:676–700

Kong SY, Remennikov A, Uy B (2010) The effect of medium strain rates on the mechanical properties of high performance steels. In: Conference: 21st Australasian conference on the mechanics of structures and materials, Melbourne, Victoria, Australia

Li L, Sun L, Dai Z, Xiong Z, Huang B, Zhang Y (2019) Experimental investigation on mechanical properties and failure mechanisms of polymer composite-metal hybrid materials processed by direct injection-molding adhesion method. J Mater Process Technol 263:385–395

Lomunov AK (1987) Methodology for studying the processes of viscoplastic deformation and material properties based on the split Hopkinson bar, Ph.D. thesis (in Russ.)

Ma L, Li Z, Liu J, Duan L, Wu J (2019) Mechanical properties of coral concrete subjected to uniaxial dynamic compression. Constr Build Mater 199:244–255

Ma L, Wu J, Wang M, Dong L, Wei H (2020) Dynamic compressive properties of dry and saturated coral rocks at high strain rates. Eng Geol 272:105615

McCormick N, Lord J (2010) Digital image correlation. Mater Today 13:52–54

Mehdikhani M, Aravand M, Sabuncuoglu B, Callens MG, Lomov SV, Gorbatikh L (2016) Full-field strain measurements at the micro-scale in fiber-reinforced composites using digital image correlation. Compos Struct 140:192–201

Nilson M (2003) Constitutive model for Armox 500T and Armox 600T at low and medium strain rates. Swedish defence research agency, Technical report

Röhrig C, Scheer T, Diebels S (2017) Mechanical characterization of a short fiber-reinforced polymer at room temperature: experimental setups evaluated by an optical measurement system. Contin Mech Thermodyn 29:1093–1111

Shi Y, Shi J, Luo Z, Liu H, Wang D, Shen H (2020) Experimental investigation on strength and deformation characteristics of red sandstone at strain rates of $10^{-2}$~55 $s^{-1}$. Adv Civ Eng, Article ID 8882976. https://doi.org/10.1155/2020/8882976

Singh Y, Udupa A, Chandrasekar S, Subbarayan G (2019) Medium to high strain-rate characterization of lead free solder alloys through metal cutting experiments. In: Proceedings of the ASME 2019 international technical conference and exhibition on packaging and integration of electronic and photonic microsystems. ASME 2019 international technical conference and exhibition on packaging and integration of electronic and photonic microsystems. Anaheim, California, USA, 7–9 Oct 2019. V001T01A008. ASME, 11p. https://doi.org/10.1115/IPACK2019-6510

Solomos G, Albertini C, Labibes K, Pizzinato V, Viaccoz B (2001) Strain rate effects of nuclear steels in room and higher temperatures. SMiRT 16, Washington DC, Paper #1356

Zabotkin K, O'Toole B, Trabia M (2003) Identification of the dynamic tensile properties of metals under moderate strain rates. In: 16th ASCE engineering mechanics conference

Zhang QB, Zhao J (2014) Quasi-static and dynamic fracture behaviour of rock materials: phenomena and mechanisms. Int J Fract 189(1):1–32

# Chapter 6
# Modeling Approaches for an Eyeball Deformation After Intravitreal Injection

**Svetlana M. Bauer, Liudmila A. Venatovskaya, Eva B. Voronkova, and Andrey B. Kachanov**

**Abstract** The mathematical models of the pressurized eyeball are studied. The eyeball is represented as two joint segments (the cornea and the sclera) of different geometric and mechanical properties. Both the cornea and sclera are considered as spherical segments of nonuniform thicknesses. Finite element simulation is performed by means of the engineering simulation software ANSYS Inc. Short time effect of fluid (medication) administration into the eye is investigated. Numerical results of two models (with and without separating strip) are compared. Results obtained with the finite element models of a composed eyeball shell are compered with the shell-based model of a pressurized spherical shell. If the strip separating the inner volume of the eye is taken into account, the difference of the pressures inside subcorneal and subscleral volume is about 1–2 mm Hg. If a single inner volume is considered, the inner pressure elevates to the average value for two subspaces from the model with the strip.

S. M. Bauer (✉) · L. A. Venatovskaya · E. B. Voronkova
St. Petersburg State University, 7/9 Universitetskaya Emb., St. Petersburg 199034, Russian Federation
e-mail: s.bauer@spbu.ru

L. A. Venatovskaya
e-mail: l.venatovskaya@spbu.ru

E. B. Voronkova
e-mail: e.voronkova@spbu.ru

S. M. Bauer · E. B. Voronkova
Research and Education Center "Materials", Don State Technical University, Pl. Gagarina 1, Rostov-on-Don 344000, Russian Federation

A. B. Kachanov
North-Western State Medical University Named After I.I. Mechnikov, Kirochnaya ul. 41, St. Petersburg 191015, Russian Federation
e-mail: andrey_kachanov@yahoo.com

St. Petersburg Branch of the S.N. Fyodorov Eye Microsurgery Federal State Institution, St. Jaroslav Hašek 21, St. Petersburg 192281, Russian Federation



H. Altenbach et al. (eds.), *Advanced Materials Modelling for Mechanical, Medical and Biological Applications*, Advanced Structured Materials 155,
https://doi.org/10.1007/978-3-030-81705-3_6

**Keywords** Ocular biomechanics · Sclera · Finite elements · Intraocular pressure (IOP) · Intravitreal injections

## 6.1 Introduction

Nowadays, intravitreal injections (IVIs) are extensively used to treat human ocular diseases (de Vries et al. 2020). An amount of fluid (medication) is brought into the eye, while performing IVIs. Administration of fluid into the eye can result in an increase in intraocular pressure (IOP) and lead to undesirable side-effects and complications such as intraocular bleeding and retinal detachment. An increase in IOP is the main risk factor for glaucoma and thinning of the retinal nerve fiber layer (de Vries et al. 2020; Kotliar et al. 2007). However, the effect of IVI-induced IOP increases on the eye is still unclear. The knowledge of the relationship between IOP and intraocular volume (IOV) in the human eye is important to make a physically correct conclusion from the data of standard measurement procedure used in ophthalmology (Moiseeva et al. 2020; Stein 2010).

The human eye is a roughly spherical organ. An outermost, fibrous, coat of the eye consists of the cornea and sclera. The main function of the fibrous coat is to protect the inner structures of the eye (the lens, the iris). The sclera forms more than 90% of fibrous eye shell and is tougher than the cornea (Lang 2000). The sclera, as well as cornea, are close to a soft membrane: the moduli of elasticity in the thickness directions are two orders of magnitude less than the tangential moduli (Iomdina 2006; Iomdina et al. 2015). For people with normal sight the eye has a shape close to spherical one. However, myopic (shortsighted) and hyperopic (longsighted) eyes have out-of-sphericity shapes (Iomdina et al. 2015; Lang 2000).

The simplest model of the eye is a spherical shell (Bauer and Voronkova 2014; Bauer et al. 2000; Kotliar et al. 2007). The stress-strain states of a pressurized spherical and ellipsoidal shells are studied by means of the exact 3D theory of elasticity and two 2D approximate shell theories of moderate thicknesses in Bauer and Voronkova (2015); Bauer et al. (2017). The approximate shell theories take into account the variation of the shell thickness, which can be important for soft materials, e.g., the human eye's tissue.

However, the presentation of the eyeball as a spherical or ellipsoidal segment is a very simple assumption. The eyeball is considered as a joined shell, consisting of two spherical segments, representing the cornea and the sclera in Bauer et al. (2020). This approach allows to describe accurately the applanation tonometry procedure and take in account not only the complex geometry of eye but also different properties of the eye's tissues (anisotropy, nonlinearity, inhomogeneity).

In the present study, two finite element models of the eyeball under IVIs-induced IOP are considered. In the first model, based on Bauer et al. (2020), we modelled the eyeball as a composite shell with a unified inner volume. To treat differently, the increments of the subcorneal and subscleral volumes the thin strip separating two chambers were added in the second model. To understand the cornea impact on the simulation results, the finite element models were compared with the shell-based model from Bauer and Voronkova (2015).

## 6.2 Problem Statement

We employed the simplified model for the eyeball previously published in Bauer et al. (2020). The corneoscleral coat of the eye was modelled as two joint segments of different geometric and mechanical properties. The outer surface of the cornea was approximated by a spherical segment with radius $R_c = 7.8$ mm (Lang 2000). We assumed that the thickness of the cornea is minimum at the apex $h_{AP} = 0.54$ mm, and increases linearly along the meridional direction and achieves its maximum at the limbus where the cornea connects to the sclera $h_L = 0.75$ mm. The sclera was represented as a spherical segment with the outer radius $R_s = 12$ mm. We proposed that scleral thickness is equal to $h_E = 0.65$ mm near the equator and $h_{PP} = 1$ mm at the posterior pole (see Fig. 6.1).

The eyeball contains different structures such as the focusing lens, iris and ciliary body. The space between the cornea and the lens is filled with aqueous humour. The posterior cavity, behind the lens, is packed with the vitreous body, a transparent jelly-like substance (Fig. 6.2).

Intravitreal injections usually is given perpendicular to the sclera, 3.5–4 mm posterior to the limbus (Rezaei and Wen 2016).

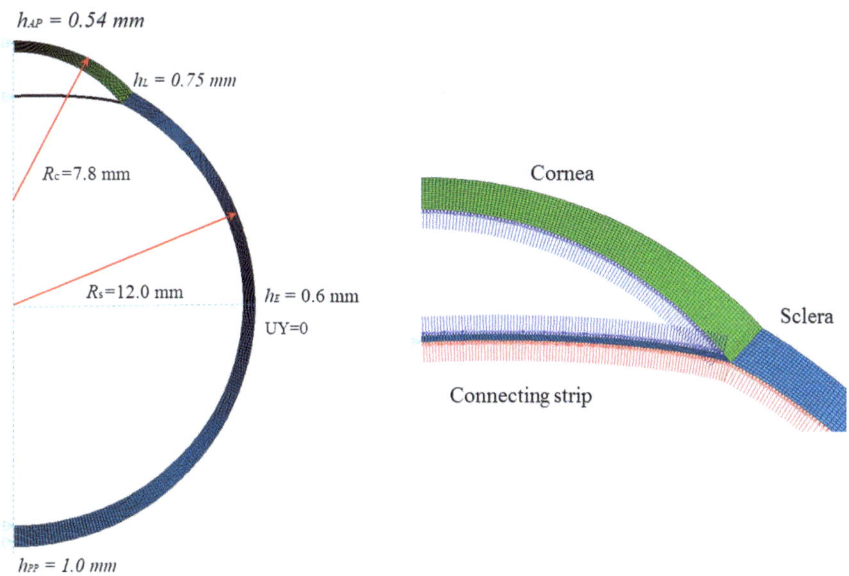


**Fig. 6.1** Finite-element simulation of the corneoscleral eye shell filled by incompressible liquid

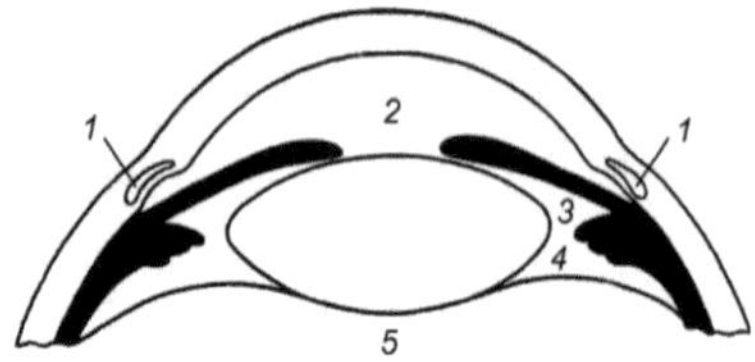


**Fig. 6.2** The hydrostatic system of the eye system: 1—Schlemm's canal, 2—the anterior chamber, 3—anterior and 4—posterior spaces of the posterior chamber, 5—the vitreous body. Adopted from Nesterov et al. (1974)

Nesterov and coauthors proposed to consider the eye as a hydrostatic system , consisting of several closed and semi-open cavities separated from each other by elastic membranes (Fig. 6.2). Each cavity has its level of pressure, which is slightly different from the pressure in other parts of the eye (Iomdina et al. 2015; Nesterov et al. 1974). The authors noted that the difference in pressure in different parts of the eye (excluding blood vessels) is small, and can be neglected in practice.

In our first, basic, model we ignored all additional components inside the eye and considered the composite shell as a cavity filled by incompressible liquid with some pressure $p$. This emulates the existence of the intraocular pressure (IOP) inside the eyeball. At the first preliminary step, the simulated composite shell was subjected to internal physiological pressure ($p = 15$ mm Hg), and then, the pressure was changed so that the internal volume of the simulated eyeball increased by the volume of the injected medication.

In the second model, we added the thin strip separating the subcorneal and scleral volumes (Fig. 6.1). This approach allows us to treat differently the increments of the scleral $V_s$ and subcorneal $V_c$ volumes. Before the emulation of the injection, initial internal pressure is applied separately to subcorneal and scleral spaces. In the second step of the simulation, the pressure that is required to change the initial scleral volume on the volume of the injected drug is applied. We assumed that internal undercorneal volume remains constant $V_c = \text{const}$, thus initial pressure in undercorneal space is also increased.

Although the cornea and sclera represent typical biological connective tissues (i.e., they are characterized by inhomogeneity, nonlinearity of physical properties), in this work we approximated the tissues of the eyeball as a transversally isotropic material with principal axes in the meridian and circumferential directions.

Nonlinear analyses were performed using engineering finite element software Ansys®, Release 18. Due to symmetry, two-dimensional (2D) modeling was carried out with finite element PLANE182. The axisymmetric problem was considered with the axis of symmetry $Y$, the $X$ axis was directed along the normal to the isotropic surface. The symmetry boundary conditions were applied along the anterior-posterior axis of the eye (at the top of the cornea and the posterior pole of the sclera), and at the equator, the displacement of the sclera in a vertical direction was prohibited.

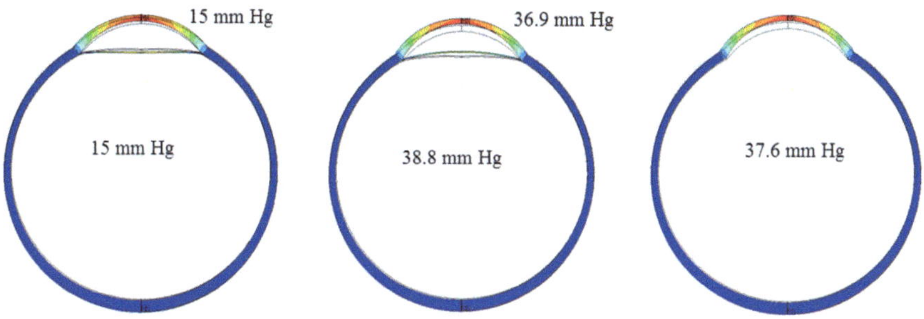


**Fig. 6.3** Deformation of the outer shell of the eye before (left) and after 0.05 ml injection of a medication for the model with the separating strip (center) and model without the strip (right)

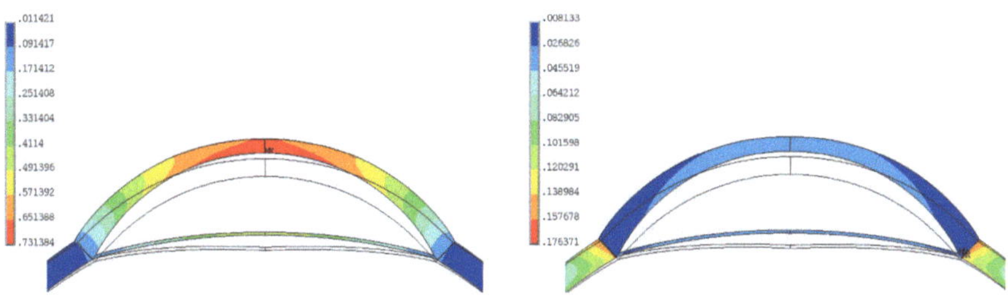


**Fig. 6.4** Displacements (left) and von Mises stress distribution (right) after 0.05 ml injection of liquid for the model with separated scleral and subcorneal spaces

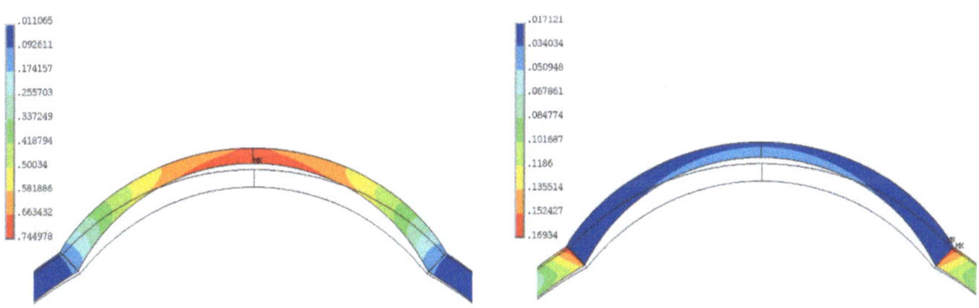


**Fig. 6.5** Displacements (left) and von Mises stress distribution (right) after 0.05 ml injection of liquid for the model without the strip

## 6.3 Results and Discussion

Figures 6.3, 6.4 and 6.5 show the profiles of the simulated eyeballs before and after IVI as well as displacements and stresses distributions in the simulated eyeballs.

Material properties are assigned as follows: Young's moduli in the isotropic surface are $E_c = 0.3$ MPa, $E_s = 15.0$ MPa, $E_{\text{strip}} = 0.9$ MPa for the cornea, sclera and connection strip, respectively; Young's moduli in the normal direction are taken $E'_c = E_c/20$, $E'_s = E_s/10$, $E'_{\text{strip}} = E_{\text{strip}}/20$ (Iomdina 2006; Iomdina et al. 2015);

**Table 6.1** IOP in the eye before and after IVIs (clinical data)

| Clinical refraction | Anterior–posterior axis | Initial IOP, mm Hg | | IOP after IVI, mm Hg | |
|---|---|---|---|---|---|
| | | Maklakoff tonometry | Pneumo-tonometry | Maklakoff tonometry | Pneumo-tonometry |
| Hyperopia | 22.5 | 21 | 24 | 32 | 39 |
| | 22.6 | 20 | 25 | 34 | 38 |

for all considered tissues Poisson's ratios in the isotropic surface are proposed to be $\nu = 0.49$, and the Poisson ratios that correspond to a contraction in the thickness direction when an extension is applied in the plane of isotropy, are $\nu' = 0.01$; in-plane and transverse shear moduli are $G_i = E_i/2(1+\nu)$, $G_i' = G_i/2$, $i = \{c, s, \text{strip}\}$. After the injection of 0.05 ml of a medication into the eye cavity, the internal pressure increased from 15 to 38.8 mm Hg in the scleral space and to 36.9 mm Hg in the subcorneal space. For model without the strip, the pressure elevates to 37.6 mm Hg which is roughly the average value for the pressure increases from the previous model. Thus, adding a strip allows us to estimate more accurately the pressure change in different parts of the eyeball.

With a lower scleral stiffness $E_s = 7.0$ MPa, $E_s' = E_s/40$ the internal pressure increases to 30 mm Hg for the scleral volume and to 29.0 mm Hg for the subcorneal volume. For the model with a signal internal space, the pressure changes to 29.2 mm Hg. The numerical simulation shows that the elastic properties of the sclera significantly affect the assessment of the evelated intraocular pressure (IOP) after intravitreal injection.

The clinical data obtained from the St. Petersburg Branch of the S. Fyodorov Eye Microsurgery Federal State Institution are presented in Table 6.1.

IOP was measured with Maklakoff 's applanation tonometry and pneumotonometry before and 3 minutes after 0.1 ml injection of a medication for the eyes with farsightedness (hyperopia). It should be noted that the eyes with hyperopia are usually smaller than the eyes with normal vision or eyes with shortsightedness (myopia). Therefore, the equal amount of a liquid delivered into the eye has a stronger effect on the eyes with hyperopia.

Our simulations show that pressure-induced deformations of the eyeball (under glaucoma disease or after intraocular injections) depend mainly on the extensibility of the sclera. To understand whether it is possible to exclude the cornea from the modeling we compared the results of finite element models presented here with the analytical models from Bauer and Voronkova (2015). Pressurized spherical shell made of transversally orthotropic material was treated by means of the 2D approximate shell theories of moderate thicknesses described in Paliy and Spiro (1977). The Palii-Spiro (PS) theory is the extension of the Timoshenko-Reissner (TR) type theory that takes into account deformations, rotation and bending of the fibers and their elongations in the direction of the thickness of the shell.

**Table 6.2** The pressure in the corneoscleral eye shell after 0.05 ml injection obtained by FEM simulation and PS shell theory, mm Hg

| Young's moduli of the cornea | Young's moduli of the sclera | FEM simulation | Results obtained by PS theory $h = 0.6$ mm |
|---|---|---|---|
| $E_c = 0.3$ MPa, | $E_s = 15$ MPa, $E_s' = E_s/10$ | 37.6 | 74.9 |
| $E_c' = E_c/20$ | $E_s = 10$ MPa, $E_s' = E_s/10$ | 35.5 | 55.3 |
| | $E_s = 7$ MPa, $E_s' = E_s/40$ | 29.2 | 42.3 |
| $E_c = 1.0$ MPa, | $E_s = 7$ MPa, $E_s' = E_s/40$ | 35.5 | 42.35 |
| $E_c' = E_c/20$ | | | |

The displacement of the midsurface of a pressurized spherical shell with inner radius $R_1$ and outer radius $R_2$ can be calculated by the following equation (Bauer and Voronkova 2015):

$$u^{PS} = \frac{w}{R} = \frac{1-\nu}{2E}\frac{p_1 - p_2}{\alpha}\left[1 - \alpha(1-\nu^\star)\frac{p_1 + p_2}{p_1 - p_2}\right],$$

where $w$ is the displacement of the middle surface, $\alpha = h/R$ is relative shell thickness, $h = R_2 - R_1$ and $R = (R_2 + R_1)/2$ are the thickness and the radius of the midsurface of the shell; $E$ and $E'$ are Young's moduli in the isotropic surface and in the normal direction, $\nu$ and $\nu'$ are corresponding Poisson's ratios and $\nu^\star = E\nu'/E'(1-\nu)$; $p_1$ and $p_2$ are the uniform pressures applied to the inner and outer surfaces of the shell.

The pressure inside the corneoscleral eye shell after an injection of liquid calculated by the finite element model without the strip and the PS theory are presented in Table 6.2. For the PS theory model, we proposed the uniform shell thickness of $h = 0.6$ mm. Different sets of material parameters for the sclera and the cornea were used to perform comparisons of simulation results. For the weak cornea ($E_c = 0.3$ MPa, $E_c' = E_c/20$) and very stiff sclera ($E_s = 15$ MPa, $E_s' = E_s/10$) the PS theory-based model result in almost twice as much inner pressure as the FEM model. For the less stiff sclera ($E_s = 10$ MPa and $E_s = 7$ MPa), the two models provide closer results although the difference between simulation predictions is about 30%. The models results are even closer when the stiffer cornea ($E_c = 1$ MPa, $E_c' = E_c/20$) is considered (see Table 6.2). The results reported above were obtained with the corneal transverse shear modulus twice smaller than the in-plane shear modulus ($G_c' = G_c/2$). If we assume $G_c' = G$, we obtain the values of the increased intraocular pressure about 1 mm Hg higher.

The simulations show that the simplified shell theory-based models of the eye can give satisfactory estimations even though they neglect the complex inner structure of the eyeball. It is also important to note that the stress-strain tests cannot be performed within the living organisms (in vivo), and most studies on mechanical properties of the eye tissues are based on experiments performed on cadaver eyes Iomdina (2006). This can lead to incorrect (overestimated) values of the material parameters as well as to a wide variation between the reported material properties due to differences in testing protocols, age and health conditions of the donors.

## 6.4 Conclusion

The mathematical models describing stress-strain state of the outermost, fibrous, coat of the eye shell subjected to initial internal pressure and pressure induced by the additional volume of fluid are presented. Numerical results of two model (with and without separating strip) are compared. If the strip separating the inner volume of the eye is taken into account, the difference of the pressures inside subcorneal and subscleral volume is about 1–2 mm Hg. If a single inner volume is considered, the inner pressure elevates to the average value for two subspaces from the model with the strip. It is showed that pressure-induced deformations of the eyeball (under glaucoma disease or after intraocular injections) depends mainly on the extensibility of the sclera. Results obtained with the finite element models of a composed eyeball shell are compered with the shell-based model of a pressurized spherical shell.

The simulations show that the simplified shell theory-based models of the eye can give satisfactory estimations even though they neglect the complex inner structure of the eyeball.

**Acknowledgements** Svetlana M. Bauer and Eva B. Voronkova acknowledge support by the grant of Goverment of Russian Federation no. 14.Z50.31.0046.

## References

Bauer SM, Voronkova EB (2014) Models of shells and plates in the problems of ophthalmology. Vestnik St. Petersburg Univ: Math 47(3):123–139 http://orcid.org/10.3103/S1063454114030029

Bauer SM, Voronkova EB (2015) Nonclassical shell theories in ocular biomechanics. In: Altenbach H, Mikhasev GI (eds) Shell and membrane theories in mechanics and biology. Advanced structured materials, vol 45. Springer, Cham, pp 81–97. https://doi.org/10.1007/978-3-319-02535-3_4

Bauer SM, Zimin BA, Tovstik PE (2000) Simplest models of plates and shells in ophthalmologys (in Russ.). SPbGU, St. Petersburg

Bauer SM, Venatovskaya LA, Voronkova EB, Ibragimova OP, Smirnov AL (2017) Scleral deformation under intravitreal injections. Russian J Biomech 21(2):107–115

Bauer SM, Venatovskaya LA, Avershina LA, Pikusova SM (2020) Scleral deformation under intravitreal injections elastotonometry for evaluation of biomechanical characteristics of the eye after surgical hyperopia correction elastotonometry for evaluation of biomechanical characteristics of the eye after surgical hyperopia correction (in Russ.). Vestnik Oftalmologii 136(5):32–38 de Vries VA, Bassil FL, Ramdas WD (2020) The effects of intravitreal injections on intraocular pressure and retinal nerve fiber layer: a systematic review and meta-analysis. Sci Rep 10(1):13248. https://doi.org/10.1038/s41598-020-70269-7

Iomdina E (2006) Mechanical properties of the human eye tissues. In: Shtein A, Tsaturyan A (eds) Modern problems of biomechanics. Moscow University Press, Moscow, vol 11: Biomechanics: achievements and prospects, pp 184–201

Iomdina EN, Bauer SM, Kotlyar KE (2015) Biomechanics of the human: theoretical aspects and clinical applications (in Russ.). Real Time, Moscow

Kotliar K, Maier M, Bauer S, Feucht N, Lohmann C, Lanzl I (2007) Effect of intravitreal injections and volume changes on intraocular pressure: clinical results and biomechanical model. Acta Ophthamologica Scandinavica 85(7):777–781

Lang GK (2000) Ophthalmology. Thieme, Stuttgart

Moiseeva IN, Stein AA, Lyubimov GA (2020) Estimation of the elastic characteristics of the eyeball from the response of the enucleated eye to the introduction of given volumes of liquid. Russ J Biomech 24(4):351–366

Nesterov A, Bunin A, Katsnelson L (1974) Intraocular pressure. Physiology and pathology (in Russ.). Nauka, Moscow

Paliy OM, Spiro VE (1977) Anisotropic Shells in Shipbuildings. Theory and Analysis (in Russ.). Sudostroenie, Leningrad

Rezaei KA, Wen JC (2016) Intravitreal injection technique. MedEdPORTAL 12. https://doi.org/10.15766/mep_2374-8265.10502

Stein AA (2010) Pressure-volume dependence for the eyeball under an external load. Fluid Dyn 45(2):177–186 http://orcid.org/10.1134/S001546281002002X

# Chapter 7
# Numerical Simulation of the Dynamics of Three-Dimensional Anisotropic Bodies Based on Non-classical Boundary Integral Equations Under the Impact of a Shock Load

**Aleksandr A. Belov and Andrey N. Petrov**

**Abstract** The classical formulation of the boundary integral equation method is successfully implemented for solving three-dimensional isotropic problems of the dynamic theory of elasticity, viscoelasticity, and poroelasticity. The extension of this method for solving dynamic anisotropic problems requires the development of special new schemes. At present, numerical schemes are being constructed based on the double application of the reciprocity theorem, which goes back to the work of D. Nardini and K. A. Brebbia or based on the integral representation of Green's matrices. The use of regular Fredholm integral equations of the first kind (integral equations on a plane wave) is an alternative to the classical formulation of the boundary integral equation method. The construction of such boundary integral equations is based on the structure of the dynamic fundamental solution. The approach employs explicit boundary integral equations and goes back to the work of Babeshko. The paper considers the application of the non-classical approach of the boundary integral equation method in combination with the integral Laplace transform in time to modeling wave processes in anisotropic elastic bodies. In this case, the inverse Laplace transform is constructed numerically using the Durbin method. A numerical solution of the dynamic problem of anisotropic elasticity theory by the boundary integral equation method in non-classical formulation is given. The boundary element scheme of the boundary integral equation method is constructed on the basis of a regular integral equation of the first kind. Numerical results prove the efficiency of using boundary integral equations on a single plane wave for solving three-dimensional anisotropic dynamic problems of elasticity theory. The achieved accuracy of calculation is not inferior to the accuracy of boundary element schemes for classical boundary integral equations.

A. A. Belov (✉) · A. N. Petrov
National Research Lobachevsky State University of Nizhny Novgorod, 23 Gagarin Avenue, Nizhny Novgorod, Russian Federation 603950
e-mail: belov_a2@mech.unn.ru

A. N. Petrov
e-mail: andrey.petrov@mech.unn.ru

 
H. Altenbach et al. (eds.), *Advanced Materials Modelling for Mechanical, Medical and Biological Applications*, Advanced Structured Materials 155,
https://doi.org/10.1007/978-3-030-81705-3_7

**Keywords** Three-dimensional problems · Boundary element method · Laplace transform inversion · Fredholm equation of the first kind

## 7.1 Introduction

The method of boundary integral equations (BIE) is a universal numerical–analytical method which is particularly useful for solving three-dimensional wave initial-boundary value problems of linear elasticity, viscoelasticity, and poroelasticity (Schanz 2001; Liu et al. 2011; Biot 1956a, b; Manolis and Dineva 2015; Manolis and Beskos 1989; Bazhenov and Igumnov 2008). The advantages of this method lie in the automatic fulfillment of the conditions for the solutions behavior at infinity when considering non-stationary processes in semi-infinite bodies and media and dimension reduction of the problem under study per unit.

Extension of the BIE method for solving dynamic anisotropic problems requires the development of new special schemes (Igumnov et al. 2014). A non-classical extension of the BIE methods and the boundary element method (BEM) to anisotropic dynamic initial-boundary value problems is presented. The considered approach is an alternative to the BIE method with a double application of the reciprocity theorem (Gaul et al. 2003; Brebbia et al. 1984) and integral representation of Green's matrices in the form proposed in Wang and Achenbach (1995). The approach employs the structure of the dynamic fundamental solution described in Babeshko (1985). The transition from the Laplace domain to the time domain is constructed numerically by the Durbin (1974) method. The present paper is a continuation of the research presented in Igumnov et al. (2015), Igumnov and Belov (2006).

## 7.2 Problem Statement and Solution Method

We consider a piecewise homogeneous body $\Omega$ with a boundary $\Gamma = \partial\Omega$ in three-dimensional Euclidean space $R^3$ in which a Cartesian coordinate system is introduced. The equations of motion for an elastically deformable medium and the boundary conditions are as follows:

$$Lu = 0, \quad L = \sigma_{ij,j} - \partial_t^2,$$
$$(l^0 u)_i \equiv u_i\big|_{\partial\Omega}, \; (l^1 u)_i \equiv \sigma_{ij}(u) m_j\big|_{\partial\Omega} \equiv c_{ijkl} m_j \, u_{l,k}, \quad i, j, k, l = 1, 2, 3,$$

where $\sigma_{ij} = c_{ijkl} u_{k,l}$; $\sigma$ is a stress tensor; $u = (u_1, u_2, u_3)$ is a displacement vector; $c_{ijkl}$ are elastic moduli; $l^0$ and $l^1$ are corresponding boundary operators; $m = (m_1, m_2, m_3)$ is outer normal vector.

The Green tensor for the linear anisotropic theory of elasticity is written in the form (Norris 1994):

$$U(x,\omega) = -c\Delta \int\limits_{S^2} \overline{u}\mathrm{d}S(n) = -c\Delta \int\limits_{S^2} G(nx,\omega,n)F^{(0)}\mathrm{d}S(n),$$

$$G(\xi) = \sum_{\alpha=1}^{N} \frac{e^{ik_\alpha|\xi|}}{2ik_\alpha\lambda_\alpha} A_\alpha(\mathrm{sgn}\xi) \otimes A_\alpha(\mathrm{sgn}\xi), \quad \xi = nx,$$

$$U(x,t) = (2\pi)^{-1} \int\limits_{-\infty}^{+\infty} U(x,\omega)\mathrm{e}^{-i\omega t}\mathrm{d}\omega,$$

$$\lambda_\alpha = \frac{1}{2} A_\alpha(\upsilon_\alpha^2 M + K)A_\alpha, \quad \upsilon_\alpha = \omega/k_\alpha,$$

where $K = L_0(n), M = \rho I, F^{(0)} = 1, g_S = \overline{u}$, $\rho$ is the density, $I$ is identity matrix, $c = -(8\pi^2)^{-1}$, $\Delta$ is Laplace operator in space $R^3$, $A_{ij}(\partial) = c_{ijkl}\partial_j\partial_l$, $k_\alpha$ are eigenvalues, $\hat{A}_\alpha$ are eigenvectors, $u = (u_1, u_2, u_3)$ is displacement vector, $\omega$ is frequency, and $n$ is outer vector determining the direction of plane waves propagation.

The following equations are valid for dynamic problems (Igumnov 2006):

$$\gamma \hat{A}_\alpha \otimes \hat{A}_\alpha \int\limits_{\Gamma} \left[l^1 \mathrm{e}^{ik_\alpha n(x-y)} l^0 \psi(y) - \mathrm{e}^{ik_\alpha n(x-y)} l^1 \psi(y)\right] d_y\Gamma = 0, \tag{7.1}$$

$$\gamma \hat{A}_\alpha \otimes \hat{A}_\alpha \int\limits_{\Gamma} \left[l^1 \delta(\tau) * l^0 \psi(y,t) - \delta(\tau) * l^1 \psi(y,t)\right] d_y\Gamma = 0, \tag{7.2}$$

$$\det A(k_\alpha) = 0, \quad A(k,\omega,n)\hat{A} = 0,$$

$$A^*(k_\alpha \mathrm{sgn}\xi) = tr A^*(k_\alpha)\hat{A}_\alpha(\mathrm{sgn}\xi) \otimes \hat{A}_\alpha(\mathrm{sgn}\xi),$$

$$\tau = t - \frac{n(x-y)}{\upsilon_\alpha},$$

$$\pm\gamma c \sum_{\alpha=1}^{N} ig_S(k_\alpha) \int\limits_{\Gamma} \left[l^1 e^{ik_\alpha n(x-y)} l^0 \psi(y) - \mathrm{e}^{ik_\alpha n(x-y)} l^1 \psi(y)\right] d_y\Gamma = 0, \tag{7.3}$$

$$\pm\gamma c \sum_{\alpha=1}^{N} ig_S(k_\alpha) \int\limits_{\Gamma} \left[l^1 \delta(\tau) * \psi(y,t) - \delta(\tau) * l^1 \psi(y,t)\right] d_y\Gamma = 0, \tag{7.4}$$

where * means convolution by $t$, $\psi = (\psi^0, \psi^1)$ are boundary functions, $\gamma = (l^0, l^1)$, $\delta$ is Dirac delta function, and $\otimes$ is tensor product.

Thus, solving the problem by the BIE method means that the corresponding relations from (7.1)–(7.4) for eigenvalues $k_\alpha$ and eigenvectors $\hat{A}_\alpha$ are satisfied. For the theory of elasticity, from Eq. (7.1) with account for the relation

$$\frac{iA^*(k_\alpha \mathrm{sgn}\xi)}{[\det A(k)]'\big|_{k=k_\alpha}} = \frac{1}{2ik_\alpha\lambda_\alpha} A_\alpha(\mathrm{sgn}\xi)\otimes A_\alpha(\mathrm{sgn}\xi)$$

an integral equation is selected that coincides (at $x = 0$) with the equation from Babeshko (1985), Vatulyan (1993).

Babeshko (1985) proposed a method for constructing new integral equations for solving boundary value problems. A BEM scheme for solving dynamic boundary value problems can be constructed based on Eqs. (7.1) and (7.2).

## 7.3 Boundary Element Discretization

The basic process of BE-discretization consists of dividing the surface $\partial\Omega$ into boundary elements: quadrangular and triangular eight-node biquadratic elements (Fig. 7.1).

Triangular elements are considered as degenerate quadrangular elements. The relationship between the local and global coordinate systems $\xi = (\xi_1, \xi_2)$, $y = (y_1(\xi), y_2(\xi), y_3(\xi))$ is written via form functions $N^i(\xi)$:

$$y_i(\xi) = \sum_{l=1}^{8} N^l(\xi) y_i^{\beta(k,l)}, \quad i = 1, 2, 3, \quad \xi \in \Delta_e,$$

where $\beta(k, l)$ is the global node number with the local number $l$ in $k$-th element.

Next, we determine the natural basis, the metric tensor, and the unit normal on the element:

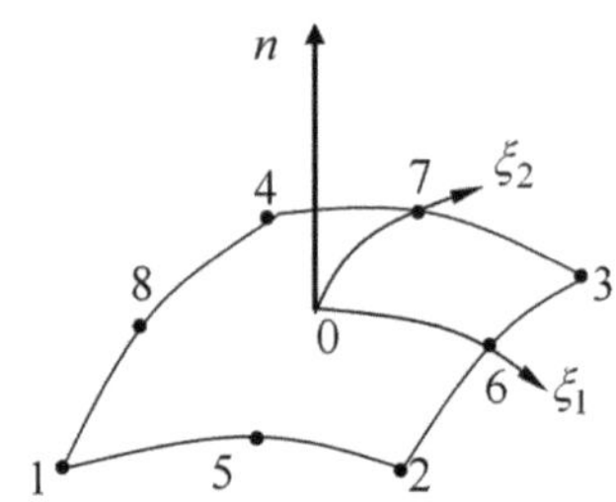


**Fig. 7.1** Quadrilateral boundary element

$$\mathrm{x}_e(\xi_1,\xi_2)=\sum_{n=1}^{8}F_n(\xi_1,\xi_2)\,\mathrm{x}_n^{\gamma(e,n)},\ \mathrm{x}\in\Gamma_e,$$

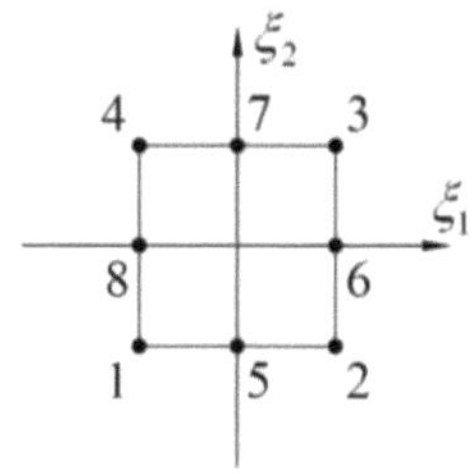


$$\bar{\mathrm{u}}(\xi_1,\xi_2,s)=\sum_{n=1}^{4}R_n(\xi_1,\xi_2)\,\bar{\mathrm{u}}_n^{\gamma(e,n)}(s),$$

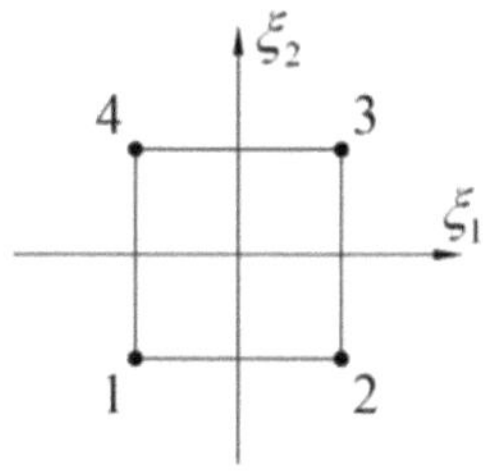


$$\bar{\mathrm{t}}(\xi_1,\xi_2,s)=\bar{\mathrm{t}}^{\gamma(e,1)}(s),$$

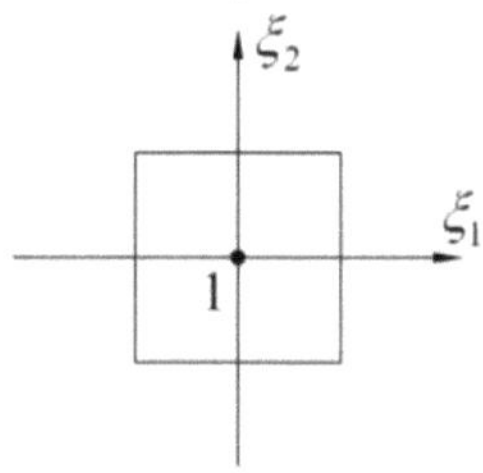


$\gamma(e, n)$ is the global node number with the local number $n$ in $e$-th element, $F_n(\xi_1, \xi_2)$ are quadratic shape functions, and $R_n(\xi_1, \xi_2)$ are linear shape function.

Unknown boundary fields are interpolated through node values. Consider a case called matched interpolation, where bilinear elements are employed to approximate boundary displacements while constant elements are used to approximate surface forces.

We apply the collocation method to obtain a BIE discrete analogue. The approximation nodes of the initial-boundary functions are selected as collocation nodes $y^m$. As a result, a system of linear algebraic equations is formed.

Numerical Gaussian integration is based on a fixed order formula. To solve the final system of linear algebraic equations $Ay = f$, the Tikhonov regularization algorithm is used.

1. A complex conjugate and transposed matrix $A^* = \left(\overline{a_{ji}}\right)$ is constructed.
2. The system $(\alpha E + A^* A)y_\alpha = A^* f$ is considered instead of the initial one.

## 7.4 Numerical Laplace Transform Inversion

To obtain the original solution, the Durbin method of the numerical Laplace transform is applied with a linear approximation of the transformant (Bazhenov and Igumnov 2008):

$$\bar{f}(s) = \int_0^\infty e^{-st} f(t)\mathrm{d}t,\ f(t) = \frac{1}{2\pi i}\int_{\alpha-i\infty}^{\alpha+i\infty} \bar{f}(s)e^{st}\mathrm{d}s,$$

$$f(0) \approx \sum_{k=1}^{n}\left[\frac{(F_{k+1}-F_k)\Delta_k}{2\pi}\right],$$

$$f(t) \approx \frac{e^{\alpha t}}{\pi t^2}\sum_{k=1}^{n}\left[\frac{F_{k+1}-F_k}{\Delta_k}(\cos(\omega_{k+1}t)-\cos(\omega_k t)) + \frac{G_{k+1}-G_k}{\Delta_k}(\sin(\omega_{k+1}t)-\sin(\omega_k t))\right],$$

where $\Delta_k = \omega_{k+1} - \omega_k$, $\tau_k = (\omega_{k+1} + \omega_k)/2$, $F_k = \mathrm{Re}[\bar{f}(\alpha + i\tau_k)]$, and$G_k = \mathrm{Im}[\bar{f}(\alpha + i\tau_k)]$.

## 7.5 Numerical Investigations

The problem was solved in an anisotropic formulation for the load acting along the normal in the form of the Heaviside function on the cube face with an edge length $l_1 = 1$ m, weakened by a cubic cavity with an edge length $l_2 = 0.45$ m (Fig. 7.2) (Roy et al. 2011). The top cube face in the direction opposite to the axis $x_3$ is subjected to a Heaviside-type load; on the opposite cube face, a rigid termination condition is modeled, and the rest of the cube surface is free. The effect of spatial sampling on the results of the BEM analysis is studied. For the cube with and without a cavity, three different uniform surface grids were used with the number of elements 216, 384, 600 and 312, 600, 984, respectively.

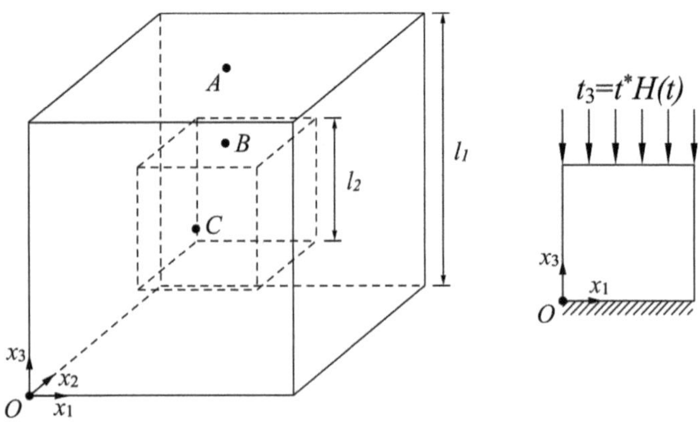


**Fig. 7.2** Physical problem statement

The cavity is located in the center of the cube. The load on $x_3 = l_1$: $t_3 = t^* H(t)$, $t^* = -100$ Pa, $H(t)$ is Heaviside function. On the face, $x_3 = 0$ m: $u_1 = u_2 = u_3 = 0$. The rest of the surface is free from surface forces. The observation points: $A$ (0,5, 0,5, 1,0), $B$ (0,5, 0,5, 0,725), $C$ (0,5, 0,275, 0,5) are located in the centers of the upper cube face, the upper face of the cavity and the front face of the cavity, respectively. Two configurations are considered: a homogeneous body and a body with a cavity with material parameters: $\rho = 3227\,\text{kg/m}^3$, matrix of orthotropic elastic material parameters

$$\mathbf{C} = \begin{bmatrix} 257 & 90 & 65 & 0 & 0 & 0 \\ 90 & 200 & 79 & 0 & 0 & 0 \\ 65 & 79 & 203 & 0 & 0 & 0 \\ 0 & 0 & 0 & 62 & 0 & 0 \\ 0 & 0 & 0 & 0 & 74 & 0 \\ 0 & 0 & 0 & 0 & 0 & 65 \end{bmatrix} \text{GPa.}$$

Boundary element solutions are compared with finite element solutions (Figs. 7.3, 7.4, and 7.5). A comparative analysis of boundary element solutions for a cube with and without a cavity is carried out. The convergence study showed that for the case of the body with a cavity, a larger number of boundary elements is required compared to the case of the body without a cavity, although in both cases acceptable results are achieved even with a significantly small number of boundary elements. A comparative analysis of the solutions obtained according to the classical and non-classical BIE schemes is carried out (Belov and Petrov 2020) (Figs. 7.6, 7.7, and 7.8). In the figures, the abbreviations ncBIE and FEM denote non-classical BIE and finite element method, respectively.

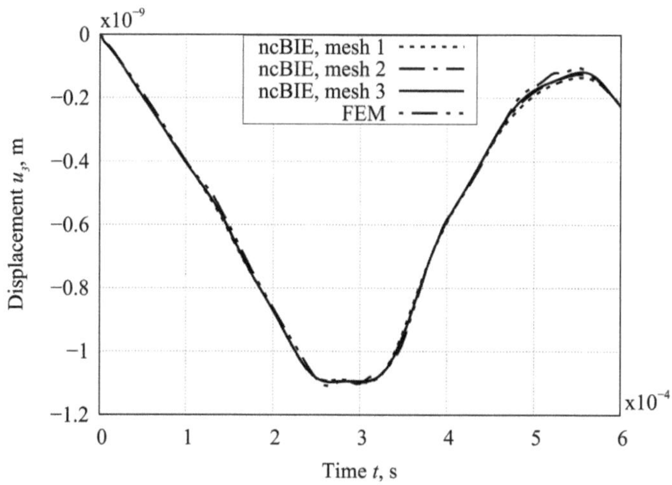


**Fig. 7.3** Displacements at point $A$ for cavity-free case

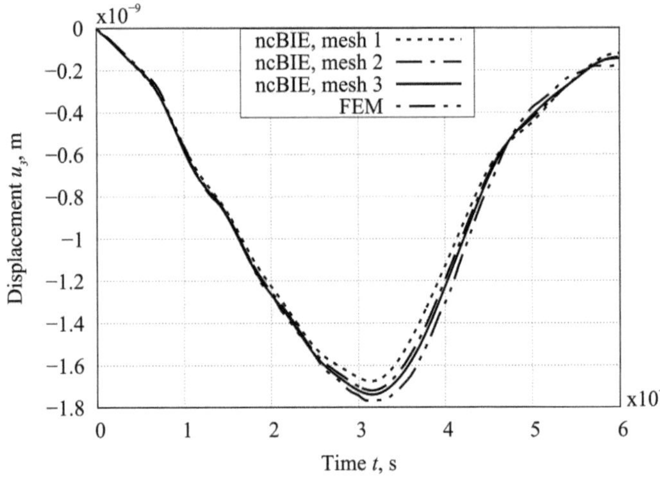


**Fig. 7.4** Displacements at point *A* for the case with cavity

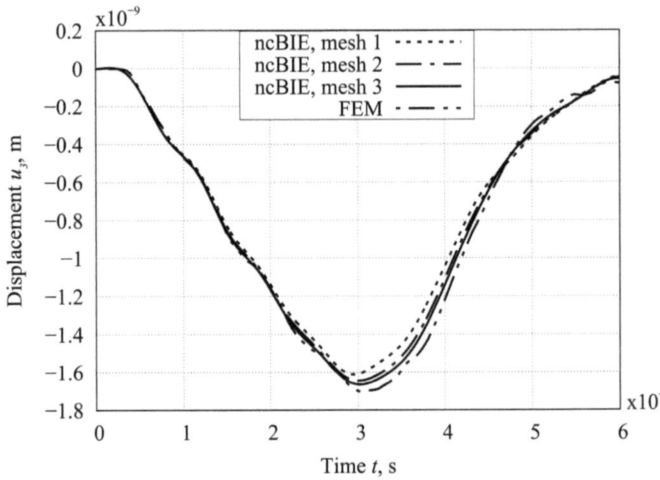


**Fig. 7.5** Displacement at point *B* for the case with cavity

## 7.6 Conclusions

The results obtained indicate that the use of BIE on a single plane wave is promising for solving three-dimensional dynamic problems of the theory of elasticity. The achieved calculation accuracy is not inferior to that of BEM schemes for classical BIE.

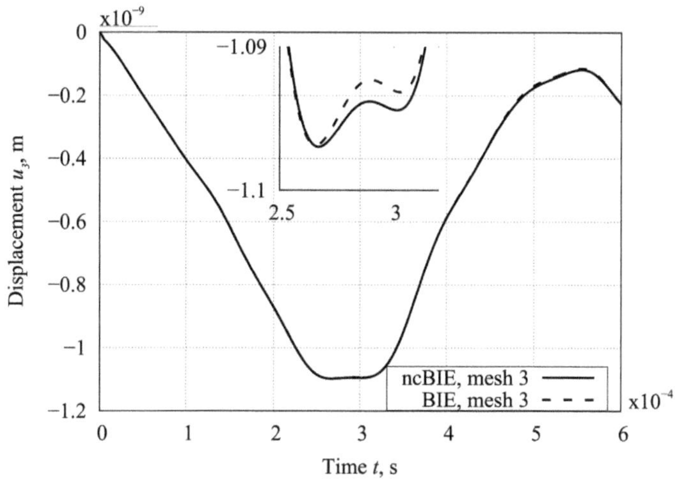


**Fig. 7.6** Displacements at point $A$ for cavity-free case

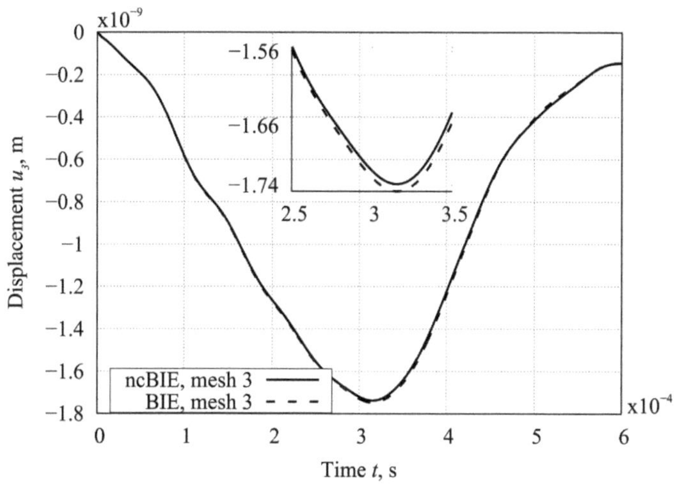


**Fig. 7.7** Displacements at point $A$ for the case with cavity

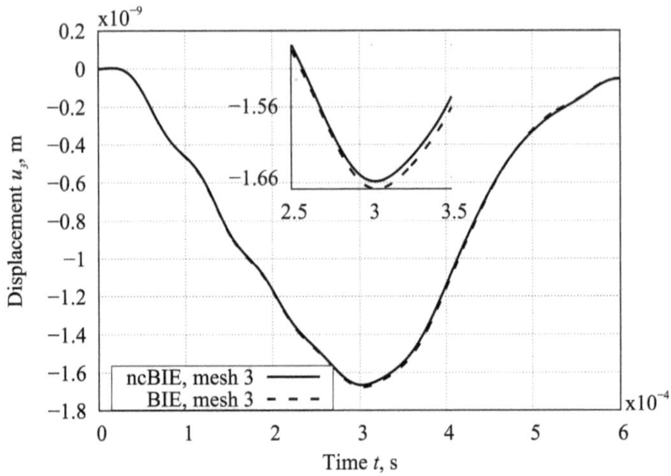


**Fig. 7.8** Displacements at point $B$ for the case with cavity

**Acknowledgements** The reported study was funded by RFBR, project number 20-08-00442.

# References

Babeshko VA (1985) A new method for solving boundary value problems in continuum mechanics and mathematical physics for non-classical domains. Doklady Akademii Nauk USSR 284(1):73–76 (in Russ)

Bazhenov VG, Igumnov LA (2008) Methods of boundary integral equations and boundary elements in solving problems of three-dimensional dynamic theory of elasticity with conjugate fields. Fizmatlit, Moscow, 352p (in Russian)

Belov AA, Petrov AN (2020) Analysis of the dynamics of anisotropic elastic bodies based on non-classical boundary integral equations. In: Proceedings of the XXVI Gorshkov international symposium, pp 44–46 (in Russ.)

Biot MA (1956a) Theory of propagation of elastic waves in a fluid-saturated porous solid. I. Low-frequency range. J Acoust Soc Am 28(2):168–178. https://doi.org/10.1121/1.1908239

Biot MA (1956) Theory of propagation of elastic waves in a fluid-saturated porous solid II. Higher frequency range. J Acoust Soc Am 28(2):179–191. https://doi.org/10.1121/1.1908241

Brebbia CA, Telles JCF, Wrobel L (1984) Boundary element techniques. Springer, Berlin

Durbin F (1974) Numerical inversion of Laplace transforms: an efficient improvement to Dubner and Abate's method. Comput J 17(4):371–376. https://doi.org/10.1093/comjnl/17.4.371

Gaul L, Kogl M, Wagner M (2003) Boundary element methods for engineers and scientists. Springer, Berlin

Igumnov LA (2006) Boundary integral equations on plane waves for 3-D problems. Dokl Akad Nauk SSSR 409(5):622–624 (In Russian)

Igumnov LA, Belov AA (2006) Numerical solution of nonstationary problems of three-dimensional theory of elasticity using method of integral equations on a single plane wave. Probl Strength Plast 68:22–26 (in Russ). https://doi.org/10.32326/1814-9146-2006-68-1-7-21

Igumnov LA, Markov IP, Rataushko YY (2014) Modeling the dynamics of 3-D elastic anisotropic solids using boundary element method. Adv Mater Res 1040:633–637. https://doi.org/10.4028/www.scientific.net/AMR.1040.633

Igumnov LA, Rataushko YY, Ipatov AA (2015) Treating coupled boundary value problems of 3D elastodynamics by means of boundary integral equations method. Adv Mater-Stud Appl 283–299

Liu YJ, Mukherjee S, Nishimura N, Schanz M, Ye W, Sutradhar A, Pan E, Dumont NA, Frangi A, Saez A (2011) Recent advances and emerging applications of the boundary element method. Appl Mech Rev 64(3):030802. https://doi.org/10.1115/1.4005491

Manolis GD, Beskos DE (1989) Integral formulation and fundamental solutions of dynamic poroelasticity and thermoelasticity. Acta Mech 76:89–104. https://doi.org/10.1007/BF01172983

Manolis GD, Dineva PS (2015) Elastic waves in continuous and discontinuous geological media by boundary integral equation methods: a review. Soil Dyn Earthq Eng 70:11–29. https://doi.org/10.1016/j.soildyn.2014.11.013

Norris AN (1994) Dynamic Green's functions in anisotropic piezoelectric, thermoelastic and poroelastic solids. Proc R Soc Lond A 447:175–188

Roy S, Gebert J-M, Stasiuk G, Piat R, Weidenmann KA, Wanner A (2011) Complete determination of elastic moduli of interpenetrating metal/ceramic composites using ultrasonic techniques and micromechanical modelling. Mater Sci Eng A-Struct 528:8226–8235. https://doi.org/10.1016/j.msea.2011.07.029

Schanz M (2001) Wave propogation in viscoelastic and poroelastic continua. Springer, Berlin

Vatulyan AO (1993) On boundary integral equations of the I-th kind in dynamic problems of anisotropic elasticity theory. Dokl Akad Nauk SSSR 333(3):312–314 (In Russian)

Wang CY, Achenbach JD (1995) Three-dimensional time-harmonic elastodynamic Green's functions for anisotropic solids. Proc R Soc Lond A 449:441–458

# Chapter 8
# Fracture Phenomena in Swarms

**Paolo D'Avanzo, Alessio Ciro Rapisarda, and Salvatore Samuele Sirletti**

**Abstract** In this paper, we describe qualitatively material deformations and fracture phenomena using a discrete 2D kinematic model. This model is based on the swarm intelligence approach, where simple laws among the agents can lead the whole group to achieve complex behaviors; it seems to be promising because it is able to not only qualitatively reproduce standard deformations and fracture phenomena but also a lot of exotically ones that other methods in literature cannot reproduce or their computational costs are too high. Instead, in our model the fracture arises more naturally. Furthermore, it has a low computational cost and it is parallelizable, allowing us to take profit of CUDA® architecture. Some numerical simulations are provided and discussed using two different kind of lattices and changing some model's parameters.

**Keywords** Discrete mechanical system · Second gradient continua · Elastic law · Swarm robotics · Numerical simulations · Exotic phenomena · Fracture phenomena · Material deformation · Deformable bodies

## 8.1 Introduction

With the massive technological development, ever more performing hardware with which to perform increasingly complex calculations, the simulations of standard and

P. D'Avanzo (✉) · A. C. Rapisarda · S. S. Sirletti
Dipartimento di Scienze Fisiche, Università di Napoli "Federico II", Compl. Univ. di Monte S. Angelo, Edificio G, Via Cinthia, 80126 Napoli, Italy
e-mail: paolo.dvnz@gmail.com

A. C. Rapisarda
e-mail: alessioxe@gmail.com

S. S. Sirletti
e-mail: salvatoresamuele.sirletti@gmail.com

S. S. Sirletti
Institut für Theoretische Physik, Johann Wolfgang Goethe-Universität Frankfurt am Main, Max-von-Laue-Str. 1, 60438 Frankfurt am Main, Germany

 
H. Altenbach et al. (eds.), *Advanced Materials Modelling for Mechanical, Medical and Biological Applications*, Advanced Structured Materials 155,
https://doi.org/10.1007/978-3-030-81705-3_8

non-standard continuous have gained a lot of interest and many models were proposed to describe the deformation of materials (Battista et al. 2018; Casale et al. 1998; dell'Erba 2015; Dell'Erba 2020c, 2021a; Müller et al. 2005, 2007; Şahin 2005) and the arise of fracture phenomena (Battista et al. 2016; Bilotta et al. 2018; Cuomo 2019; De Angelo et al. 2019a; Spagnuolo et al. 2017; Turco et al. 2016b). In particular, new numerical approaches have been developed in order to obtain better and quicker simulations. On this trail, we developed an algorithm, presented in D'Avanzo et al. (2021), able to obtain qualitative interesting simulations of deformation. In this paper, we present an enrichment for the proposed algorithm implementing some new features, like the fracture. An important aspect of our model is its purely kinematic basis, i.e., there are not forces or masses; indeed, it is based on the theory of the swarm intelligence (Dell'Erba 2018, 2019a, 2020a, b, 2021a; Moriconi and Dell'Erba 2014; Şahin 2005), so in our model there are only points and interaction laws that rule their motion based on their position. The algorithm developed is parallelizable and runs with CUDA® programming language, so we can obtain numerical simulations in few seconds, also if they are very complex, using a modest desktop PC.

We want to remark that our model does not have any physical meaning yet. It concerns only a set of variables, that are points of a Cartesian plane, which change coordinates with the succession of the algorithm's iterations. Therefore, physical concepts such as "*motion*", "*time*", or "*velocity*" do not exist in our model. We will explore in further papers if and how our model is linked with the physics of deformations, so here we will use some terms only through analogy as already explained and done in D'Avanzo et al. (2021).

Particular interest will be spent on what we will define as "exotic behaviors" (like in D'Avanzo et al. (2021), see Sects. 8.3.2.2 and 8.3.3.2) and "plastic deformations" (see Sect. 8.3.1) because one of our aim is to simulate non-standard continua behavior. Models that can handle non-standard continua are being developed more and more nowadays because standard deformation models, like Cauchy, are unable to describe the behavior of modern composite (Avella et al. 1995, 2000; Casale et al. 1998; Dell'Erba 1995) and structured materials (Dell'Erba 2001).

We are aware that a lot of mathematical results are demanding for proving the computation scheme proposed but we believe that it can become somehow equivalent to those used in continuum mechanics once it has been developed. This hypothesis requires a deep investigation in this direction. But it is our opinion the model can be useful in a large variety of mechanical application. In particular, we believe that some of the following problems:

1. Second gradient continuum theory (Abdoul-Anziz et al. 2019; Barchiesi et al. 2019a; Epstein and Smelser 2020; Germain 2020; Scerrato et al. 2016; Sciarra et al. 2007; Turco et al. 2016a; Yang and Misra 2012).
2. Peridynamics (dell'Isola et al. 2015a; Javili et al. 2019).
3. Bone growth phenomena (George et al. 2018, 2020; Giorgio et al. 2016, 2017a, 2019b, 2020b; Rapisarda et al. 2018, 2020).

4. Generalized shell theories (Altenbach and Eremeyev 2015, 2008a, b, 2011; Altenbach et al. 2010; Andreaus et al. 2018; Bilotta et al. 2019; Desmorat et al. 2020; Giorgio 2020b; Giorgio et al. 2021; Spagnuolo et al. 2020; Steigmann 2018; Steigmann and dell'Isola 2015) and composite laminates (Altenbach et al. 2015; Browning and Askes 2019; Cazzani and Ruge 2016; Cazzani et al. 2018a, b; Giorgio et al. 2021).
5. Beams in large deformations (Barchiesi et al. 2020a, b, 2021a; Baroudi et al. 2019; Giorgio and Del Vescovo 2018, 2019; Spagnuolo and Andreaus 2019).
6. Micro-morphic (Abali et al. 2012; Franciosi et al. 2019; Gagneux and Millet 2016; Giorgio and Scerrato 2017; Giorgio et al. 2019a; Misra and Poorsolhjouy 2016a, b; Misra et al. 2020; Rickert et al. 2019; Scerrato et al. 2014) and micro-polar continuum theories (Altenbach and Eremeyev 2015; Eremeyev and Pietraszkiewicz 2016; Eugster et al. 2014; Giorgio 2020a; Greco 2020; Olive 2019; Turco 2018b; Turco et al. 2019).

can be dramatically simplified if the algorithm presented in D'Avanzo et al. (2021) and enriched here will prove to be an efficient integration scheme for all mentioned continuum models (see, e.g., also Wiech et al. 2018). We remark that generalized continua already demanded the introduction of complex integration schemes, among which we believe that isogeometric ones are of particular relevance (Cazzani et al. 2016a, b, c; Cuomo et al. 2014; Greco et al. 2018; Niiranen et al. 2016; Schulte et al. 2020; Tran and Niiranen 2020; Yildizdag et al. 2020b). Therefore, we believe that the algorithm we are developing can represent a powerful predictive tool in metamaterial theory (Alibert et al. 2003; Barchiesi et al. 2019c, 2021b; Cuomo et al. 2016; Del Vescovo and Giorgio 2014; dell'Isola et al. 2015b, 2019; Eremeyev and Turco 2020; Eugster et al. 2019; Giorgio et al. 2017b, 2019c, 2020a; Scerrato and Giorgio 2019; Seppecher et al. 2011; Spagnuolo 2020; Spagnuolo and Scerrato 2020; Vangelatos et al. 2019; Yang et al. 2019; Yildizdag et al. 2019, 2020a) and in general in the theory of deformable bodies. Furthermore, we believe that it can represent a heuristic tool in helping in search of micro-macro identification results (Abali et al. 2016; Andreaus et al. 2016; Barchiesi et al. 2019b; Bolzon et al. 2002; Boutin et al. 2017; De Angelo et al. 2019a, 2020; Giorgio et al. 2018; Nejadsadeghi et al. 2019; Placidi et al. 2017; Rosi et al. 2018; Yang et al. 2018). We highlight that the presented algorithm can be used for calculating families of equilibrium shapes by using variable load parameters, in a very similar way as done in continuum mechanics (Boutin et al. 2017; Turco 2018a, 2019; Turco and Barchiesi 2019; Turco et al. 2020). When damage phenomena occur in standard continuum mechanics some Karush–Kuhn–Tucker (KKT) conditions (Placidi 2016; Placidi et al. 2018a, b; Timofeev et al. 2020) need to be used in order to take into account for plastic and damage phenomena or crack onset (Bilotta et al. 2018; Cuomo 2019; De Angelo et al. 2019b). All quoted dissipation phenomena can be more easily modeled in the present algorithm, employing the purely deformation-based criterion already in Spagnuolo et al. (2017) and Turco et al. (2016b). We introduced a similar threshold in our algorithm (see Sect. 8.2.2.2), and we believe that with further developments and research it will be possible to predict not only crack onset and damage progression but also plastic phenomena (Contrafatto and Cuomo 2002; Cuomo 2017; Del Piero 2018; Vaiana et al. 2018).

The paper is organized as follows: In Sect. 8.2, we recall the models introduced in D'Avanzo et al. (2021) and we construct step-by-step new version of model. Section 8.3 is dedicated to show some numerical simulations obtained; we divide them in two groups, the standard ones and the exotic ones. In Sect. 8.4, conclusions with advantages and disadvantages are discussed.

## 8.2 Review and Development of the Model Used

The model here used has been introduced in D'Avanzo et al. (2021) for the first time. The model uses the "swarm intelligence" (**SI**) concept to simulate qualitatively various kinds of material deformations. The SI concept is a method well known in complex system research area where is necessary to simulate how a high number of objects interacting each other: behavior of biological system (animals' swarms), robotics (swarm robotics) and the behavior of elastic materials (Battista et al. 2016, 2018; deir'Erba and Moriconi 2015; dell'Erba 2015; Dell'Erba 2018, 2019a, b, 2020a, c, 2021a, b; Vangelatos et al. 2019) are some examples. It is important to note that the interaction among the elements is locally defined and from it the macroscopic behavior arises.

### *8.2.1 A Brief Review of the Previous Model*

In the previous model, the swarm elements occupy all the points of a chosen lattice with step $\sigma$ and labeled with an index $i$ in a Cartesian plane. The lattices used are the squared lattice and the hexagonal lattice. These swarm elements inherited the concept of neighborhood and order of a neighbor from the graph theory that is recalled intuitively here in Fig. 8.1.

Formed the swarm, a role is assigned to its elements between two different ones:

1. **Leaders**. In analogy with biological swarms, these elements lead the evolution and the deformation of the swarm, so in this case the leaders are points that have a "motion" a priori imposed and are not influenced by other swarm's elements.
2. **Followers**. These points "follow" the motion of the leaders and the deformation of the swarm; the motion of a follower depends only by the interaction with its considered neighbors.

Therefore, the Leaders are moved every iteration of the algorithm according to their imposed motion, after that the followers are moved according with the provided interaction laws. So, the algorithm takes only a collection of points in the plane and with some interaction's laws elaborates a new set of positions; the latter is considered the evolution of the previous configuration. Setting a numerical value to

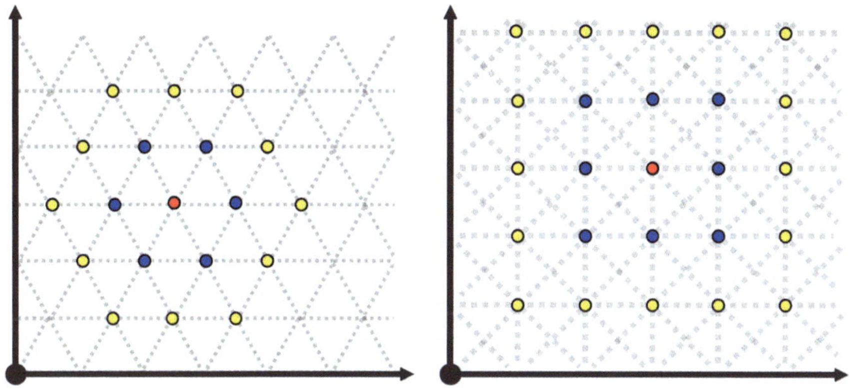

**Fig. 8.1** Example of a swarm's portion on a hexagonal lattice (left) and on the squared lattice (right). Considering the red point: The blue points are the first neighbors (first order) of the red one, the yellows are its second neighbors (second order)

each configuration (algorithm steps), starting imposing that the initial configuration has value 0, is possible to study the evolution of the swarm with this "*pseudo-time*" parameter; so, to each "*pseudo-time-step*" (**PTS**) value is associated with a configuration of the swarm. The "*displacement*" or the "*motion*" of a point of the swarm is intended as the changing of the position of the point for each change of the pseudo-time-step and the "*velocity*" as the displacement in a single pseudo-time-step. These model's laws are:

$$\mathbf{r}_i(t_{m+1}) = \mathbf{r}_i(t_m) + \mathbf{Int}_i(t_m) \tag{8.1}$$

$$\mathbf{Int}_i(t_m) = \sum_{h=1}^{k}\left(\left[\sum_{j\in\mathcal{N}_h(i)}(\mathbf{A}_{ij}(t_m))\right] + \left[\sum_{l\in\mathcal{P}_h(i)}(\mathbf{B}_{il}(t_m))\right]\right) \tag{8.2}$$

$$\mathbf{A}_{ij}(t_m) = -\frac{\alpha_j}{N_i}\left(R_{ij}(t_m) - R_{ij}(t_0)\right)\frac{\mathbf{r}_i(t_m) - \mathbf{r}_{ij}(t_m)}{|\mathbf{r}_i(t_m) - \mathbf{r}_{ij}(t_m)|} \tag{8.3}$$

$$\mathbf{B}_{il}(t_m) = -\frac{\beta_l}{N_i}\left(\mathbf{r}_i(t_m) - \langle\rho_{il}(t_m)\rangle\right) \tag{8.4}$$

For each PTS ($t_m$), each follower position is updated to ($\mathbf{r}_i(t_{m+1})$) by a term ($\mathbf{Int}_i(t_m)$) that considers two types of interactions, for example considering the interaction on a point $i$ with its $j$th neighbor:

- **The elastic interaction $\mathbf{A}_{ij}$.** It is called in this way because of its similarity with the Hooke's law; indeed, this interaction is proportional to a coefficient $\alpha_j$ times to the variation of current distance between the point $i$ and the $j$ point ($R_{ij}(t_m)$)

**Table 8.1** List of quantities in the equations of the model

| | |
|---|---|
| $\mathbf{r}_i(t_{m+1})$ | Is the updated-position of the point $P_i$ at $p$-time step $t_{m+1}$ |
| $\mathbf{Int}_i(t_m)$ | Is the "interaction" among the point $P_i$ and all its neighbors in the $h$th-order shell until the $k$th order |
| $N_i$ | Is the number of all the considered neighbors of the point $P_i$ |
| $\mathbf{r}_{ij}(t_m)$ | Is the updated position of the $j$th neighbor of the point $P_i$ at the $p$-time step $t_m$ |
| $R_{ij}(t_m)$ | Is the Euclidean distance between point $P_i$ and its $j$th neighbor at the $p$-time step $t_m$ |
| $R_{ij}(t_0)$ | Is the Euclidean distance between point $P_i$ and its $j$th neighbor in $C_0$ |
| $\langle\rho_{il}(t_m)\rangle$ | Is the average point of the pair $\rho_{il}$ considering the position of the points at time $t_m$ |
| $\alpha_j$ | Is a coefficient that modulates the intensity of the "elastic" interaction ($\mathbf{A}_{ij}$) and the convergence velocity. It depends by the $j$th neighbor's shell order |
| $\beta_l$ | Is a coefficient that modulates the intensity of the "alignment" interaction ($\mathbf{B}_{il}$). It depends by the $j$th neighbor's shell order; the $l$ is the label of the couple $\rho_{il}$ |
| $\sigma$ | Is the step of the lattice |

respect to the equilibrium distance ($R_{ij}(t_0)$), it has direction along the line that join $i$ and $j$.

- **The alignment interaction $\mathbf{B}_{ij}$**. Its own aim is to avoid the points superimposition problem forcing every point to be near the centroid of its neighborhood. This interaction is proportional to a coefficient $\beta_j$ times to the difference of the position of $i$ point and the average point of the couple of the opposite neighbors $\langle\rho_{il}\rangle$. Every couple $\rho_{il}$ belongs to a set $\mathcal{P}_h(i)$ that collects all the existing couples around $i$ of order $h$.

The interactions are divided with a factor $N_i$, in this way the average on all the neighbors of the $i$ point up to $k$ order is considered. Here we would remind that all the neighbors of an $i$ point are labeled (or collected) in shell $N_h(i)$; in this way, the points at the edge of the swarm are not missing of points with to interact. More details are provided in Table 8.1.

Define the interaction law among the followers and assign the leaders motion; the algorithm is provided to calculate the swarm evolution with respect to the deformation carried out by leaders, pseudo-time-step by pseudo-time-step. Each swarm configuration is made by the algorithm considering only the configuration at the previous PTS ($t_{m-1}$), and this happens in two different phases:

1. **The interaction phase**, where all the interactions of the followers are calculated.
2. **Spread phase**, where the points are moved.

The leaders' velocities are posed to zero after a pre-established number of time steps, so the whole simulation process is divided into two phases:

1. **Active deformation phase**, where the leaders are moved to deform the swarm.
2. **Settling phase**, where the leaders are held in place while the followers continue to interact pseudo-time-step by pseudo-time-step, until they achieve the "*equilibrium position*".

**Table 8.2** The algorithm operation scheme of the model proposed in D'Avanzo et al. (2021) ($\mathbf{v}_j$ is the imposed "velocity" of the $j$th leader)

| | | |
|---|---|---|
| 1) | **for-all** p-time step $i$: | |
| 2) | **for-all** Leader $j$: | |
| 3) | **do**: $\mathbf{x}_j^{(i)} = \mathbf{x}_j^{(i-1)} + \mathbf{v}_j^{(i)}$ | (The Leaders are translated) |
| 4) | **for-all** Follower $k$: | **Interaction phase** |
| 5) | **do**: $\mathbf{Int}_k^{(i)} = \mathbf{Function}_k^{(i)}\left(neighbors_k^{(i-1)}\right)$ | (The interactions are evaluated) |
| 6) | **for-all** Follower $k$: | **Spread phase** |
| 7) | **do**: $\mathbf{x}_k^{(i)} = \mathbf{x}_k^{(i-1)} + \mathbf{Int}_k^{(i)}$ | (The Followers are translated) |
| 8) | **end** | |

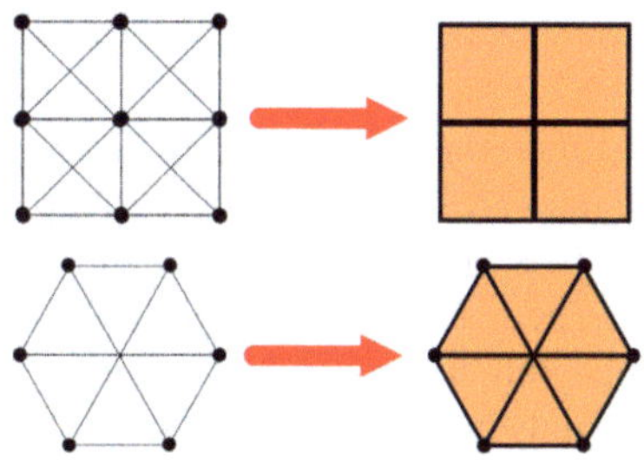

**Fig. 8.2** Here we provided examples of EAUs defined on the two kinds of lattice used. They are defined on the initial configuration of the swarm but are studied through all the pseudo-time-steps

It is important to note that with the terms "*equilibrium position*" or "*final configuration*" we mean that with the sequence of algorithm's cycles, the position of the swarm points is unchanged within a tolerance $\varepsilon$.

How the algorithm works is showed in Table 8.2.

To give a visual interpretation of the results given by the algorithm and taking as much information as possible from the final configuration, the concept of "*elementary area unit*" (**EAU**) of the swarm was introduced: Small geometrical figures are defined, where their vertices are swarm points; these figures can create, in a topological sense, a partition of the swarm's area in the initial configuration. We want to empathize that this is only a graphical representation of the deformation. The geometrical figures of the EAUs depend by the lattice used:

- *Triangles* for the hexagonal lattice;
- *Quadrilaterals* for the squared one.

An example of the EAUs is provided in Fig. 8.2.

The areas of the EAUs are analyzed with the evolution of the swarm. The variation of the areas of the EAUs compared with the initial configuration adds information about the deformation state of the EAUs. This information is represented on a chromatic scale which color values are evaluated as:

$$\text{EAUColor} = \frac{|A_0 - A|}{A_0} \tag{8.5}$$

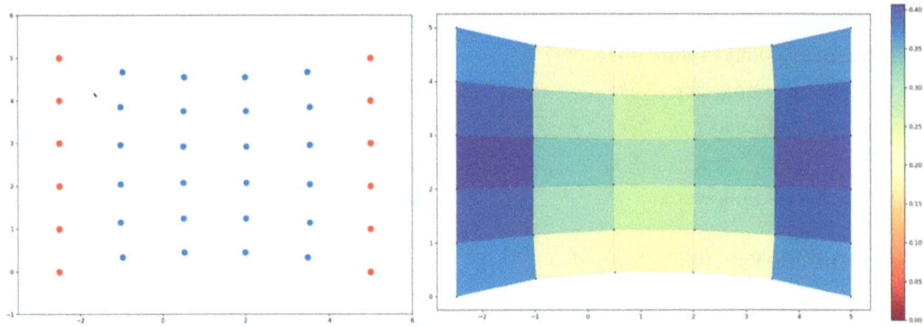

Fig. 8.3 On the left we provided an example of the row algorithm's output; it is possible to note the leaders in red and the followers in blue. On the right, instead, we provided an example of colored version of the algorithm's output with its color bar. The colors result a helpful medium to better understand the swarm deformation

Therefore, the final configuration is colored in such a way that the most deformed regions are more evident, an example is provided in Fig. 8.3. Note that in the following (Sect. 8.3) we used Eq. 8.5 without the module.

The results of the paper (D'Avanzo et al. 2021) are promising, and we believe that with the proper developments it can be transform in a quantitative model useful to study bodies deformations.

## 8.2.2 *The Model Developments Here Introduced*

In D'Avanzo et al. (2021), some development ideas were just introduced; in this article, we have considered the most important one: the fracture of swarm concept. To introduce the fracture, some improvements have been made: In the updated model, not only the final configuration is important, but also how this configuration is reached, this is because the fracture is a dynamical phenomenon where is important also the non-equilibrium configurations; if the latter changes, also the fracture phenomena could change. Here we introduce the news in order to introduce the model's scheme used.

### 8.2.2.1 The $\gamma$ Parameter

Here we propose an alternative to the alignment term $\mathbf{B}_{ij}$ that we call $\gamma$ parameter, to avoid the overlap problem in a much better way and that permits to manage the swarm's fracture. Furthermore, as just observed in some similar models (Battista et al. 2016, 2018; Bernalte Sínchez et al. 2020; D'Avanzo et al. 2021; deir'Erba and Moriconi 2015; Dell'Erba 2019b; Marini et al. 2020; Ramárez et al. 2021), the deformation information in the swarm is propagated through its elements when the

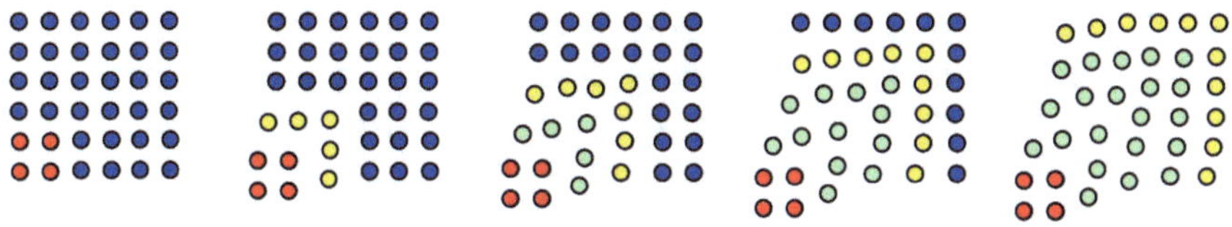

**Fig. 8.4** Here we provided an example of propagation of deformation's information in the swarm. We show four consecutive swarm configurations, ordered from the left (initial configuration with pseudo-time-step 0) to the right (configuration at fourth pseudo-time-step). The information about the initial leaders' movement (red dots) is in yellow

pseudo-time-steps go on. So, it is possible to say that "*if the whole neighborhood of an i point (until the k order) have not been moved, then no interaction is on the point because for this no deformation has taken place*". Here is important to remark a swarm characteristic: Each swarm's element does not have any knowledge of the global swarm's state, but only of the local swarm's state, that is the state of its neighbors. The deformation's information will propagate across the swarm proceeding in stripes that are proportional to the neighbors $k$-order considered in the interaction. In Fig. 8.4, we show an explicit example of what we have said, considering only the interaction to the first neighbors.

In the previous model, it was necessary to take care about the leaders' velocity, assign them extremely low value to avoid the superposition problem (especially in the compression simulations) and to do enough time to the followers to suit themselves basing on the deformations. We have introduced $\gamma$ algorithm's cycles between two pseudo-time-steps, called intermediate frame or $\gamma$-frame, in which the leaders are stopped meanwhile the follower goes on. Respect to the pseudo-time these $\gamma$-frames are instantaneous. It results a highly "*rigidity*" of the swarm when it is deformed. With "*rigidity*" of the swarm we mean the response speed of the swarm. Therefore, a "*rigid*" swarm has a quicker response compared to a "*soft*" one to the same leader-induced deformation. An example is provided comparing Fig. 8.5 with the previous Fig. 8.4.

It must be noted that using a parameter $\gamma = (k - 1)$ with first neighbors' interaction or using a parameter $\gamma = 0$ with interaction to $k$-order neighbors, is not the same; we have found two crucial differences.

- The first difference is that, using the $\gamma$-frame, the followers are moved because their neighbors are moved in the previous intermediate frames; in this way they have more steps to better suit the deformation and they come near the local equilibrium configuration. Otherwise, without the $\gamma$-frame, because of the followers interact for only one pseudo-time-step, they suit the deformation slower.
- The second difference is that the interaction to the $k$-order neighbors has a different final configuration from the one that would be obtained using only the first neighbor's interaction. The $\gamma$-frames do not alter the final configuration but only how this is reached (if fracture or critical overlap does not occur). This propriety is fundamental for the fracture and to avoid the overlap problem.

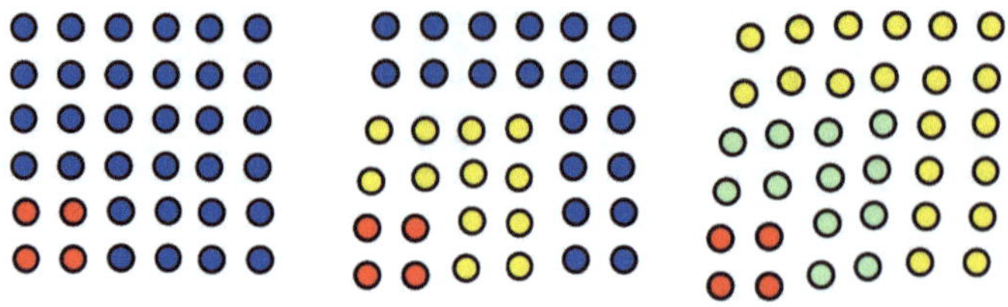

**Fig. 8.5** Here there are three consecutive swarm configurations, ordered from the left (initial configuration with pseudo-time-step 0) to the right (configuration at second pseudo-time-step). These configurations are an example of propagation of deformation's information in the swarm: the information about the initial leaders' movement (red dots) is in yellow. Here, nevertheless the leaders' velocity is the same (of the previous Fig. 8.4), and the interaction is extended until the first neighbors, setting $\gamma = 1$, two shell are involved in every $p$-time step, so the apparent propagation velocity of the deformation is now doubled respect to the previous image, so the material is less "soft" and more reactive (i.e., all the swarm is involved in the deformation in less PTSs)

With considerations on the coasts and the benefits, we have decided to show simulations obtained without considering the alignment interaction $\mathbf{B}_{il}$, replacing it with a certain number of intermediate steps. We would like to note the alignment term has also an effect on the final swarm's configuration, so it is not definitively deleted, but a new tool is added to the model. In Fig. 8.6, we show an example about how the swarm evolution depends on the $\gamma$ parameter, showing the difference between initial, intermediate, and final configurations.

With $\gamma$ parameter we can modulate how much shells are involved in one pseudo-time-step, and so it is related to the deformation propagation velocity in the swarm; this concept is analogous to the concept of correlation length in reticular models (like the Ising model) and to the concept of the persistence length in polymeric models. Hence, the $\gamma$ parameter could be linked to the rigidity of the swarm. Here $\gamma$ is intended as the number of bands of points that are propagating the information of deformation in each pseudo-time-step, but this is only theoretically because the $\alpha_j$ must be considered. This because that the "*rigidity*" depends also how much the swarm's points move in an intermediate frame.

### 8.2.2.2 The Fracture

The fracture of a material is the main phenomenon that we want to simulate, this is because the other models, like the ones based on FEM, are so slowly in this kind of simulation; in our model the fracture phenomena are simply introduced with a threshold on the distance between two points: If two points have distance less than the threshold value $T_B$, these two points are considered linked and they continue to interact each other normally; otherwise if in a certain pseudo-time-step, the distance between them is more than $T_B$, the bound between them is considered broken and the two points no more interact (then after a fracture simply the points interact with

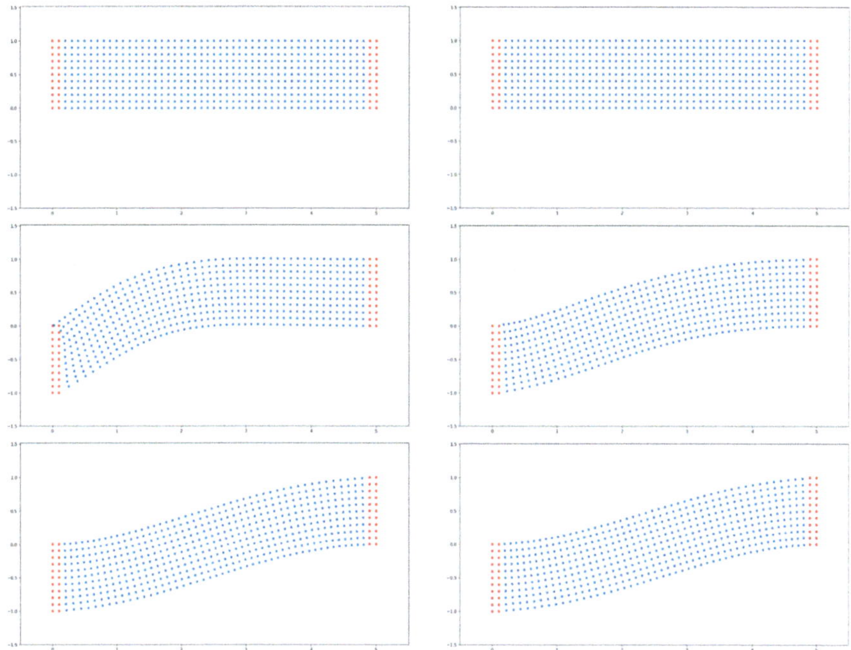

**Fig. 8.6** Different algorithm responses to the same conditions but with and without intermediate steps. The left images are configurations relative to a same PTSs during the swarm's evolution but on the top without the $\gamma$ intermediate frames while on the tight with a high number of $\gamma$ intermediate frames. On the top is possible to observe the same initial condition, in the middle the different responses, on the bottom the final configurations that correspond each other

fewer neighbors). So considering the interaction on *i* by a neighbor *j* and calling their distance $d_{ij}$, we have:

$$\text{Interaction}_{i-j} = \begin{cases} 0 & \text{if } d_{ij} \leq T_B \\ \mathbf{A}_{ij} & \text{otherwise} \end{cases} . \tag{8.6}$$

In Fig. 8.7, we have showed the simple algorithm's flux diagram about the fracture phenomenon.

Nevertheless, to introduce this simple rule in the simulations, we had to redefine the EAU: to visualize of the fracture phenomenon, during the EAU coloration process only the EAUs that do not have broken bonds are excluded. In this way, the broken bonds leave white holes or slits that can be interpreted like fractures.

In the case of hexagonal lattice, there are no particular problems and the exclusion can be carried out simply creating a good result. Instead, in the case of squared lattice, problems raise up because of the diagonal bounds which are included in the EAU (see Fig. 8.2) and, in addition, with the same amount of considered points in the two different types of lattices, the EAU in the squared one are fewer and bigger than the

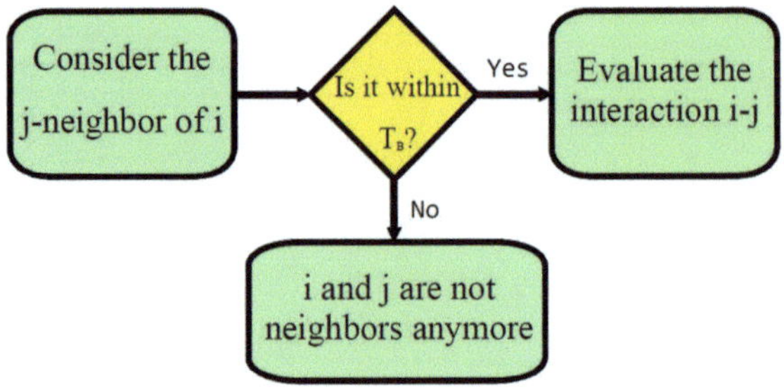


**Fig. 8.7** Algorithm's flux diagram about the fracture phenomenon

same in the hexagonal one; the EAU break can leave big holes in the colorized swarm although there are still many unbroken bonds. All these mentioned facts cause a big loss of information; see Fig. 8.8.

We have avoided this problem by reformulating the squared partition (for the squared lattice) with the so-called *mini-EAU*. Each new mini-EAU represents the area of a triangle in the squared lattice, as is possible to see in Fig. 8.9.

So, in this article the EAUs are defined like in Fig. 8.10 and in Figs. 8.11 and 8.12 we provided an example on how the mini-EAUs improve the algorithm's output as an image resolution improvement.

Furthermore, besides the UAEs in the future simulations, we will show also the intact bounds. This addition is useful to get how the swarm is deformed also when EAUs fail (e.g., understand if different pieces of the swarm are still connected). Therefore, in this way, during the fracture phenomena, more details are preserved, also in the squared lattice; for example, see Fig. 8.13. All those precautions will be needed particularly in Sects. 8.3.2.2 and 8.3.3.2 where the deformations will cause massive losses of EAUs.

So, an example of breaking phenomena for swarm on the hexagonal lattice is provided in Fig. 8.14, meanwhile an example of breaking phenomena for swarm on squared lattice is provided in Fig. 8.15. We remark that also the squared lattice colored can suffer the problem of apparent mini-EAUs isolated like the hexagonal one shown in Fig. 8.8 on the bottom and the bounds can help to avoid the problem as shown in Fig. 8.13.

#### 8.2.2.3 The Interaction Law Modifications

Another important change respect to D'Avanzo et al. (2021) is the management of the boundary points via a modification of interaction law.

In D'Avanzo et al. (2021) is shown that the combination of the parameters $\alpha_j$ can cause "*exotic*" behaviors of the swarm where it reaches particular equilibrium configurations. Some of those have qualitative similarities with particular kinds of phenomena, as formation of density patterns like stripes or holes. We noticed that the

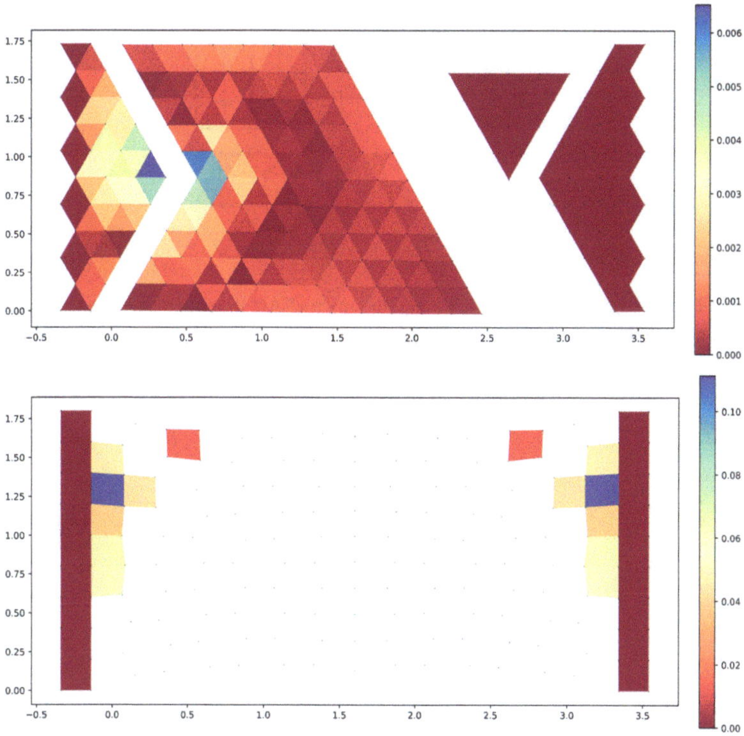


**Fig. 8.8** Here we provided two snapshot of the fracture for the two lattices. On the top we have the mean problem for the hexagonal lattice, on the bottom for the squared one. The problem of the hexagonal lattice is that some EAUs can appear isolated as if the swarm is broken in pieces, but actually they are still bounded. The squared lattice has two mean problems; the first is the same of the hexagonal lattice, the second is the fact that often, at least for traction simulations, the break of few bounds could be enough for destroy a whole region of UAEs causing a big loss of information

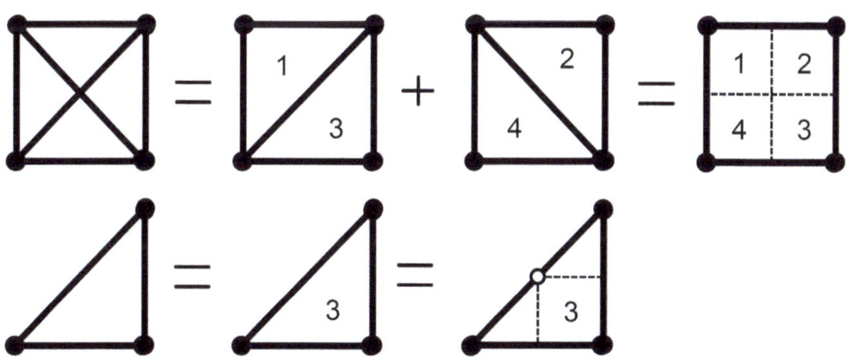


**Fig. 8.9** Every squared complete cell has four mini-EAU. The intersection point of the dotted lines is the centroid of the four vertices. If one of the vertices miss, for example the high left one, the only mini-EAU that can be created is the number 3 and the white dot is the middle point of the diagonal. If two or three vertices miss, no mini-EAU can be created

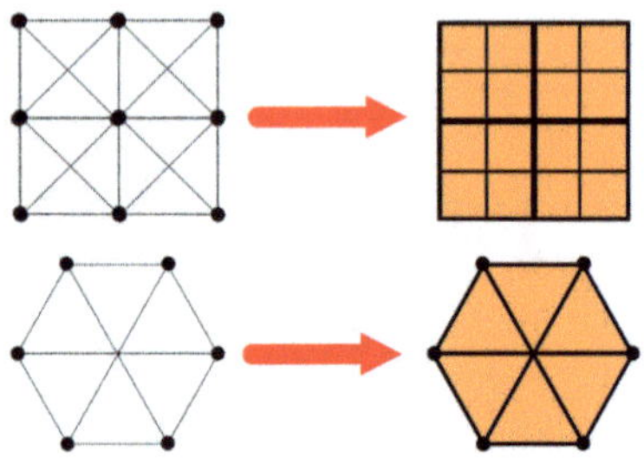

**Fig. 8.10** The new definition of the EAU and the micro-EAU in orange (the triangular EAUs are defined exactly in the same way of Fig. 8.2)

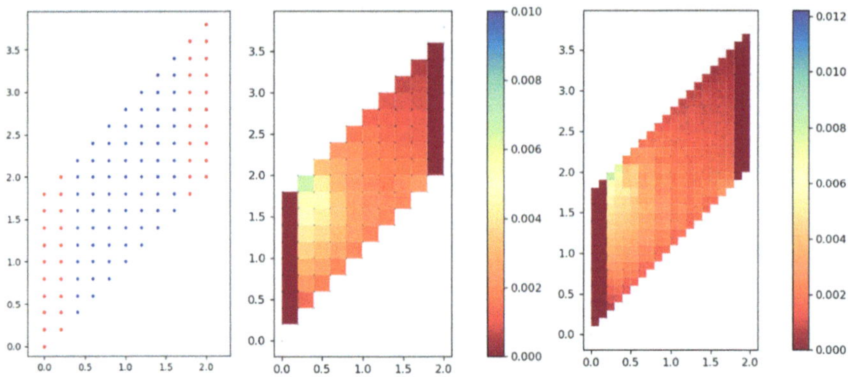


**Fig. 8.11** Here we show the graphical difference between squared EAU and mini-EAU. On the left we provided the initial condition of the swarm on which is applied a small traction on the left side, in the middle there is the old EAU colorized and, on the right, the new version of mini-EAU colorized of the same final configuration. It is easy to see an improvement of the quality information that can be taken from the colors

edges of the swarm are particularly subject to show local overlap and oscillations. In general, the points at borders of the swarm suffer the incompleteness of the neighbors' shells, that is why some research use "frame" points around the swarm (Battista et al. 2016, 2018; Dell'Erba 2019b). For those reasons in this paper, we have also changed slightly the swarm evolution law to better manage the points on the edges of the swarm.

As it is possible to see from Eq. 8.3, the interaction of every shell order is averaged with the same weight $1/N_i$; furthermore, the boundary elements, that have incomplete shell, can have an average displacement bigger than the inner followers that have complete shell. This is due to the lack of points at edge of swarm that does not balance the interaction. These features cause an exotic behavior at the edge of the swarm (i.e., local overlap and oscillations). See Fig. 8.16 on the top.

So, we have introduced a modification to manage this behavior and for make the bound more stable and less sensitive to the overlap. We have split the interaction for shell order, and each one has its weight $\omega_k$ that depends on the number of neigh-

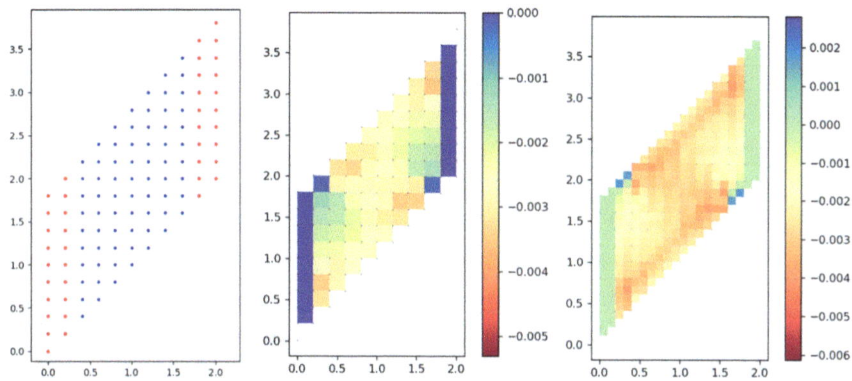


**Fig. 8.12** Here we show the graphical difference between squared EAU and mini-EAU. On the left we provided the initial condition of the swarm on which is applied a small shear thought the leader, in the middle there is the old EAU colorized and, on the right, the new version of mini-EAU colorized of the same final configuration. It is easy to see an improvement of the quality information that can be taken from the colors. Here we used Eq. 8.5 without module

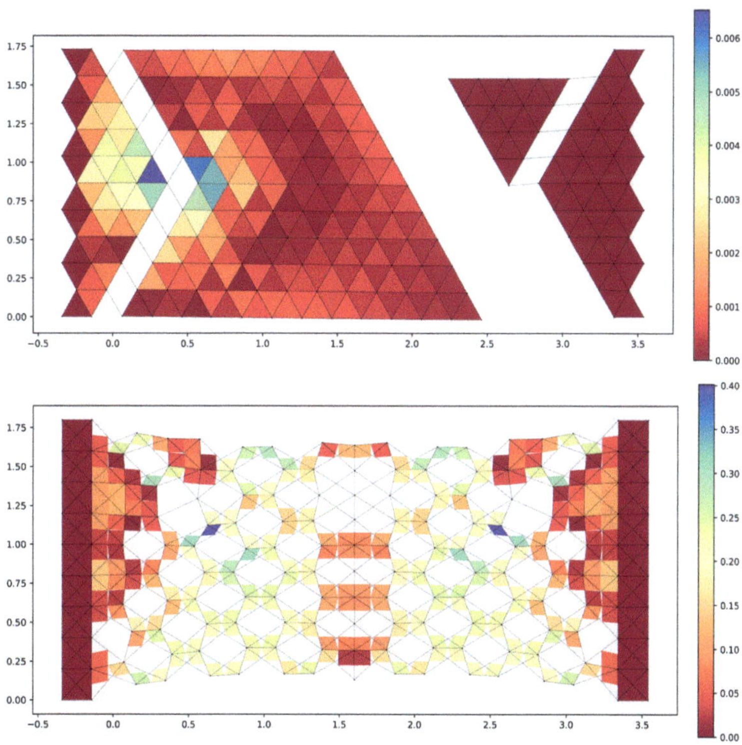


**Fig. 8.13** Here we provided the same simulations shown in Fig. 8.8 to resume the developments done for the EAUs: On the top, it is possible to note how the bounds are useful to understand if portions of the swarm are still connected while on the bottom it is possible to note the nice result of the mini-EAU

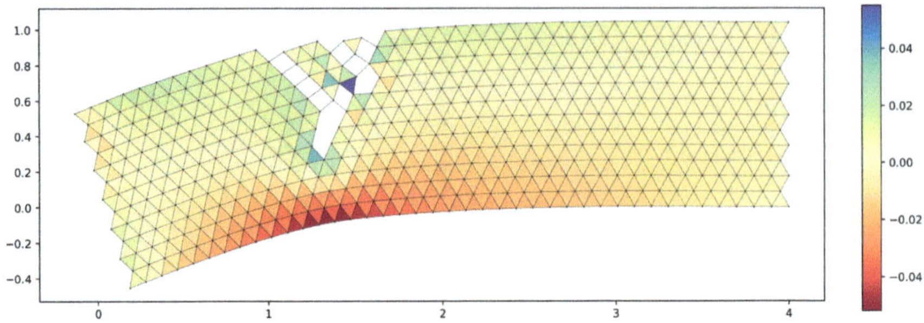


**Fig. 8.14** This figure shows the swarm configuration at a chosen pseudo-time-step. The figure has the only aim to show the fracture on hexagonal lattice, and the swarm is taken little so it possible to appreciate the triangles. Here we used Eq. 8.5 without module

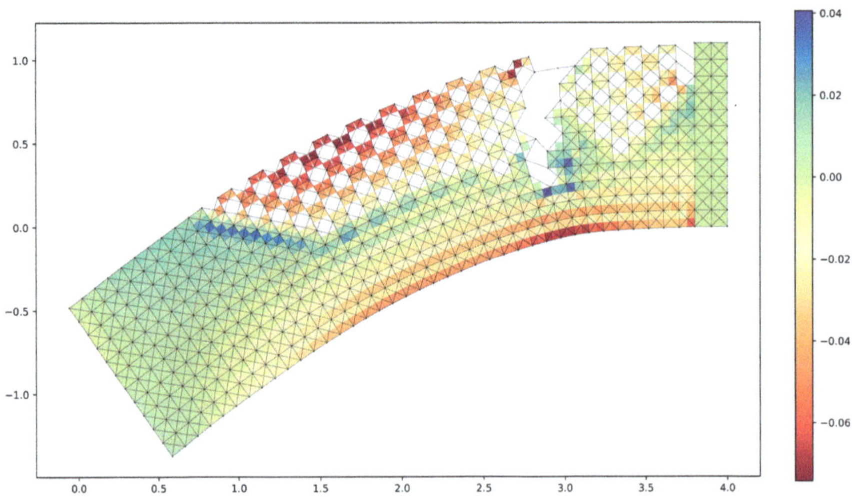


**Fig. 8.15** This figure shows the swarm configuration at a chosen pseudo-time-step. The figure has the only aim to show the fracture on squared lattice. Here we used Eq. 8.5 without module

bors in the shell. Hence, these weights are a "measure" of how much the shells are "complete". A shell's weight is so defined as

$$\omega_k = \frac{1}{2Z_k - N_k(i)} \tag{8.7}$$

where $Z_k$ is the number of points that the $k$-order shell would have if complete, $N_k$ is the number of the $i$'s $k$-order neighbors. So, if the shell is complete the weight is the same used in the previous work, otherwise the weight is less, reducing the interaction on $i$ due to all shell's points of the same order.

Without this correction, exotic phenomena raise up first at the edges; they extend themselves to all the lattice increasing $\alpha_j$. With the proposed correction, the exotic

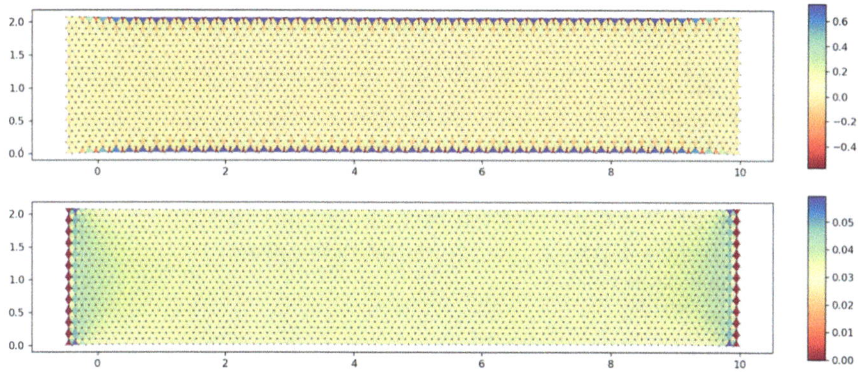


**Fig. 8.16** Here we show an example of the done modification. This is the same final configuration with (bottom) and without (top) the modification. We would like to emphasize the edge's behavior between the two figures. The modification gives more stability

phenomena arise on the whole swarm for $\alpha_j$ values in exotic range. In Fig. 8.16 is possible to see the effects of this correction.

This modification introduces the possibility to modulate the interaction on the edge's points with another opportune modification on the weight $\omega_k$, and in this way it is possible to define the final configuration's "*stability*" and the swarm resistance to the deformation and fracture. This aspect will be developed in future research.

#### 8.2.2.4 The Model Here Proposed

Here we propose a model quite similar to the one presented in D'Avanzo et al. (2021), except for what we have just said in the previous:

- the aliment term $\mathbf{B}_{ij}$ isn't considered;
- the introduction of the $\gamma$-frames;
- fracture introduction;
- different manage of the swarm's edge.

The interaction formulas are reduced to:

$$\mathbf{r}_i(t_{m+1}) = \mathbf{r}_i(t_m) + \mathbf{Int}_i(t_m) \tag{8.8}$$

$$\mathbf{Int}_i(t_m) = \sum_{h=1}^{k}\left(\sum_{j\in\mathcal{N}_h(i)} \mathbf{A}_{ij}(t_m)\right) \tag{8.9}$$

$$\mathbf{A}_{ij}(t_m) = -\frac{\alpha_j}{2Z_h - N_h(i)}\left(R_{ij}(t_m) - R_{ij}(t_0)\right)\frac{\mathbf{r}_i(t_m) - \mathbf{r}_{ij}(t_m)}{|\mathbf{r}_i(t_m) - \mathbf{r}_{ij}(t_m)|} \tag{8.10}$$

**Table 8.3** Algorithm operation scheme of the model here proposed ($\mathbf{v}_j$ is the imposed "*velocity*" of the *j*th leader)

| | |
|---|---|
| **for-all** p-time step $i$: | |
| **for-all** Leader $j$: | |
| **do**: $\mathbf{x}_j^{(i)} = \mathbf{x}_j^{(i-1)} + \mathbf{v}_j^{(i)}$ | (The Leaders are translated) |
| **for-all** Follower $k$: | **$\gamma$-frame initialization** |
| **do**: $\mathbf{x}_k^{(i-1)(-1)} = \mathbf{x}_k^{(i-1)}$ | |
| **for-all** ($\gamma$ *frame* + 1) $p$: | **Interaction phase** |
| **for-all** Follower $k$: | |
| $\mathbf{Int}_k^{(i)(p)} = 0$ | |
| **for-all** $neighbors_{kj}j^{(i-1)(p-1)}$ $j$: | |
| **if** $d(k, j) < T_B$: | Check distance $k - i$ |
| **do**: $\mathbf{Int}_{k-1}^{(i)(p)} = \mathbf{A}(k, j)$ | The interaction is evaluated |
| **do**: $\mathbf{Int}_k^{(i)(p)} += \mathbf{Int}_{k-1}^{(i)(p)}$ | |
| **else**: | |
| **do**: $\mathcal{N}(i) = \mathcal{N}(i) - neighbor_{kj}$ | $j$ and $k$ are not neighbors anymore |
| **for-all** Follower $k$: | Spread phase in $\gamma$−frame |
| **do**: $\mathbf{x}_k^{(i-1)(p)} = \mathbf{x}_k^{(i-1)(p-1)} + \mathbf{Int}_k^{(i)(p)}$ | |
| **for-all** Follower $k$: | **Spread phase** |
| **do**: $\mathbf{x}_k^{(i)} = \mathbf{x}_k^{(i-1)(p)}$ | The Followers are translated |
| **end** | |

The notation is the same of Table 8.1, and obviously the alignment term lacks; the working scheme is slightly different, as is possible to see in Table 8.3 comparing it with Table 8.2.

Of course, in Table 8.3 a simple version of the algorithm is presented due to a being it more understandable; the real algorithm is bigger and complicated and does not use many nested "for" cycles. We would like to note that the modifications done (the $\mathbf{B}_{il}$ removing, the edges correction and the $\gamma$-frame) are model tools; this means that we are providing more tools to the model in order to simulate deformations and fractures enriching it.

When it will be possible to simulate quantitatively material's deformations, it will be possible to understand which model's tools keep the physical reality and which not; it could be useful to find other model's tools for "develop" new physical realities, e.g., metamaterials.

## 8.3 Numerical Simulations

In this section, we will show some simulations varying the interaction constants $\alpha_j$ to analyze how fracture occur in normal and exotic ranges. We have chosen to show traction simulations (to be brief) and for these we have used the same systems inspired by the ASTM D638 sample, which initial conditions are shown in Fig. 8.17. The swarm's size is 2312 points for the hexagonal lattice, 2504 points for the squared one. This choice is due to the hardware limitations of the PC used to get the maximum efficiency; however, it can manage bigger swarms. Each simulation that will be shown takes from 30 s to 3 min to be done, this is thanks to the CUDA® architecture that is used here.

Also in this paper, like in D'Avanzo et al. (2021), we will use equal values for $\alpha_j$ for each point in the shell of the same order, so in the following only two interaction parameters will be defined:

- $\alpha_1$ for the first neighbors;
- $\alpha_2$ for the second neighbors.

In the following, we will show some simulations with parameters $\alpha_1$ and $\alpha_2$ in exotic range but we want to underline that it is the combination of the two parameters that determines the exoticity of the range and so the swarm's behavior. In the model here proposed, there are some parameters that could be explored ($\alpha_1$, $\alpha_2$, $T_B$, $\gamma$, the lattice, the number of elements and $\omega_k$). In the following, we will focus on analyzing the swarm response to the variation of only $\alpha_1$, $\alpha_2$, $T_B$ parameters and to the lattice itself.

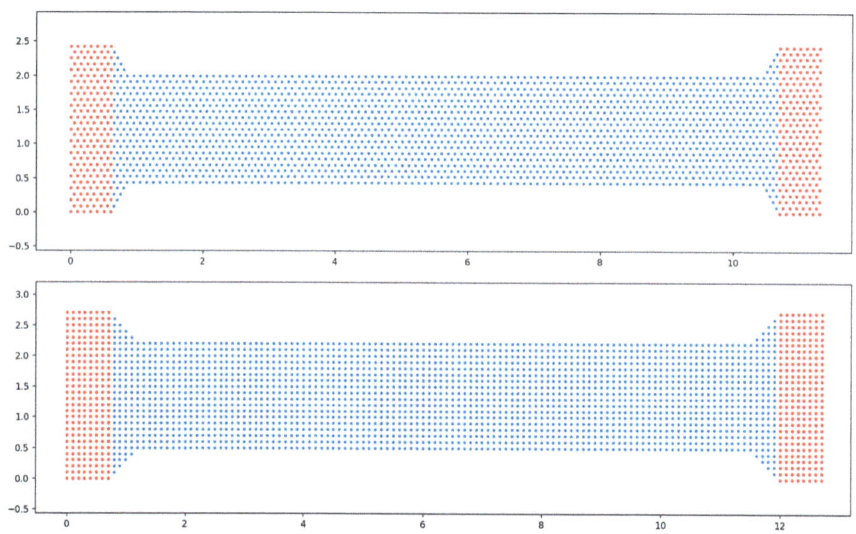


**Fig. 8.17** Here we provided the initial configurations of the swarm used in the following simulations, on the top for the hexagonal lattice on the bottom for the squared one

### 8.3.1 Notations and Fracture Thresholds

For the sake of simplicity, we will analyze the behaviors of the fracture in the swarms in graphic analogy with fracture experimentally observed in real materials. So, we define three kinds of fracture phenomena:

- "*Brittle*" (**B**) when the fracture is smooth and clear;
- "*Ductile*" (**D**) when the regions of the swarm near the fracture keep the deformation;
- "*Completely ductile*" (**CD**) when the swarm structure is heavily compromised.

A graphic example of those is provided in Fig. 8.18.

We are interested in to verify if our model can qualitatively achieve these three kinds of fracture listed above. If the swarm's response is ductile or completely ductile, it will be called a "*plastic*" response.

To get the fracture phenomenon, we introduced a breaking threshold $T_B$: if the distance between two neighbors of the swarm is greater than $T_B$, their bound is break and they do not interact each other anymore. Generally, we can define different thresholds $T_{B_k}$ for neighbors of order $k$. So, the value of $T_B$ is an important parameter of the model that has a strong influence on the swarm's evolution and fracture. We choose two values for $T_B$: $1.05R_0$ and $1.20R_0$, where $R_0$ is the equilibrium distance between the considered points. Such choice is heuristic; indeed, there are a lot of values to explore and so we selected a "small" threshold and a "large" threshold. Because a bound between two points can be viewed as a little spring, the thresholds correspond to a limit on the possible strain: the small one ($\varepsilon \approx 0.05$) and the large one ($\varepsilon \approx 0.20$). In fact

$$\varepsilon = \frac{T_B - R_0}{R_0} = 0.20 \implies T_B = 1.20R_0. \tag{8.11}$$

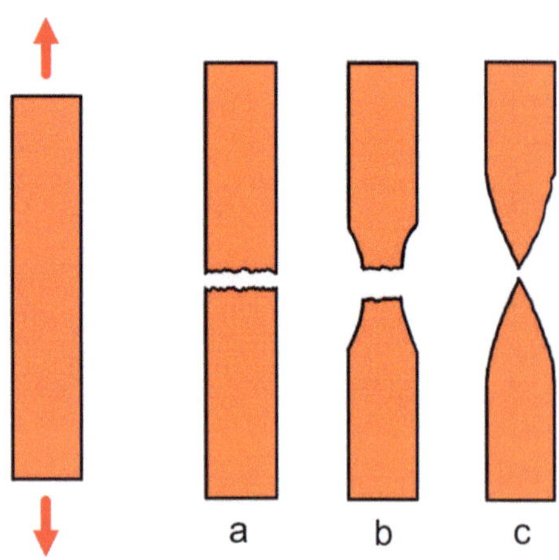


**Fig. 8.18** Here we provided three examples of fracture phenomena that we observed in our traction tests: **a** brittle, **b** ductile, **c** completely ductile

We want to underline that the responses of those "springs" are always linear with the distance (as the Hooke law, see Eq. 8.10). So, we consider the "plastic" response of the swarm an **emergent phenomenon** probably due to a large amount of breaking bonds in a swarm's region that compromises its structure. In particular, the "plastic" response does not depend only by the breaking threshold but also by the number of locally breaking bonds.

In future research, we will introduce a "plastic" interaction between the swarm's points, i.e., a nonlinear interaction for the "springs" over a defined threshold before the break of the bonds.

Following we will show some simulations done. We would like to empathize that we don't have the claim that our algorithm is really describing the physical reality of material's fractures and deformations; we want only to show some interesting results obtained from our developed algorithm and model, however similar to some experimental results. We want to recall that our model right now does not have a physical interpretation, so it is not possible to give precise information about the orders of magnitude in length and in time, furthermore we cannot give any explanation about the asymmetry that appears in the simulations. In the following, we will provide six images for each simulation to show different phases of the breaking phenomena and the evolution of the swarm: from immediately before the arise of the fracture to once that the fracture occurred, until the final configuration.

### *8.3.2 Fractures with First Neighbors' Interaction*

In this section, we show some simulations done with only the first neighbors' interaction, varying the fracture threshold $T_B$. The following parameters have been set:

| Parameter | Value used |
|---|---|
| $\alpha_2$ | 0 |
| $\gamma$ | 99 |
| $v_L$ | −0.002 |
| $v_R$ | 0 |

$v_L$ and $v_R$ represent the leaders' velocity on the left and right side of the swarm, respectively. So, the swarm is clamped on the right side and pulled on the left side for simulate a tensile test. The $\gamma$ value has been chosen in a way that the deformation's information is transmitted across the used swarm samples; these have about 100 points along the $x$-axis, the precise number depends by the lattice. What said means that we have simulated materials with a quick response, quite stiff. We want to underline that in this section we are considering only first neighbors' interaction, that is why $\alpha_2$ is zero meanwhile the values used for $\alpha_1$ are specified next.

Because the high number of simulations, we have provided a practical table (Table 8.4) that may be useful to resume ranges, parameters used and results. The range kind are "Normal" (**N**) and "Exotic" (**E**),

**Table 8.4** Schematic view of the simulations with first neighbors' interaction

| Lattice type | $\alpha_1$ | $T_B$ | Range kind | Fracture kind | Figure number |
|---|---|---|---|---|---|
| H | 1.0 | $1.05R_0$ | N | B | 8.19 |
| H | 1.0 | $1.20R_0$ | N | B | 8.20 |
| S | 1.0 | $1.05R_0$ | N | B | 8.21 |
| S | 1.0 | $1.20R_0$ | N | B | 8.22 |
| H | 2.0 | $1.05R_0$ | E | B | 8.24 |
| H | 2.1 | $1.05R_0$ | E | B | 8.25 |
| H | 2.3 | $1.20R_0$ | E | D | 8.26 |
| H | 2.6 | $1.20R_0$ | E | CD | 8.27 |
| S | 2.0 | $1.05R_0$ | E | B | 8.28 |
| S | 2.4 | $1.05R_0$ | E | B | 8.29 |
| S | 2.1 | $1.20R_0$ | E | D | 8.30 |
| S | 2.2 | $1.20R_0$ | E | D | 8.31 |
| H | 2.1 | $1.20R_0$ | E | B/D | 8.32 |
| H | 3.0 | $1.20R_0$ | E | CD | 8.33 |
| S | 2.6 | $1.20R_0$ | E | CD | 8.34 |
| S | 2.7 | $1.20R_0$ | E | CD | 8.35 |

#### 8.3.2.1 $\alpha_1$ in the Normal Range

We observed that, until $\alpha_1$ changes in the normal range, the obtained fractures, from simulations on the same lattices, do not show differences. For this, the simulations provided in this section are all performed with a fixed value of $\alpha_1 = 1$. We now show the lattice response to the fracture threshold variation, in Figs. 8.19 and 8.20 the lattice is hexagonal, in Figs. 8.21 and 8.22 for the squared one. The fracture in hexagonal lattice is characterized by a smooth "V" shape so we say it brittle. It seems that the fracture threshold does not influence the fracture phenomenon.

For the squared lattice, the breaking threshold influences the fracture (Figs. 8.21 and 8.22). Also, the squared lattice exhibits a brittle fracture although many bonds are broken near it. However, the number of broken bonds is insufficient to cause a plastic response. The squared lattice is characterized by regions that contain some broken bonds; these usually give rise to a *graphical* "chessboard pattern".

The fracture begins from the swarm's inner side (e.g., in Figs. 8.19b and 8.20b) and propagates itself to the bound (this could be seen especially in Fig. 8.19c where some bounds still exist). The same phenomenology in the squared lattice occurs (Figs. 8.21b and 8.22b) but with the difference that here "chessboard pattern" appear (Figs. 8.21a, b and 8.22a, b).

In the corners, there are areas in tensile state (in blue), especially in the hexagonal lattice (Figs. 8.19a and 8.20b), instead in the squared one these areas are less evident, or some bounds are broken in these areas (Figs. 8.21a and 8.22a).

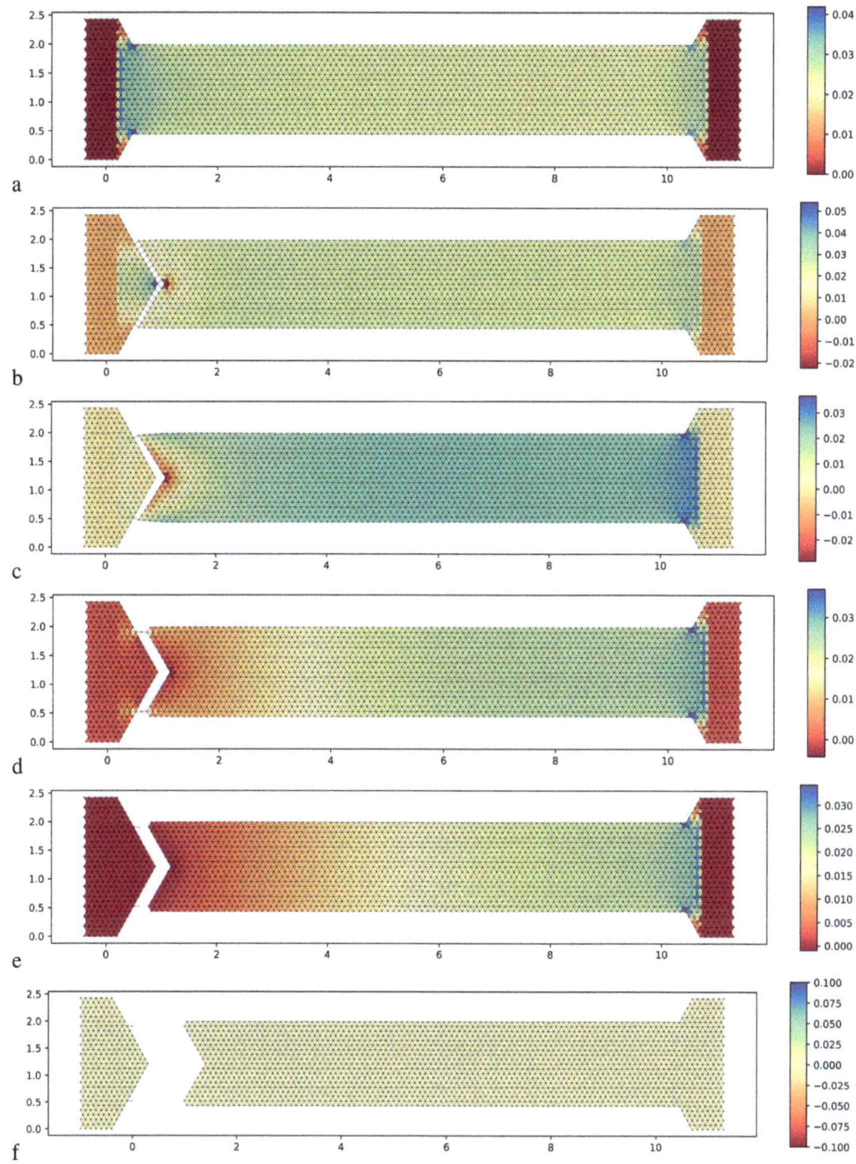


**Fig. 8.19** Hexagonal lattice with $\alpha_1 = 1.0$ and $T_B = 1.05R_0$. PTSs here used are: (1920, 1930, 1940, 1960, 2000, 6000). We show some PTS of the simulation: **a** before the fracture; **b–d** during the fracture; **e** after the fracture; **f** the final configuration

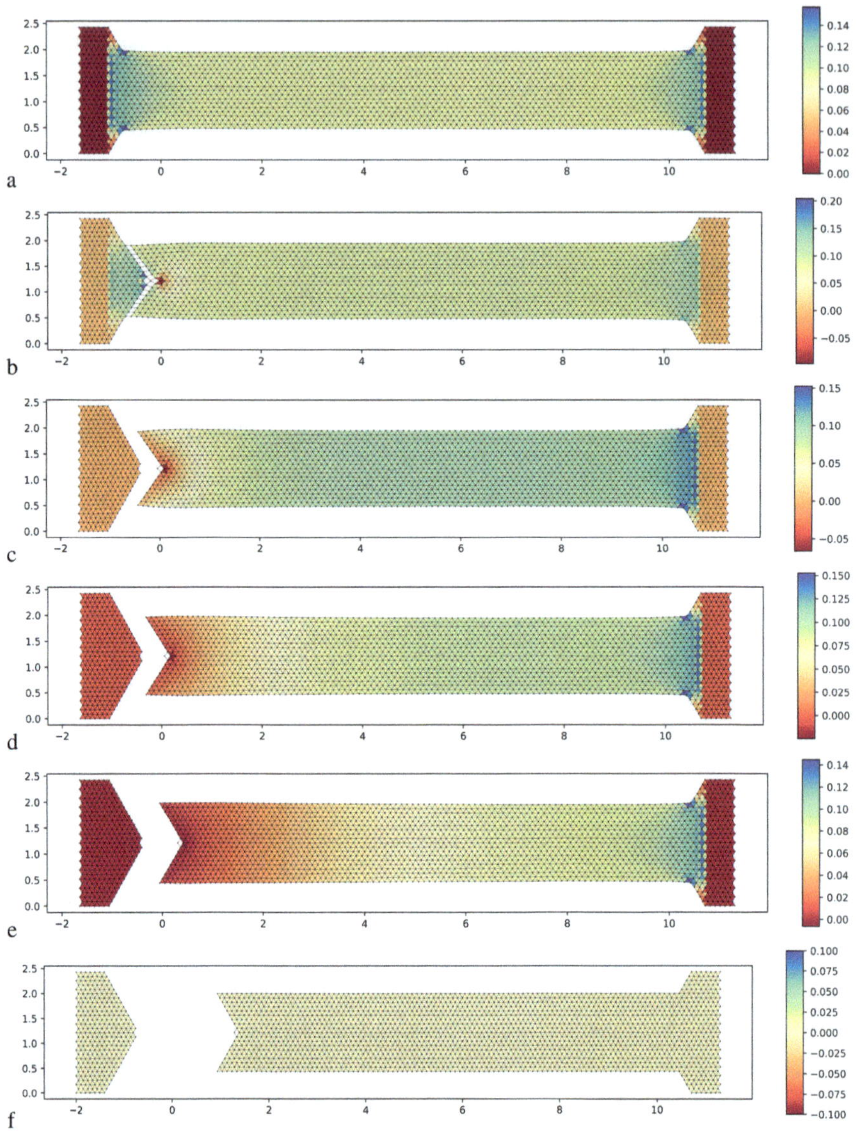


**Fig. 8.20** Hexagonal lattice with $\alpha_1 = 1.0$ and $T_B = 1.20R_0$. PTSs here used are: (8180, 8190, 8200, 8210, 8250, 15,000). We show some PTSs of the simulation: **a** before the fracture; **b** during the fracture; **c**–**e** after the fracture; **f** the final configuration

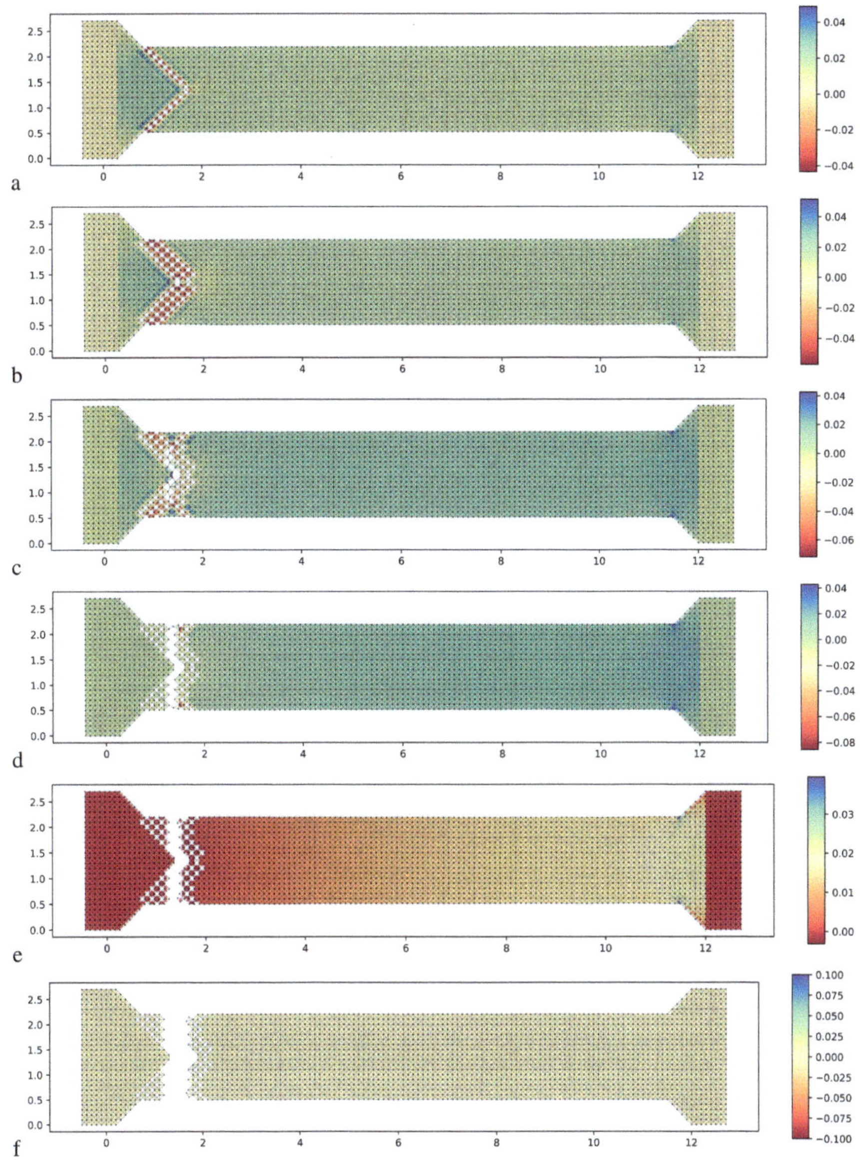

**Fig. 8.21** Squared lattice with $\alpha_1 = 1.0$ and $T_B = 1.05R_0$. PTSs here used are: (2070, 2100, 2110, 2130, 2180, 7000). We show some PTS of the simulation: **a**, **b** before the fracture; **c**, **d** during the fracture; **e** after the fracture; **f** the final configuration

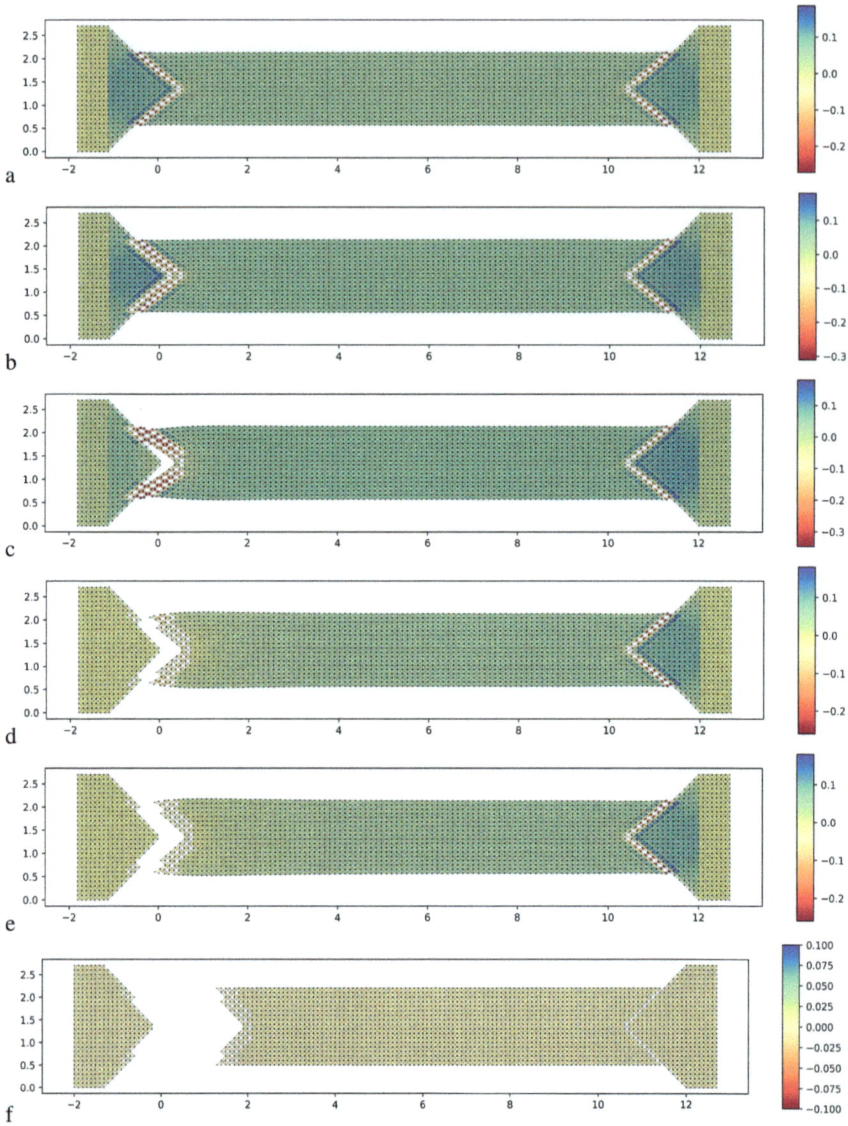


**Fig. 8.22** Squared lattice with $\alpha_1 = 1$ and $T_B = 1.20R_0$. PTSs here used are: (9010, 9020, 9030, 9040, 9050, 15,000). We show some PTS of the simulation: **a**, **b** before the fracture; **c** during the fracture; **d**, **e** after the fracture; **f** the final configuration

There is the Poisson effect in every simulation, but here is hardly evident maybe due to the shape and the number of elements of the swarm. However, we can observe it comparing Figs. 8.21a and 8.22a: in the latter, a greater elongation causes a visible thinning.

#### 8.3.2.2 $\alpha_1$ in the Exotic Range

When $\alpha_1$ changes remaining in the exotic range, a large amount of inner bounds breaks. Analyzing this phenomenon, we have found that in the first PTSs of the simulation high-deformation areas appear with bubble shapes (see Fig. 8.23a, b). If the fracture is not considered, this bubble expands along the whole lattice giving rise exotic phenomena (see Fig. 8.23g); otherwise, considering the bounds' break, when a bound in the bubble breaks, the latter start to propagate along the lattice as a wave (see Fig. 8.23a–f): interference-like and reflection-like phenomena are observed. The bubble propagation breaks many bounds. This phenomenon occurs also with the interaction to second neighbors.

This phenomenon raises up in a similar way also for the squared lattice. Basing on the lattice, holes patterns are formed and, growing $\alpha$, this phenomenon becomes more and more evident. In some simulations that we will show, an equilibrium configuration, as defined before with $\varepsilon$, can't be reached because some swarm's points continue to oscillate indefinitely. In these cases, we consider the equilibrium configuration reached when the most part of the swarm is in equilibrium, while the parts still oscillating do not cause modification to this configuration. We will not discuss these phenomena deeper in this paper because here we were focused on the fracture phenomena.

Hexagonal Lattice

In the exotic range, the final fracture depends on the $\alpha_1$ parameter and on the breaking threshold $T_B$ because from the combination of these two parameters different pattern of initial break bonds arises, and so a different swarm evolution. In this article, we have chosen to show the more interesting simulations showing what our algorithm and model can produce.

In Figs. 8.24 and 8.25, a similar fracture phenomenon can be observed: First holes are formed and from these mini fractures start (Figs. 8.24a–d and 8.25a–d). The difference between the two simulations, that have very close parameters, is the number of initial broken bonds; in both cases, the pattern is the same. So, for a low $T_B$ and $\alpha_1$ around the inferior limit in the exotic ranges the response is fragile.

In Figs. 8.26 and 8.27, higher values of $T_B$ and $\alpha$ have been used; here the fracture is ductile. In the first PTSs, many bounds are broken; this compromises the swarm's structure that cannot keep its shape.

In the normal range, the fractures could be "*predictable*": For every $\alpha_1$ value, the break starts from the inner swarm's and expands itself on the edges every time in the same swarm's region (near the leaders, on the left). On the other hand, in the

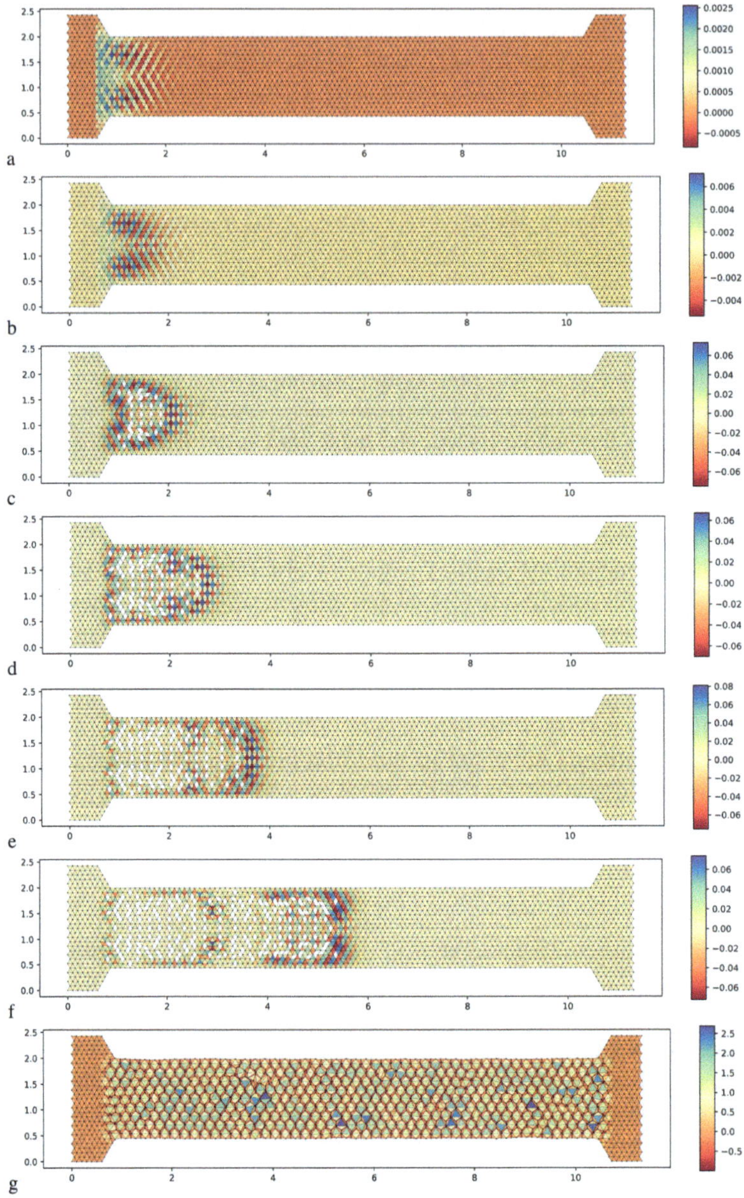


**Fig. 8.23** Hexagonal lattice with threshold $1.05R_0$. Here we show some snapshot of the swarm evolution with $\alpha_1$ in the exotic range. In **a, b** the formation of two areas with high deformation could be seen; these in **c** merge and in **d–f** propagate across the swarm. During the propagation, the high-deformation areas leave behind them self break bounds patterns that are common in these kinds of simulations. In **g**, the exotic phenomenon rises without bounds break

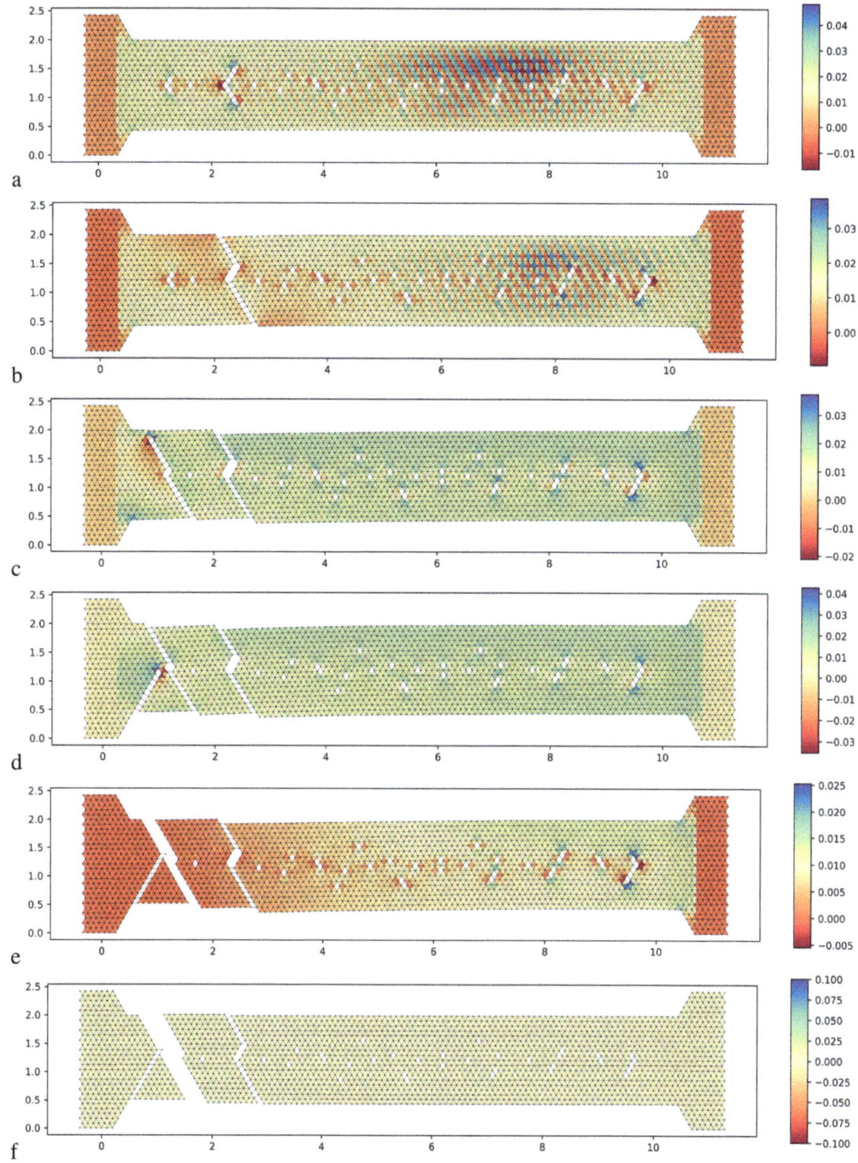


**Fig. 8.24** Hexagonal lattice with $\alpha_1 = 2.0$ and $T_B = 1.05R_0$. PTSs here used are: (1300, 1310, 1610, 1650, 1750, 20,000). We show some PTS of the simulation: **a** before the fracture, **b**–**d** during the fracture, **e** while the fracture is ending and **f** the final configuration. In **a**, **b** is visible a residue of the high-deformation zone described in 8.3.2.2

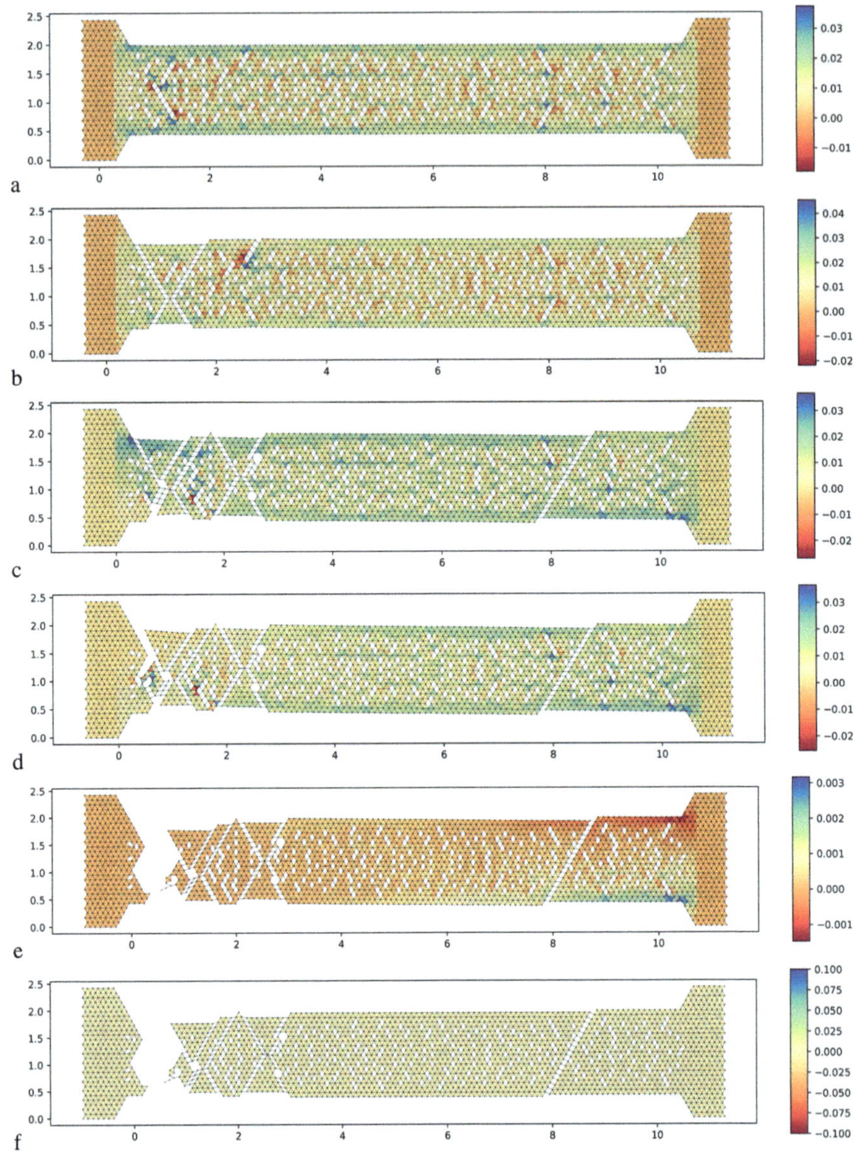


**Fig. 8.25** Hexagonal lattice with $\alpha_1 = 2.1$ and $T_B = 1.05R_0$. PTSs here used are: (1500, 2000, 3000, 3060, 4500, 15,000). We show some PTS of the simulation: **a** before the fracture, **b–d** during the fracture, **e** while the fracture is ending and **f** the final configuration

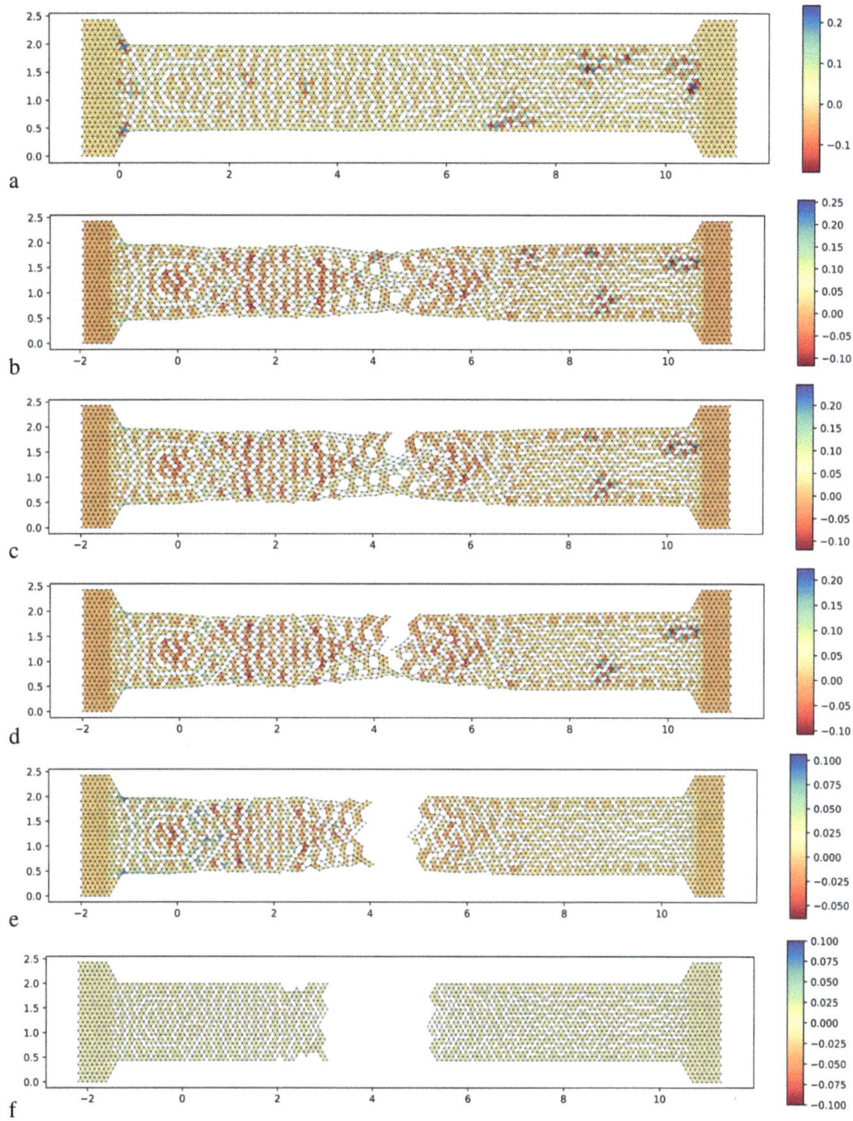


**Fig. 8.26** Hexagonal lattice with $\alpha_1 = 2.3$ and $T_B = 1.20R_0$. PTSs here used are: (3500, 9750, 9950, 10,000, 10,050, 20,000). We show some PTS of the simulation: **a** before the fracture, **b–d** during the fracture, **e** while the fracture is ending and **f** the final configuration

exotic range the fracture type and where the fracture occur depends on the $\alpha_1$ and $T_B$ parameters, in particular the ductile fractures are preferred when $T_B$ is high.

Squared Lattice

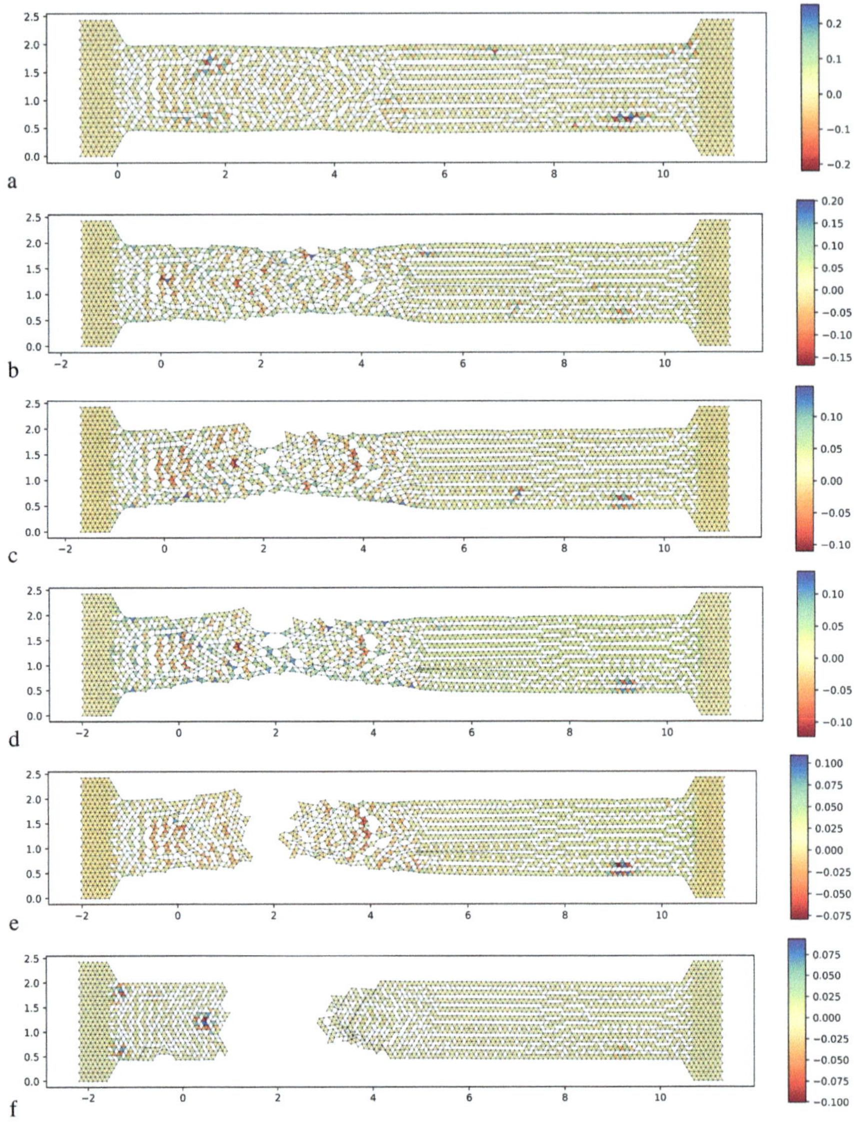

**Fig. 8.27** Hexagonal lattice with $\alpha_1 = 2.6$ and $T_B = 1.20R_0$. PTSs here used are: (3500, 8000, 8500, 10,000, 10,100, 15,000). We show some PTS of the simulation: **a** before the fracture, **b–d** during the fracture, **e** while the fracture is ending and **f** the final configuration

At the same way of the hexagonal lattice, also on the squared one in the exotic range the final fracture depends on the $\alpha_1$ parameter and on the breaking threshold $T_B$ due to the different pattern of initial break bonds. In this article, we have chosen to show the more interesting simulations showing what our algorithm and model can produce.

In Fig. 8.28, the fracture evolution is quite like that shown on Fig. 8.21 with the only difference on the number of patterns, that anyway are of the same type; this is due to the $\alpha_1$ value on the in lower limit of the exotic range.

In Fig. 8.29 is shown a simulation in which a fracture resembling a ripped fabrics occur. Although many bonds are broken, the swarm can keep its shape at equilibrium configuration. From this and other simulations, we can deduce that the squared lattice seems to be resilient to ductility: This is due to the higher number of coordination and to the cross bonds that strengthen the structural integrity of the swarm.

In Figs. 8.30 and 8.31, there are fractures with similar evolution: There are extraordinarily strong local deformations and an evident chessboard structure arises; this is specific for the squared lattice. The fracture is hardly ductile: in Figs. 8.30d and 8.31d two collapsing flaps are formed, the collapse is due the bound's lack.

Comparing the simulations in Figs. 8.28, 8.29, 8.30 and 8.31, it is possible to understand that growing $\alpha_1$, the number of broken inner bounds grows up, particularly it can be seen in Fig. 8.29.

Extreme Behaviors

In this paragraph, we have assorted some interesting exotic simulations different from the previous shown.

For the hexagonal lattice, the extreme behaviors broke the swarm in more and more small pieces, like in Fig. 8.32; the swarm has a lot of micro-fractures that together compromise the whole structure. In Fig. 8.33, the inner fragmentation damaged drastically the swarm, but the edges are not jagged like in Fig. 8.32. The final form is compromised, and the fracture is ductile. Notice the horizontal lines pattern of the hexagonal lattice.

In the case of the squared lattice instead, the swarm resembles soft materials or fabrics. In Figs. 8.34 and 8.35 indeed, there are strong local deformations and fractures resembling ripped fabrics occur again in a more evident way. We considered these fractures completely ductile and noted that they have raised in the highest deformation regions. We can observe that in Fig. 8.35, a second region of high deformation on the right side of the swarm appears, but it recovered the shape at last (this shows again the resilience to the ductility of the squared lattice).

### 8.3.3 Fractures with Second Neighbors' Interaction

In this section, we will provide simulations in which the interaction is extended also to the second neighbors. So, here we analyze the responses of the swarm varying the

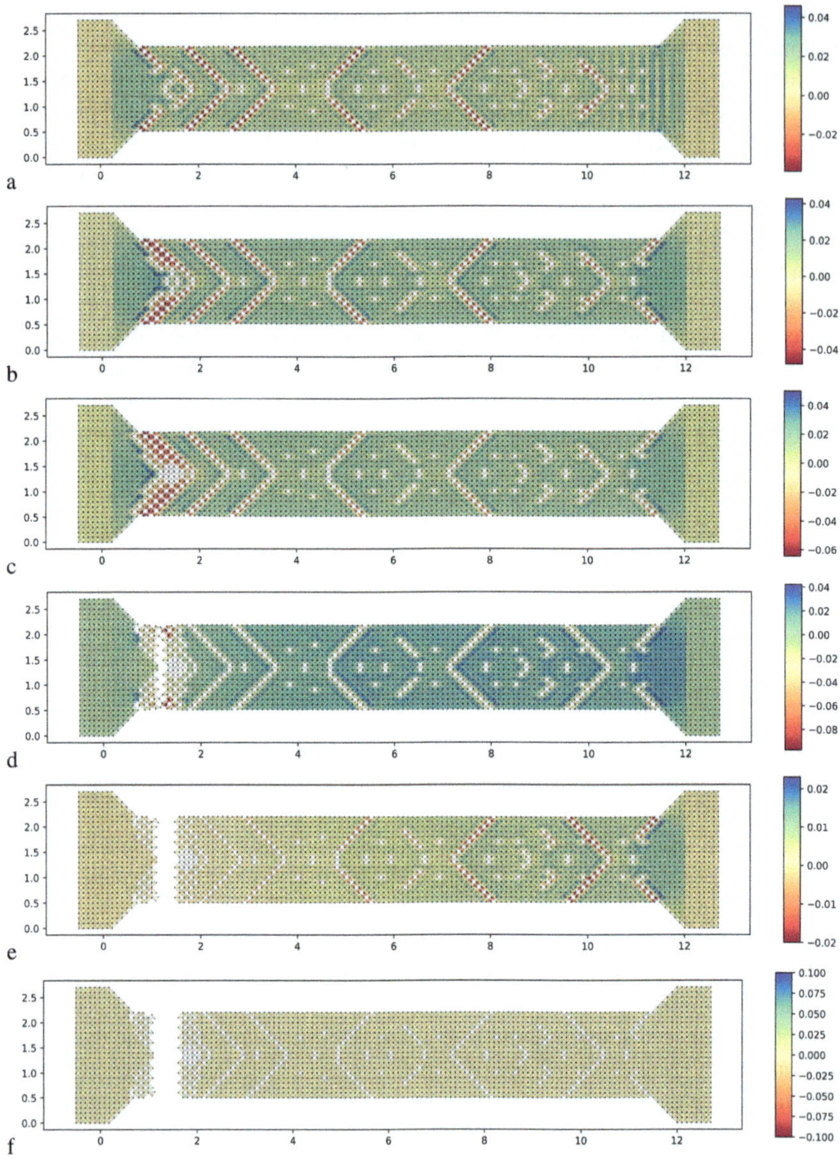

**Fig. 8.28** Squared lattice with $\alpha_1 = 2.0$ and $T_B = 1.05R_0$. PTSs here used are: (2300, 2360, 2470, 2480, 2600, 5000). We show some PTS of the simulation: **a**, **b** before the fracture, **c**, **d** during the fracture, **e** after the fracture; **f** the final configuration

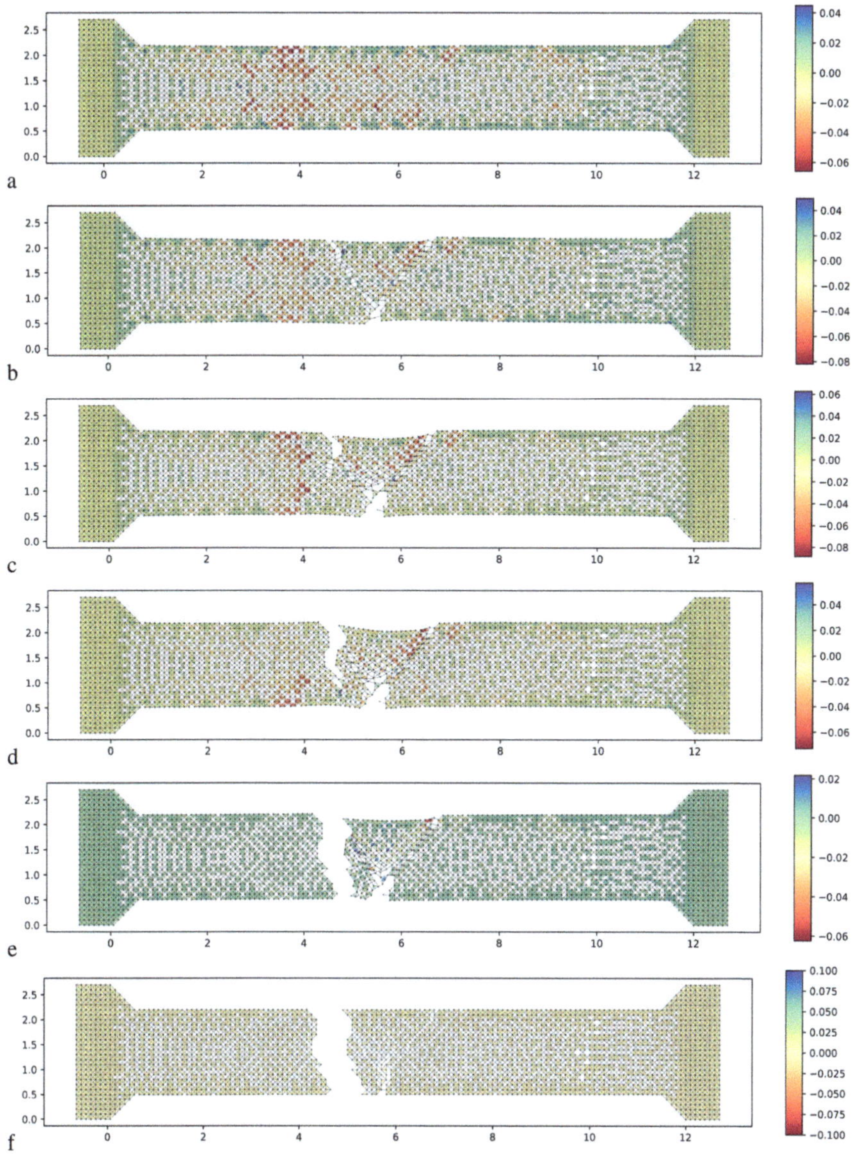

**Fig. 8.29** Squared lattice with $\alpha_1 = 2.4$ and $T_B = 1.05R_0$. PTSs here used are: (2500, 2820, 3040, 3070, 3200, 10,000). We show some PTS of the simulation: **a** before the fracture, **b–d** during the fracture, **e** after the fracture; **f** the final configuration

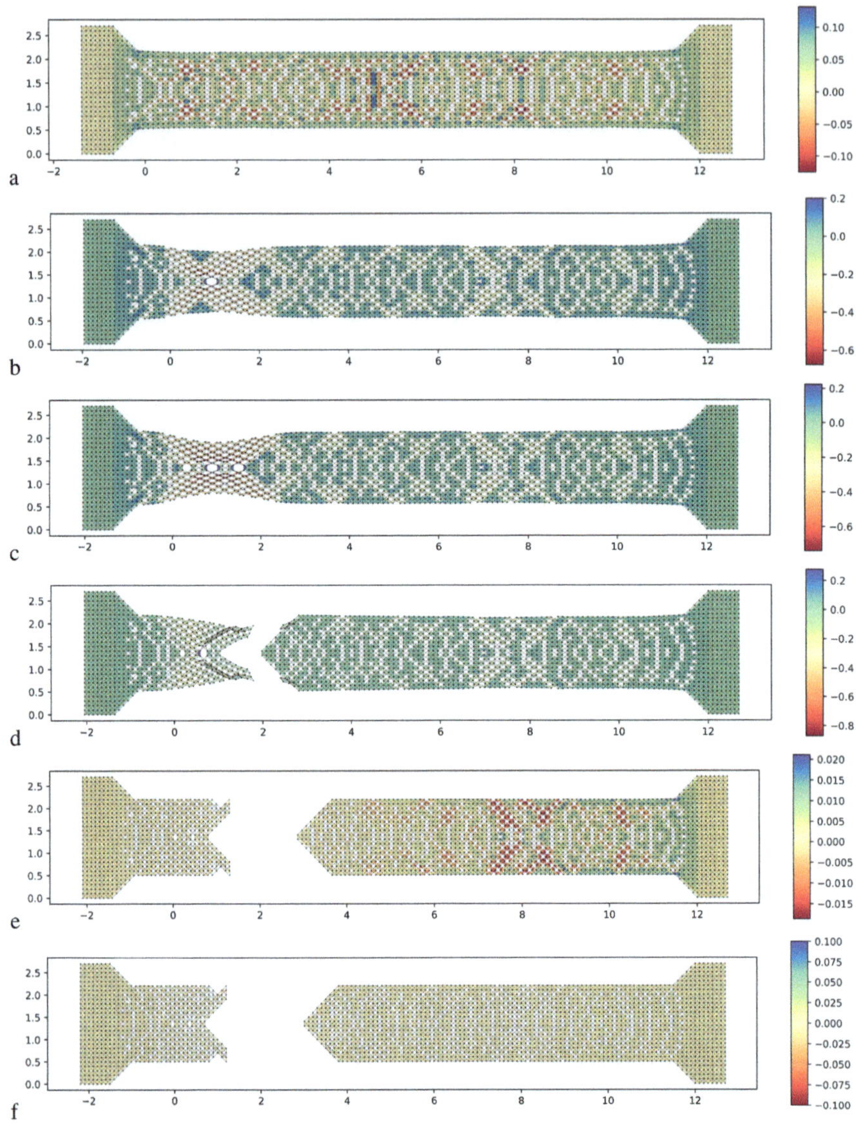

**Fig. 8.30** Squared lattice with $\alpha_1 = 2.1$ and $T_B = 1.20R_0$. PTSs here used are: (7000, 9800, 10,200, 10,250, 10,500, 15,000). We show some PTS of the simulation: **a** before the fracture, **b**, **c** during the fracture, **d**, **e** after the fracture; **f** the final configuration

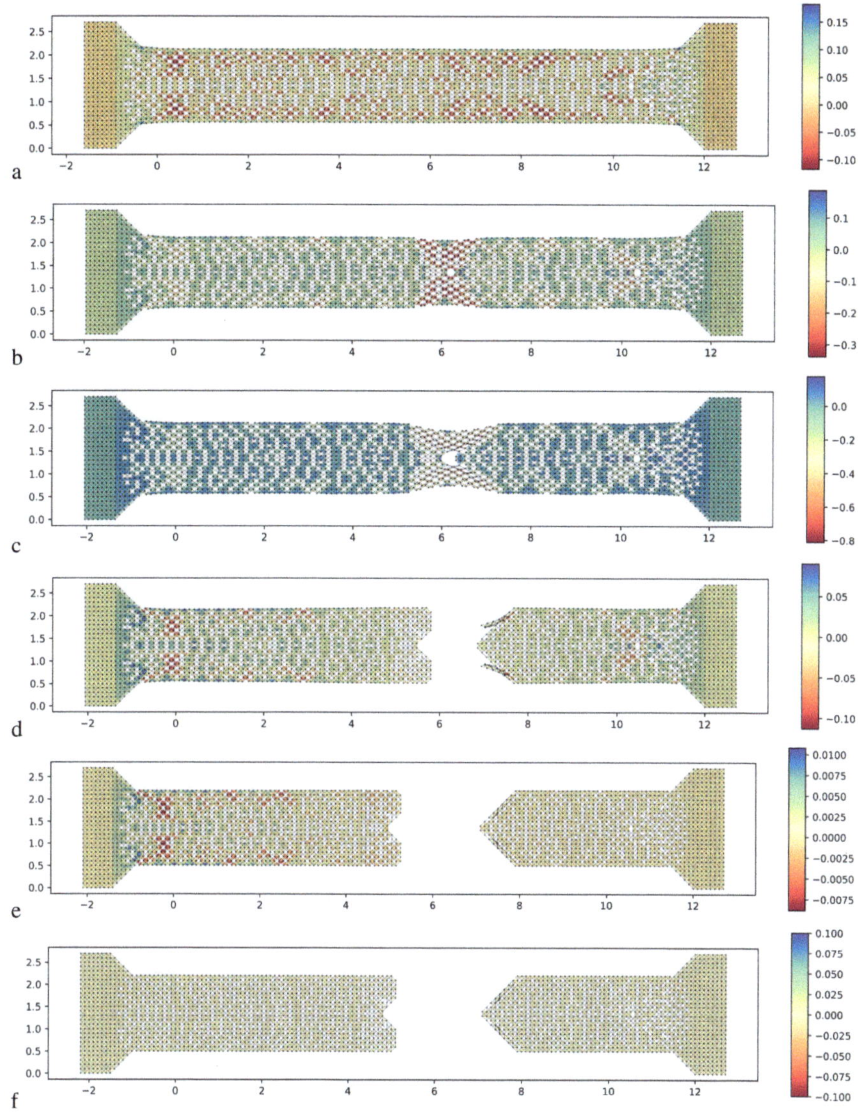


**Fig. 8.31** Squared lattice with $\alpha_1 = 2.2$ and $T_B = 1.20R_0$. PTSs here used are: (8000, 9800, 10,250, 10,300, 10,500, 15,000). We show some PTS of the simulation: **a** before the fracture, **b**, **c** during the fracture, **d**, **e** after the fracture; **f** the final configuration

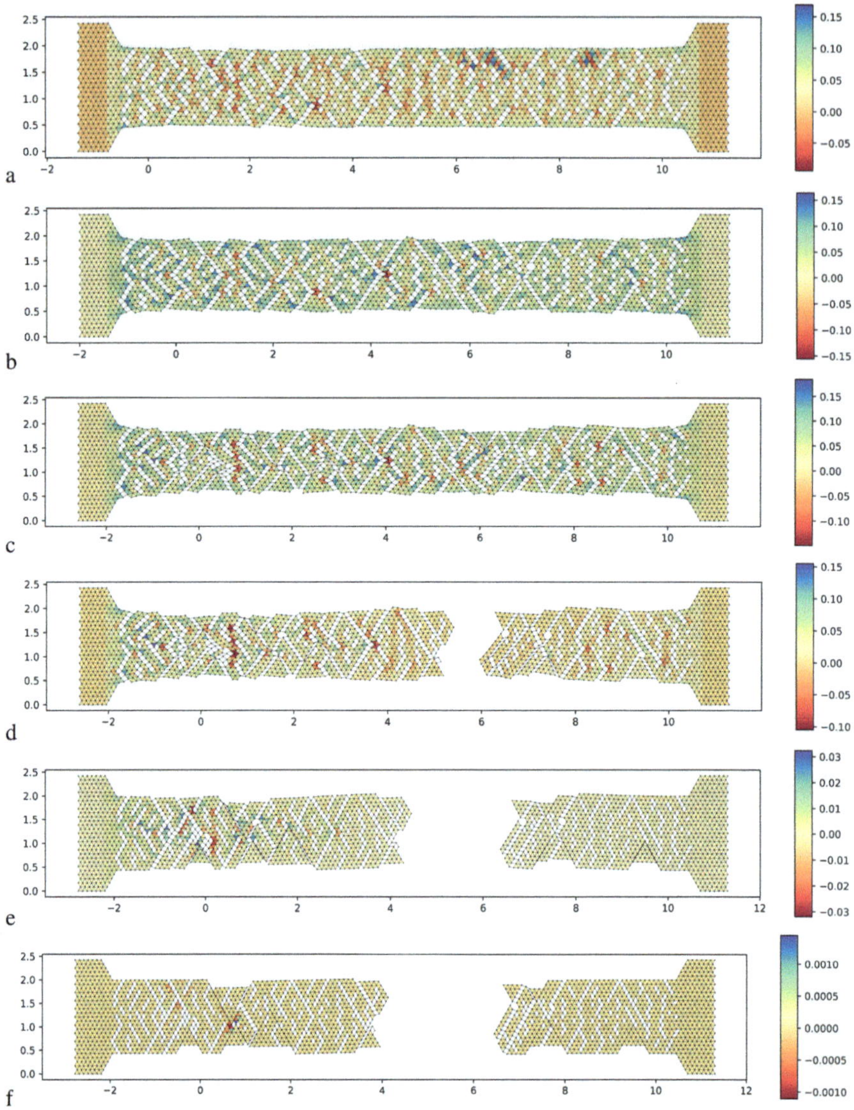

**Fig. 8.32** Hexagonal lattice with $\alpha_1 = 2.1$ and $T_B = 1.20R_0$. PTSs here used are: (7000, 10,000, 13,100, 13,200, 14,000, 30,000). We show some PTS of the simulation: **a** before the fracture, **b, c** during the fracture, **d, e** after the fracture; **f** the final configuration

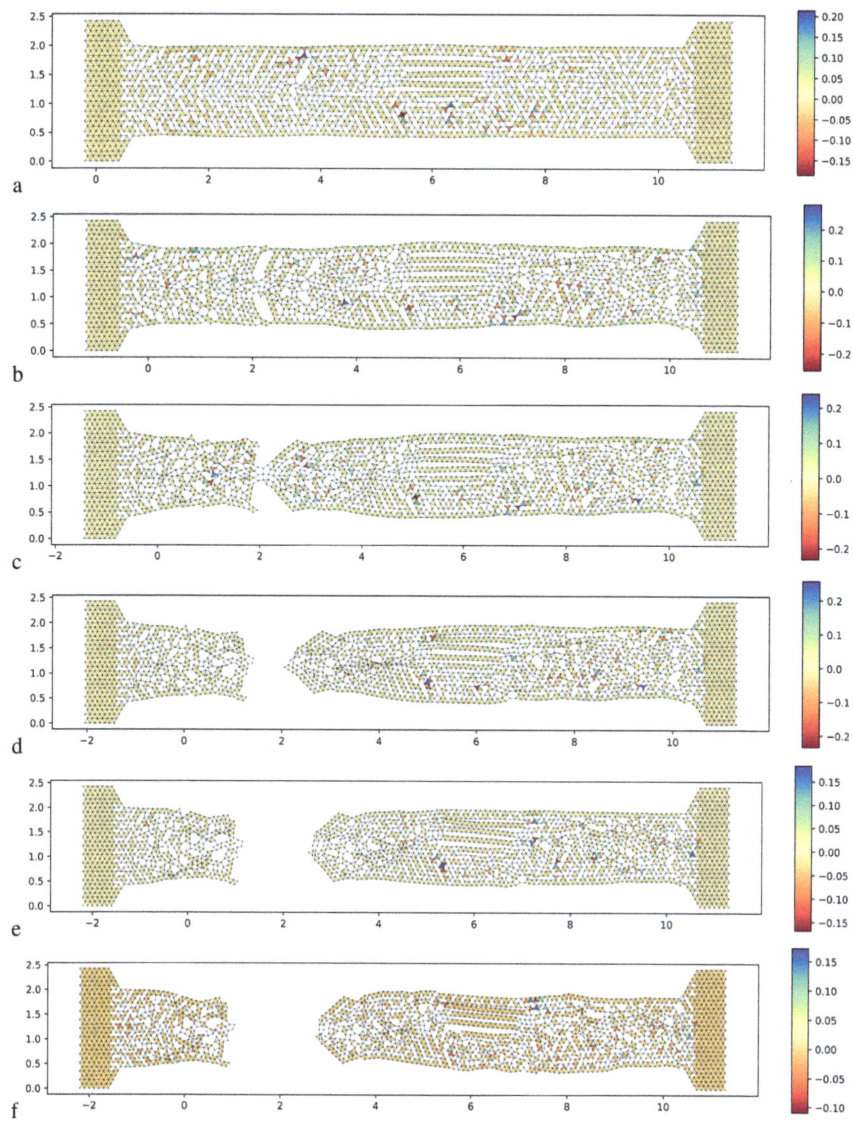


**Fig. 8.33** Hexagonal lattice with $\alpha_1 = 3.0$ and $T_B = 1.20R_0$. PTSs here used are: (1000, 6000, 7200, 10,200, 11,000, 40,000). We show some PTS of the simulation: **a** before the fracture, **b**, **c** during the fracture, **d**, **e** after the fracture; **f** the final configuration

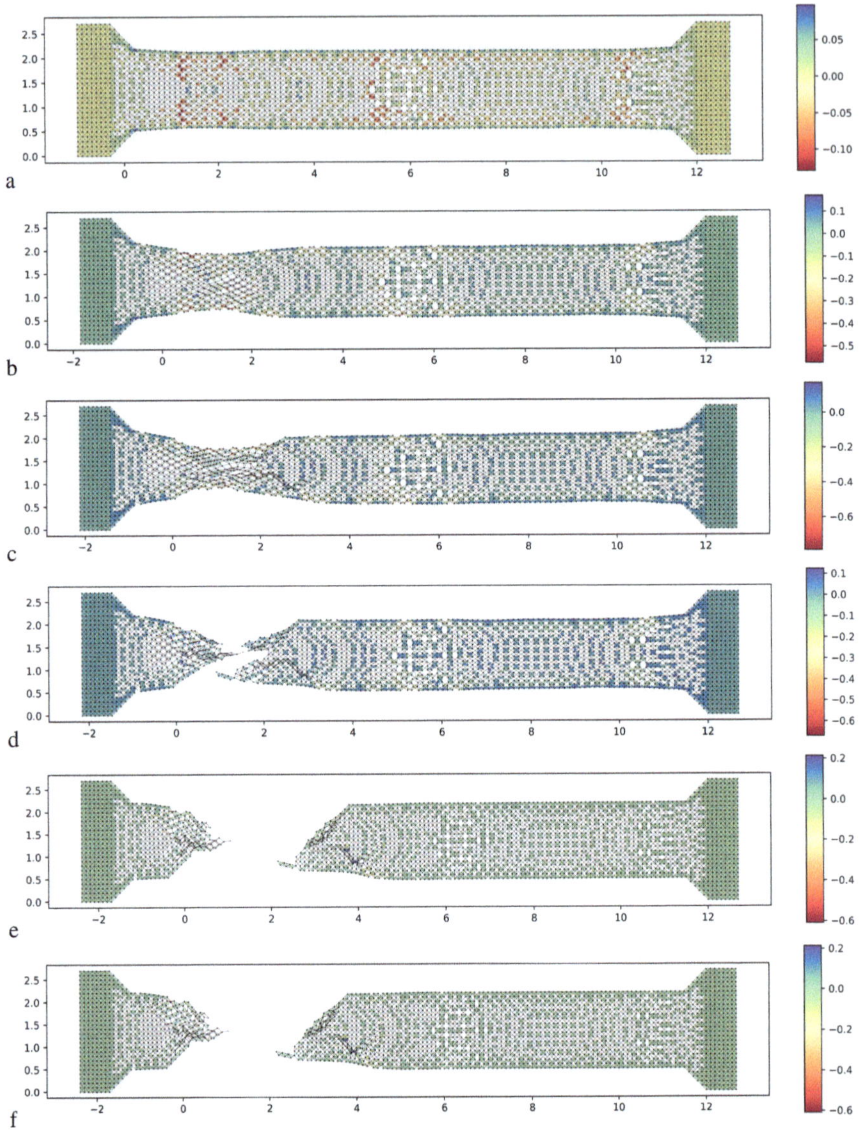


**Fig. 8.34** Squared lattice with $\alpha_1 = 2.6$ and $T_B = 1.20R_0$. PTSs here used are: (5000, 9200, 10,600, 10,800, 12,000, 20,000). We show some PTS of the simulation: **a** before the fracture, **b**, **c** during the fracture, **d**, **e** after the fracture; **f** the final configuration

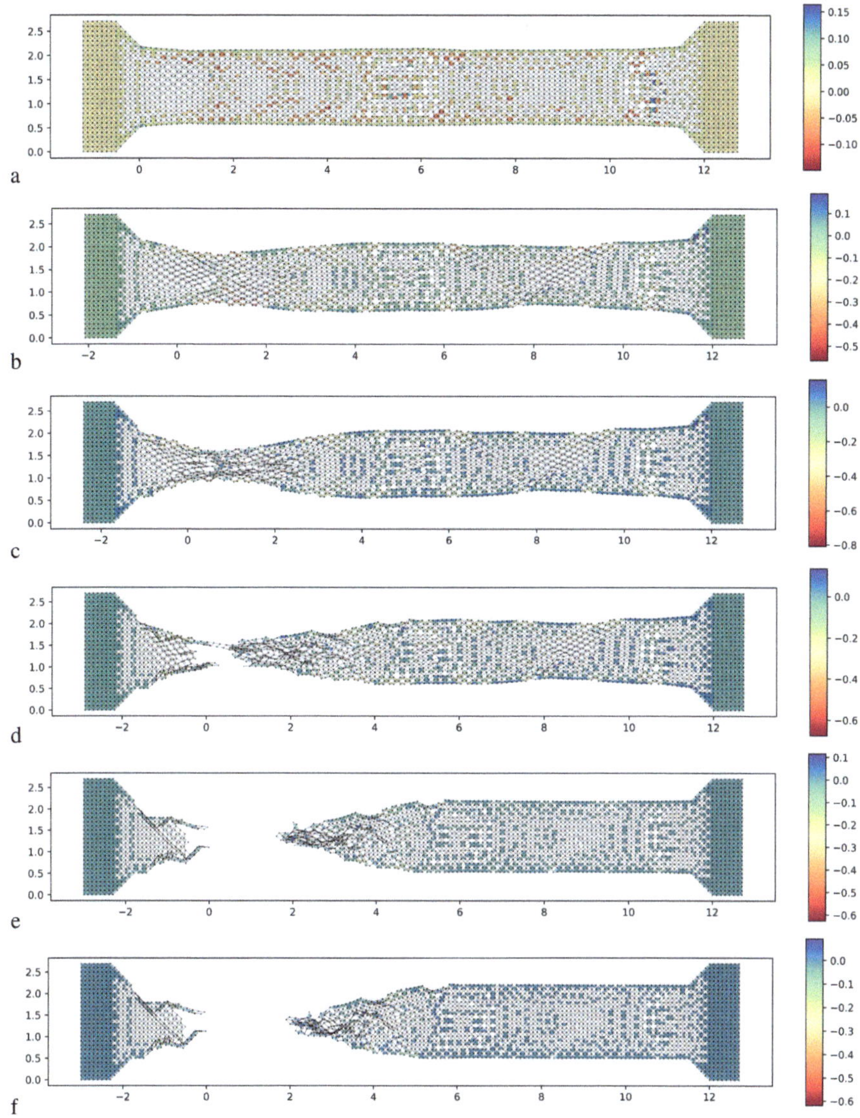


**Fig. 8.35** Squared lattice with $\alpha_1 = 2.7$ and $T_B = 1.20R_0$. PTSs here used are: (6000, 10,400, 12,200, 14,400, 15,000, 30,000). We show some PTSs of the simulation: **a** before the fracture, **b**, **c** during the fracture, **d**, **e** after the fracture; **f** the final configuration

**Table 8.5** Schematic view of the simulations with first neighbors' interaction

| Lattice type | $\alpha_1$ | $\alpha_2$ | $T_{B_1}$ | $T_{B_2}$ | Range kind | Fracture kind | Figure number |
|---|---|---|---|---|---|---|---|
| H | 1.0 | 0.5 | $1.05R_0$ | $1.05R_0$ | N | B | 8.36 |
| H | 1.0 | 0.5 | $1.20R_0$ | $1.20R_0$ | N | B | 8.37 |
| S | 1.0 | 0.5 | $1.05R_0$ | $1.05R_0$ | N | B | 8.38 |
| S | 1.0 | 0.5 | $1.20R_0$ | $1.20R_0$ | N | B | 8.39 |
| H | 1.9 | 1.1 | $1.05R_0$ | $1.05R_0$ | E | B | 8.40 |
| H | 1.3 | 1.5 | $1.05R_0$ | $1.05R_0$ | E | B | 8.41 |
| H | 1.9 | 0.3 | $1.20R_0$ | $1.20R_0$ | E | B | 8.42 |
| H | 2.5 | 0.1 | $1.20R_0$ | $1.20R_0$ | E | B/D | 8.43 |
| S | 1.1 | 1.5 | $1.05R_0$ | $1.05R_0$ | E | B | 8.44 |
| S | 2.1 | 0.5 | $1.05R_0$ | $1.05R_0$ | E | B | 8.45 |
| S | 1.5 | 1, 1 | $1.20R_0$ | $1.20R_0$ | E | B | 8.46 |
| S | 1.9 | 0.3 | $1.20R_0$ | $1.20R_0$ | E | B/D | 8.47 |
| H | 3.1 | 0.05 | $1.20R_0$ | $1.05R_0$ | E | B/D | 8.48 |
| H | 2.6 | 0.1 | $1.20R_0$ | $1.05R_0$ | E | CD | 8.49 |
| S | 2.2 | 0.05 | $1.20R_0$ | $1.05R_0$ | E | D/CD | 8.50 |
| S | 2.9 | 0.05 | $1.20R_0$ | $1.05R_0$ | E | CD | 8.51 |

breaking thresholds $T_{B_1}$ for the first neighbors and $T_{B_2}$ for the second ones and the parameters $\alpha_1$ and $\alpha_2$. We set the other parameters shown in table.

| Parameter | Value used |
|---|---|
| $\gamma$ | 49 |
| $v_L$ | −0.002 |
| $v_R$ | 0 |

With $v_L$ and $v_R$, we mean respectively the velocity of the leaders on the left and right side of the swarm. So, the swarm is clamped on the right side and pulled on the left side for simulate a tensile test like in Sect. 8.3.2, but this time the $\gamma$ value is halved. This because, as told in Sect. 8.2.2.1, consider the second neighbor means to double the propagation velocity of the deformation, so we halved $\gamma$ value to keep it the same.

We provided Table 8.5 to collect and recap the simulations shown in this section and aid the comprehension of the parameters involved.

#### 8.3.3.1 Normal Range

For the simulations with the second neighbors and the parameters $\alpha_1$ and $\alpha_2$ in the normal range, the swarm behaviors are very similar to those shown in Sect. 8.3.2.1. This can be observed for the hexagonal lattice comparing Fig. 8.36 with Fig. 8.19 and

Fig. 8.37 with Fig. 8.20. This is somewhat less valid for the squared one comparing Figs. 8.38 and 8.39 with Fig. 8.21.

Furthermore, also in these simulations, with $\alpha_1$ and $\alpha_2$ is the normal range, there are not differences in the final configuration (for chosen breaking thresholds and lattice). For this reason, we used the same fixed values $\alpha_1 = 1$ and $\alpha_2 = 0.5$ for the simulations in normal range. We want to emphasize that the invariance of the final configuration is a big difference with the results got in D'Avanzo et al. (2021), but we want to recall that in this paper the interaction laws are different as told in Sect. 8.2.2.4.

The most interesting observation can be done on the "*fracture velocity*", i.e., it happens in less PTSs, and in the squared lattice behavior that breaks itself similar the hexagonal lattice, i.e., the fracture has a more evident "V" shape respect to the same simulations with the interaction to the first neighbors.

#### 8.3.3.2 Exotic Range

In the exotic range, the final fracture depends on the parameters $\alpha_1$, $\alpha_2$, $T_{B_1}$ and $T_{B_2}$. This because from the combination of these parameters different pattern of initial break bonds arises, and so different swarm evolution. In this article, we have chosen to show the more interesting simulations showing what our algorithm can produce.

Hexagonal Lattice

The simulations shown in Figs. 8.40 and 8.41 have a similar response, i.e., a fragile fracture that remember stiff materials or bones. In each case, there is big bounds lack. In Fig. 8.40, there is the typical line pattern (on the right), that is due to high $\alpha_1$ value. The place where the fracture occurs changes, so sometimes the latter occur on the left side, sometimes on the center, with a no predictable way. This show that the algorithm can produce a very rich phenomenology.

The simulation shown in Fig. 8.42 is similar to the first neighbors exotic simulations, this is because $\alpha_2$ is low. In this case, despite the presence of a large amount of micro-fractures, the swarm's structure is not compromised, and the break is fragile.

The simulation in Fig. 8.43 is interesting because it exhibits two different break patterns: On the left the swarm fragments itself, on the right the common hexagonal lattice's horizontal line pattern arises. So, there are two different swarm's behaviors: On the left side, there is a localized Poisson effect, while on the left side there isn't. However, the fracture remains fragile.

Squared Lattice

The first thing to be noted is that a high $\alpha_2$ inhibits the chessboard pattern, as is possible to see in Figs. 8.44 and 8.46 where $\alpha_2 > 1$; on the other hand, in Figs. 8.45 and 8.46, where $\alpha_2 < 1$, the chessboard pattern still remains noticeable.

The simulation shown in Fig. 8.44 is like the ones with the interaction to the first neighbors, indeed there is a fragile shaped "V" fracture. The first broken inner bonds make a geometric pattern not so extended.

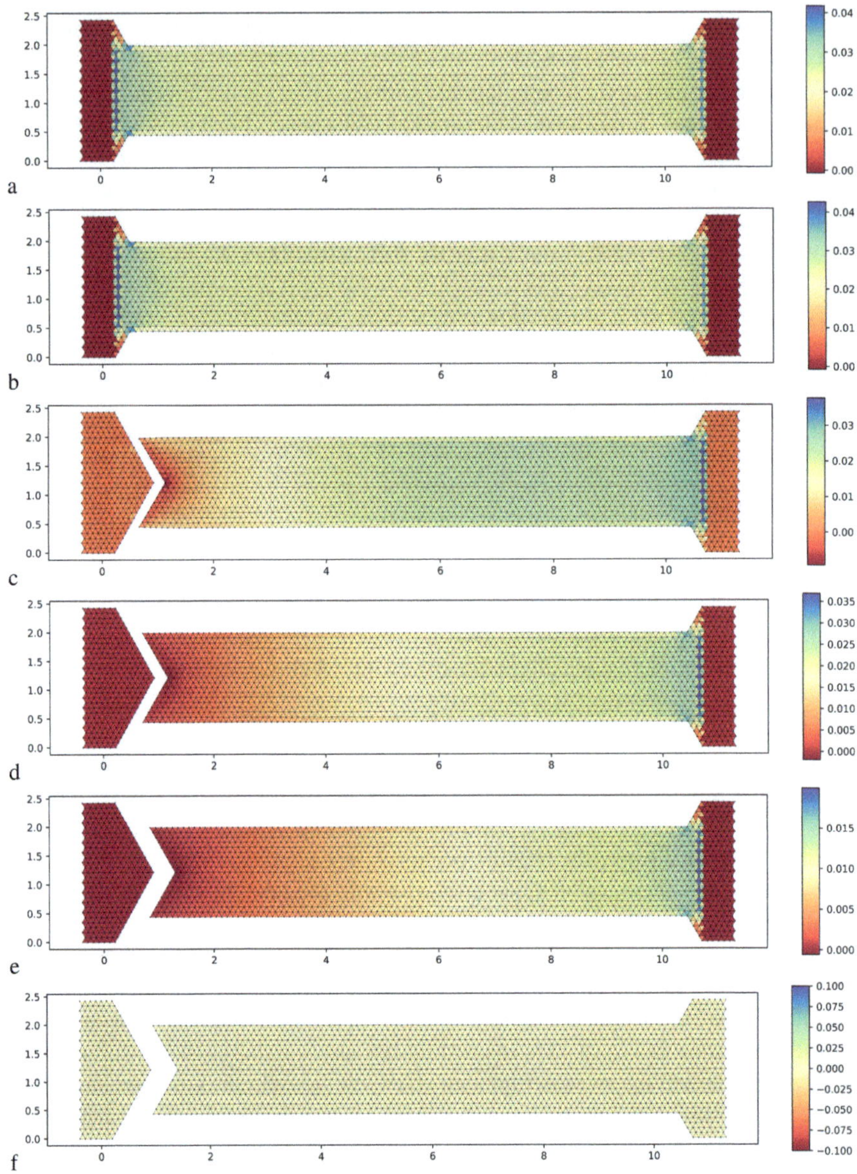


**Fig. 8.36** Hexagonal lattice with $\alpha_1 = 1.0, \alpha_2 = 0.5, T_{B_1} = 1.05R_0$ and $T_{B_2} = 1.05R_0$. PTSs here used are: (900, 920, 930, 940, 1000, 2000). We show some PTS of the simulation: **a**, **b** before the fracture, **c** during the fracture, **d**, **e** after the fracture; **f** the final configuration

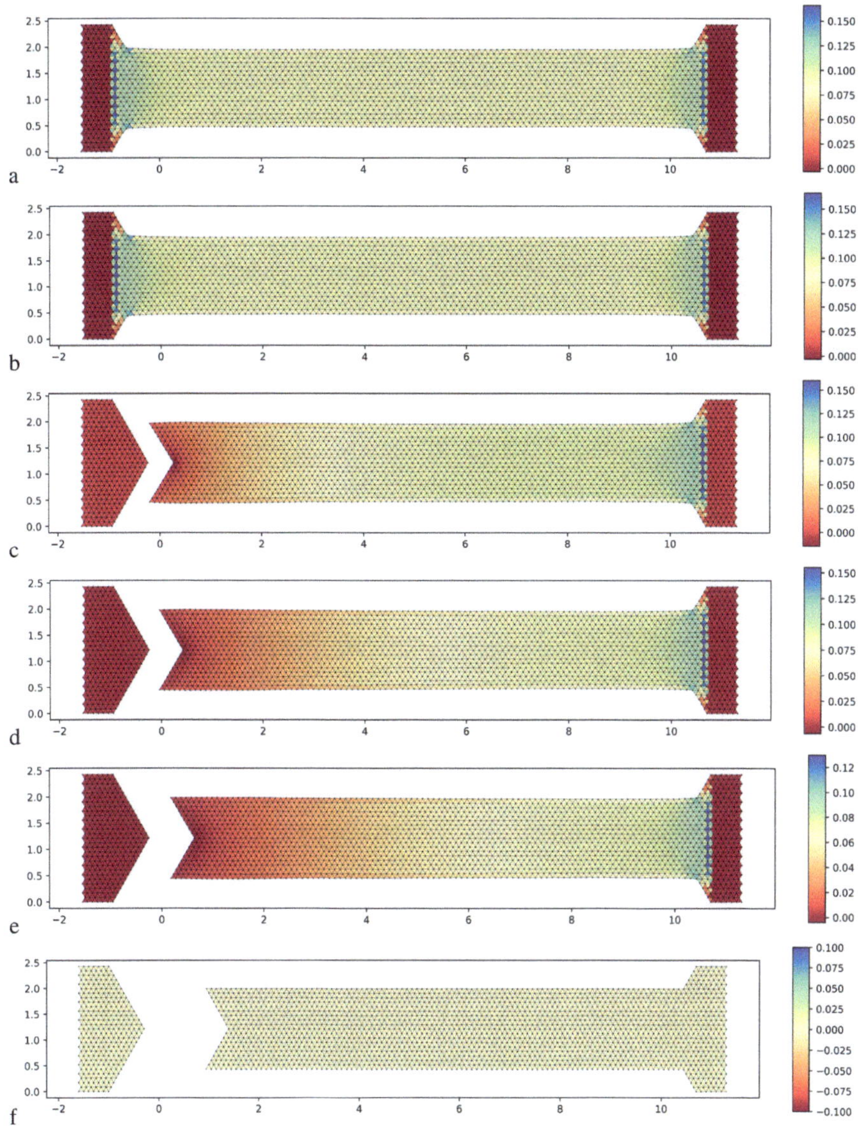

**Fig. 8.37** Hexagonal lattice with $\alpha_1 = 1.0, \alpha_2 = 0.5, T_{B_1} = 1.20R_0$ and $T_{B_2} = 1.20R_0$. PTSs here used are: (3830, 3840, 3850, 3860, 3880, 6000). We show some PTS of the simulation: **a**, **b** before the fracture, **c** during the fracture, **d**, **e** after the fracture; **f** the final configuration

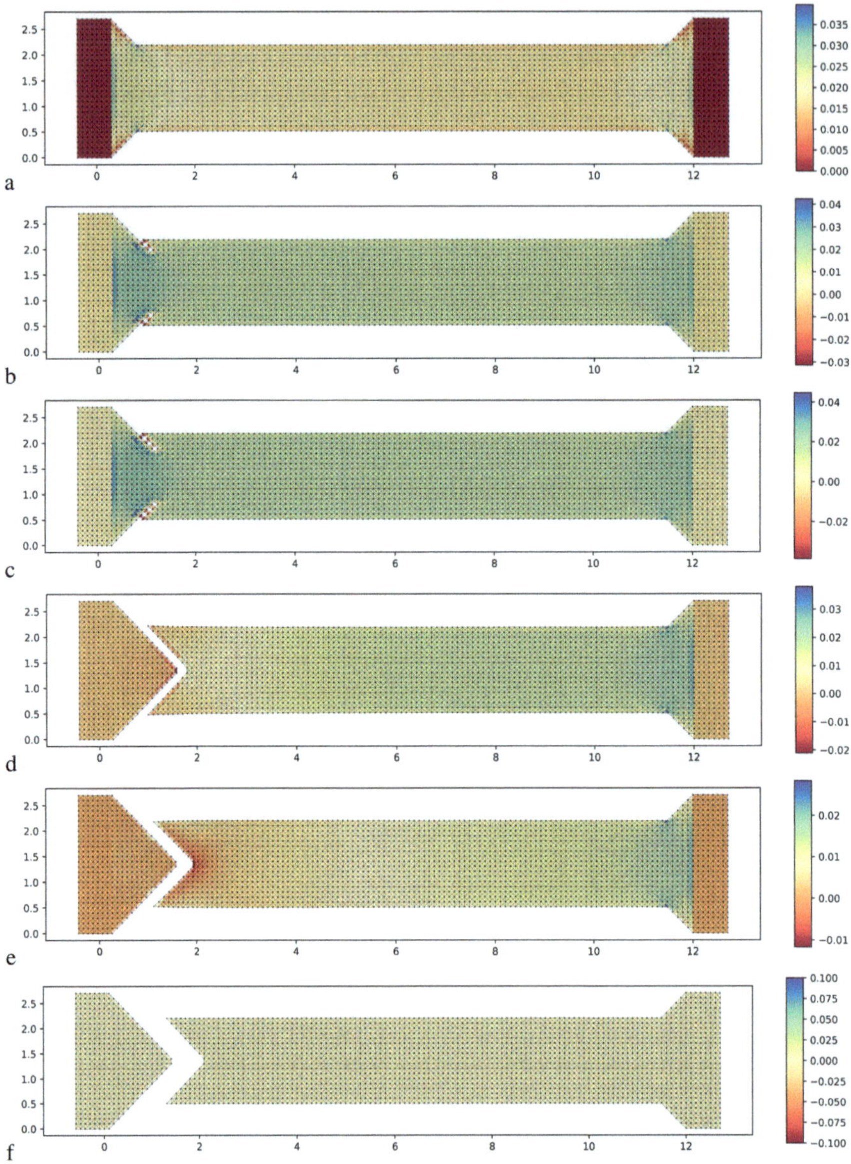


**Fig. 8.38** Squared lattice with $\alpha_1 = 1.0$, $\alpha_2 = 0.5$, $T_{B_1} = 1.05R_0$ and $T_{B_2} = 1.05R_0$. PTSs here used are: (1000, 1040, 1050, 1060, 1100, 3000). We show some PTS of the simulation: **a–c** before the fracture, **d** during the fracture, **e** after the fracture; **f** the final configuration

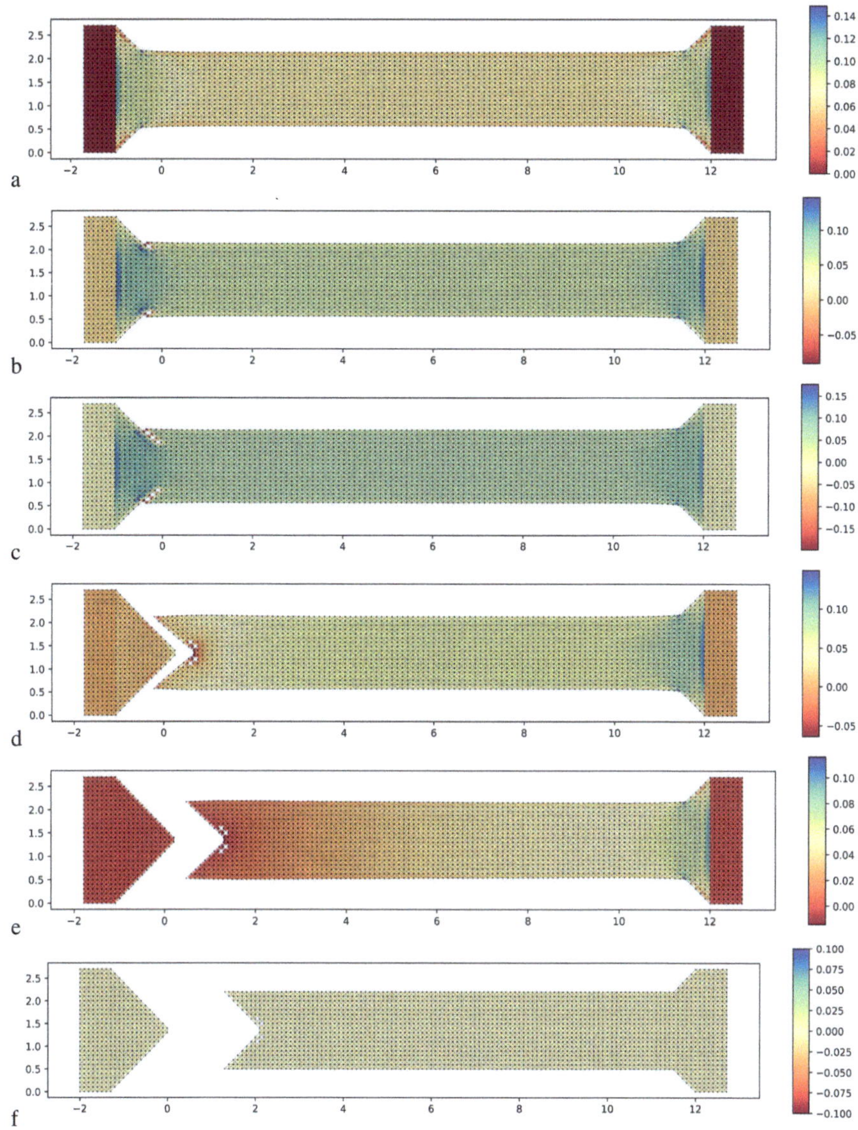


**Fig. 8.39** Squared lattice with $\alpha_1 = 1.0$, $\alpha_2 = 0.5$, $T_{B_1} = 1.20R_0$ and $T_{B_2} = 1.20R_0$. PTSs here used are: (4280, 4320, 4430, 4440, 4480, 7000). We show some PTS of the simulation: **a–c** before the fracture, **d** during the fracture, **e** after the fracture; **f** the final configuration

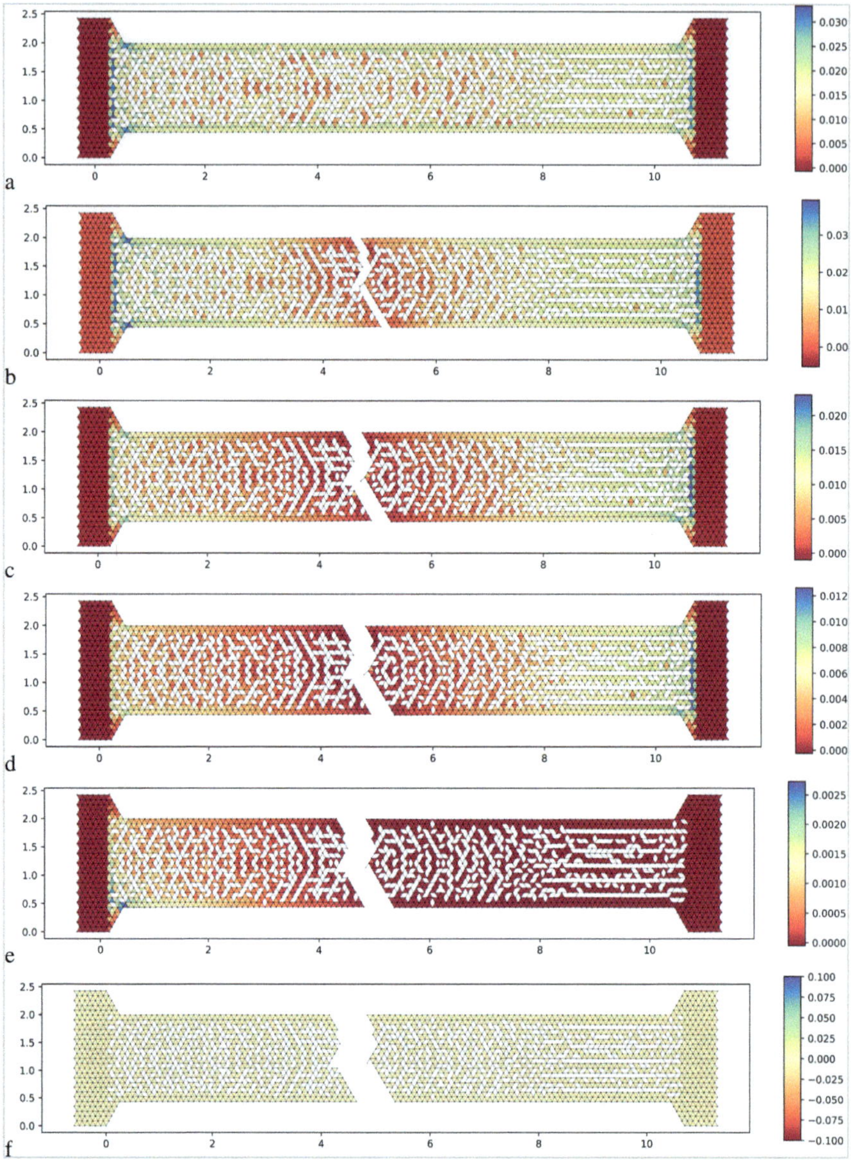


**Fig. 8.40** Hexagonal lattice with $\alpha_1 = 1.9, \alpha_2 = 1.1, T_{B_1} = 1.05R_0$ and $T_{B_2} = 1.05R_0$. PTSs here used are: (800, 940, 950, 960, 1100, 2500). We show some PTS of the simulation: **a** before the fracture, **b**, **c** during the fracture, **d**, **e** after the fracture; **f** the final configuration

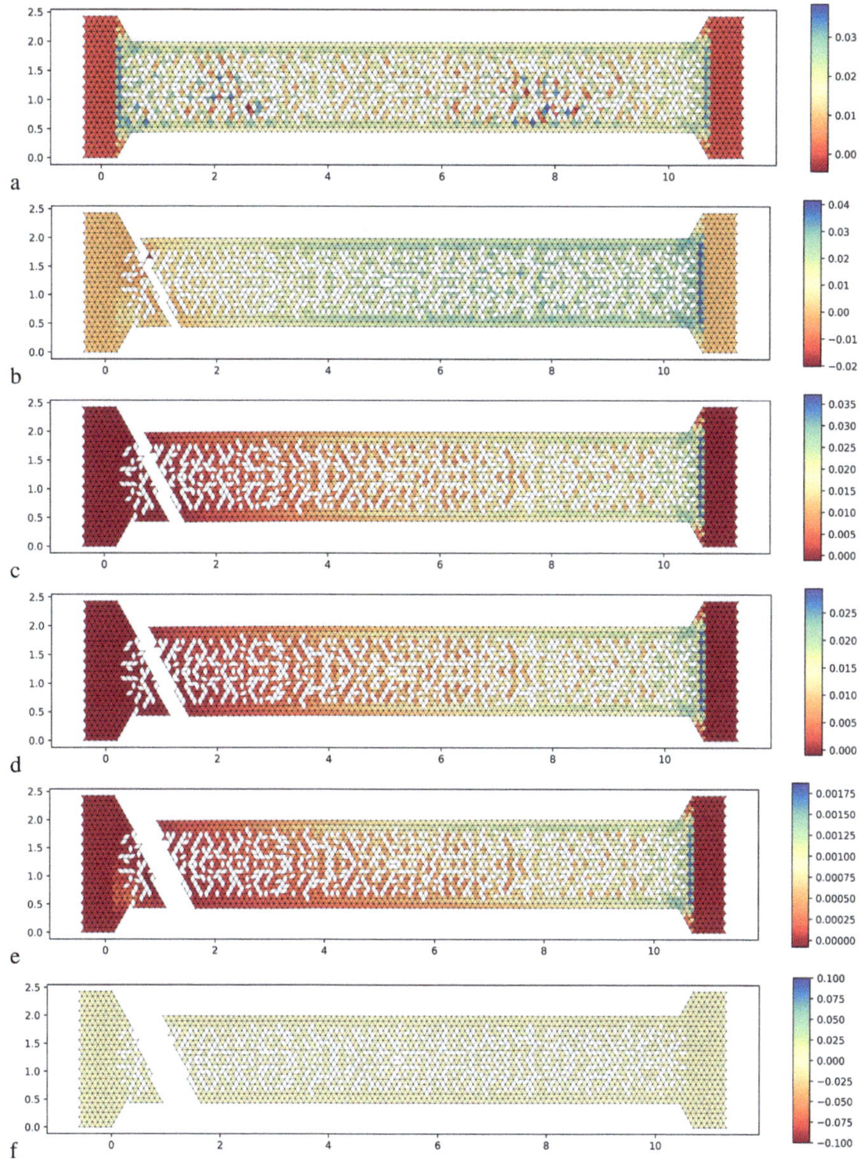


**Fig. 8.41** Hexagonal lattice with $\alpha_1 = 1.3, \alpha_2 = 1.5, T_{B_1} = 1.05R_0$ and $T_{B_2} = 1.05R_0$. PTSs here used are: (800, 970, 980, 990, 1100, 2500). We show some PTS of the simulation: **a** before the fracture, **b** during the fracture, **c**–**e** after the fracture; **f** the final configuration

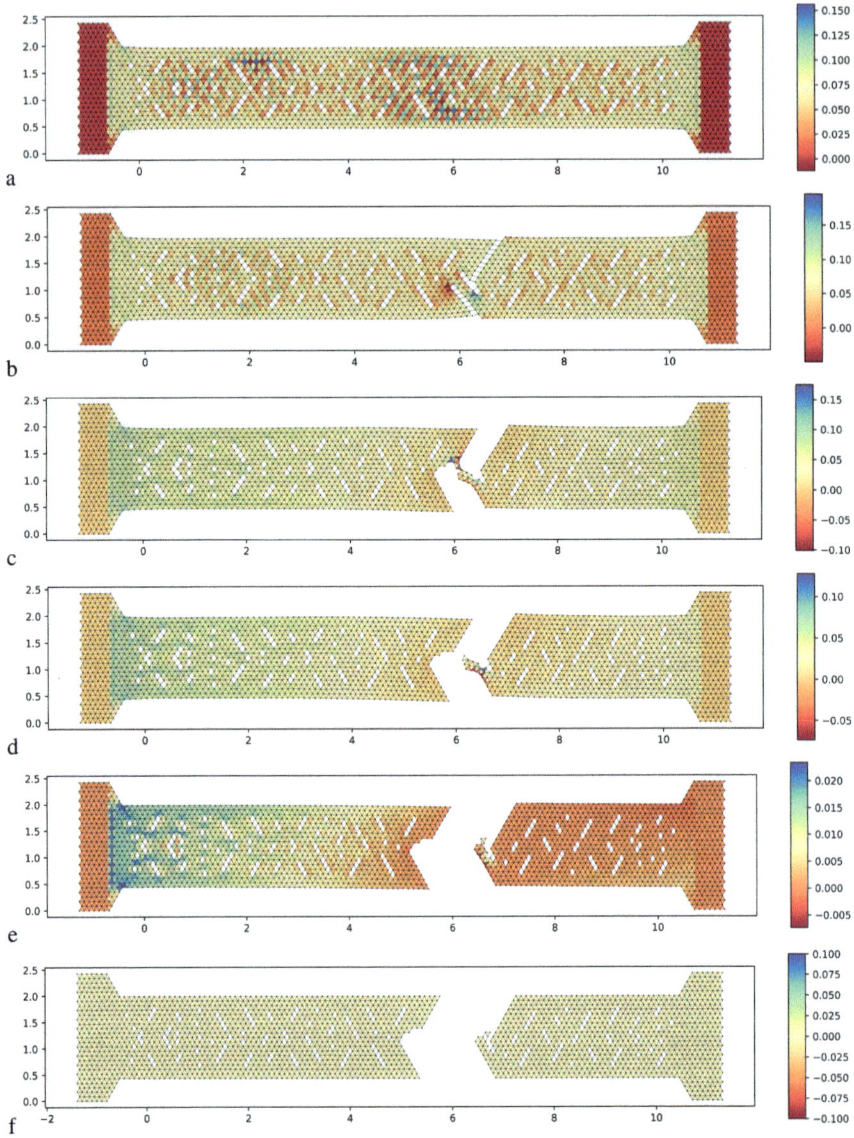


**Fig. 8.42** Hexagonal lattice with $\alpha_1 = 1.9, \alpha_2 = 0.3, T_{B_1} = 1.20R_0$ and $T_{B_2} = 1.20R_0$. PTSs here used are: (3000, 3170, 3180, 3190, 3250, 7000). We show some PTS of the simulation: **a**, **b** before the fracture, **c** during the fracture, **d**, **e** after the fracture; **f** the final configuration

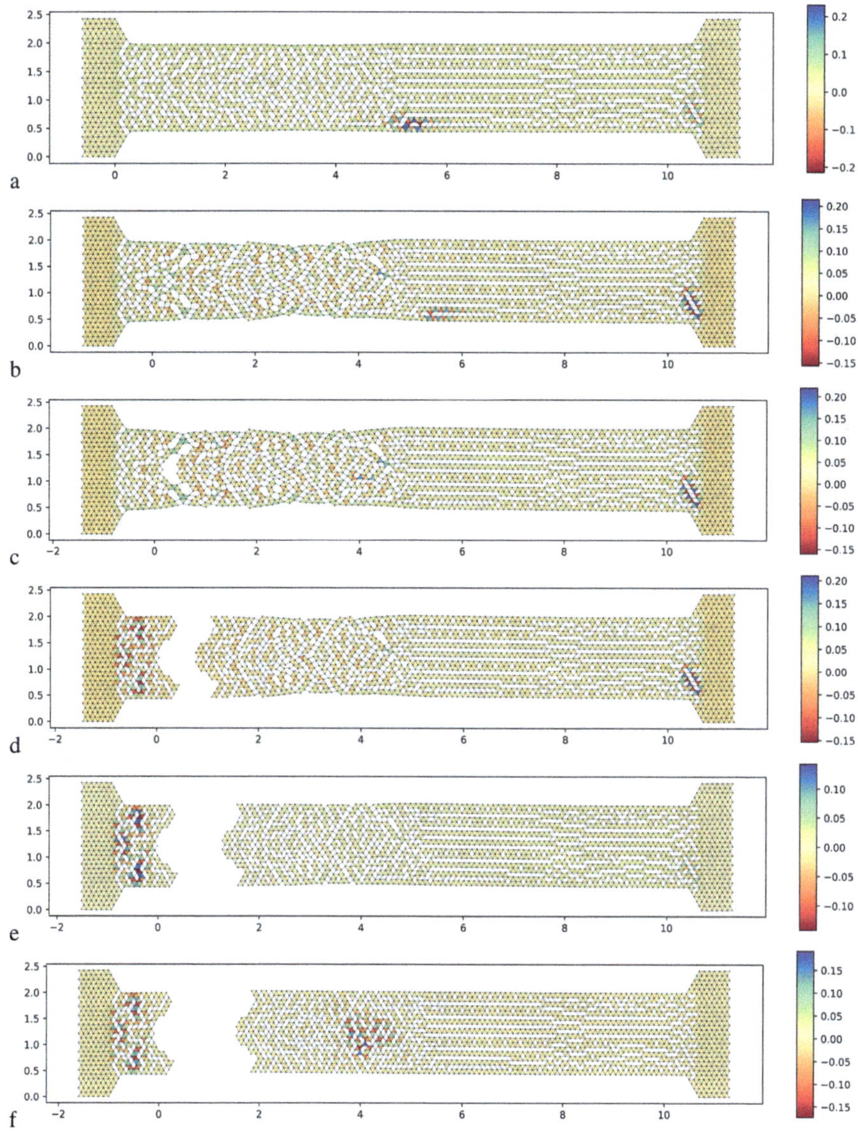


**Fig. 8.43** Hexagonal lattice with $\alpha_1 = 2.5$, $\alpha_2 = 0.1$, $T_{B_1} = 1.20R_0$ and $T_{B_2} = 1.20R_0$. PTSs here used are: (1500, 3400, 3600, 3650, 3800, 7000). We show some PTS of the simulation: **a** before the fracture, **b**, **c** during the fracture, **d**, **e** after the fracture; **f** the final configuration

Also, in the simulations in Figs. 8.43 and 8.45, different patterns are made on the left and right side; in the place around the fracture, an expansive chessboard pattern is formed (Fig. 8.45b–d). This kind of fracture reminds a yielding due to micro-fractures (Fig. 8.45c).

The fracture in Fig. 8.46 is very similar to the one in Fig. 8.39, except for the initial broken bound pattern; it is surprising that the fracture has occurred on the swarm's right side. The fracture is fragile.

In the last simulation, Fig. 8.47, the swarm evolution reminds an elastic/soft material. A high-deformation area with a strong Poisson effect occurs (Fig. 8.47b) where the fracture from the inner side to the outside is made (Fig. 8.47c). Anyway, in the end the swarm's structure is not compromised; so, the fracture is fragile.

Extreme Behaviors

In the previous two paragraphs, no ductile or completely ductile simulations have been shown; this is due to the used parameters: We have observed that the second neighbor's interaction tends to strengthen the structural integrity. Considering this, we have performed some simulation with low $\alpha_2$ and $T_{B_2}$ values, trying to obtain ductile fractures. Below, we will show some interesting results, in which extreme behaviors occur.

The simulations in Figs. 8.48 and 8.49 show a break pattern as in Fig. 8.43, i.e., different from the right to the left swarm's sides, with the presence of horizontal lines, so the evolution is very similar to that; evidently, in these traction tests, the left side is more inclined to the fracture respect to the right one with horizontal lines. The highest $\alpha_2$ parameter in Fig. 8.48 causes a greater fragmentation respect to Fig. 8.49. In both simulations arise a completely ductile fracture but is interesting to note that in Fig. 8.49, the swarm assume a curved shape in the equilibrium configuration.

The simulation shown in Fig. 8.50 is very interesting because it reminds the ASTM's sample fracture. The fracture is ductile and occurs in the area where there is most local deformation.

The simulation shown in Fig. 8.51 is like those in Figs. 8.34 and 8.35, i.e., it reminds a ripped fabric. It is too interesting that, although high local deformation occurs (Fig. 8.50c), the swarm partially recovers the shape, showing again his resilience to ductility of the squared lattice. Anyway, the fracture is completely ductile.

## 8.4 Conclusions

From tensile simulation provided, we can deduce that the two lattices analyzed show different behaviors: Each one is characterized by a specific breaking pattern.

The hexagonal lattice is characterized by a breaking pattern in which horizontal lines appear, and it easily undergoes a ductile fracture; it usually does not recover its original shape (Figs. 8.25, 8.27, 8.32, 8.33, 8.48 and 8.49). Instead, the squared lattice is characterized by a breaking pattern in which "chessboard pattern" occurs.

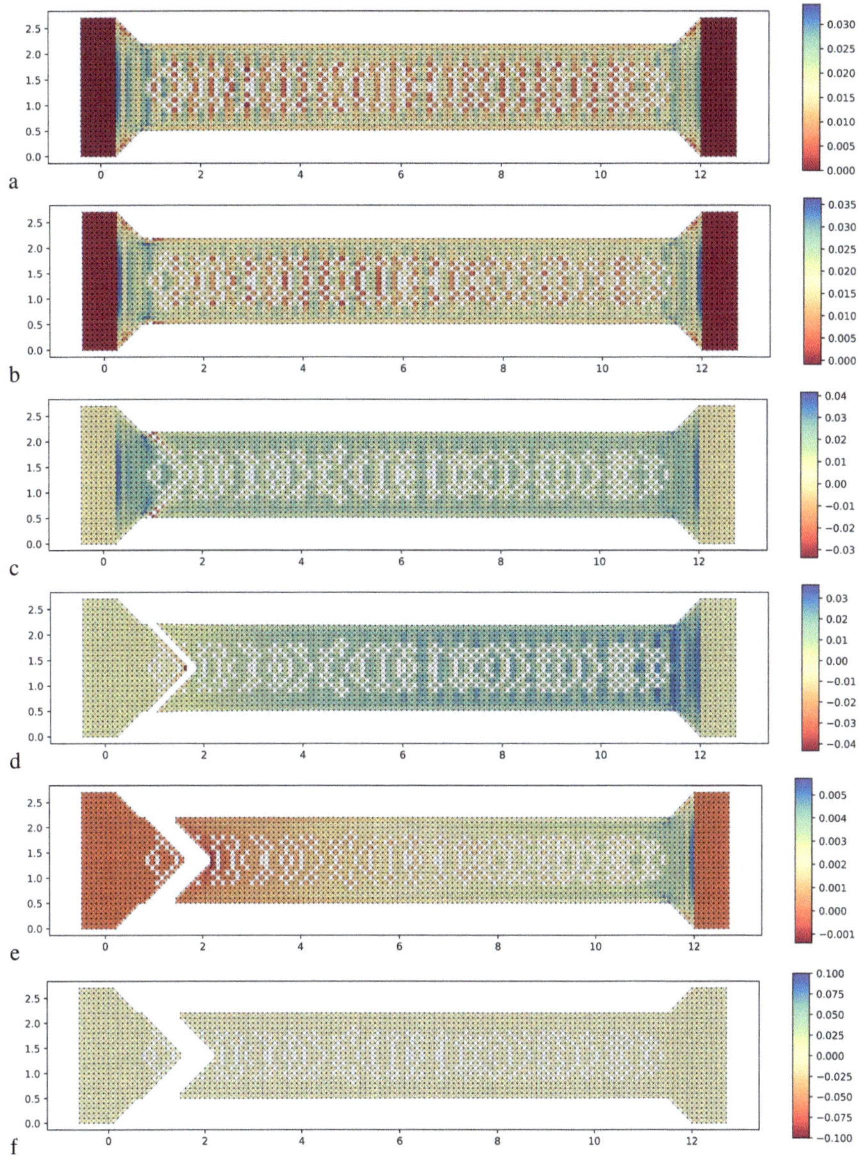


**Fig. 8.44** Squared lattice with $\alpha_1 = 1.1$, $\alpha_2 = 1.5$, $T_{B_1} = 1.05R_0$ and $T_{B_2} = 1.05R_0$. PTSs here used are: (1000, 1100, 1110, 1120, 1200, 2500). We show some PTS of the simulation: **a–c** before the fracture, **d** during the fracture, **e** after the fracture; **f** the final configuration

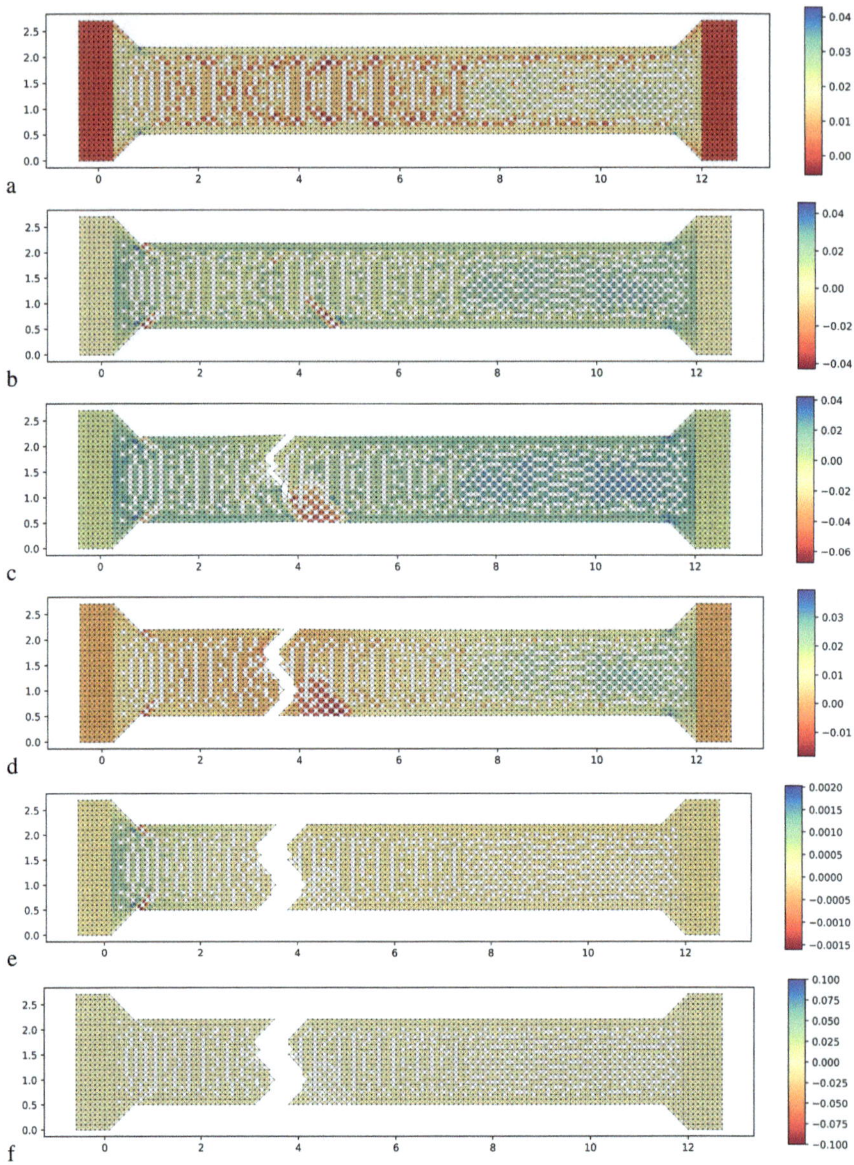


**Fig. 8.45** Squared lattice with $\alpha_1 = 2.1$, $\alpha_2 = 0.5$, $T_{B_1} = 1.05R_0$ and $T_{B_2} = 1.05R_0$. PTSs here used are: (1000, 1120, 1130, 1140, 1400, 2500). We show some PTS of the simulation: **a**, **b** before the fracture, **c** during the fracture, **d**, **e** after the fracture; **f** the final configuration

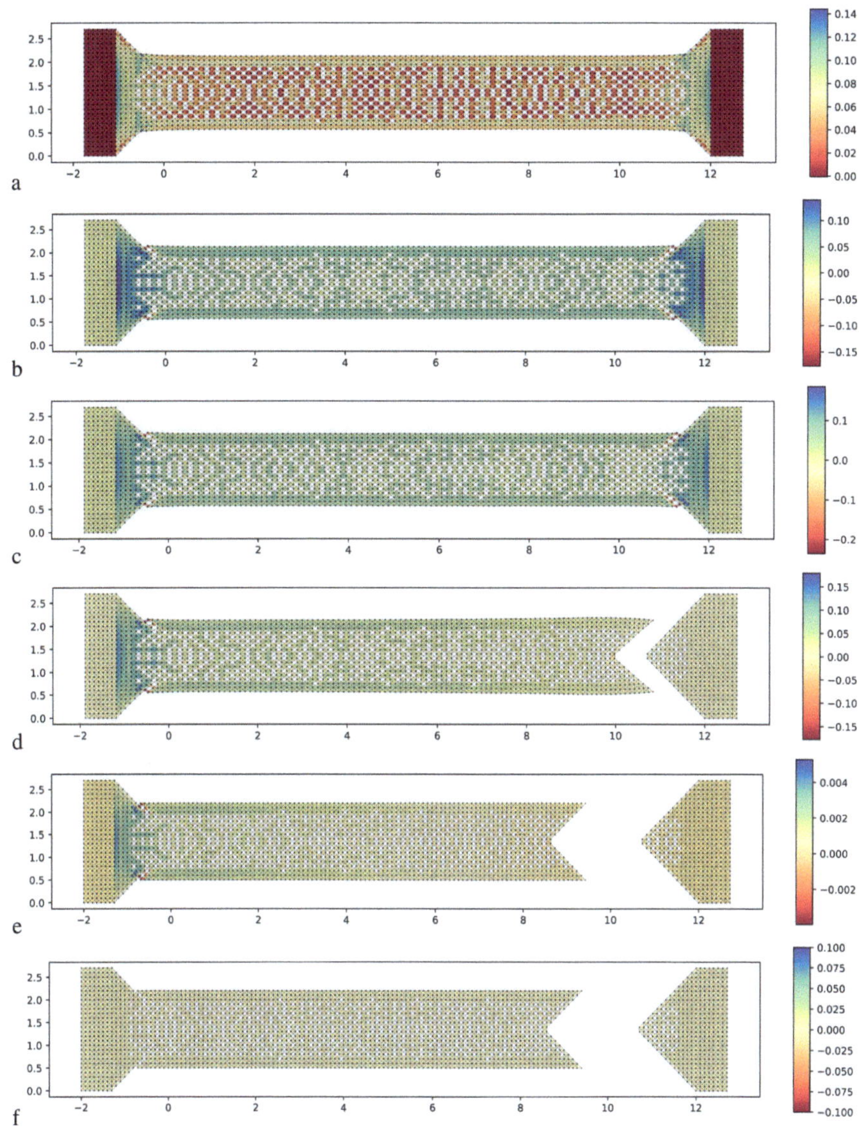


**Fig. 8.46** Squared lattice with $\alpha_1 = 1.5$, $\alpha_2 = 1.1$, $T_{B_1} = 1.20R_0$ and $T_{B_2} = 1.20R_0$. PTSs here used are: (4400, 4510, 4700, 4710, 5000, 7500). We show some PTS of the simulation: **a–c** before the fracture, **d** during the fracture, **e** after the fracture; **f** the final configuration

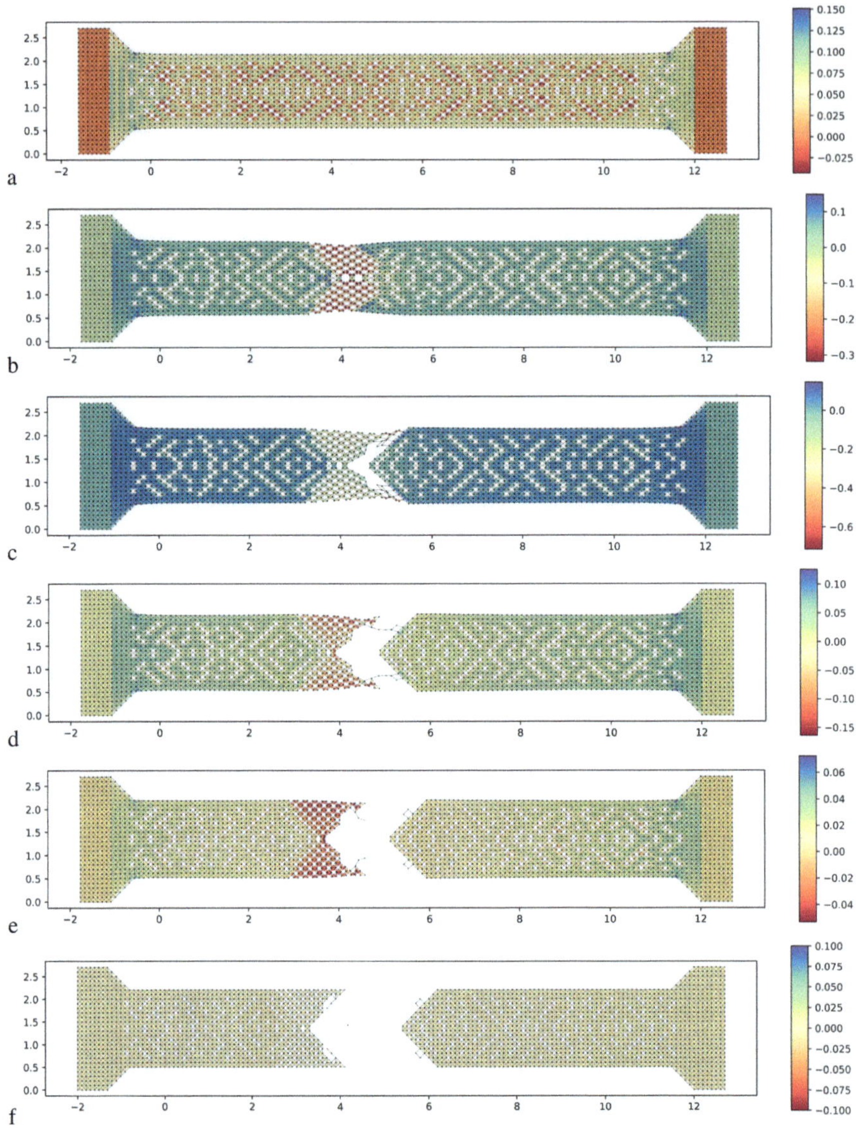


**Fig. 8.47** Squared lattice with $\alpha_1 = 1.9$, $\alpha_2 = 0.3$, $T_{B_1} = 1.20R_0$ and $T_{B_2} = 1.20R_0$. PTSs here used are: (4040, 4380, 4400, 4410, 4430, 7500). We show some PTS of the simulation: **a–c** before the fracture, **d** during the fracture, **e** after the fracture; **f** the final configuration

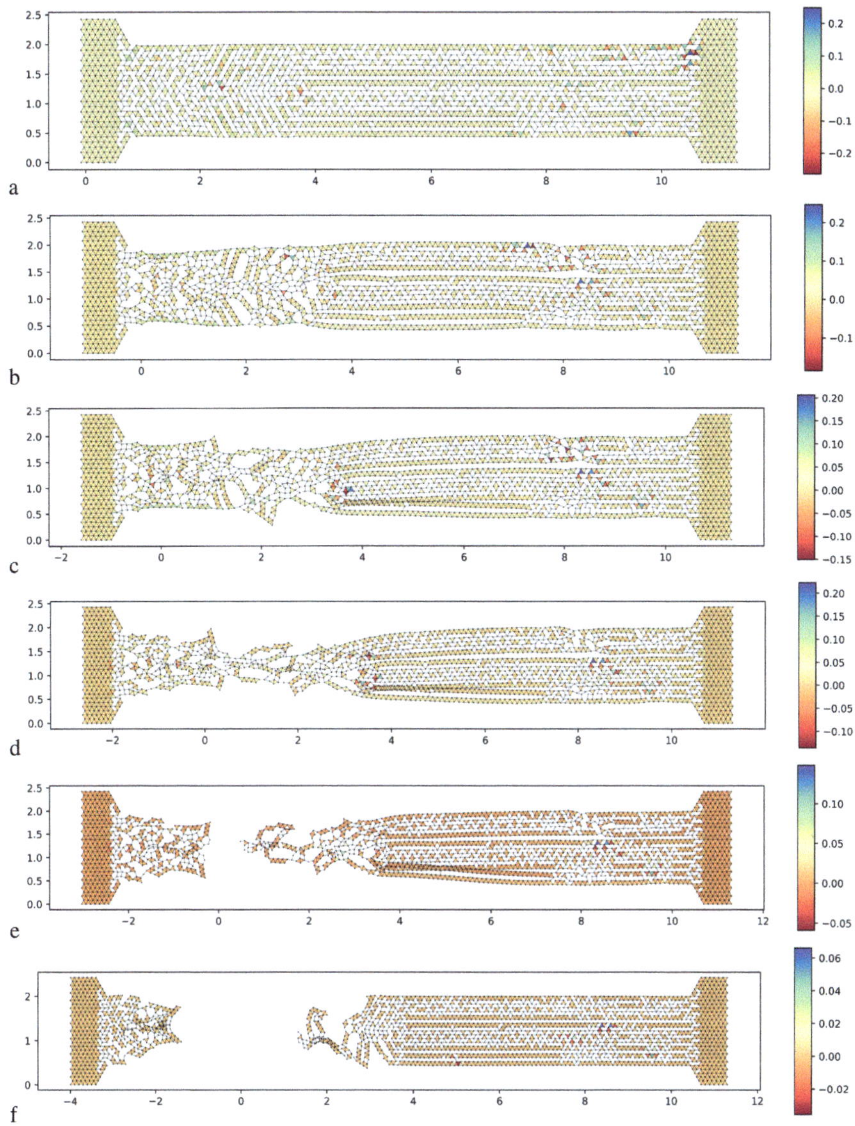


**Fig. 8.48** Hexagonal lattice with $\alpha_1 = 3.1$, $\alpha_2 = 0.05$, $T_{B_1} = 1.20R_0$ and $T_{B_2} = 1.05R_0$. PTSs here used are: (200, 2800, 4000, 6600, 7600, 15,000). We show some PTS of the simulation: **a** before the fracture, **b–d** during the fracture, **e** after the fracture, **f** the final configuration

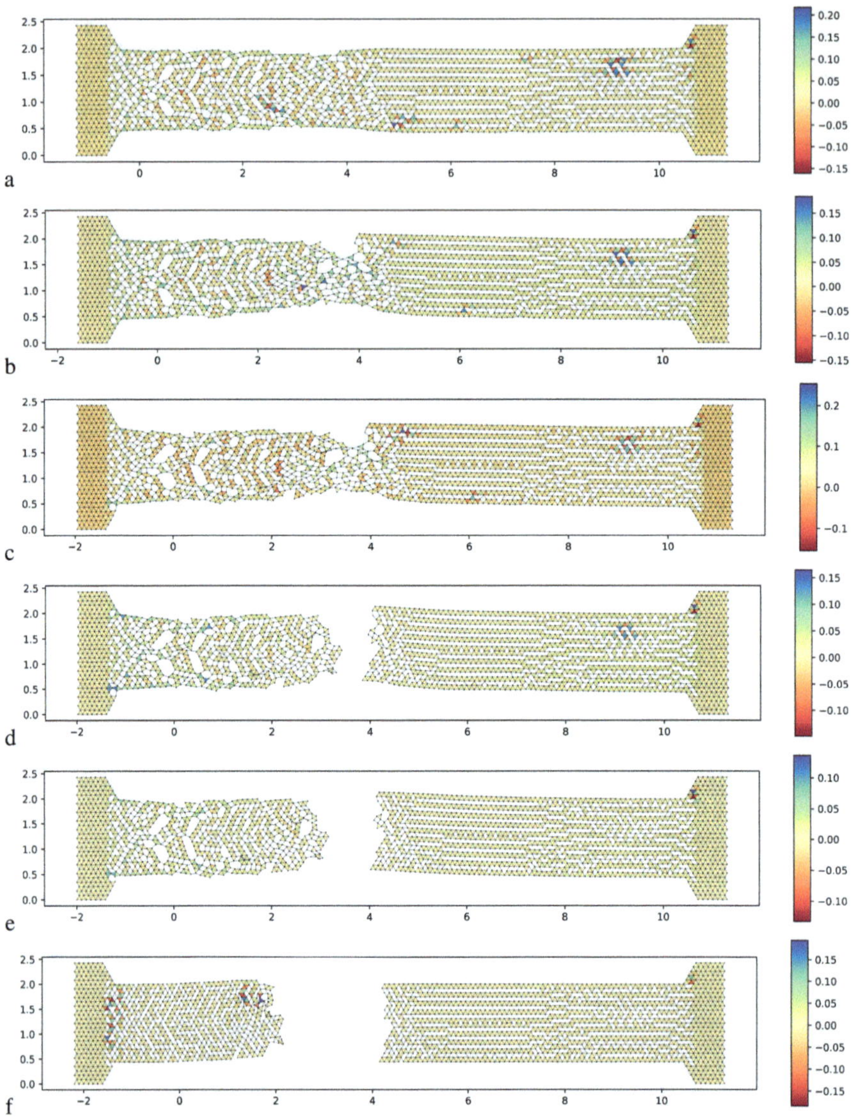

**Fig. 8.49** Hexagonal lattice with $\alpha_1 = 2.6$, $\alpha_2 = 0.1$, $T_{B_1} = 1.20R_0$ and $T_{B_2} = 1.05R_0$. PTSs here used are: (3100, 4000, 4900, 4950, 5000, 15,000). We show some PTS of the simulation: **a** before the fracture, **b**, **c** during the fracture, **d**, **e** after the fracture, **f** the final configuration

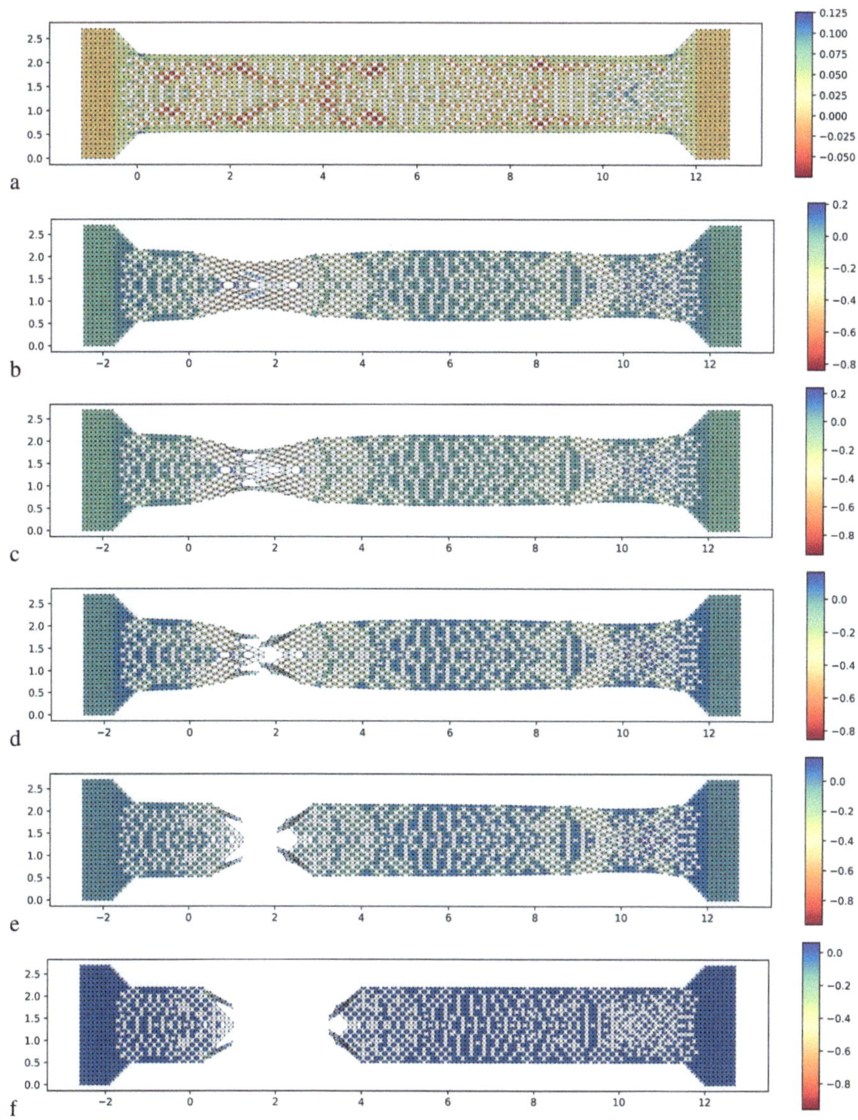


**Fig. 8.50** Squared lattice with $\alpha_1 = 2.2$, $\alpha_2 = 0.05$, $T_{B_1} = 1.20R_0$ and $T_{B_2} = 1.05R_0$. PTSs here used are: (3000, 6100, 6170, 6180, 6200, 10,000). We show some PTS of the simulation: **a** before the fracture, **b**, **c** during the fracture, **d**, **e** after the fracture, **f** the final configuration

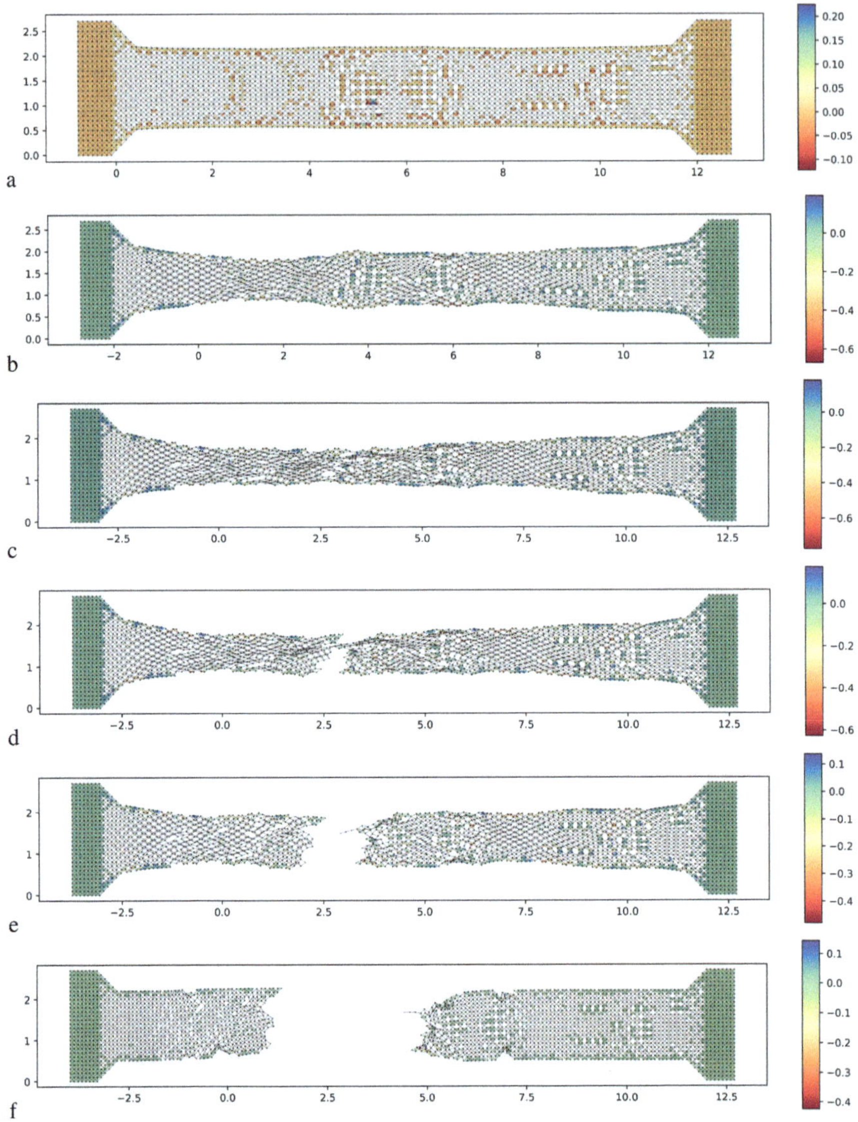


**Fig. 8.51** Squared lattice with $\alpha_1 = 2.9$, $\alpha_2 = 0.05$, $T_{B_1} = 1.20R_0$ and $T_{B_2} = 1.05R_0$. PTSs here used are: (2000, 7000, 9200, 9340, 9400, 15,000). We show some PTS of the simulation: **a** before the fracture, **b**, **c** during the fracture, **d**, **e** after the fracture, **f** the final configuration

Furthermore, although it undergoes high local deformation that cause the breaking of many bonds, the squared lattice seems to be resilient to ductile fracture because it can efficiently recover its original shape (Figs. 8.29, 8.30, 8.31, 8.35, 8.47, 8.50 and 8.51).

This resilience to ductility of the squared lattice is probably due to the following reasons:

- It has a coordination number higher than the hexagonal one, so the points are more linked each other and a more resistant structure arises.
- It has the cross-diagonal bounds that sustain horizontal and vertical bounds.
- Diagonal bounds are longer than the horizontal and vertical ones, so in the traction testes the horizontal ones break first and in average only the vertical and diagonal bounds alive (making the chessboard pattern). The diagonal bounds flatten out on the traction direction allowing the swarm to sustain higher deformation and to recover its original shape efficiently.

In simulations provided in Sect. 8.3, in which second neighbor interaction is considered, we can note that does not easily arise a ductile fracture. This is probably due to second neighbor interaction that strengthens the structural integrity of the lattice, also with low value for $\alpha_2$. This is true for both lattices analyzed.

We can sum up the advantage of our model in:

- **The algorithm is fast (despite the fracture introduction)**
  The fracture introduction does not affect the algorithm speed, compared to the one presented in D'Avanzo et al. (2021) some optimizations have been done. The velocity that it can reaches depend on the number of the neighbors considered in the computation of the interactions: In the better case, i.e., with hexagonal lattice and first neighbors' interaction (maximum 6 neighbors), it can compute about 100,000 step per second; in the worst case, i.e., with squared lattice and second neighbors' interaction (maximum 24 neighbors), it can compute about 25,000 step per second. The time taken by the simulations provided is from 0.5 to 3 min.
- **The algorithm complexity is linear**
  The algorithm complexity is linear, indeed for every point (for a swarm that contains $N$ points) the interaction is evaluated with a set of neighbors which number is fixed once the lattice is chosen. So, if the number of the mean considered neighbors is $N_k$, every step the algorithm evaluates $N \times N_k$ interactions and so its complexity is linear $O(N)$.
- **It can qualitatively reproduce a lot of phenomena**
  Our model is high suitable in different kind of situations: In fact, changing the model's parameters (i.e., $\alpha_1$, $\alpha_2$, $\beta$, $\gamma$, $\omega_k$, $N$, the kind of lattice) is possible to obtain a high variety of different responses.
- **The superposition is more manageable**
  The $\gamma$-frame can prevent the superposition problem more efficiently than the alignment term $\mathbf{B}_{ij}$. As just anticipated, the $\gamma$ parameter modulates the information's propagation velocity along the lattice it could be useful to characterize the kind of material that one wants to simulate.

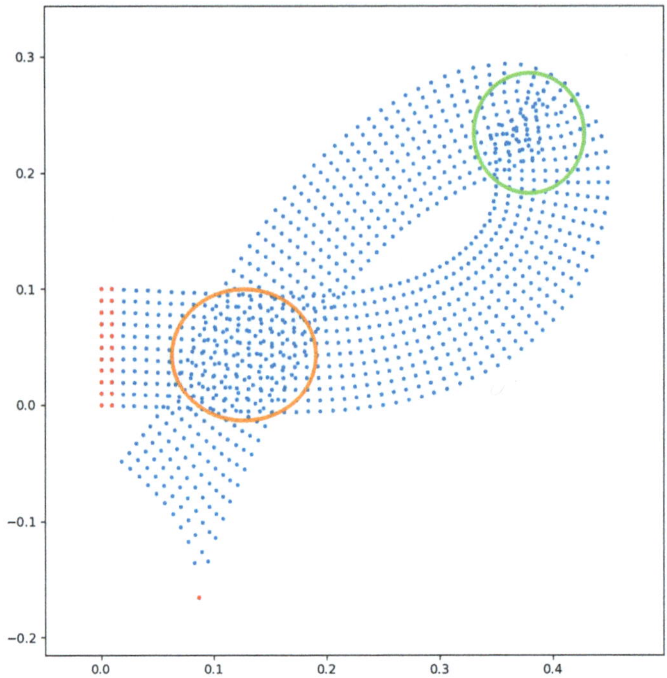


**Fig. 8.52** In orange circle, the auto-interpenetration problem occurs, meanwhile in green circle we can see the superposition problem

However our model, as every other model, has also some disadvantages that can be summarized in:

- **Now it can only supply qualitative results**
  There isn't yet a connection between the constitutive parameters of continua and the parameters of the model. This connection will be investigated in future works.
- **The problem of auto-interpenetration still remains**
  In this paper, there is no attempt to avoid auto-interpenetration of the swarm's pieces far each other, e.g., if in a beam the two opposite edges are stressed in a way that they occupy the same region of the space (see Fig. 8.52).

## References

Abali BE, Völlmecke C, Woodward B, Kashtalyan M, Guz I, Müller WH (2012) Numerical modeling of functionally graded materials using a variational formulation. Continuum Mech Thermodyn 24(4–6):377–390

Abali BE, Wu C-C, Müller WH (2016) An energy-based method to determine material constants in nonlinear rheology with applications. Continuum Mech Thermodyn 28(5):1221–1246

Abdoul-Anziz H, Seppecher P, Bellis C (2019) Homogenization of frame lattices leading to second gradient models coupling classical strain and strain-gradient terms. Math Mech Solids 24(12):3976–3999

Alibert J-J, Seppecher P, dell'Isola F (2003) Truss modular beams with deformation energy depending on higher displacement gradients. Math Mech Solids 8(1):51–73

Altenbach H, Eremeyev V (2015) On the constitutive equations of viscoelastic micropolar plates and shells of differential type. Math Mech Complex Syst 3(3):273–283

Altenbach H, Eremeyev VA (2008a) Direct approach-based analysis of plates composed of functionally graded materials. Arch Appl Mech 78(10):775–794

Altenbach H, Eremeyev VA (2008b) Analysis of the viscoelastic behavior of plates made of functionally graded materials. ZAMM-Z Angew Math Mech 88(5):332–341

Altenbach H, Eremeyev VA (2011) On the shell theory on the nanoscale with surface stresses. Int J Eng Sci 49(12):1294–1301

Altenbach H, Eremeyev VA, Naumenko K (2015) On the use of the first order shear deformation plate theory for the analysis of three-layer plates with thin soft core layer. ZAMM-Z Angew Math Mech 95(10):1004–1011

Altenbach J, Altenbach H, Eremeyev VA (2010) On generalized Cosserat-type theories of plates and shells: a short review and bibliography. Arch Appl Mech 80(1):73–92

Andreaus U, dell'Isola F, Giorgio I, Placidi L, Lekszycki T, Rizzi NL (2016) Numerical simulations of classical problems in two-dimensional (non) linear second gradient elasticity. Int J Eng Sci 108:34–50

Andreaus U, Spagnuolo M, Lekszycki T, Eugster SR (2018) A Ritz approach for the static analysis of planar pantographic structures modeled with nonlinear Euler-Bernoulli beams. Continuum Mech Thermodyn 30(5):1103–1123

Avella M, Dell'Erba R, d'Orazio L, Martuscelli E (1995) Influence of molecular weight and molecular weight distribution on crystallization and thermal behavior of isotactic polypropylene. Polym Netw Blends 5:47–54

Avella M, Dell'Erba R, Martuscelli E, Partch R (2000) Thermosetting based composites reinforced with silicon carbide whiskers. J Polym Mater 17:445–458

Barchiesi E, Eugster SR, Placidi L, dell'Isola F (2019a) Pantographic beam: a complete second gradient 1D-continuum in plane. Z Angew Math Phys 70(5):135

Barchiesi E, Ganzosch G, Liebold C, Placidi L, Grygoruk R, Müller WH (2019b) Out-of-plane buckling of pantographic fabrics in displacement-controlled shear tests: experimental results and model validation. Continuum Mech Thermodyn 31(1):33–45

Barchiesi E, Spagnuolo M, Placidi L (2019c) Mechanical metamaterials: a state of the art. Math Mech Solids 24(1):212–234

Barchiesi E, dell'Isola F, Hild F, Seppecher P (2020a) Two-dimensional continua capable of large elastic extension in two independent directions: asymptotic homogenization, numerical simulations and experimental evidence. Mech Res Commun 103:103466

Barchiesi E, Eugster SR, dell'Isola F, Hild F (2020b) Large in-plane elastic deformations of bi-pantographic fabrics: asymptotic homogenization and experimental validation. Math Mech Solids 25(3):739–767

Barchiesi E, dell'Isola F, Bersani AM, Turco E (2021a) Equilibria determination of elastic articulated duoskelion beams in 2D via a Riks-type algorithm. Int J Non-Linear Mech 128:1–24

Barchiesi E, dell'Isola F, Hild F (2021b) On the validation of homogenized modeling for bi-pantographic metamaterials via digital image correlation. Int J Solids Struct 208:49–62

Baroudi D, Giorgio I, Battista A, Turco E, Igumnov LA (2019) Nonlinear dynamics of uniformly loaded elastica: experimental and numerical evidence of motion around curled stable equilibrium configurations. ZAMM-Z Angew Math Mech 99(7):e201800121

Battista A, Rosa L, Dell'Erba R, Greco L (2016) Numerical investigation of a particle system compared with first and second gradient continua: deformation and fracture phenomena. Math Mech Solids 22. https://doi.org/10.1177/1081286516657889

Battista A, D'Avanzo P, Laudato M (2018) Discrete systems with geometry-driven evolution: application to 1D elasticity and granular media. Mech Res Commun 92:107–110. ISSN 0093-6413. https://doi.org/10.1016/j.mechrescom.2018.08.006. URL https://www.sciencedirect.com/science/article/pii/S009364131830106X

Bernalte Sínchez P, Papaelias M, Garcáa Mírquez FP (2020) Autonomous underwater vehicles: instrumentation and measurements. IEEE Instrum Meas Mag 23:105–114. https://doi.org/10.1109/MIM.2020.9062680

Bilotta A, Morassi A, Turco E (2018) The use of quasi-isospectral operators for damage detection in rods. Meccanica 53(1):319–345

Bilotta A, Morassi A, Rosset E, Turco E, Vessella S (2019) Numerical size estimates of inclusions in Kirchhoff-Love elastic plates. Int J Solids Struct 168:58–72

Bolzon G, Fedele R, Maier G (2002) Parameter identification of a cohesive crack model by Kalman filter. Comput Methods Appl Mech Eng 191(25–26):2847–2871

Boutin C, dell'Isola I, Giorgio F, Placidi L (2017) Linear pantographic sheets: asymptotic micro-macro models identification. Math Mech Complex Syst 5(2):127–162

Browning F, Askes H (2019) Analytical solutions for the natural frequencies of rectangular symmetric angle-ply laminated plates. Math Mech Complex Syst 7(1):51–61

Casale L, Avella M, Dell'Erba R, Martuscelli E (1998) Broom fibres as reinforcements for thermoplastic matrices. Macromol Symp 127(1):211–218. https://doi.org/10.1002/masy.19981270128. URL https://hal.archives-ouvertes.fr/hal-01998469

Cazzani A, Ruge P (2016) Stabilization by deflation for sparse dynamical systems without loss of sparsity. Mech Syst Signal Process 70:664–681

Cazzani A, Malagù M, Turco E (2016a) Isogeometric analysis of plane-curved beams. Math Mech Solids 21(5):562–577

Cazzani A, Malagù M, Turco E, Stochino F (2016b) Constitutive models for strongly curved beams in the frame of isogeometric analysis. Math Mech Solids 21(2):182–209

Cazzani A, Stochino F, Turco E (2016c) An analytical assessment of finite element and isogeometric analyses of the whole spectrum of Timoshenko beams. ZAMM-Z Angew Math Mech 96(10):1220–1244

Cazzani A, Rizzi NL, Stochino F, Turco E (2018a) Modal analysis of laminates by a mixed assumed-strain finite element model. Math Mech Solids 23(1):99–119

Cazzani A, Serra M, Stochino F, Turco E (2018b) A refined assumed strain finite element model for statics and dynamics of laminated plates. Continuum Mech Thermodyn 1–28. https://doi.org/10.1007/s00161-018-0707-x

Contrafatto L, Cuomo M (2002) A new thermodynamically consistent continuum model for hardening plasticity coupled with damage. Int J Solids Struct 39(25):6241–6271

Cuomo M (2017) Forms of the dissipation function for a class of viscoplastic models. Math Mech Complex Syst 5(3):217–237

Cuomo M (2019) Continuum damage model for strain gradient materials with applications to 1D examples. Continuum Mech Thermodyn 31(4):969–987

Cuomo M, Contrafatto L, Greco L (2014) A variational model based on isogeometric interpolation for the analysis of cracked bodies. Int J Eng Sci 80:173–188

Cuomo M, dell'Isola F, Greco L (2016) Simplified analysis of a generalized bias test for fabrics with two families of inextensible fibres. Z Angew Math Phys 67(3):61

D'Avanzo P, Rapisarda AC, Sirletti SS (2021) From the swarm robotics to material deformations. Math Appl Continuum Struct Mech 127:105–114

De Angelo M, Barchiesi E, Giorgio I, Abali BE (2019a) Numerical identification of constitutive parameters in reduced-order bi-dimensional models for pantographic structures: application to out-of-plane buckling. Arch Appl Mech 89(7):1333–1358

De Angelo M, Spagnuolo M, D'Annibale F, Pfaff A, Hoschke K, Misra A, Dupuy C, Peyre P, Dirrenberger J, Pawlikowski M (2019b) The macroscopic behavior of pantographic sheets depends mainly on their microstructure: experimental evidence and qualitative analysis of damage in metallic specimens. Continuum Mech Thermodyn 31(4):1181–1203

De Angelo M, Placidi L, Nejadsadeghi N, Misra A (2020) Non-standard Timoshenko beam model for chiral metamaterial: identification of stiffness parameters. Mech Res Commun 103:103462
deir'Erba R, Moriconi C (2015) High power LEDs in localization of underwater robotics swarms. IFAC-PapersOnLine 48(10):117–122. ISSN 2405-8963. In: 2nd IFAC conference on embedded systems, computer intelligence and telematics CESCIT 2015. https://doi.org/10.1016/j.ifacol.2015.08.118. URL https://www.sciencedirect.com/science/article/pii/S2405896315009854
Del Piero G (2018) The variational structure of classical plasticity. Math Mech Complex Syst 6(3):137–180
Del Vescovo D, Giorgio I (2014) Dynamic problems for metamaterials: review of existing models and ideas for further research. Int J Eng Sci 80:153–172
Dell'Erba R, Dell'Isola F, Rotoli G (1995) The influence of the curvature dependence of the surface tension on the geometry of electrically charged menisci. Continuum Mech Thermodyn 11:47–54
Dell'Erba R (2001) Rheo-mechanical and rheo-optical characterisation of ultra high molecular mass poly(methylmethacrylate) in solution. Polymer 42:2655–2663. https://doi.org/10.1016/S0032-3861(00)00636-4
dell'Erba R (2015) Determination of spatial configuration of an underwater swarm with minimum data. Int J Adv Robot Syst 12(7):97. https://doi.org/10.5772/61035
Dell'Erba R (2018) Position-based dynamic of a particle system: a configurable algorithm to describe complex behaviour of continuum material starting from swarm robotics. Continuum Mechanics and Thermodynamics 30. https://doi.org/10.1007/s00161-018-0663-5
Dell'Erba R (2019a) On how swarm robotics can be used to describe particle system's deformation. Continuum Mech Thermodyn 11. https://doi.org/10.1007/s00161-019-00845-4
Dell'Erba R (2019b) Swarm robotics and complex behaviour of continuum material. Continuum Mech Thermodyn 31: 1–26. https://doi.org/10.1007/s00161-018-0675-1
Dell'Erba R (2020a) A tool to describe particle system evolution from swarm robotics behavior, pp 191–217. https://doi.org/10.1007/978-3-030-50460-1_13
Dell'Erba R (2020b) A plausible description of continuum material behavior derived by swarm robot flocking rules
Dell'Erba R (2020c) The distances measurement problem for an underwater robotic swarm: a semi-experimental trial, using power LEDs, in unknown sea water conditions. Continuum Mech Thermodyn 1–9. https://doi.org/10.1007/s00161-020-00923-y
Dell'Erba R (2021a) Flocking rules governing swarm robot as tool to describe continuum deformation, pp 223–243. https://doi.org/10.1007/978-3-030-53755-5_14
Dell'Erba R (2021b) Distance estimations in unknown sea underwater conditions by power LED for robotics swarms. Continuum Mech Thermodyn. https://doi.org/10.1007/s00161-020-00889-x. URL https://hal.archives-ouvertes.fr/hal-03119198
dell'Isola F, Andreaus U, Placidi L (2015a) At the origins and in the vanguard of peridynamics, non-local and higher-gradient continuum mechanics: an underestimated and still topical contribution of Gabrio Piola. Math Mech Solids 20(8):887–928
dell'Isola F, Lekszycki T, Pawlikowski M, Grygoruk R, Greco L (2015b) Designing a light fabric metamaterial being highly macroscopically tough under directional extension: first experimental evidence. Z Angew Math Phys 66(6):3473–3498
dell'Isola F, Seppecher P, Alibert J-J et al (2019) Pantographic metamaterials: an example of mathematically driven design and of its technological challenges. Continuum Mech Thermodyn 31(4):851–884
Desmorat B, Spagnuolo M, Turco E (2020) Stiffness optimization in nonlinear pantographic structures. Math Mech Solids 25(12):2252–2262
Epstein M, Smelser R (2020) An appreciation and discussion of Paul Germain's "the method of virtual power in the mechanics of continuous media, I: second-gradient theory." Math Mech Complex Syst 8(2):191–199
Eremeyev VA, Pietraszkiewicz W (2016) Material symmetry group and constitutive equations of micropolar anisotropic elastic solids. Math Mech Solids 21(2):210–221

Eremeyev VA, Turco E (2020) Enriched buckling for beam-lattice metamaterials. Mech Res Commun 103:103458
Eugster S, Hesch C, Betsch P, Glocker C (2014) Director-based beam finite elements relying on the geometrically exact beam theory formulated in skew coordinates. Int J Numer Methods Eng 97(2):111–129
Eugster S, dell'Isola F, Steigmann D (2019) Continuum theory for mechanical metamaterials with a cubic lattice substructure. Math Mech Complex Syst 7(1):75–98
Franciosi P, Spagnuolo M, Salman OU (2019) Mean Green operators of deformable fiber networks embedded in a compliant matrix and property estimates. Continuum Mech Thermodyn 31(1):101–132
Gagneux G, Millet O (2016) Modeling capillary hysteresis in unsatured porous media. Math Mech Complex Syst 4(1):67–77
George D, Allena R, Remond Y (2018) A multiphysics stimulus for continuum mechanics bone remodeling. Math Mech Complex Syst 6(4):307–319
George D, Allena R, Bourzac C, Pallu S, Bensidhoum M, Portier H, Rémond Y (2020) A new comprehensive approach for bone remodeling under medium and high mechanical load based on cellular activity. Math Mech Complex Syst 8(4):287–306
Germain P (2020) The method of virtual power in the mechanics of continuous media, I: second-gradient theory. Math Mech Complex Syst 8(2):153–190
Giorgio I (2020a) A discrete formulation of Kirchhoff rods in large-motion dynamics. Math Mech Solids 25(5):1081–1100
Giorgio I (2020b) Lattice shells composed of two families of curved Kirchhoff rods: an archetypal example, topology optimization of a cycloidal metamaterial. Continuum Mech Thermodyn. https://doi.org/10.1007/s00161-020-00955-4
Giorgio I, Del Vescovo D (2018) Non-linear lumped-parameter modeling of planar multi-link manipulators with highly flexible arms. Robotics 7(4):60
Giorgio I, Del Vescovo D (2019) Energy-based trajectory tracking and vibration control for multilink highly flexible manipulators. Math Mech Complex Syst 7(2):159–174
Giorgio I, Scerrato D (2017) Multi-scale concrete model with rate-dependent internal friction. Eur J Environ Civ Eng 21(7–8):821–839
Giorgio I, Andreaus U, Scerrato D, dell'Isola F (2016) A visco-poroelastic model of functional adaptation in bones reconstructed with bio-resorbable materials. Biomech Model Mechanobiol 15(5):1325–1343
Giorgio I, Andreaus U, dell'Isola F, Lekszycki T (2017a) Viscous second gradient porous materials for bones reconstructed with bio-resorbable grafts. Extreme Mech Lett 13:141–147
Giorgio I, Rizzi NL, Turco E (2017b) Continuum modelling of pantographic sheets for out-of-plane bifurcation and vibrational analysis. Proc R Soc A Math Phys Eng Sci 473(2207):20170636
Giorgio I, Harrison P, dell'Isola F, Alsayednoor J, Turco E (2018) Wrinkling in engineering fabrics: a comparison between two different comprehensive modelling approaches. Proc R Soc A Math Phys Eng Sci 474(2216):20180063
Giorgio I, De Angelo M, Turco E, Misra A (2019a) A Biot–Cosserat two-dimensional elastic nonlinear model for a micromorphic medium. Continuum Mech Thermodyn 1–13
Giorgio I, dell'Isola F, Andreaus U, Alzahrani F, Hayat T, Lekszycki T (2019b) On mechanically driven biological stimulus for bone remodeling as a diffusive phenomenon. Biomech Model Mechanobiol 18(6):1639–1663
Giorgio I, Rizzi NL, Andreaus U, Steigmann DJ (2019c) A two-dimensional continuum model of pantographic sheets moving in a 3D space and accounting for the offset and relative rotations of the fibers. Math Mech Complex Syst 7(4):311–325
Giorgio I, Ciallella A, Scerrato D (2020a) A study about the impact of the topological arrangement of fibers on fiber-reinforced composites: some guidelines aiming at the development of new ultra-stiff and ultra-soft metamaterials. Int J Solids Struct 203:73–83

Giorgio I, Spagnuolo M, Andreaus U, Scerrato D, Bersani AM (2020b) In-depth gaze at the astonishing mechanical behavior of bone: a review for designing bio-inspired hierarchical metamaterials. Math Mech Solids. https://doi.org/10.1177/1081286520978516

Giorgio I, Varano V, dell'Isola F, Rizzi NL (2021) Two layers pantographs: a 2D continuum model accounting for the beams' offset and relative rotations as averages in SO(3) Lie groups. Int J Solids Struct 216:43–58

Greco L (2020) An iso-parametric $G^1$-conforming finite element for the nonlinear analysis of Kirchhoff rod. Part I: the 2D case. Continuum Mech Thermodyn 1–24

Greco L, Cuomo M, Contrafatto L (2018) A reconstructed local B formulation for isogeometric Kirchhoff-Love shells. Comput Methods Appl Mech Eng 332:462–487

Javili A, Morasata R, Oterkus E, Oterkus S (2019) Peridynamics review. Math Mech Solids 24(11):3714–3739

Marini S, Gjeci N, Govindaraj S, But A, Sportich B, Ottaviani E, Mírquez FPG, Bernalte Sanchez PJ, Pedersen J, Clausen CV, Madricardo F, Foglini F, Bonofiglio F, Barbieri L, Antonini M, Montenegro Camacho YS, Weiss P, Nowak K, Peer M, Gobert T, Turetta A, Chatzidouros E, Lee D, Zarras D, Steriotis T, Charalambopoulou G, Yamas T, Papaelias M (2020) Enduruns: an integrated and flexible approach for seabed survey through autonomous mobile vehicles. J Mar Sci Eng 8(9). ISSN 2077-1312. https://doi.org/10.3390/jmse8090633. URL https://www.mdpi.com/2077-1312/8/9/633

Misra A, Poorsolhjouy P (2016a) Granular micromechanics based micromorphic model predicts frequency band gaps. Continuum Mech Thermodyn 28(1–2):215–234

Misra A, Poorsolhjouy P (2016b) Elastic behavior of 2D grain packing modeled as micromorphic media based on granular micromechanics. J Eng Mech 143(1):C4016005

Misra A, Nejadsadeghi N, De Angelo M, Placidi L (2020) Chiral metamaterial predicted by granular micromechanics: verified with 1D example synthesized using additive manufacturing. Continuum Mech Thermodyn 1–17

Moriconi C, Dell'Erba R (2014) Social dependability: a proposed evolution for future robotics. https://doi.org/10.12910/EAI2014-63

Müller M, Heidelberger B, Teschner M, Gross M (2005) Meshless deformations based on shape matching. ACM Trans Graph 24:471–478. https://doi.org/10.1145/1186822.1073216

Müller M, Heidelberger B, Hennix M, Ratcliff J (2007) Position based dynamics. J Vis Commun Image Represent 18:109–118. https://doi.org/10.1016/j.jvcir.2007.01.005

Nejadsadeghi N, De Angelo M, Drobnicki R, Lekszycki T, dell'Isola F, Misra A (2019) Parametric experimentation on pantographic unit cells reveals local extremum configuration. Exp Mech 59(6):927–939

Niiranen J, Khakalo S, Balobanov V, Niemi AH (2016) Variational formulation and isogeometric analysis for fourth-order boundary value problems of gradient-elastic bar and plane strain/stress problems. Comput Methods Appl Mech Eng 308:182–211

Olive M (2019) Effective computation of SO(3) and O(3) linear representation symmetry classes. Math Mech Complex Syst 7(3):203–237

Placidi L (2016) A variational approach for a nonlinear one-dimensional damage-elasto-plastic second-gradient continuum model. Continuum Mech Thermodyn 28(1–2):119–137

Placidi L, Andreaus U, Giorgio I (2017) Identification of two-dimensional pantographic structure via a linear D4 orthotropic second gradient elastic model. J Eng Math 103(1):1–21

Placidi L, Barchiesi E, Misra A (2018a) A strain gradient variational approach to damage: a comparison with damage gradient models and numerical results. Math Mech Complex Syst 6(2):77–100

Placidi L, Misra A, Barchiesi E (2018b) Two-dimensional strain gradient damage modeling: a variational approach. Z Angew Math Phys 69(3):56

Ramárez IS, Bernalte Sínchez PJ, Papaelias M, Mírquez FPG (2021) Autonomous underwater vehicles and field of view in underwater operations. J Mar Sci Eng 9(3). ISSN 2077-1312. https://doi.org/10.3390/jmse9030277. URL https://www.mdpi.com/2077-1312/9/3/277

Rapisarda A, Della Corte A, Drobnicki R, Cosmo F, Rosa L (2018) A model for bone mechanics and remodeling including cell populations dynamics. Z Angew Math Phys 70. https://doi.org/10.1007/s00033-018-1055-1
Rapisarda A, Almasi M, Almasi N, Barchiesi E, Della Corte A, Scerrato D (2020) Bone mechanics and cell populations: mathematical description and parametric study of the model, pp 107–126. ISBN 978-3-030-50463-2. https://doi.org/10.1007/978-3-030-50464-9_7
Rickert W, Vilchevskaya E, Müller W (2019) A note on Couette flow of micropolar fluids according to Eringen's theory. Math Mech Complex Syst 7(1):25–50
Rosi G, Placidi L, Auffray N (2018) On the validity range of strain-gradient elasticity: a mixed static-dynamic identification procedure. Eur J Mech A Solids 69:179–191
Şahin E (2005) Swarm robotics: from sources of inspiration to domains of application, pp 10–20
Scerrato D, Giorgio I (2019) Equilibrium of two-dimensional cycloidal pantographic metamaterials in three-dimensional deformations. Symmetry 11(12):1523
Scerrato D, Giorgio I, Madeo A, Limam A, Darve F (2014) A simple non-linear model for internal friction in modified concrete. Int J Eng Sci 80:136–152
Scerrato D, Zhurba Eremeeva IA, Lekszycki T, Rizzi NL (2016) On the effect of shear stiffness on the plane deformation of linear second gradient pantographic sheets. ZAMM-Z Angew Math Mech 96(11):1268–1279
Schulte J, Dittmann M, Eugster S, Hesch S, Reinicke T, Dell'Isola F, Hesch C (2020) Isogeometric analysis of fiber reinforced composites using Kirchhoff-Love shell elements. Comput Methods Appl Mech Eng 362:112845
Sciarra G, dell'Isola F, Coussy O (2007) Second gradient poromechanics. Int J Solids Struct 44(20):6607–6629
Seppecher P, Alibert JJ, dell'Isola F (2011) Linear elastic trusses leading to continua with exotic mechanical interactions. J Phys Conf Ser 319:012018. IOP Publishing
Spagnuolo M (2020) Circuit analogies in the search for new metamaterials: phenomenology of a mechanical diode. In: Nonlinear wave dynamics of materials and structures. Springer, pp 411–422
Spagnuolo M, Andreaus U (2019) A targeted review on large deformations of planar elastic beams: extensibility, distributed loads, buckling and post-buckling. Math Mech Solids 24(1):258–280
Spagnuolo M, Scerrato D (2020) The mechanical diode: on the tracks of James Maxwell employing mechanical–electrical analogies in the design of metamaterials. In: Developments and novel approaches in biomechanics and metamaterials. Springer, pp 459–469
Spagnuolo M, Barcz K, Pfaff A, dell'Isola F, Franciosi P (2017) Qualitative pivot damage analysis in aluminum printed pantographic sheets: numerics and experiments. Mech Res Commun 83:47–52
Spagnuolo M, Yildizdag ME, Andreaus U, Cazzani AM (2020) Are higher-gradient models also capable of predicting mechanical behavior in the case of wide-knit pantographic structures? Math Mech Solids. https://doi.org/10.1177/1081286520937339
Steigmann DJ (2018) Equilibrium of elastic lattice shells. J Eng Math 109(1):47–61
Steigmann DJ, dell'Isola F (2015) Mechanical response of fabric sheets to three-dimensional bending, twisting, and stretching. Acta Mech Sin 31(3):373–382
Timofeev D, Barchiesi E, Misra A, Placidi L (2020) Hemivariational continuum approach for granular solids with damage-induced anisotropy evolution. Math Mech Solids 1081286520968149
Tran LV, Niiranen J (2020) A geometrically nonlinear Euler-Bernoulli beam model within strain gradient elasticity with isogeometric analysis and lattice structure applications. Math Mech Complex Syst 8(4):345–371
Turco E (2018a) Discrete is it enough? The revival of Piola-Hencky keynotes to analyze three-dimensional elastica. Continuum Mech Thermodyn 30(5):1039–1057
Turco E (2018b) In-plane shear loading of granular membranes modeled as a Lagrangian assembly of rotating elastic particles. Mech Res Commun 92:61–66
Turco E (2019) Numerically driven tuning of equilibrium paths for pantographic beams. Continuum Mech Thermodyn 31(6):1941–1960
Turco E, Barchiesi E (2019) Equilibrium paths of Hencky pantographic beams in a three-point bending problem. Math Mech Complex Syst 7(4):287–310

Turco E, dell'Isola F, Cazzani A, Rizzi NL (2016a) Hencky-type discrete model for pantographic structures: numerical comparison with second gradient continuum models. Z Angew Math Phys 67(4):85

Turco E, dell'Isola F, Rizzi NL, Grygoruk R, Müller WH, Liebold C (2016b) Fiber rupture in sheared planar pantographic sheets: numerical and experimental evidence. Mech Res Commun 76:86–90

Turco E, dell'Isola F, Misra A (2019) A nonlinear Lagrangian particle model for grains assemblies including grain relative rotations. Int J Numer Anal Methods Geomech 43(5):1051–1079

Turco E, Barchiesi E, Giorgio I, dell'Isola F (2020) A Lagrangian Hencky-type non-linear model suitable for metamaterials design of shearable and extensible slender deformable bodies alternative to Timoshenko theory. Int J Non-Linear Mech 123:103481

Vaiana N, Sessa S, Marmo F, Rosati L (2018) A class of uniaxial phenomenological models for simulating hysteretic phenomena in rate-independent mechanical systems and materials. Nonlinear Dyn 93(3):1647–1669

Vangelatos Z, Gu GX, Grigoropoulos CP (2019) Architected metamaterials with tailored 3D buckling mechanisms at the microscale. Extreme Mech Lett 33:100580

Wiech J, Eremeyev VA, Giorgio I (2018) Virtual spring damper method for nonholonomic robotic swarm self-organization and leader following. Continuum Mech Thermodyn 30(5):1091–1102

Yang H, Ganzosch G, Giorgio I, Abali BE (2018) Material characterization and computations of a polymeric metamaterial with a pantographic substructure. Z Angew Math Phys 69(4):1–16

Yang H, Abali BE, Timofeev D, Müller WH (2019) Determination of metamaterial parameters by means of a homogenization approach based on asymptotic analysis. Continuum Mech Thermodyn 1–20

Yang Y, Misra A (2012) Micromechanics based second gradient continuum theory for shear band modeling in cohesive granular materials following damage elasticity. Int J Solids Struct 49(18):2500–2514

Yildizdag ME, Tran CA, Barchiesi E, Spagnuolo M, dell'Isola F, Hild F (2019) A multi-disciplinary approach for mechanical metamaterial synthesis: a hierarchical modular multiscale cellular structure paradigm. In: State of the art and future trends in material modeling. Springer, pp 485–505

Yildizdag ME, Barchiesi E, dell'Isola F (2020a) Three-point bending test of pantographic blocks: numerical and experimental investigation. Math Mech Solids 25(10):1965–1978

Yildizdag ME, Demirtas M, Ergin A (2020b) Multipatch discontinuous Galerkin isogeometric analysis of composite laminates. Continuum Mech Thermodyn 32(3):607–620

# Chapter 9
# Integral Equations in Semi-inverse Boundary Value Problems for an Elastic Strip

**Alexander N. Galybin**

**Abstract** This study deals with four incorrectly posed boundary value problems for an elastic strip. The formulations assume that three scalar conditions are given on one side of the strip and one on the other side. The Fourier transform is used to find analytical solutions in all four problems. However, the direct inverse transform is not applicable, which necessitates the reduction to integral equations of the Fredholm type of the first kind. Numerical approaches associated with non-stable solutions of the derived integral equation are discussed.

## 9.1 Introduction

This study continues the developments of mathematical background for incorrectly posed problems for layered elastic media. It is assumed that some experimental data on strain/displacements can be measured on a part of stress-free surface. This constitutes the main difference from classical formulation. The latter are well developed and do not present essential difficulties, e.g. Ufljand (1968) or Shevljakov (1977) who presents a general method capable of dealing with 2D and 3D-layered structures involving the Fourier transform and matrix algorithms and also some solutions for the standard formulations of the boundary value problems, BVP, of elasticity. On the other hand, there are inverse formulations that can be reduced to the Cauchy initial value problem; see Shvab (1989) who refers such problems as the (**u**, **p**) problem, which indicates that both displacement, **u**, and stress, **p**, vectors are known on a part of the boundary and nothing is known outside of it. This problem has a unique solution and, therefore, is conditionally ill-posed. There are some other formulations that can be reduced to the (**u**, **p**) problem; see Kubo (1988), Gao and Mura (1991), Yeih et al. (1993), Hsieh and Mura (1993), Bonnet and Constantinescu (2005).

---

A. N. Galybin (✉)
The Schmidt Institute of Physics of the Earth (IPE RAS), Moscow, Russia
e-mail: a.n.galybin@gmail.com

Don State Technical University, Rostov-on-Don, Russia

 
H. Altenbach et al. (eds.), *Advanced Materials Modelling for Mechanical, Medical and Biological Applications*, Advanced Structured Materials 155,
https://doi.org/10.1007/978-3-030-81705-3_9

Both mentioned types of problems present certain interest for applications especially for the problem of debonding detection on the interfaces between dissimilar layers. Such problems are often observed in engineering (e.g. coatings or FGM structures), geomechanics (layered rocks), biomechanics (live tissues) and other areas. There exist several routine methods that can be used for monitoring surface deformations or strains; however, in some situations measurement of all displacements on the surface of a body can be impractical. Thus, optical methods can easily measure deflection of the stress-free surface while the distribution of lateral strain over the surface becomes unavailable as any other devises as strain gauges obscure optical access for the surface. Bearing this in mind, a new formulation has been proposed in Galybin (2020) in which three conditions have been set of the surface of the coating and one condition on the interface between the coating and a substrate. Such formulations have been referred to as a semi-inverse problem. In this study, we investigate similar formulations for an elastic strip. It has been shown by Galybin and Rogerson (2019) that the case of a single strip provides information for the stress–displacement distributions of the interfaces of the layers and can be used for fracture detection.

Here we consider four possible formulations of semi-inverse BVP for a single strip. In all cases, the Fourier transforms are used to find out analytical solutions, and then we reduce inversion of the Fourier transform to integral equations.

## 9.2 Preliminaries

### 9.2.1 *Kolosov–Muskhelishvili Formulas*

The Kolosov–Muskhelishvili formulas for the stress and displacement components in a Cartesian coordinate system *Oxy* are as follows (Muskhelishvili 1963):

$$
\begin{aligned}
P &\equiv \frac{\sigma_{xx} + \sigma_{yy}}{2} = \Phi(z) + \overline{\Phi(z)}, \\
D &\equiv \frac{\sigma_{yy} - \sigma_{xx}}{2} + i\sigma_{xy} = \overline{z}\Phi'(z) + \Psi(z), \\
2GW &\equiv 2G(u_x + iu_y) = \kappa\phi(z) - z\overline{\Phi(z)} - \overline{\psi(z)}.
\end{aligned} \tag{9.1}
$$

Here $\sigma_{xx}, \sigma_{yy}, \sigma_{xy}$ are stress components, $P$ is mean stress, $D$ is complex stress deviator, $W = u_x + iu_y$ is complex displacement vectors with the components $u_x$, $u_y$ along the $x$ and $y$ axes, respectively, $G$ is shear modulus, $\kappa = (3 - 4\nu)$ for the plain stress and $\kappa = (3 - \nu)/(1 + \nu)$ for the plane stress conditions, $\nu$ is Poisson's ratio, and $\Phi(z) = \varphi'(z)$, $\Psi(z) = \psi'(z)$ are complex potentials (holomorphic functions of a complex variable $z = x + iy$).

For straight boundaries, it is convenient to replace the potential $\Psi(z)$ by a new unknown holomorphic function $\Omega(z)$ as follows:

$$
\Omega(z) = z\Phi'(z) + \Phi(z) + \Psi(z). \tag{9.2}
$$

Then the expression for the complex stress deviator takes the form

$$D(z, \overline{z}) = (\overline{z} - z)\Phi'(z) - \Phi(z) + \Omega(z). \tag{9.3}$$

By differentiating the third formula in (9.1) with respect to the variable $x$ and making use of (9.2) and (9.3), one obtains the following expressions for the derivative of the displacement vector, $W' = w = u + iv$, and its complex conjugate, $\overline{W'} = \overline{w} = u - iv$, in terms of the holomorphic functions $\Phi(z)$ and $\Omega(z)$ as follows:

$$\begin{aligned} 2Gw &= \kappa\Phi(z) - \overline{\Omega(z)} + (\overline{z} - z)\overline{\Phi'(z)}, \\ 2G\overline{w} &= \kappa\overline{\Phi(z)} - \Omega(z) - (\overline{z} - z)\Phi'(z). \end{aligned} \tag{9.4}$$

The sum of the first and the second formulas in (9.1) followed by complex conjugation results in the following two equivalent relationships:

$$\begin{aligned} \sigma \equiv \sigma_{yy} + i\sigma_{xy} &= \overline{\Phi(z)} + \Omega(z) + (\overline{z} - z)\Phi'(z), \\ \overline{\sigma} = \sigma_{yy} - i\sigma_{xy} &= \Phi(z) + \overline{\Omega(z)} - (\overline{z} - z)\overline{\Phi'(z)}. \end{aligned} \tag{9.5}$$

Now the expressions for the complex potentials $\Phi(z)$ and $\Omega(z)$ in terms of the stress and displacement vectors are found from (9.4) and (9.5) as follows:

$$(\kappa + 1)\Phi(z) = \overline{\sigma} + 2Gw. \tag{9.6}$$

$$(\kappa + 1)\Omega(z) = \kappa\sigma - 2G\overline{w} + (\kappa + 1)(z - \overline{z})\Phi'(z). \tag{9.7}$$

The sum of (9.6) and (9.7) results in

$$\Phi(z) + \Omega(z) - (z - \overline{z})\Phi'(z) = \frac{\kappa\sigma + \overline{\sigma} + 2G(w - \overline{w})}{\kappa + 1}. \tag{9.8}$$

If $y = 0$, then (9.8) takes the form

$$\Phi(x) + \Omega(x) = \sigma_{yy}(x, 0) + i\big[\tfrac{\kappa - 1}{\kappa + 1}\sigma_{xy}(x, 0) + \tfrac{4G}{\kappa + 1}v(x, 0)\big]. \tag{9.9}$$

It follows from (9.9) that if the right-hand side of (9.9) is known, then one arrives to the initial (Cauchy) problem for finding the holomorphic function $\Phi(z) + \Omega(z)$. This fact will be used further on for the formulation of four boundary value problems.

### 9.2.2 *Fourier Transforms*

The Fourier transform, $\mathbf{F}(.)$ (abbreviated as the FT hereafter), and its inverse, $\mathbf{F}^{-1}(.)$, are specified as follows: (Bateman and Erdelyi 1954)

$$\mathbf{F}(Y(x)) \equiv \hat{Y}(s) = \int_{-\infty}^{\infty} Y(x)\mathrm{e}^{-isx}\mathrm{d}x,$$

$$\mathbf{F}^{-1}\left(\hat{Y}(s)\right) \equiv Y(x) = \frac{1}{2\pi}\int_{-\infty}^{\infty} \hat{Y}(s)\mathrm{e}^{isx}\mathrm{d}s, \tag{9.10}$$

where $s$ is a real parameter, $\mathrm{Im}(s) = 0$. The FT of the derivative is

$$\mathbf{F}\left(Y'(x)\right) = is\hat{Y}(s). \tag{9.11}$$

The FT for complex conjugated functions is

$$\mathbf{F}\left(\overline{Y(x)}\right) = \overline{\hat{Y}(s)} = \int_{-\infty}^{\infty} \overline{Y(x)}\mathrm{e}^{-isx}\mathrm{d}x = \overline{\int_{-\infty}^{\infty} Y(s)\mathrm{e}^{isx}\mathrm{d}s} = \overline{\hat{Y}(-s)}. \tag{9.12}$$

If $Y_0(x)$ is a real-valued function, then $\overline{\hat{Y}_0(s)} = \hat{Y}_0(-s)$.

Application of the FT with respect to the variable $x$ to any holomorphic function $H(z) = H(x, y)$ results (due to the Cauchy–Riemann equations) in the following relationships for the transforms of the complex potentials

$$\begin{pmatrix} \hat{\Phi}(s, y) \\ \hat{\Omega}(s, y) \end{pmatrix} = \mathrm{e}^{-ys}\begin{pmatrix} \hat{\Phi}(s, 0) \\ \hat{\Omega}(s, 0) \end{pmatrix}. \tag{9.13}$$

Applying complex conjugation to (9.13), one finds from (9.11) that

$$\begin{pmatrix} \overline{\hat{\Phi}(-s, y)} \\ \overline{\hat{\Omega}(-s, y)} \end{pmatrix} = \mathrm{e}^{ys}\begin{pmatrix} \overline{\hat{\Phi}(-s, 0)} \\ \overline{\hat{\Omega}(-s, 0)} \end{pmatrix}. \tag{9.14}$$

Relationships (9.13) and (9.14) are frequently used furthere on.

## 9.3 Problem Formulation

A strip of the thickness $h$ and with elastic constants $G$ and $\kappa$ is considered. Let us introduce a Cartesian coordinate frame $Oxy$ such that the $x$-axis coincides with the low boundary $\Gamma_1$ of the layer, and the $y$-axis is directed towards the upper boundary $\Gamma_2$. Thus, the elastic strip occupies the domain $\{-\infty < x < \infty, 0 < y < 2\,h\}$.

It is further assumed that three boundary conditions are given on $\Gamma_1$:

$$\sigma_{yy}(x, 0) = \sigma_1(x), \quad \sigma_{xy}(x, 0) = \tau_1(x), \quad v(x, 0) = v_1(x), \tag{9.15}$$

and one of the following boundary condition is given on $\Gamma_2$:

$$\sigma_{xy}(x, 2h) = \tau_2(x) \tag{9.16}$$

$$\sigma_{yy}(x, 2h) = \sigma_2(x), \tag{9.17}$$

$$v(x, 2h) = v_2(x), \tag{9.18}$$

$$u(x, 2h) = u_2(x). \tag{9.19}$$

Therefore, we further consider four boundary value problems, BVPs, for an elastic strip as follows:

- BVP-T employs conditions (9.15) and (9.16),
- BVP-N employs (9.15) and (9.17),
- BVP-V employs (9.15) and (9.18),
- BVP-U employs (9.15) and (9.19).

It is also assumed that the boundary $\Gamma_1$ is loaded by concentrated forces only applied at the points $t_j$ ($j = 1, \ldots, J$). Therefore, the complex stress vector on $\Gamma_1$ has the form

$$\sigma_1(x) + i\tau_1(x) = \sum_{j=0}^{J} P_j\delta(x - t_j), \quad x, t_j \in \Gamma_1, \tag{9.20}$$

where $P_j$ are complex intensities of the applied loads, and $\delta(t)$ is the Dirac delta function.

The assumption above ensures that the deflection of the lower boundary can be measured over the entire surface, and therefore, the tangential derivative of the normal displacements is also known on the boundary with exception of the points $t_j$, where it is discontinuous and tends to infinity.

It is understood that monitoring of the tangential displacements (or surface strain) is also possible. However, it is more difficult that deflection monitoring can be done by optical methods. For this reason, we do not employ any data regarding measurements of the $u_x(x, 0)$ field. Instead, we use one of the conditions (9.16)–(9.19) on the other side of the strip. Therefore, it is evident that such formulation differs from the well-known correct formulations and also from the Cauchy problem considered by Galybin and Rogerson (2019). It is somewhat related to the problem for the detection of a sliding zone on the interface between a coating and a substrate considered in Galybin (2020). The difference from the latter mentioned problem lays in the fact that no relationships are introduced between the stress and displacement vectors on the upper boundary as it was used in Galybin (2020) due to continuity conditions on the interface.

It is worth noting that the measurements of the deflection are subject to errors and can be modelled as follows:

$$v_1(x) = u_y^{\text{ideal}}(x, 0) + \text{Noise}(x), \quad x \in \Gamma_1, \tag{9.21}$$

where $u_y^{\text{ideal}}(x, 0)$ is the function of ideal deflections calculated from the solution of a correctly posed BVP and Noise($x$) is a function that generates for every point $x$ a random noise of certain intensity $\varepsilon$ that can be modelled by uniformly distributed random numbers or by normally distributed errors with zero mean or by other means. This means that in the case of ill-posed formulations, the solutions may possess large errors as in the Cauchy problem, which indicates that some special methods could be necessary to apply in order to stabilize the solutions (Tikhonov and Arsenin 1977). This will be discussed further.

## 9.4 Fourier Solutions for the Semi-Inverse Boundary Value Problems for an Elastic Strip

This section presents solutions of for semi-inverse boundary value problems for an elastic strip with the boundaries $\Gamma_1 = \{z = t : -\infty < t < \infty\}$ and $\Gamma_2 = \{z = t + 2ih : -\infty < t < \infty\}$. We present full solutions in terms of the FT of the complex potentials, which enable one to analyse the stress state at any point of the strip.

In all four problems, we use boundary conditions (9.15), from which it follows that the boundary value of the holomorphic function $\Phi(z) + \Omega(z)$ is known on the boundary $\Gamma_1$ where $y = 0$. Let us apply the FT to (9.8) and use the properties of the FT to obtain

$$\begin{aligned}(1 + 2sy)\hat{\Phi}(s, y) + \hat{\Omega}(s, y) &= \hat{\sigma}_{yy}(s, y)\\ &+ i\left[\tfrac{\kappa-1}{\kappa+1}\hat{\sigma}_{xy}(s, y) + \tfrac{4G}{\kappa+1}\hat{v}(s, y)\right].\end{aligned} \tag{9.22}$$

On the boundary $\Gamma_1$ ($y = 0$), one can present (9.22) through the FT of conditions (9.15) as follows:

$$\hat{\Phi}(s, 0) + \hat{\Omega}(s, 0) = f(s). \tag{9.23}$$

where for briefness, it was denoted

$$f(s) = \hat{\sigma}_1(s) + i\left[\tfrac{\kappa-1}{\kappa+1}\hat{\tau}_1(s) + \tfrac{4G}{\kappa+1}\hat{v}_1(s)\right]. \tag{9.24}$$

By using (9.13) and (9.14), one transforms (9.23) to the form

$$\hat{\Omega}(s, y) = e^{-ys} f(s) - \hat{\Phi}(s, y). \tag{9.25}$$

Due to (9.12) complex conjugation of (9.25) results in

$$\overline{\hat{\Omega}(s, y)} = e^{ys}\overline{f(-s)} - \overline{\hat{\Phi}(-s, y)}. \tag{9.26}$$

By applying the FT to the stress and displacement vectors and excluding the FT of $\Omega$ via (9.25) and (9.26), one finds

$$\begin{aligned}\hat{\sigma}_{yy}(s, y) + i\hat{\sigma}_{xy}(s, y) &= \mathrm{e}^{ys}\overline{\hat{\Phi}(-s, 0)} - (1 - 2ys)\mathrm{e}^{-ys}\hat{\Phi}(s, 0)\\ &\quad + \mathrm{e}^{-ys} f(s),\\ \hat{\sigma}_{yy}(s, y) - i\hat{\sigma}_{xy}(s, y) &= \mathrm{e}^{-ys}\hat{\Phi}(s, 0) - (1 + 2ys)\mathrm{e}^{ys}\overline{\hat{\Phi}(-s, 0)}\\ &\quad + \mathrm{e}^{ys}\overline{f(-s)},\end{aligned} \tag{9.27}$$

$$\begin{aligned}2G\big(\hat{u}(s, y) + i\hat{v}(s, y)\big) &= \kappa\mathrm{e}^{-ys}\hat{\Phi}(s, 0) + (1 + 2ys)\mathrm{e}^{ys}\overline{\hat{\Phi}(-s, 0)}\\ &\quad - \mathrm{e}^{ys}\overline{f(-s)},\\ 2G\big(\hat{u}(s, y) - i\hat{v}(s, y)\big) &= \kappa\mathrm{e}^{ys}\overline{\hat{\Phi}(-s, 0)} + (1 - 2ys)\mathrm{e}^{-ys}\hat{\Phi}(s, 0)\\ &\quad - \mathrm{e}^{-ys} f(s).\end{aligned} \tag{9.28}$$

Taking the difference of the equations in expressions (9.27) and (9.28) on $\Gamma_1$, one finds

$$\begin{aligned}&i\hat{\tau}_1(s) = \overline{\hat{\Phi}(-s, 0)} - \hat{\Phi}(s, 0) + \tfrac{1}{2} f(s) - \tfrac{1}{2}\overline{f(-s)},\\ &4iG\hat{v}_1(s, 0) = (\kappa - 1)\hat{\Phi}(s, 0) - (\kappa - 1)\overline{\hat{\Phi}(-s, 0)} + f(s) - \overline{f(-s)}.\end{aligned} \tag{9.29}$$

The following relationships are obtained from (9.29)

$$\begin{aligned}\overline{\hat{\Phi}(-s, 0)} &= \hat{\Phi}(s, 0) + i\hat{\tau}_1(s) - \tfrac{1}{2}\Big[f(s) - \overline{f(-s)}\Big],\\ \overline{\hat{\Phi}(-s, 0)} &= \hat{\Phi}(s, 0) - \tfrac{4G}{\kappa-1}i\hat{v}_1(s) + \tfrac{1}{\kappa-1}\Big[f(s) - \overline{f(-s)}\Big].\end{aligned} \tag{9.30}$$

These relationships are equivalent as it is evident from (9.22)–(9.26).

Substitution of either of Eq. (9.30) into (9.28), (9.27) and (9.28) allows one to express the stress and displacement vectors (and their conjugates) through the only unknown $\Phi(s, 0)$. The latter is found from one of the conditions (9.16)–(9.19) on $\Gamma_2$.

Let us find the solutions for BVPs in terms of the FT.

**BVP-T**. Three boundary conditions (9.15) on $\Gamma_1$ are complemented by a single condition (9.16), i.e. the shear stresses are known on $\Gamma_2$.

Consider the difference of equations in (9.27)

$$i\hat{\sigma}_{xy}(s,y) = (1+ys)e^{ys}\overline{\hat{\Phi}(-s,0)} - (1-ys)e^{-ys}\hat{\Phi}(s,0) + \frac{e^{-ys}f(s) - e^{ys}\overline{f(-s)}}{2} \tag{9.31}$$

Substitution of the first equation in (9.30) into (9.31) results in (9.32)

$$\begin{aligned} i\hat{\sigma}_{xy}(s,y) &= \left[(1+ys)\mathrm{e}^{ys} - (1-ys)\mathrm{e}^{-ys}\right]\hat{\Phi}(s,0) + \frac{e^{-ys}f(s) - e^{ys}\overline{f(-s)}}{2} \\ &+ (1+ys)\mathrm{e}^{ys}\left(i\hat{\tau}_1(s) - \frac{f(s) - \overline{f(-s)}}{2}\right). \end{aligned} \tag{9.32}$$

Putting $y = 2h$ and using the FT of $\tau_2(x)$ on the boundary $\Gamma_2$, one finds the solution for the unknown FT of $\Phi(x, 0)$ on $\Gamma_1$ in the form

$$\hat{\Phi}(s,0) = \frac{i\hat{\tau}_2(s) - (1+2hs)\mathrm{e}^{2hs}\left(i\hat{\tau}_1(s) - \mathrm{Re} f(s)\right) - \mathrm{Re}\left(\mathrm{e}^{-2hs}f(s)\right)}{2\sinh(2hs) + 4hs\cosh(2hs)} \tag{9.33}$$

If one assumes the absence of shear stresses on the boundaries, then (9.33) takes the following form:

$$\hat{\Phi}(s,0) = \frac{\sinh(2hs)\hat{\sigma}_1(s) + i\left(\sinh(2hs) + 2hs\mathrm{e}^{2hs}\right)\frac{4G}{\kappa+1}\hat{v}_1(s)}{2\sinh(2hs) + 4hs\cosh(2hs)} \tag{9.34}$$

It is evident from (9.33) that the numerator should vanish when $s$ tends to zero (to compensate the first-order singularity in the denominator at $s = 0$). This should also provide the global balance of the whole strip. It is rather evident that this condition is fulfilled for (9.34).

**BVP-N**. Three boundary conditions (9.15) on $\Gamma_1$ are complemented by a single condition (9.17), i.e. the normal stresses are known on $\Gamma_2$.

Consider the sum of equations in (9.27)

$$\hat{\sigma}_{yy}(s,y) = ys\left[\mathrm{e}^{-ys}\hat{\Phi}(s,0) - \mathrm{e}^{ys}\overline{\hat{\Phi}(-s,0)}\right] + \frac{\mathrm{e}^{-ys}f(s) + \mathrm{e}^{ys}\overline{f(-s)}}{2}. \tag{9.35}$$

Substitution of the first equation in (9.30) into (9.35) results in

$$\begin{aligned} \hat{\sigma}_{yy}(s,y) &= -2ys\sinh(ys)\hat{\Phi}(s,0) + \frac{e^{-ys}f(s) + e^{ys}\overline{f(-s)}}{2} \\ &- ys\mathrm{e}^{ys}\left(i\hat{\tau}_1(s,0) - \frac{f(s) - \overline{f(-s)}}{2}\right) \end{aligned} \tag{9.36}$$

Employing the FT of condition (9.17) on the boundary $\Gamma_2$ ($y = 2h$), one finds the following solution:

$$\hat{\Phi}(s,0) = \frac{\mathrm{Re}\left(\mathrm{e}^{-2hs}f(s)\right) - \hat{\sigma}_2(s) - 2ihs\mathrm{e}^{2hs}\left(\hat{\tau}_1(s) - \mathrm{Im} f(s)\right)}{4hs\sinh(2hs)} \tag{9.37}$$

**BVP-V**. Three boundary conditions (9.15) on $\Gamma_1$ are complemented by a single condition (9.18), i.e. the tangential derivative of the normal displacements is given on $\Gamma_2$.

The difference of equations in (9.28) has the form

$$4iG\hat{v}(s,y) = (\kappa - 1 + 2ys)e^{-ys}\hat{\Phi}(s,0) + (1-\kappa+2ys)e^{ys}\overline{\hat{\Phi}(-s,0)} \\ + e^{-ys}f(s) - e^{ys}\overline{f(-s)}, \tag{9.38}$$

Substitution of the second equation in (9.30) in (9.38)

$$4iG\hat{v}(s,y) = (\kappa - 1 + 2ys)e^{-ys}\hat{\Phi}(s,0) + (1-\kappa+2ys)e^{ys}\hat{\Phi}(s,0) \\ - (1-\kappa+2ys)e^{ys}\left(\tfrac{4G}{\kappa-1}i\hat{v}_1(s) + \tfrac{f(s)-\overline{f(-s)}}{\kappa-1}\right) \\ + e^{-ys}f(s) - e^{ys}\overline{f(-s)}. \tag{9.39}$$

Employing the FT of condition (9.18) on the boundary $\Gamma_2$ ($y = 2h$), one finds the following solution:

$$\hat{\Phi}(s,0) = i\frac{2G\hat{v}_2(s) - (1-\frac{4hs}{\kappa-1})e^{2hs}\big(2G\hat{v}_1(s) + \mathrm{Im} f(s)\big) - \mathrm{Im}\big(e^{-2hs}f(s)\big)}{4hs\cosh(2hs) - (\kappa-1)\sinh(2hs)} \tag{9.40}$$

**BVP-U**. Three boundary conditions (9.15) on $\Gamma_1$ are complemented by a single condition (9.19), i.e. the lateral strain is given on $\Gamma_2$.

The sum of equations in (9.28) gives

$$4G\hat{u}(s,y) = (1+\kappa-2ys)e^{-ys}\hat{\Phi}(s,0) + (1+\kappa+2ys)e^{ys}\overline{\hat{\Phi}(-s,0)} \\ - e^{ys}\overline{f(-s)} - e^{-ys}f(s) \tag{9.41}$$

Substitution of the second equation in (9.30) into (9.41) results in

$$4G\hat{u}(s,y) = (1+\kappa-2ys)e^{-ys}\hat{\Phi}(s,0) + (1+\kappa+2ys)e^{ys}\hat{\Phi}(s,0) \\ - \tfrac{1+\kappa+2ys}{\kappa-1}\left(4iGe^{ys}\hat{v}_1(s) + f(s) - \overline{f(-s)}\right) \\ - e^{ys}\overline{f(-s)} - e^{-ys}f(s). \tag{9.42}$$

Putting $y = 2h$ in (9.42) and using the FT of (9.19), one finds the following solution:

$$\hat{\Phi}(s,0) = \frac{2G\hat{u}_1(s) + i\frac{1+\kappa+4hs}{\kappa-1}e^{2hs}\big(2G\hat{v}_1(s) - \mathrm{Im} f(s)\big) + \mathrm{Re}\big(e^{-2hs}f(s)\big)}{(1+\kappa)\cosh(2hs) + 2hs\sinh(2hs)} \tag{9.43}$$

This section presents analytical solutions for the FT of the complex potential $\Phi(z)$ on the boundary $\Gamma_1$ for four semi-inverse BVPs. The results for the second potential $\Omega(z)$ on the boundary $\Gamma_1$ are found by formula (9.23). In order to find the FT of $\Phi(z)$ and $\Omega(z)$ everywhere in the strip, one should apply formulas (9.13), i.e. by multiplying the right-hand sides of (9.33), (9.37), (9.40) and (9.43) and similar expressions for $\hat{\Omega}(s, 0)$ by $\exp(-ys)$.

## 9.5 Integral Equations

It is evident from the structure of the solutions for $\hat{\Phi}(s, 0)$ obtained in the previous section that $\Phi(x, 0)$ can be directly found by inversion of the FT.

$$\begin{aligned}\begin{pmatrix}\Phi(x,0)\\ \Omega(x,0)\end{pmatrix} &= \frac{1}{2\pi}\int\limits_{-\infty}^{\infty}\begin{pmatrix}\hat{\Phi}(s,0)\\ \hat{\Omega}(s,0)\end{pmatrix}\mathrm{e}^{isx}\mathrm{d}s\\ &= \frac{1}{2\pi}\int\limits_{-\infty}^{\infty}\begin{pmatrix}1\\ -1\end{pmatrix}\hat{\Phi}(s,0)\mathrm{e}^{isx}\mathrm{d}s + \frac{1}{2\pi}\int\limits_{-\infty}^{\infty}\begin{pmatrix}0\\ 1\end{pmatrix}f(s)\mathrm{e}^{isx}\mathrm{d}s\\ &= \frac{1}{2\pi}\int\limits_{-\infty}^{\infty}\begin{pmatrix}1\\ -1\end{pmatrix}\hat{\Phi}(s,0)\mathrm{e}^{isx}\mathrm{d}s + \begin{pmatrix}0\\ 1\end{pmatrix}f(x).\end{aligned} \tag{9.44}$$

However, direct inversion of the solution $\hat{\Phi}(s, y)$ cannot be performed because some integrals are divergent due to the presence the factor $\exp(-sy)$ in their integrands. Therefore, it is suggested, following the approaches reported in Galybin (2020), Galybin and Rogerson (2019), to reduce the FT inversion to integral equations of the first kind by using the following manipulations.

Let us present the exponent $\exp(-sy)$ as the difference of the hyperbolic functions

$$\mathrm{e}^{-ys} = \cosh(ys) - \sinh(ys). \tag{9.45}$$

followed by substitution of (9.45) into (9.13). Thus, one finds

$$\begin{aligned}\begin{pmatrix}\hat{\Phi}(s,y)\\ \hat{\Omega}(s,y)\end{pmatrix} &= (\cosh(ys) - \sinh(ys))\begin{pmatrix}\hat{\Phi}(s,0)\\ \hat{\Omega}(s,0)\end{pmatrix}\\ &= \cosh(ys)(1 - \tan(ys))\begin{pmatrix}\hat{\Phi}(s,0)\\ \hat{\Omega}(s,0)\end{pmatrix}.\end{aligned} \tag{9.46}$$

One can divide (9.46) by $\cosh(ys) > 0$ and use expression (9.44) to obtain

$$\frac{1}{\cosh(ys)}\begin{pmatrix}\hat{\Phi}(s,y)\\ \hat{\Omega}(s,y)\end{pmatrix} = (1-\tan(ys))\begin{pmatrix}1\\ -1\end{pmatrix}\hat{\Phi}(s,0) - (1-\tan(ys))\begin{pmatrix}0\\ f(s)\end{pmatrix} \tag{9.47}$$

Let us further use the table integrals from Bateman and Erdelyi (1954) to derive the following inverse transforms:

$$\mathbf{F}^{-1}\left(\cosh^{-1}(\pi s)\right) = \frac{1}{2\pi}\frac{1}{\cosh\frac{x}{2}}, \quad \mathbf{F}^{-1}(\tanh(\pi s)) = \frac{-1}{2\pi i}\frac{1}{\sinh\frac{x}{2}} \tag{9.48}$$

As soon as the inversion of $\hat{\Omega}(s, y)$ is linearly expressed through inversion of $\hat{\Phi}(s, y)$, we can further consider the first line from (9.47) only. Assuming $y = 2h$ and applying the convolution theorem, one arrives at the following integral equation for the determination of unknown $\Phi(z)$

$$\frac{1}{2\pi}\int_{-\infty}^{\infty}\frac{\Phi(t,y)}{\cosh\frac{\pi}{2}\frac{t-x}{y}}dt = y\Phi(x,0) + \frac{1}{2\pi i}\int_{-\infty}^{\infty}\frac{\Phi(t,0)}{\sinh\frac{\pi}{2}\frac{t-x}{y}}dt, \quad 0 < y \le 2h \tag{9.49}$$

Equation (9.49) presents the fundamental integral equations for all BVP considered. It belongs to the class of Fredholm equations of the first kind. Therefore, its solution can be unstable. In such case, the application of special regularization techniques, e.g. as Tikhonov's regularization (Tikhonov and Arsenin 1977) or SVD truncation, is necessary. It is worth noting that the right-hand side of (9.49) is subjected to measurement errors as $\Phi(x, 0)$ contains data of deflection monitoring, which again emphasizes the importance of regularization. It has been shown numerically that the SVD regularization provides reasonable results in somewhat similar problem; see more details in Galybin and Rogerson (2019).

## 9.6 Conclusions

Four non-classical BVPs are considered in this paper. Such problems can be applied for the detection of delamination and sliding zones on the interfaces of layered materials or coating and substrates.

The BVPs are considered for an elastic strip. They are referred to the class of semi-inverse problems because they use three scalar conditions of one side of the strip and one on the other side. As a result, the BVPs become incorrect and possess a non-stable solution.

We present analytical solutions for the FT of complex potentials for four semi-inverse BVP and a fundamental integral equation to which these problems can be reduced. The importance of using regularization techniques for this integral equation is outlined, in particular, the SVD regularization is recommended.

**Acknowledgements** This work was supported by the grant of the Government of the Russian Federation No. 14.Z50.31.0046.

## References

Bateman H, Erdelyi A (1954) Tables of integral transforms. McGraw-Hill, New York
Bonnet M, Constantinescu A (2005) Inverse problems in elasticity. Inverse Prob 21:R1–R50
Galybin AN (2020) Integral equations in direct and inverse problems of elastostatics for crack detection in layered structures. Math Mech Solids 25(5):1140–1154
Galybin AN, Rogerson GA (2019) Ill-posed problem of the Cauchy type for an elastic strip. Math Mech Solids 24(9):2986–2998
Gao Z, Mura T (1991) Elasticity problems with partially overspecified boundary conditions. Int J Eng Sci 29(6):685–692
Hsieh SC, Mura T (1993) Nondestructive cavity identification in structures. J Solids Struct 30(12):1579–1587
Kubo S (1988) Inverse problems related to the mechanics and fracture of solids and structures. JSME Int J 31:157–166
Muskhelishvili NI (1963) Some basic problems of the mathematical theory of elasticity. P. Noordhoff, Groningen, The Netherlands
Shevljakov YA (1977) Matrix algorithms in the theory of elasticity of inhomogeneous media. Vischa Schola, Kiev-Odessa, in Russian
Shvab AA (1989) Incorrectly posed static problems of elasticity. Solid Mech 24:98–106
Tikhonov AN, Arsenin VY (1977) Solution of ill-posed problems. Winston, Wiley, New York
Ufljand JS (1968) Integral transforms in problems of the elasticity theory. Nauka, Leningrad (in Russian)
Yeih WC, Koya T, Mura T (1993) An inverse problem in elasticity with partially overspecified boundary conditions. I. Theoretical approach. Trans ASME J Appl Mech 60:595–600

# Chapter 10
# Surface Wave Propagation in Elastic Half-Spaces with Periodic Coatings

**Evgeny Glushkov, Natalia Glushkova, and Sergey Fomenko**

**Abstract** Contact problems for functionally graded (FG) materials of complex structure are of traditional research interest due to various practical applications in the chemical industry, materials science, etc. In this article, we consider the dynamic response of such materials with FG or stepwise layered coatings. Numerical analysis is carried out based on the explicit integral and asymptotic representations obtained in terms of Green's matrix for elastic half-spaces with periodic coverages. Their soft interlayers form internal waveguide channels, which leads to unusual wave and energy transfer effects in response to dynamic surface loading.

**Keywords** Laminate composites · Functionally graded materials · Periodic coatings · Guided waves · Mode repulsion · Resonance response

## 10.1 Introduction

It is a pleasure for us to contribute to the book dedicated to the anniversary of our long-time friend, Professor Sergey Aizikovich. We (EG and NG) have known him since our students' days at the Mechanics and Mathematics Department of the Rostov State University (RSU) at the beginning of the 1970s. Then, we all stayed at the Institute of Mechanics and Applied Mathematics RSU, headed in those years by the founder of the Rostov School of Mechanics, Prof. I.I. Vorovich (in the 1980s, he became a full member of the Soviet Academy of Science, the first academician in the field of mechanics in the South Russia region, outside of the capitals).

E. Glushkov (✉) · N. Glushkova · S. Fomenko
Institute for Mathematics, Mechanics and Informatics, Kuban State University, Krasnodar, Russia
e-mail: evg@math.kubsu.ru

N. Glushkova
e-mail: nvg@math.kubsu.ru

S. Fomenko
e-mail: sfom@yandex.ru

 
H. Altenbach et al. (eds.), *Advanced Materials Modelling for Mechanical, Medical and Biological Applications*, Advanced Structured Materials 155,
https://doi.org/10.1007/978-3-030-81705-3_10

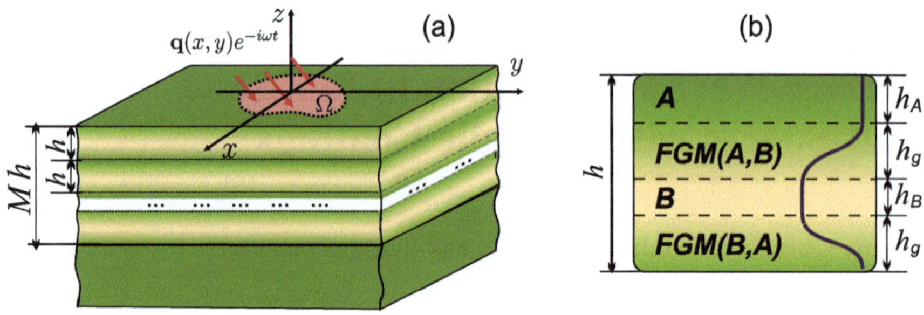


**Fig. 10.1** Geometry of the problem; **a**) stratified elastic base, **b**) an FGM cell

Although we worked in the same research area: integral equation methods in contact problems of elasticity, we were supervised by different scientific leaders, Vorovich's pupils and followers V. M. Alexandrov and V. A. Babeshko. Therefore, we were in various departments of the Institute. Sergey was in the Soil Mechanics Department, dealing with the static contact problems for civil engineering foundations and material indentation testing, while we were in the Department of Wave Processes, engaged in the dynamic contact problems of geophysics and seismic prospecting, one of the leading research topics of Babeshko's team at those years. Nevertheless, in both cases, the mathematical framework of the numerical studies was the same: the integral equation based approach developed in Vorovich et al. (1974), Vorovich and Babeshko (1979).

Within this approach, the vector $\mathbf{u}$ of the displacements caused by a surface load $\mathbf{q}$ in an elastic substrate (a stack of layers, a stratified half-space, etc.; Fig. 10.1) is sought as the convolution of the substructure's Green's matrix $k$ with the load vector $\mathbf{q}$, or equivalently, as the inverse Fourier transform of their Fourier symbols $K = \mathcal{F}_{xy}[k]$ and $\mathbf{Q} = \mathcal{F}_{xy}[\mathbf{q}]$:

$$\mathbf{u}(\mathbf{x}) = (k * \mathbf{q})(\mathbf{x}) = \frac{1}{(2\pi)^2} \int\limits_{\Gamma_1}\int\limits_{\Gamma_2} K(\alpha_1, \alpha_2, z)\mathbf{Q}(\alpha_1, \alpha_2) e^{-i(\alpha_1 x + \alpha_2 y)} \mathrm{d}\alpha_1 \mathrm{d}\alpha_2. \tag{10.1}$$

The columns $\mathbf{k}_j$ of the $3 \times 3$ Green's matrix $k = (\mathbf{k}_1 \vdots \mathbf{k}_2 \vdots \mathbf{k}_3)$ are the displacement vectors corresponding to the point loads $\boldsymbol{\tau} = \delta(x, y)\mathbf{i}_j$ applied to the substructure's surface $z = 0$ along the coordinate vectors $\mathbf{i}_j$, $j = 1, 2, 3$; $\delta$ is Dirac's delta-function; $\mathcal{F}_{xy}$ is the integral Fourier transform with respect to the horizontal coordinates $x$ and $y$. The integration paths $\Gamma_1$ and $\Gamma_2$ go along the real axes rounding real poles of $K$ elements in accordance with the principle of limiting absorption. Hereinafter, we adhere to the traditional notations of the Vorovich's school [for more details see Vorovich and Babeshko (1979), Glushkov et al. (2012, 2015) and the works cited therein].

Representation (10.1) is valid for any elastic base with piecewise-constant or piecewise-continuous dependence of mechanical properties on the depth $z$, i.e., for layered and functionally graded materials (FGM). The only difference is in the $K$ matrix, which is determined by the specific laws of $z$-dependence. With classical homogeneous media (free or clamped layer, half-space), the $K$ elements are derived in an explicit analytical form. However, to obtain them for multilayer or FG structures, robust numerical algorithms are to be developed. This was what we had been doing at the Institute, Sergey for static indentation in FGM half-spaces (Aizikovich 1982; Aizikovich and Alexandrov 1982) and we for their dynamic response (Glushkov 1978; Glushkova 1981).

In the most general FGM case, the $K$ matrix calculation is reduced to the numerical evaluation of a system of linear ordinary differential equations (ODE) for the $z$ variable. The solution to such systems is featured by exponential and oscillating behavior of its components, resulting in ill-conditioning and rapid loss of accuracy during numerical integration. The same kinds of difficulties arise in the implementation of the transfer matrix methods for layered media. This is the most challenging problem in developing efficient algorithms of Green's matrix calculation for stratified substrates. Starting from the pioneering works (Haskell 1953; Petrashen 1952; Thomson 1950), at those years, the most efforts were spent to suppress such numerical instability (Alekseev and Mikhailenko 1976; Knopoff 1964).

Over time, it became clear to us that the best way to ensure numerical stability is to explicitly extract the exponential and oscillating parts of the solution so that only smooth modulating functions remain to be obtained numerically (Babeshko et al. 1987). This idea has been implemented in solving various specific problems and stood the test of time. In recent years, a significant contribution to its implementation and active development has been made by S. Fomenko, the third co-author of the present paper (Glushkov et al. 2012, 2015).

An interesting type of stratification is a non-monotonic change of properties in alternating interlayers or FGMs with sign-variable gradients (Fig. 10.2). It occurs in such critical applications as multilayer polymer composites used for hydrogen storages (Gentilleau et al. 2011), or hardening coatings of periodic structure. Sergey Aizikovich and his team have thoroughly investigated the features of indentation of such materials (Aizikovich et al. 2008, 2011; Vasiliev et al. 2014, 2016; Volkov et al. 2013).

This and similar images appear first in any Google request for Aizikovich's works. Hence, they could be considered the emblem of his team, and this is not accidental. Rotated 90 degrees, they represent nothing more than a mountain landscape - the second or, perhaps, the main Sergey's interest in life. We (EG and NG) even met each other at the hike that he organized and led through the Caucasus Mountains on the summer vacation far in 1971. Since that, he seems to have visited the main mountain regions of all continents (the Caucasus, Alps, Pamir, Kamchatka, Himalayas, Andes, etc.), and climbed the highest peak of the Soviet Union - Communism Peak (7495 m). Till present, year after year, he invariably alternates winter alpine skiing and summer hiking and mountaineering.

He not only attracted us to mountains but also drew our attention to structures with periodic coatings. Alternating elastic properties form internal waveguide channels providing interesting effects of wave energy transfer (Glushkov et al. 2015).

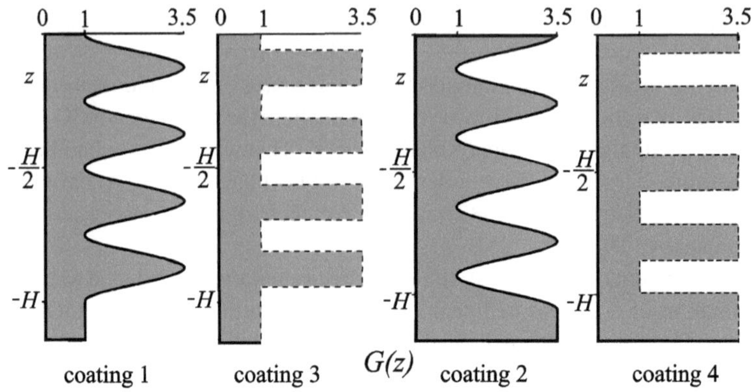


**Fig. 10.2** Depth dependences of elastic properties for periodic gradient and layered coatings with alternating soft/hard and hard/soft interlayer cells (Vasiliev et al. 2016)

Therefore, it should be clear why we have chosen our recent results on the features of traveling wave propagation in such structures for this book. Namely, the numerical examples below illustrate the effect of low-frequency splitting of dispersion curves into $M$ branches, where $M$ is the number of interlayer cells in the periodic coverage. The influence of the law of periodic stratification (stack of layers or FGM) on the source energy distribution among the traveling waves is also analyzed and discussed. It should be mentioned that Green's matrix-based representation (1) is inapplicable with non-vertical inhomogeneity, which requires other approaches [e.g., wave propagation in media with roughness coatings Mikhasev et al. (2021)].

## 10.2 Mathematical Framework

Consider time-harmonic oscillation $\mathbf{u}(\mathbf{x}, \omega)e^{-i\omega t}$ of an elastic stratified half-space caused by a load $\mathbf{q}(x, y)$ applied to its surface $z = 0$ in a load area $\Omega$ (Fig. 10.1a); $\omega$ is angular frequency. In Cartesian coordinates $\mathbf{x} = (x, y, z)$, $\mathbf{u} = (u_x, u_y, u_z) = (u_1, u_2, u_3)$ is the complex displacement amplitude, and $\boldsymbol{\tau} = (\tau_{xz}, \tau_{yz}, \sigma_z)$ is a stress vector. The upper part of the half-space is a periodic coverage of total thickness $H = Mh$ consisting of $M$ sublayers (cells) of thickness $h$. The material is isotropic, and the dependence of the cells' mechanical properties on the depth $z$ can be stepwise or gradient. In both cases, the displacement and stress fields $\mathbf{u}(\mathbf{x})$ and $\boldsymbol{\tau}(\mathbf{x})$ are continuous in the whole elastic volume. The vector $\mathbf{u}$ obeys the equations of motion

$$\sigma_{ij,j} + \rho\omega^2 u_i = 0, \quad i = 1, 2, 3,$$

in which it enters via the stress tensor elements

$$\sigma_{ij} = \lambda \operatorname{div} \mathbf{u} + \mu(u_{i,j} + u_{j,i}), \quad i, j = 1, 2, 3.$$

In the numerical examples below, we use a general law of depth dependence schematically sketched in Fig. 10.1b. Sublayers $A$ and $B$ are homogeneous, while the material parameters of interlayers FGM($A$, $B$) and FGM($B$, $A$) continuously vary from the values for material $A$ to those for material $B$ and vice versa. The law of FG variation is specified by a cubic spline, providing continuity together with the first derivatives. If the FGM interlayers are of zero thickness ($h_g = 0$), the dependence becomes stepwise discontinuous, depicting a cell composed of two homogeneous sublayers of thicknesses $h_A = h_B = h/2$ and materials $A$ and $B$. Conversely, if $h_g = h/2$ and $h_A = h_B = 0$, there is a fully FG cell with the material properties gradually varying from $A$ to $B$ and back. The half-space part beneath the coating ($-\infty < z \le -H$) is homogeneous with the properties of material $A$.

Using the integral transform $\mathcal{F}_{xy}$ applied to the governing elastodynamic equations and boundary conditions of the corresponding boundary value problem, the wave field $\mathbf{u}(\mathbf{x})$ is obtained in the form of double path integral (10.1) (Glushkov et al. 2012, 2015; Vorovich and Babeshko 1979). This explicit representation is valid for both isotropic and anisotropic stratified substrates. In the isotropic case, transition to the polar coordinates

$$\begin{aligned} x &= r\cos\varphi, \quad y = r\sin\varphi, \quad r = \sqrt{x^2 + y^2} \\ \alpha_1 &= \alpha\cos\gamma, \quad \alpha_2 = \alpha\sin\gamma, \quad \alpha = \sqrt{\alpha_1^2 + \alpha_2^2} \end{aligned}$$

reduces the double integral to a single path integral, evaluated using the residue technique. It leads to the far-field asymptotic representation of the source-generated wave field $\mathbf{u}$ in terms of traveling GWs $\mathbf{u}_n$:

$$\mathbf{u}(\mathbf{x}) = \sum_{n=1}^{N} \mathbf{u}_n(\mathbf{x})[1 + O(\zeta_n r)^{-1}], \quad \zeta_n r \to \infty \tag{10.2}$$

$$\mathbf{u}_n(\mathbf{x}) = \mathbf{a}_n(\varphi, z)e^{i\zeta_n r}/\sqrt{\zeta_n r},$$

$$\mathbf{a}_n = \sqrt{\frac{i}{2\pi}}\zeta_n \operatorname{res} K(-\alpha, \varphi, z)|_{\alpha=\zeta_n}\mathbf{Q}(-\zeta_n, \varphi).$$

Here, $N$ is the number of positive real poles $\zeta_n$ of the matrix $K$ elements; in the infrequent case of backward GW associated with a pole $\zeta_n$, the negative counterpart $-\zeta_n$ gets into the closed contour and contributes into expansion (10.2) instead of $\zeta_n$, (Vorovich and Babeshko 1979).

The poles $\alpha = \pm\zeta_n$ are roots of the characteristic (dispersion) equation

$$\Delta(\alpha, \omega) = 0, \tag{10.3}$$

where $\Delta$ is a common denominator of the $K$ matrix elements. The roots continuously depend on the regular frequency $\omega$ specifying dispersion curves $\zeta = \zeta_n(\omega)$ in the frequency-wavenumber.

The term $\mathbf{u}_n$ of expression (10.2) associated with real poles $\zeta_n$, $n = 1, 2, ..., N$ describe cylindrical traveling GWs. In this context, $\zeta_n$ are their wavenumbers, and the residues yield amplitude factors $\mathbf{a}_n(\varphi, z)$ providing their modal eigenforms as functions of the depth $z$ and direction of propagation $\varphi$. With a homogeneous elastic half-space, Eq. (10.1) has only one real root ($N = 1$) that yields the classical Rayleigh wave; with a free layer, the terms $\mathbf{u}_n$ are Lamb waves, etc. For the coated half-space under consideration, the GWs $\mathbf{u}_n$ are surface acoustic waves (SAWs), propagating from the source to infinity with the phase velocities $c_n = \omega/\zeta_n$, slownesses $s_n = \zeta_n/\omega$, and group velocities $v_n = \mathrm{d}\zeta_n/\mathrm{d}\omega$. The complex roots $\zeta_n = \mathrm{Re}\,\zeta_n + i\mathrm{Im}\,\zeta_n$, $\mathrm{Im}\,\zeta_n > 0$, $n \geq N + 1$, give the pseudo-surface acoustic waves (PSAW) that propagate with the additional exponential decay, controlled by the imaginary part $\mathrm{Im}\,\zeta_n$: $|\mathbf{u}_n| \sim O(e^{-|\mathrm{Im}\,\zeta_n|r}/\sqrt{|\zeta_n r|})$ as $|\zeta_n r| \to \infty$.

The GWs $\mathbf{u}_n$ transfer to infinity the wave energy supplied by the source load $\mathbf{q}$. In the time-harmonic wave field $\mathbf{u}e^{-i\omega t}$, the energy carried by GWs is quantified by the time-averaged over the period of oscillation $T = 2\pi/\omega$ amount of wave energy transferred through the lateral cylindrical surface $r = \mathrm{const}$, $0 \leq \varphi < 2\pi$, $-\infty < z \leq 0$ containing the source. The closed analytical expressions for the total source energy $E_0$ and its parts $E_V$ and $E_n$, $n = 1, 2, ..., N$, carried by the volume and traveling waves have been derived in terms of the matrix $K$ elements (Glushkov 1983; Glushkov et al. 2015). They have proven to be a convenient tool for analyzing the influence of coating structure on the source energy partition among the excited SAWs, which is performed below in parallel with analyzing their dispersion properties.

## 10.3 Numerical Examples

In this section, the numerical results are given in dimensionless form. Since each cell of the coating forms a waveguide channel, we chose its thickness $h$ as the unit of length $l_0$. Such a choice makes the results better comparable in the frequency domain. The other two characteristic values are the velocity of shear waves $c_0 = c_s$ in the harder material $A$ and its density $\rho_0 = \rho_A$. In these units, the dimensionless angular frequency is $\omega = 2\pi f l_0/c_0 = 2\pi f h/c_s$, and the unit of time-averaged energy (power) is $e_0 = \rho_0 c_0^3 l_0^2 = \rho_A c_s^3 h^2$.

The figures below illustrate the features of waveguide properties of the coatings with alternating hard and soft interlayers. For definiteness, we took the dimensionless parameters of alumina ($Al_2O_3$) and aluminum (Al) as the hard material $A$: $c_p = 1.691, c_s = 1, \rho = 1$, and soft material $B$: $c_p = 0.972, c_s = 0, 490, \rho = 0.675$, respectively. Their dimensional parameters can be obtained with the characteristic values $c_0 = 6373$ m/s and $\rho_0 = 4000$ kg/m$^3$.

The numerical examples differ by the number of cells $M$ in the coatings. To see the effect of specific stratification laws ($z$ dependences), every coating is considered

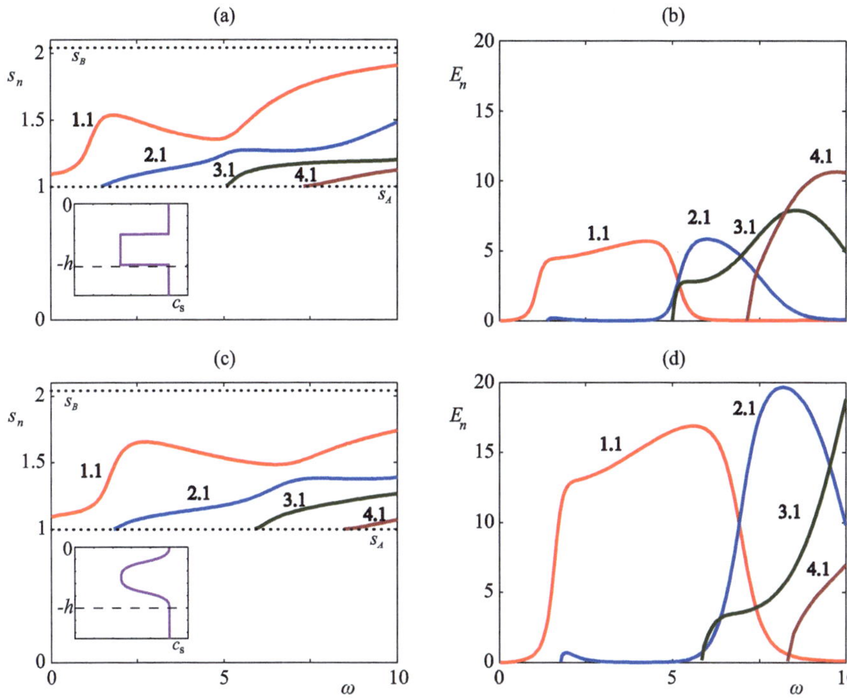


**Fig. 10.3** Slowness dispersion curves (left) and source energy partition among the excited traveling waves (right) with a single-cell laminate (top) or FGM (bottom) coating ($M = 1$)

for two extreme cases: stepwise lamination ($h_g = 0$) and full FGM ($h_g = h/2$). To exclude the influence of the load type and area, and to focus on studying the effect of coating structure, the numerical analysis is performed for the vertical point load $\mathbf{q} = (0, 0, \delta(x, y))$, which Fourier symbol is $\mathbf{Q} = (0, 0, 1)$. Subplots (a) and (c) of each figure depict the dispersion properties of the SAWs generated with such $M$-cell laminate and FGM coatings, while subplots (b) and (d) show the source energy partition among these waves depending on the frequency. For visual clarity, we plot slowness dispersion curves $s_n(\omega)$, which vary within a limited range between the $S$-wave slownesses $s_A = 1$ and $s_B = 2.041$ in the materials $A$ and $B$ in the case; unlike the conventionally used wavenumber curves $\zeta_n(\omega)$ that grow indefinitely with $\omega$, or the phase velocity curves $c_n(\omega)$ that descend from infinity in the limiting cases of stress-free or bottom-clamped coatings (super-soft or super-hard substrates).

The dispersion curves in Fig. 10.3a,c show that the inner waveguide channel, appearing in the case of one-cell coating ($M = 1$), supports a set of traveling wave modes. Like Lamb waves, their number increases with frequency, and they exhibit dispersion via increasing slowness (i.e., decreasing phase velocity) from the level $s_n = s_A = 1$ at the cutoff frequencies (except for the first mode). At lower frequencies, the poles are complex with the real part below this level, giving damped PSAWs.

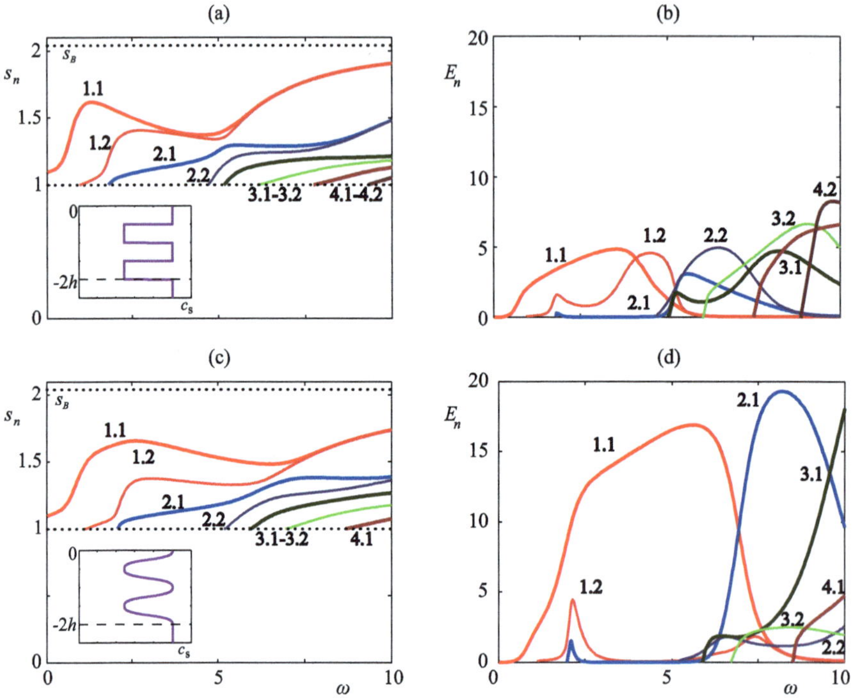


**Fig. 10.4** Same as in Fig. 10.3, but for two-cell coatings ($M = 2$)

Comparing the curves in subplots (a) and (c) indicates a small effect of the type of stratification on the dispersion properties. At the same time, in the FGM case (Fig. 10.3d), the amount of energy supplied from the source to the GW modes is more significant than with the laminated coating (Fig. 10.3b). This can be explained by the increased wave reflection from the contrast interface between the hard and soft materials at the boundary $z = -h/2$, while the FGM coating is without sharp reflecting interfaces.

The energy of each mode is localized in a limited frequency range. These ranges successively replace one another with the emergence of each new traveling wave. We can say that the source energy is sequentially transferred from one mode to another as they appear. This is the effect of mode-to-mode energy transfer inherent in multimodal waveguides (Glushkov et al. 2015).

With two cells ($M = 2$, Fig. 10.4), there are two inner channels supporting GWs. Since the geometry, mechanical properties, and environment of the soft interlayers are locally the same, the velocities and eigenforms of GWs propagating along the different channels should be similar. At least at high frequencies, when the wavelength is comparable with the channel's cross section. In the dispersion curve patterns, this manifests in the fact that the curves of the counterpart GWs approach each other with increasing frequency, practically merging and following along the trajectories

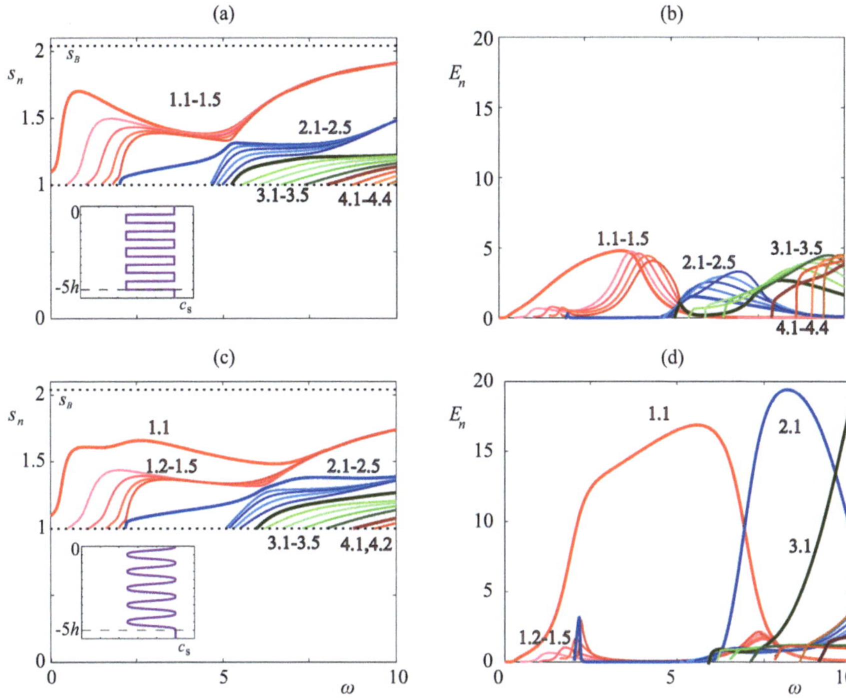


**Fig. 10.5** Same as in Figs. 10.3 and 10.4, but for five-cell coatings ($M = 5$)

for the corresponding GWs in the single-channel structure shown in Fig. 10.3. It looks like curves' splitting if to follow them reversely, decreasing frequency (Fig. 10.4a,c). Such splitting is especially pronounced in a multichannel coverage when the curves look like sheaves of lines merging with increasing frequency ($M = 5$, Fig. 10.5a,c). Therefore, in these figures, the curves are numbered with two digits: 1.1, 1.2, 2.1, etc. The first is the number of the corresponding parent curve in the single-channel case ($M = 1$, Fig. 10.3), and the second digit is the channel number.

With multichannel coatings, the effect of mode-to-mode energy transfer holds. But its manifestation with laminate or FGM channels is different. In a stepwise coverage, each GW mode of the same family carries an approximately equal amount of energy (e.g., the pairs of modal energy curves 1.1–1.2, 2.1–2.2, and 3.1–3.2 in Fig. 10.4b). On the contrary, in the FGM case (Fig. 10.4d), the first modes of these pairs carry the most energy in their frequency ranges. With increasing the number of inner channels, this difference in energy distribution among the family set modes becomes even more pronounced (Fig. 10.5b, d). Moreover, with FGM coatings, the first modes' energy in the family sets remain almost unchanged (compare curves 1.1, 2.1, and 3.1 in subplots (d) of Figs. 10.3, 10.4 and 10.5).

A qualitative difference in the indentation of laminates and FGM coatings has already been noted (Aizikovich 1982). The above examples show that dynamic

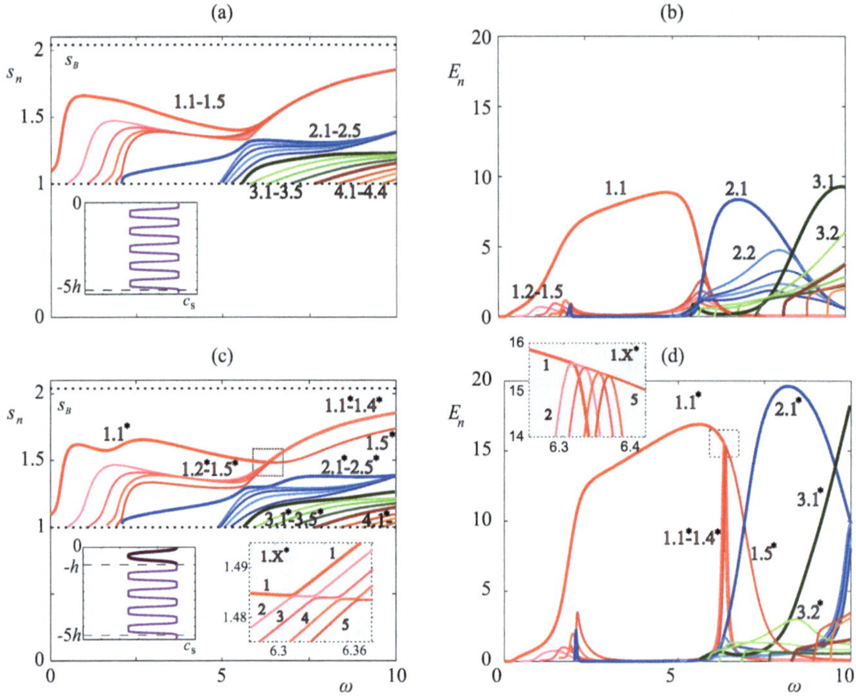


**Fig. 10.6** Five-cell coatings ($M = 5$) as in Fig. 10.5, but with another FG law (top), and with a slight change in the FG law for the top cell (c); the latter illustrates the occurrence of the mode repulsion (d) with resonance peaks of modal energy (right) shown in the enlarged inserts

response and wave energy transfer can also be qualitatively different in these structures. And wave processes are even more sensitive to stratification patterns than static indentation. Examples of such sensitivity are shown in Fig. 10.6. Its top subplots (a) and (b) are for a five-cell FGM coating (as in Fig. 10.5c, d), but with another law of continuous dependence on z: $h_g = h_A = h_B = 0.25$ instead of the fully FG variation with $h_g = 0.5$. This small variation in the FG dependence does not change the dispersion curves, but the modal energy almost halved, decreasing toward the limiting case of a purely laminate coating with $h_g = 0$ (Fig. 10.5b). However, the first mode of each family still carries the most of energy.

Further, changing the properties of only one cell (namely, the top cell is replaced here with the fully FG cell ($h_g = 0.5$)) again almost does not change the dispersion curves but increases the modal energy to the previous level (Fig. 10.6c, d). This indicates that the total level of energy supplied by the source is primarily determined by the top part of the substructure containing the upper channel.

Besides, this cell replacement leads to a qualitatively new phenomenon of mode repulsion (Zharnikov and Syresin 2015). The seemingly intersecting dispersion curves of the first family do not actually intersect but converge closely and then diverge as if repelling each other (the enlarged inset in Fig. 10.6c). In addition, sharp resonance peaks of wave energy carried by these modes arise at the repulsion frequencies (the enlarged inset in Fig. 10.6d).

## 10.4 Concluding Remarks

Numerical analysis has shown the influence of the law of periodic stratification on the frequency dynamic response of a coated elastic half-space. In particular, while the dispersion curve patterns change slightly with the transition from layered to FGM coverages, the amount of source energy increases significantly. The effects of dispersion curve splitting and mode-to-mode energy transfer in a multi-channel coating and the mode repulsion with resonance increasing of modal energy are also numerically demonstrated and discussed.

**Acknowledgements** The work is supported by the Russian Science Foundation (Project No. 17-11-01191).

## References

Aizikovich SM (1982) Asymptotic solutions of contact problems of elasticity theory for media inhomogeneous in depth. J Appl Math Mech 46:116–124
Aizikovich SM, Alexandrov VM (1982) Properties of compliance functions for a layered and continuously nonuniform halfspace. Sov Phys Doklady 27:765–767
Aizikovich SM, Krenev LI, Trubchik IS (2008) The deformation of a half-space with a gradient elastic coating under arbitrary axisymmetric loading. J Appl Math Mech 72:461–467
Aizikovich SM, Vasil'ev AS, Krenev LI et al (2011) Contact problems for functionally graded materials of complicated structure. Mech Compos Mater 47(5):539–548
Alekseev AS, Mikhailenko BG (1976) Solution of Lamb's problem for a vertically inhomogeneous elastic half-space. Izv Akad Nauk SSSR Fiz Zemli 12:11–25
Babeshko VA, Glushkov EV, Glushkova NV (1987) Methods of constructing Green's matrix of a stratified elastic half-space. USSR Comput Math Math Phys 27(1):60–65
Gentilleau B, Bertin M, Touchard F et al (2011) Stress analysis in specimens made of multi-layer polymer/composite used for hydrogen storage application: comparison with experimental results. Compos Struct 93:2760–2767
Glushkov EV (1978) Dynamic contact problems for arbitrarily shaped stamps on inhomogeneous media. Dis. Cand. fiz.-mat.nauk, 01.02.04, 134 p. Rostov State University
Glushkov EV (1983) Energy distribution of a surface source in an inhomogeneous half-space. PMM U.S.S.R. 47(1):70–75
Glushkov EV, Glushkova NV, Fomenko SI, Zhang C (2012) Surface waves in materials with functionally gradient coatings. Acoust Phys 58(3):339–353
Glushkov E, Glushkova N, Fomenko S (2015) Wave energy transfer in elastic half-spaces with soft interlayers. J Acoust Soc Am 137(4):1802–1812
Glushkova NV (1981) Propagation of elastic waves in stratified media and their interaction with surface objects. Dis. Cand. fiz.-mat.nauk, 01.02.04, 204 p. Rostov State University
Haskell NA (1953) The dispersion of surface waves on multilayered media. Bull Seismol Soc Am 43:17–34
Knopoff L (1964) A matrix method for elastic waves problems. Bull Seismol Soc Am 54:431–438

Mikhasev GI, Botogova MG, Eremeyev VA (2021) On the influence of a surface roughness on propagation of anti-plane short-length localized waves in a medium with surface coating. Int J Eng Sci 158:103428

Petrashen' GI (1952) Propagation of elastic waves in an isotropic-layered media, divided by parallel planes. Uch Zap LGU, Leningrad 26:162

Thomson WT (1950) Transmission of elastic waves through a stratified medium. J Appl Phys 21:89–93

Vasiliev A, Volkov S, Aizikovich S et al (2014) Axisymmetric contact problems of the theory of elasticity for inhomogeneous layers. ZAMM Zeitschrift fur Angewandte Mathematik und Mechanik 94:705–712

Vasiliev AS, Swain MV, Aizikovich SM, Sadyrin EV (2016) Torsion of a circular punch attached to an elastic half-space with a coating with periodically depth-varying elastic properties. Arch Appl Mech 86:1247–1254

Volkov S, Aizikovich S, Wang Y-S et al (2013) Analytical solution of axisymmetric contact problem about indentation of a circular indenter into a soft functionally graded elastic layer. Acta Mechanica Sinica/Lixue Xuebao 29:196–201

Vorovich II, Babeshko VA (1979) Dynamic mixed problems of elasticity for non-classical domains. Nauka, Moscow [in Russian]

Vorovich II, Aleksandrov VM, Babeshko VA (1974) Non-classical mixed problems of elasticity theory. Nauka, Moscow [in Russian]

Zharnikov TV, Syresin DE (2015) Repulsion of dispersion curves of quasidipole modes of anisotropic waveguides studied by finite element method. J Acoust Soc Am 137(6):EL396–EL402

# Chapter 11
# Features of High-Speed Deformation and Fracture of Fine-Grained Concrete Under Tensile Stress

**Mikhail E. Gonov, Anatoly M. Bragov, Aleksandr Yu. Konstantinov, Andrey K. Lomunov, and Andrey R. Filippov**

**Abstract** This paper analyzes an experimental study of high-rate deformation and fracture of fine-grained concrete under tensile stresses. A number of sources of foreign and domestic authors are analyzed. Based on the analysis, it was concluded that some properties of concrete have not been fully investigated. In connection with the above, an experimental study of the dynamic properties of concrete materials is relevant today. Experimental studies were carried out on the basis of modifications of the Kolsky method. These modifications make it possible to determine the strength and time characteristics of concrete deformation under high-speed loading. To analyze the nature and time of the destruction, experiments were carried out using high-speed photography. Splitting (Brazilian test) and straight tensile experiments have two speed modes. On the basis of the performed experimental work, tensile strain diagrams and stress versus time diagrams during splitting were obtained. The experimental data indicate the effect of the strain rate on the ultimate tensile strength characteristics. The dynamic tensile strength of fine-grained concrete was about 8 MPa. The value of the coefficient of dynamic tensile hardening is in the range from 4 to 6 MPa. This factor depends on the strain rate. The opposite effect of the influence of the loading rate was obtained both during splitting and stretching. This means that with an increase in the strain rate, the maximum breaking stresses decrease. On the basis of high-speed photography, the features of high-speed destruction of fine-grained concrete under tensile stresses are revealed. The found characteristics can be used to create mathematical models necessary to determine the strength of concrete structures subjected to dynamic effects.

M. E. Gonov (✉) · A. M. Bragov · A. Yu. Konstantinov · A. K. Lomunov · A. R. Filippov
National Research Lobachevsky State University of Nizhny Novgorod, 23 Gagarin Avenue, Nizhny Novgorod, Russian Federation 603950
e-mail: gonov_mikhail@mech.unn.ru

A. M. Bragov
e-mail: bragov@mech.unn.ru

A. K. Lomunov
e-mail: lomunov@mech.unn.ru

A. R. Filippov
e-mail: filippov@mech.unn.ru



H. Altenbach et al. (eds.), *Advanced Materials Modelling for Mechanical, Medical and Biological Applications*, Advanced Structured Materials 155,
https://doi.org/10.1007/978-3-030-81705-3_11

**Keywords** Dynamic strength · Stress · Deformation · Strain rate · Concrete · Dynamic tests · Kolsky method · Mechanics

## 11.1 Introduction

The relevance of studying the dynamic properties of brittle materials, in particular concretes, is due to the fact that concrete is a widespread structural material all over the world. Concrete is used in the construction of buildings and structures for both civil, industrial, and military purposes. During operation, as a result of emergency situations, concrete structures can experience intense dynamic effects. Such loads include natural disasters in the form of typhoons, tsunamis, landslides, earthquakes; man-made disasters of civil or industrial origin; terrorist threats in the form of detonating explosive devices of various types.

The variability of concrete materials leads to the need for a comprehensive study of the characteristics of concrete and their influence on the dynamic properties of the material. First of all, this is necessary for the construction of mathematical models describing its deformation and destruction during calculations. To calculate the strength of concrete structures in emergency situations, it is necessary to know how dynamic loads affect the physical and mechanical properties of concrete.

The strength characteristics of the materials are required for the calculation and design engineer. The study of the dynamic properties of concrete under tensile dynamic loads allows one to obtain experimental data, various parameters and constants. This data are required to equip dynamic deformation and fracture models. Subsequently, mathematical models, based on experimental data, replace the behavior of a real material when calculating the stress–strain state and strength of structures in such software and computing systems as ANSYS, LS-DYNA, ABAQUS, LOGOS, etc.

Today, the world scientific community is actively working on the study of the dynamic properties of brittle materials. Agar Ozbek et al. (2013) investigate the dynamic properties of porous concrete, which, due to its characteristics, breaks down into small fragments when exposed to an impact load due to the formation of many cracks. In this case, porosity is one of the main factors affecting the nature of dynamic deformation and fracture. In Hao et al. (2013a; b); Flores-Johnson (2017), a number of factors are investigated that affect the dynamic strength of concrete during impact tests. Erzar and Forquin (2010) are presented an experimental method for determining the tensile strength of concrete at high strain rates, in which a significant increase in dynamic strength was observed under the influence of the strain rate. Pedersen et al. (2013), Al-Salloum et al. (2015), Kwak and Gang (2015), Navas et al. (2018), Pereira et al. (2018) describe the modeling of the dynamic properties of concrete under tension, which says that the loading rate has a significant effect on the fracture processes and overall strength. In Xiao et al. (2015), the characteristics of concrete made from recycled aggregate under the influence of shock loading, the effect of the amount of aggregate, and its moisture content on strength indicators are studied. Guo et al. (2017) investigate the reaction of high-strength concrete to dynamic compressive load using the Kolsky method. In Kettenbeil and Ravichandran (2018), an experimental study of the dynamic behavior of an innovative concrete composite designed to reduce structural damage confirmed by high-speed loads is presented. Based on the analysis of the above sources, it can be concluded that some of the properties of concrete have not been fully investigated, and today, the experimental study of this direction is relevant. The work is devoted to the study of the features of high-speed deformation and fracture of fine-grained concrete under tension.

## 11.2 Experimental Methods

There are a number of methods for studying the dynamic properties of concrete, among which the most frequently used and effective method is the Kolsky method (Kolsky 1949). Modifications of this method make it possible to obtain a complex system of basic experiments (Bragovet al. 2020; Bragov et al. 2015, 2018a, b). Fine-grained concrete of class B22.5 was tested, the composition of which is shown in Table 11.1.

An experimental setup according to the Kolsky method includes a system of measuring rods between which a test specimen is installed, a gas gun (for accelerating a cylindrical striker), strain gauges, a speed meter, recording, and computing equipment with a software package (Fig. 11.1).

The study of the properties of concrete under dynamic tension was carried out using the following schemes: direct stretching using a modification of the Kolsky method (Bragov et al. 2018a, b); splitting ("Brazilian test" or indirect tension) (Rodriguez et al. 1994). The above experimental methods allow dynamic testing of concrete in a wide range of deformation rates ($10^2$–$10^3$ s$^{-1}$).

The experimental setup for direct tension (Fig. 11.2) differs from the traditional compression setup according to the Kolsky method in that the tubular striker is accelerated in the barrel by means of a gas gun, after which it strikes the anvil, which

**Table 11.1** Composition of fine-grained concrete

| S. No. | Material | Consumption per 1 m$^3$ (kg) |
|---|---|---|
| 1 | Cement grade 500D | 430 |
| 2 | Coarse sand (module 3 mm) | 1435 |
| 3 | Fine sand (module 1.6 mm) | 355 |
| 4 | Plasticizing additive Murasan BWA 16 | 1.6 |
| 5 | Water | 120 |

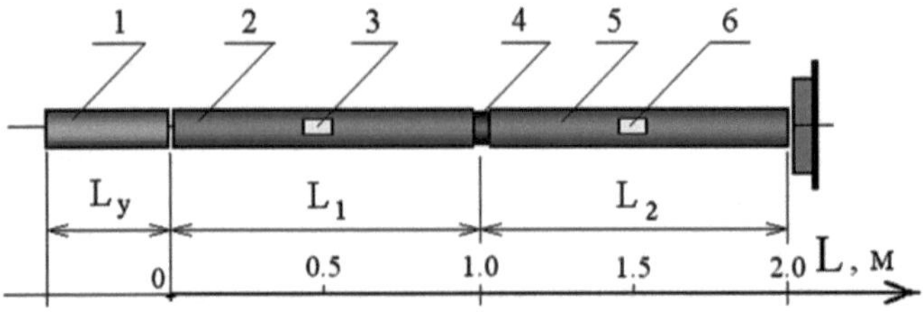


**Fig. 11.1** Schematic of the Kolsky compression method. *1*—striker, *2*—loading rod, *3* and *6*—strain gauges, *4*—specimen, *5*—support rod (Bragov and Lomunov 2017)

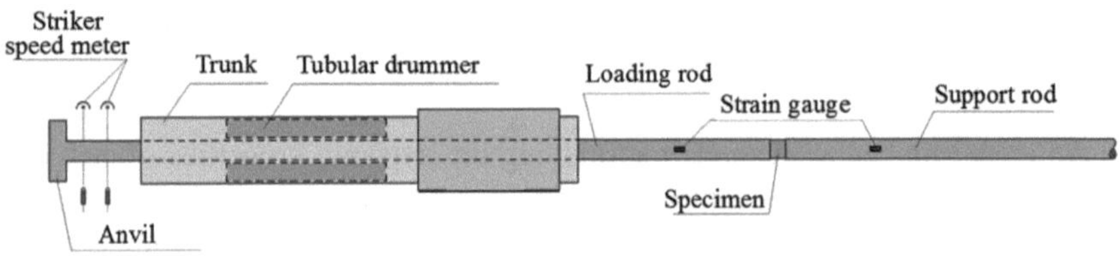


**Fig. 11.2** Schematic of the experimental setup for straight tension

is fixed at the end of the loading measuring rod, thereby exciting an elastic wave in it stretching. The specimen, in turn, is either glued to measuring rods or to special threaded nozzles, which are screwed into measuring rods with a diameter of 20 mm made of high-strength steel with an elastic modulus of $1.85 \times 10^5$ MPa.

Direct tensile concrete specimens with a diameter of just under 20 mm and a length of 10 mm were drilled using a bench drill. After that, the specimen was glued with epoxy glue to the threaded nozzles. To increase the strength of the glue line, tungsten powder was added to the epoxy glue.

The splitting experiments were carried out using a traditional compression apparatus according to the Kolsky method with a slight modification. The difference lies in the fact that the tested cylindrical specimen is installed along the cylindrical generatrix. During testing, the specimen is split along the diametrical plane into two halves. The test setup is shown in Fig. 11.3.

In this modification, duralumin measuring rods with a diameter of 20 mm of solid (loading) and hollow (support) sections with an elastic modulus of $0.71 \times 10^5$ MPa were used. A tubular support rod was used for more accurate recording of elastic

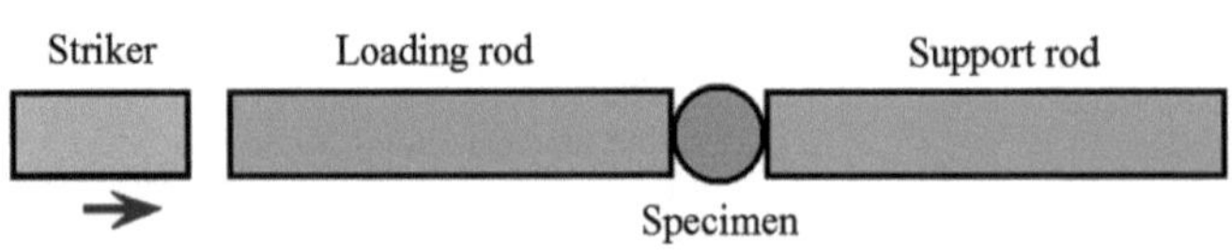


**Fig. 11.3** Schematic splitting test

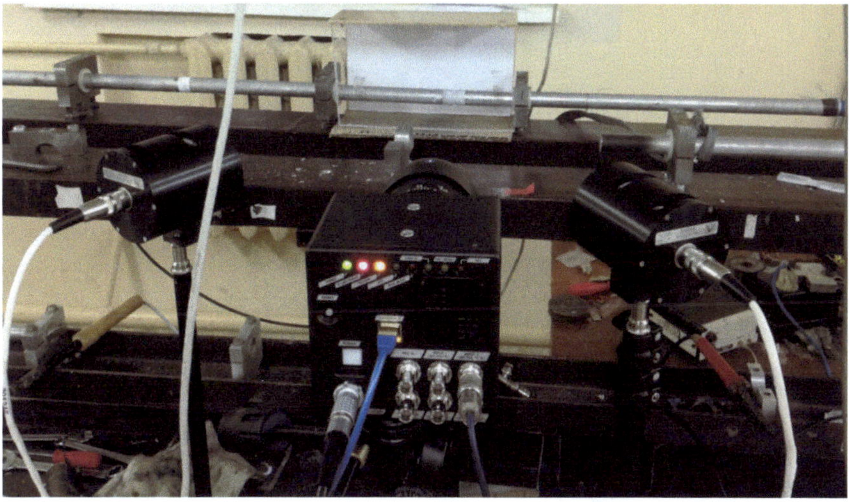

**Fig. 11.4** FASTCAM Mini U×100 high-speed camera

pulses. Concrete splitting test specimens with a diameter of 20 mm and a length of 10 mm were also drilled using a bench drill.

All tests were carried out using a FASTCAM Mini UX100 high-speed camera for subsequent analysis of the time and nature of the destruction of the specimen (Fig. 11.4). For the qualitative detection of initial cracks and deformation, the specimens were pre-painted white with a spot layer of black paint.

## 11.3 Experimental Results Under Dynamic Direct Tension

In direct tension tests, two speed modes were implemented. Experimental data and some parameters determined from their results are shown in Table 11.2.

In the first mode (Fig. 11.5), five experiments were carried out (01–05). The average speed of the striker was 7 m/s, the average deformation rate was 500 $s^{-1}$, and the average maximum tensile stress was 12 MPa.

Figures 11.5, 11.6, and 11.8 show diagrams of tests $\sigma(\varepsilon)$ and $\dot{\varepsilon}(\varepsilon)$ of concrete specimens for direct tension according to the modified Kolsky method. In this series of experiments, in all the figures, the solid lines show the stress–strain diagrams, and the dashed lines of the same color show the strain rate–strain diagrams (the corresponding axis is located on the right).

In the second mode (Fig. 11.6), five experiments were also performed (06–10). The average speed of the striker was 11 m/s, the average strain rate was 700 $s^{-1}$, and the average maximum tensile stress was 8 MPa. Figure 11.7 shows the nature of the specimen destruction.

Figure 11.8 shows the average tensile test diagrams of concrete.

**Table 11.2** Parameters of direct tension experiments

| Mode | No. experiment | Loading parameters | | | Specimen parameters | | | Experiment results | | | | |
|---|---|---|---|---|---|---|---|---|---|---|---|---|
| | | Striker length (mm) | Striker material | Striker speed (m/s) | $D$ (mm) | $S_{c.s.}$ (mm $^2$) | $L$ (mm) | Maximum stress (MPa) | Ultimate deformation (%) | Strain rate (1/s) | Lifetime (μs) | |
| | | | | | | | | | | | Diagram | Photography |
| 1 | t637-01 | 200 | Duralumin | 7.11 | 18.64 | 272.7 | 10.51 | 13.43 | 1.31 | 539.07 | 43 | 37.5 |
| | t637-02 | 200 | Duralumin | 7.87 | 18.51 | 268.9 | 10.52 | 13.79 | 1.10 | 567.69 | 34.5 | – |
| | t637-03 | 200 | Duralumin | 7.64 | 18.65 | 273.0 | 10.35 | 11.20 | 0.89 | 525.76 | 35 | 37.5 |
| | t637-04 | 200 | Duralumin | 6.38 | 18.69 | 274.2 | 10.41 | 13.04 | 0.831 | 380.64 | 34 | 37.5 |
| | t637-05 | 200 | Duralumin | 5.94 | 18.7 | 274.5 | 10.39 | 13.01 | 1.00 | 373.56 | 39 | 37.5 |
| 2 | t637-06 | 200 | Duralumin | 10.85 | 18.28 | 262.3 | 10.37 | 8.94 | 0.92 | 723.52 | 25.5 | 25 |
| | t637-07 | 200 | Duralumin | 11.23 | 18.53 | 269.5 | 9.96 | 8.24 | 1.11 | 844.88 | 26.5 | – |
| | t637-08 | 200 | Duralumin | 11.39 | 18.47 | 267.7 | 10.35 | 10.75 | 0.95 | 750.21 | 21 | 25 |
| | t637-09 | 200 | Duralumin | 10.54 | 18.2 | 260.0 | 10.42 | 6.44 | 0.66 | 545.00 | 24 | – |
| | t637-10 | 200 | Duralumin | 10.95 | 18.09 | 256.8 | 9.63 | 8.49 | 0.93 | 727.69 | 29.5 | 25 |

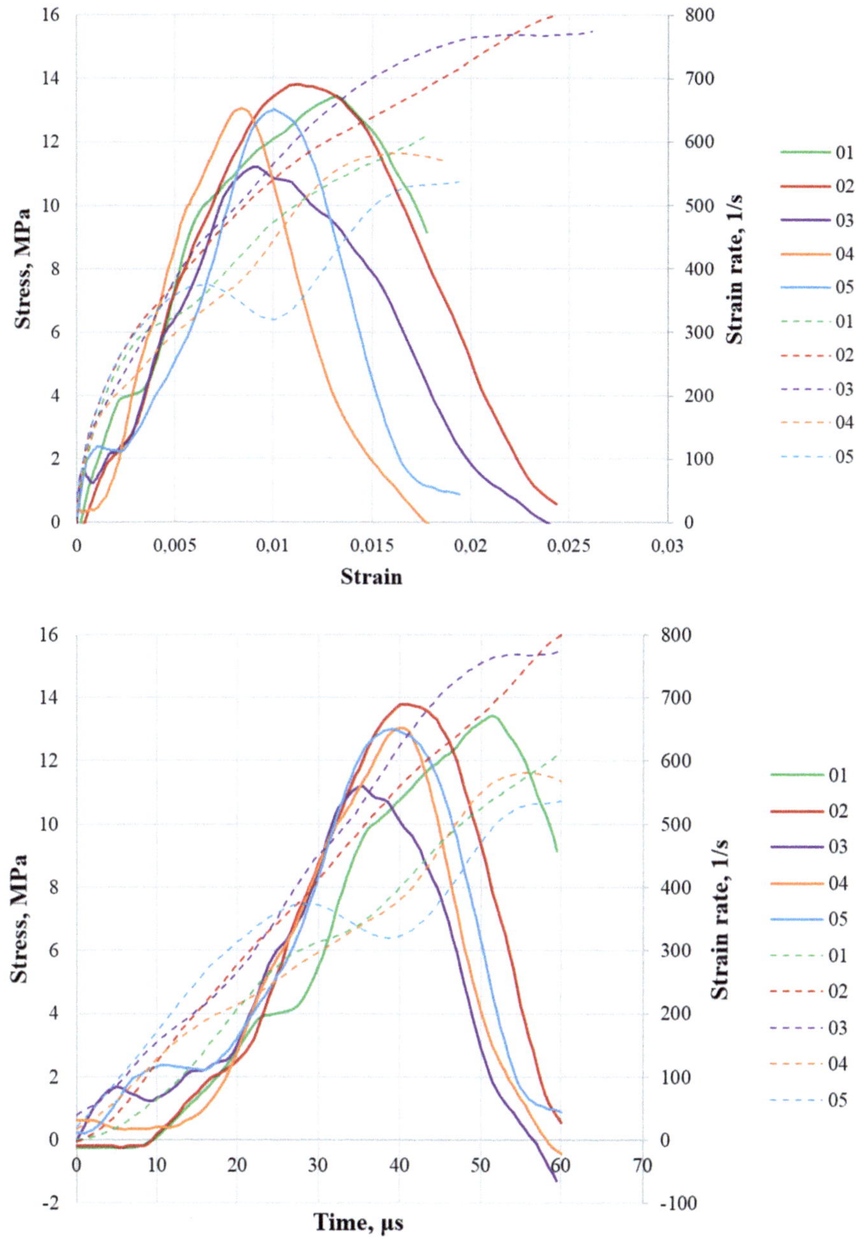


**Fig. 11.5** Diagrams of tensile test mode No. 1

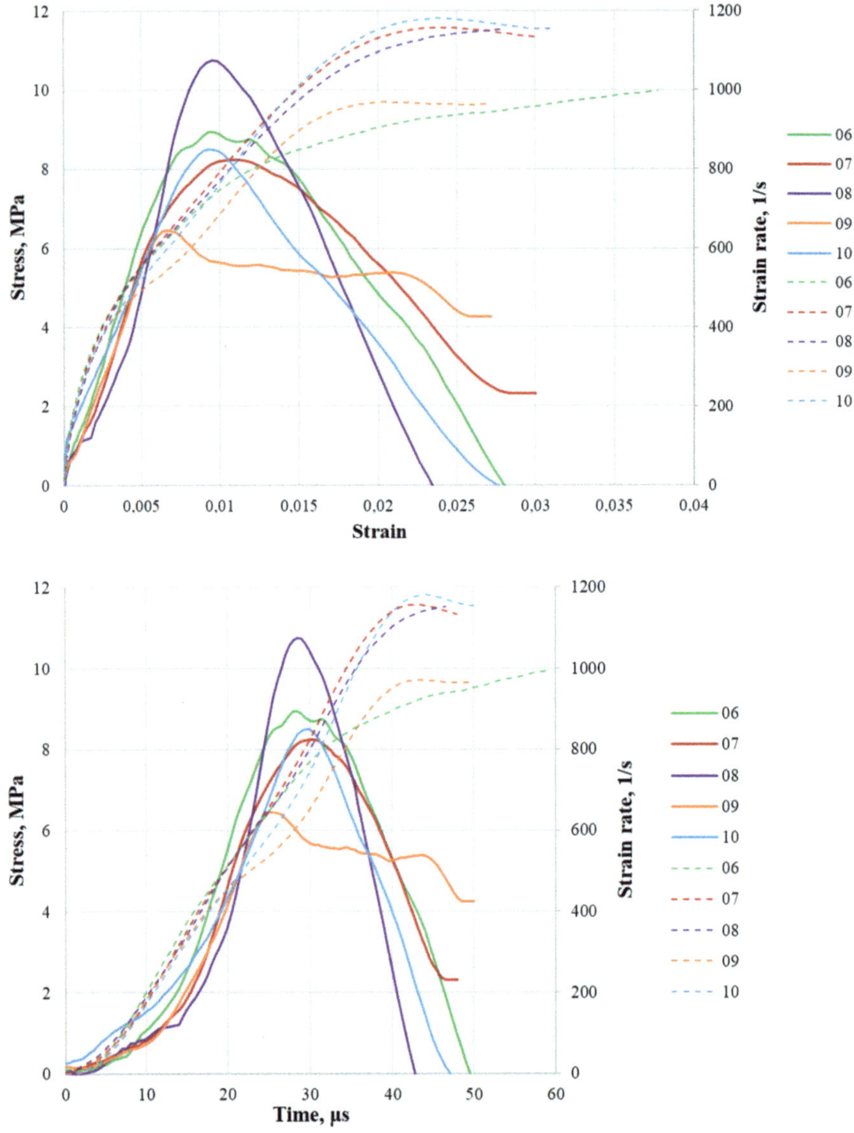


**Fig. 11.6** Diagrams of tensile test mode No. 2

**Fig. 11.7** Nature of destruction of the specimen

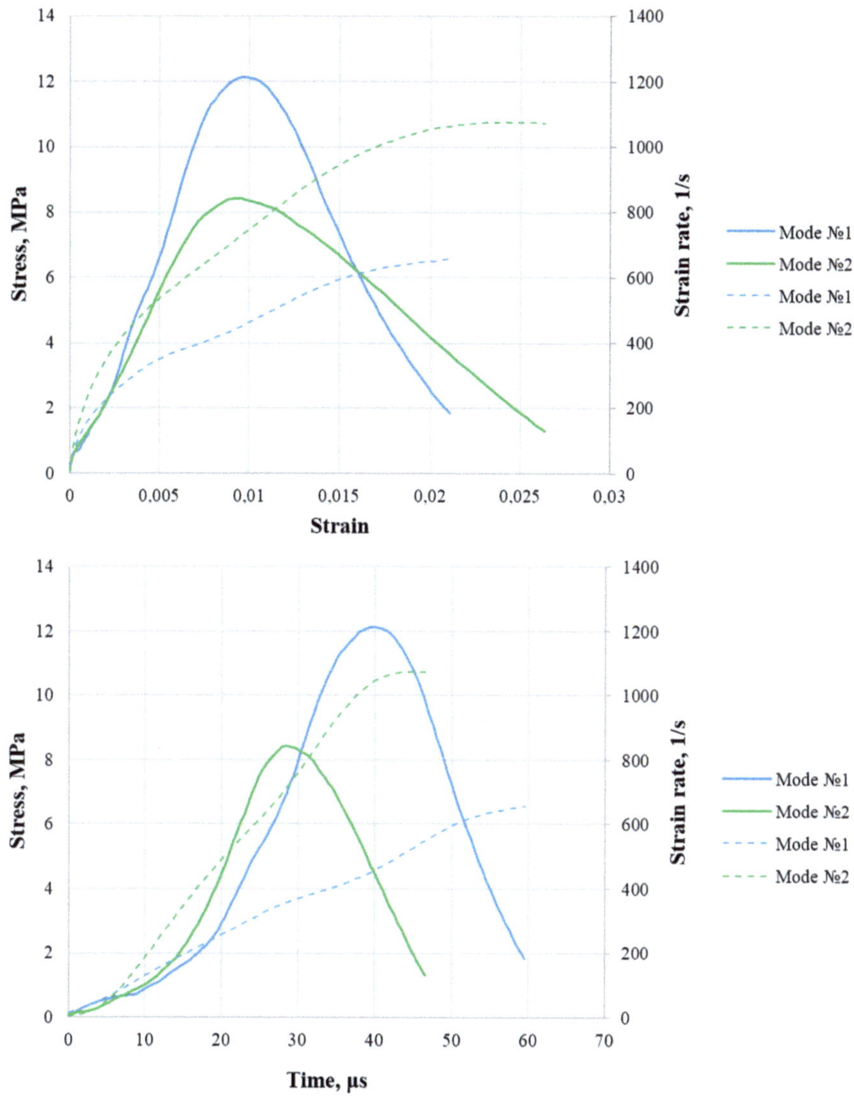


**Fig. 11.8** Averaged diagrams of stress versus time and stress versus deformation modes No. 1 and No. 2

Figure 11.9 shows diagrams of the dependence of the time to the onset of fracture and the maximum stress on the strain rate.

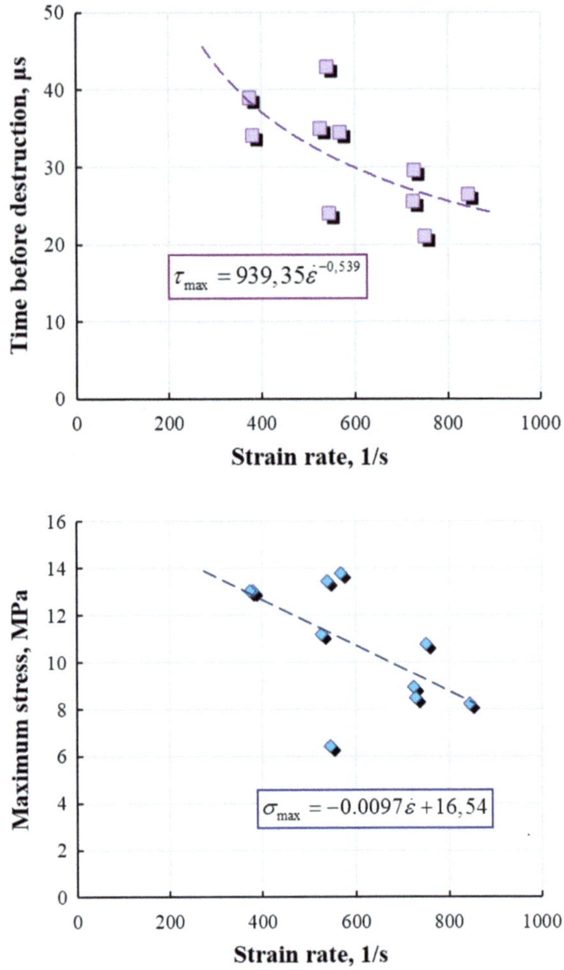


**Fig. 11.9** Diagrams of the dependence of the time to the onset of fracture and the maximum stress on the strain rate

Thus, the direct tensile test results show a negative strain rate sensitivity when the dynamic strength decreases at a higher strain rate. Figure 11.10 shows the chronology of high-speed photography of experiment No. 03 of the first high-speed mode. Figure 11.11 shows the chronology of high-speed photography of experiment No. 10 of the second high-speed mode.

The analysis of high-speed photography confirms the time to the beginning of the destruction of the specimen, shown in the diagrams above and in Table 11.2. On this basis, the maximum stress at the moment of failure can be considered reliable.

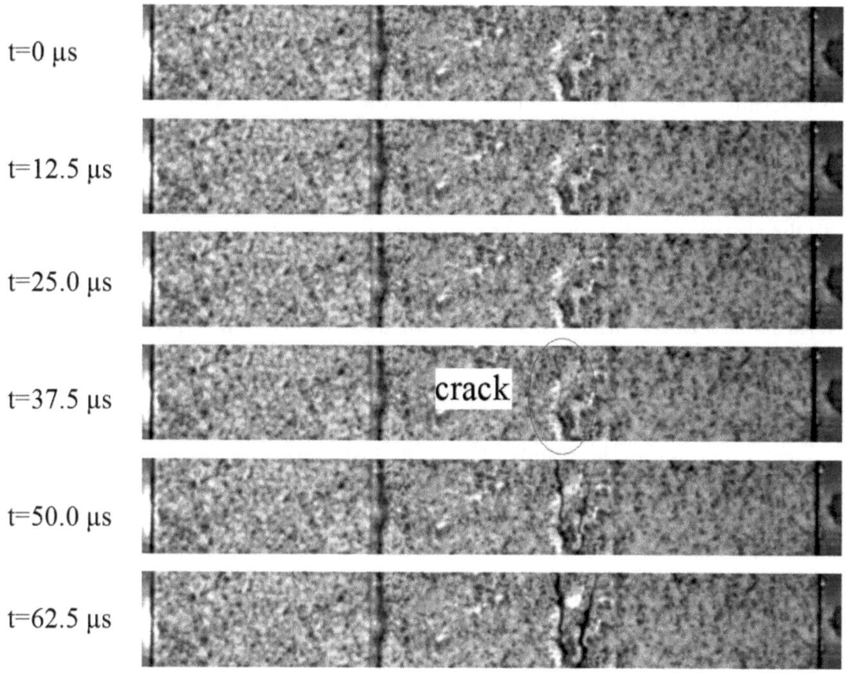


**Fig. 11.10** Chronology of high-speed photography of experiment No. 03

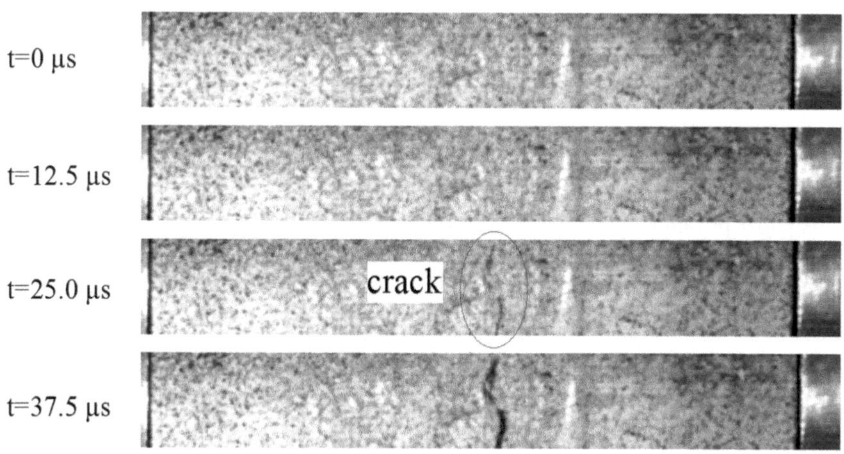


**Fig. 11.11** Timeline of high-speed photography of experiment No. 10

## 11.4 Experimental Results for Dynamic Splitting

To determine the ultimate tensile stress of fine-grained concrete during splitting ("Brazilian test"), two speed modes are implemented. Experimental data and some parameters determined from their results are shown in Table 11.3.

In the first mode (Fig. 11.12), five experiments were performed (r637-11–r637-15); the average speed of the striker was 7 m/s, the average maximum splitting stress was 8 MPa, and the average time to the onset of fracture was 30 μs. In the second speed mode (Fig. 11.13), six experiments were performed (r637-16–r637-21); the average speed of the striker was 17 m/s, the average maximum splitting stress was 6 MPa, and the average time to the onset of fracture was 22 μs.

Figure 11.14 shows the nature of the destruction of the specimen, Fig. 11.15 shows the averaged diagrams of concrete splitting tests.

Thus, according to the results of splitting tests, a decrease in dynamic strength is observed at a higher loading rate. Figure 11.16 shows the chronology of high-speed photography of experiment No. 14 of the first high-speed mode. Figure 11.17 shows the chronology of high-speed photography of experiment No. 19 of the second high-speed mode.

Analysis of high-speed photography showed that the destruction of the specimen occurred before reaching the maximum stresses in the diagrams shown above. Note that there are two voltage peaks in the diagrams. According to high-speed photography, the first stress peak corresponds to the formation of the first crack, and the maximum stresses are achieved due to the resistance of the specimen split into two parts.

## 11.5 Conclusions

This paper presents the results of a study of the dynamic properties of fine-grained concrete under tensile dynamic loads. The test procedure, the parameters of the specimens, and the modes of the experiments performed are presented. Based on the results obtained, the following conclusions can be drawn.

1. The system of experiments based on modifications of the Kolsky method allows one to determine the range of strength and time characteristics of concrete under shock loading.
2. Ultimate dynamic strength of concrete in splitting tests was in the range from 6 to 8 MPa; in tensile tests—from 8 to 12 MPa at strain rates from 500 to 700 $s^{-1}$.
3. The dynamic increase factor (DIF) of concrete in tension is in the range from 4 to 6 and depends on the rate of deformation. DIF was calculated as the ratio of the maximum stress achieved in the experiment to the static tensile strength of concrete (2.25 MPa).
4. The experimental data obtained indicate the effect of the strain rate on the ultimate tensile strength characteristics.

**Table 11.3** Parameters of splitting experiments

| Mode | No. experiment | Loading parameters | | | Specimen parameters | | | Experiment results | | | |
|---|---|---|---|---|---|---|---|---|---|---|---|
| | | Striker length (mm) | Striker material | Striker speed (m/s) | *D* (mm) | $S_{c.s.}$ (mm$^2$) | *L* (mm) | Maximum stress (MPa) | Stress growth rate (GPa/s) | Lifetime (μs) | |
| | | | | | | | | | | Diagram | Diagram |
| 1 | r637-11 | 200 | Дюраль | 7.88 | 18.62 | 272.1 | 14.48 | 7.61 | 431 | 25 | – |
| | r637-12 | 200 | Дюраль | 6.49 | 18.57 | 270.7 | 14.41 | 7.60 | 394 | 30 | 30 |
| | r637-13 | 200 | Дюраль | 6.76 | 18.62 | 272.1 | 14.34 | 7.50 | 265 | 35 | 35 |
| | r637-14 | 200 | Дюраль | 7.74 | 18.64 | 272.7 | 14.21 | 8.14 | 506 | 29 | 30 |
| | r637-15 | 200 | Дюраль | 7.89 | 18.61 | 271.8 | 14.21 | 7.74 | 458 | 28 | 30 |
| 2 | r637-16 | 200 | Дюраль | 16.46 | 18.59 | 271.2 | 14.18 | 8.48 | 532 | 25 | 25 |
| | r637-17 | 200 | Дюраль | 16.96 | 18.59 | 271.2 | 14.16 | 4.60 | 367 | 23 | 25 |
| | r637-18 | 200 | Дюраль | 16.33 | 18.61 | 271.8 | 14.11 | 5.01 | 352 | 24 | 25 |
| | r637-19 | 200 | Дюраль | 16.65 | 18.63 | 272.4 | 14 | 4.33 | 352 | 22 | 25 |
| | r637-20 | 200 | Дюраль | 16.49 | 18.62 | 272.1 | 13.88 | 4.53 | 329 | 20 | 20 |
| | r637-21 | 200 | Дюраль | 16.6 | 18.64 | 272.7 | 13.92 | 5.72 | 475 | 18 | 20 |

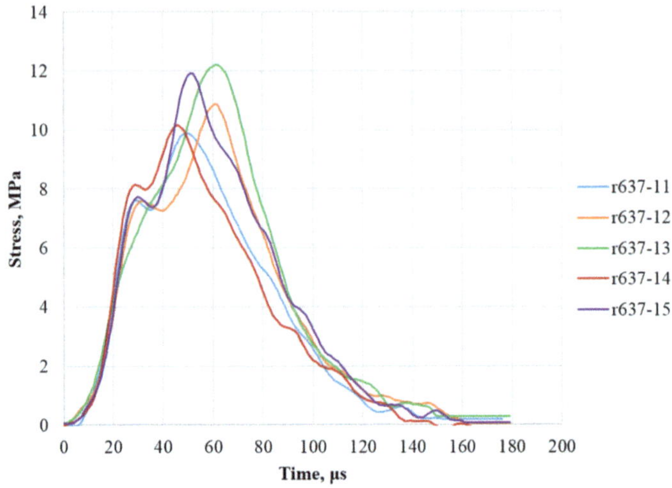


**Fig. 11.12** Diagrams of splitting tests for mode No. 1

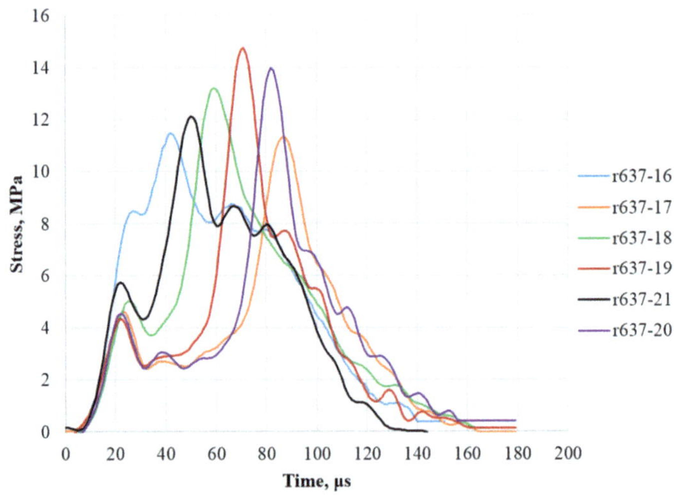


**Fig. 11.13** Diagrams of splitting tests for mode No. 2

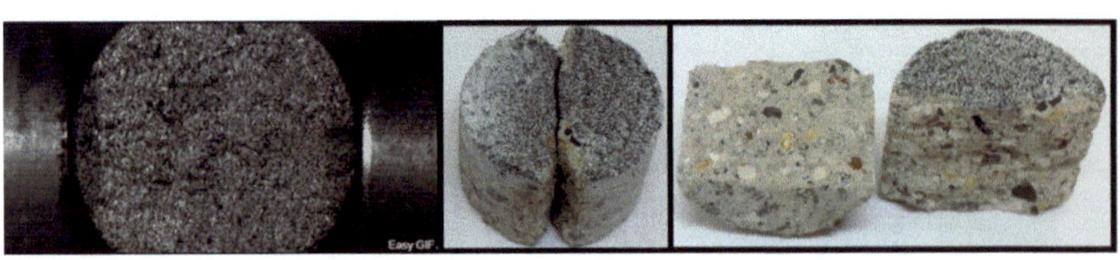

**Fig. 11.14** Nature of destruction of specimen

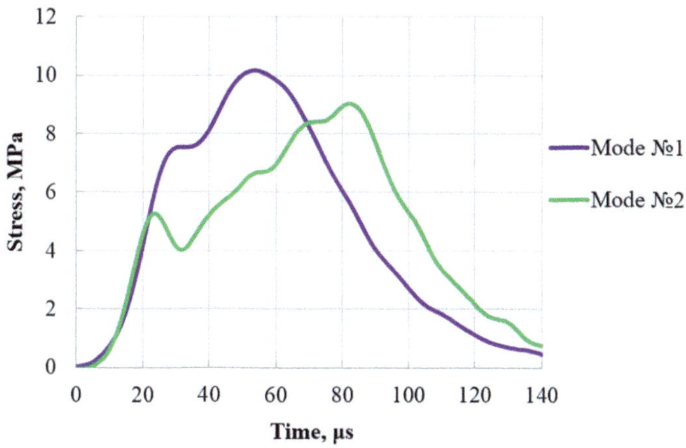


**Fig. 11.15** Averaged diagrams of voltage versus time for experimental modes No. 1 and No. 2

5. In tests for direct tension and splitting, the opposite effect of the influence of the loading rate was obtained: with an increase in the speed of the striker, and, consequently, the rate of deformation, the maximum attained stresses at which the destruction of the specimens begins, decrease.
6. Analysis of high-speed photography allows more accurate determination of the time of the destruction of the specimen, and, consequently, the corresponding breaking stress.

Thus, it can be concluded that the tensile and splitting tests carried out made it possible to obtain the dynamic characteristics of fine-grained concrete for the compilation of mathematical models and calculations of the stress–strain state. The obtained characteristics can be used in applied scientific research to assess the strength of structures subjected to dynamic influences in various software and computer systems.

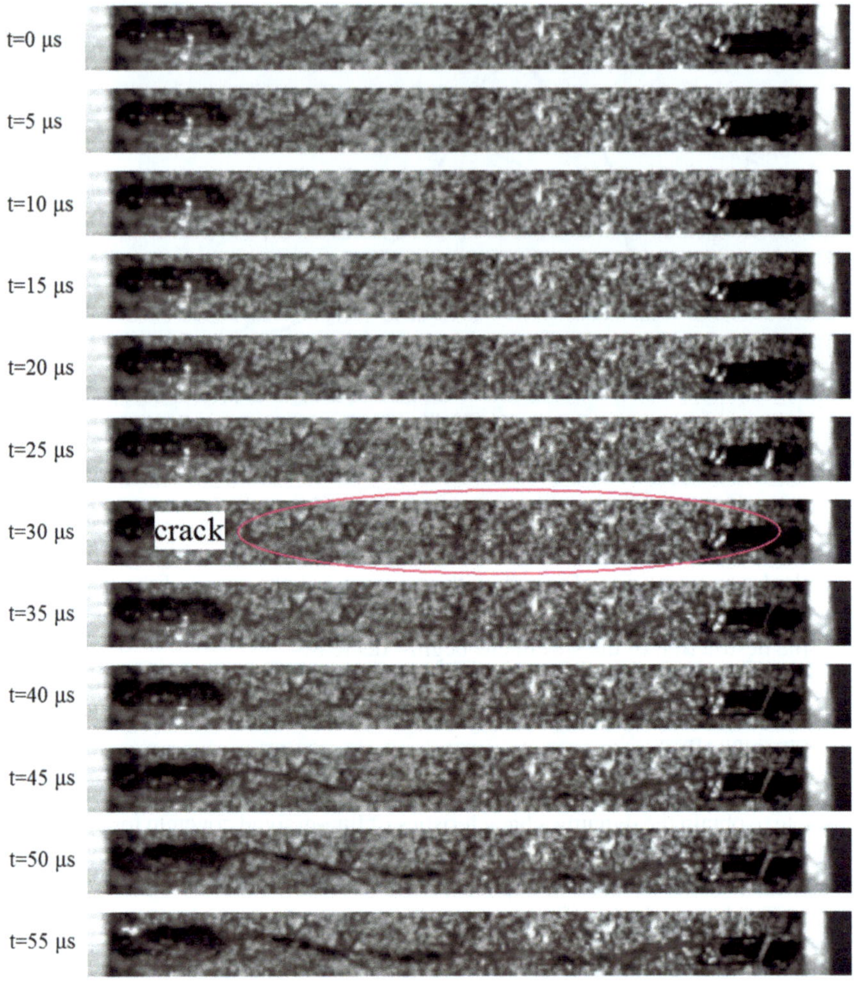


**Fig. 11.16** Timeline of high-speed photography of experiment No. 14

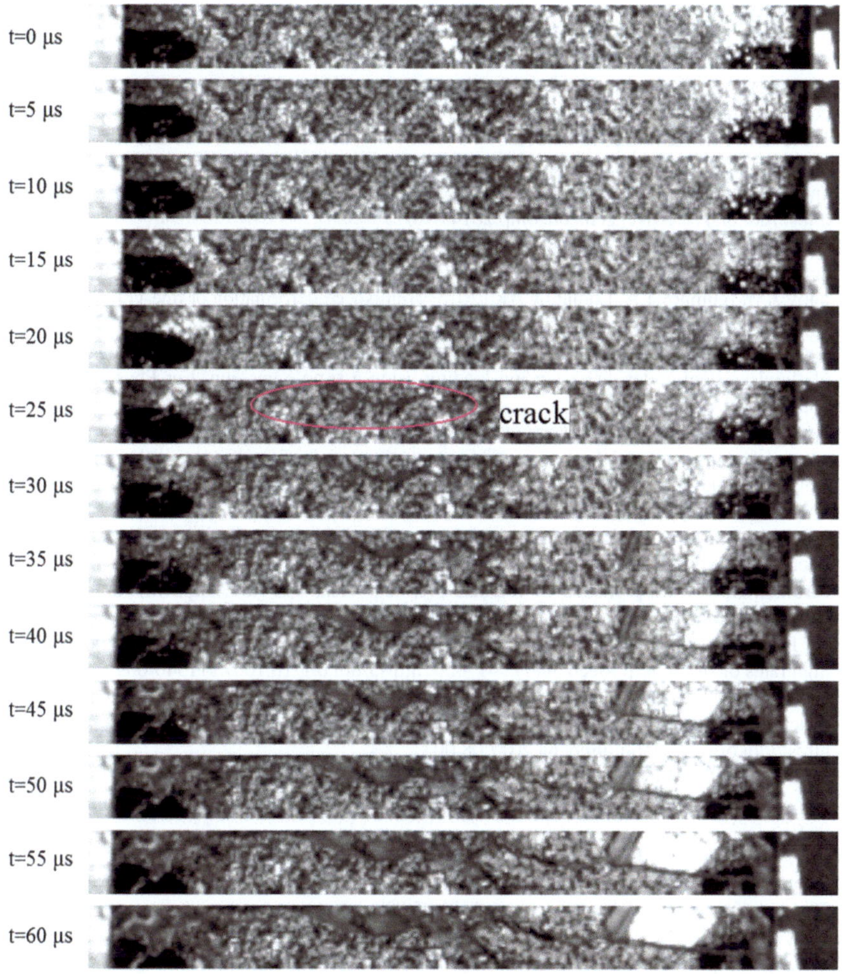


**Fig. 11.17** Timeline of high-speed photography of experiment No. 19

**Acknowledgements** The experimental results under dynamic direct tension were conducted with financial support from RFFI (Project 19-38-90225). The experimental results for dynamic splitting were conducted with financial support from the Ministry of Science and Higher Education of the Russian Federation (Project 0729-2020-0054).

## References

Agar Ozbek AS, Weerheijm J, Schlangen E, van Breugel K (2013) Dynamic behavior of porous concretes under drop weight impact testing. Cement Concrete Compos 39:1–11. https://doi.org/10.1016/j.cemconcomp.2013.03.012
Al-Salloum Y, Almusallam T, Ibrahim SM, Abbas H, Alsayed S (2015) Rate dependent behavior and modeling of concrete based on SHPB experiments. Cement Concrete Compos 55:34–44. https://doi.org/10.1016/j.cemconcomp.2014.07.011
Bragov AM, Lomunov AK (2017) Using the Kolsky method to study the mechanisms of high-rate deformation of materials of physical nature. N. Novgorod: Izd-vo NNGU (in Russ.)
Bragov AM, Igumnov LA, Lomunov AK (2015) High-speed deformation of fine-grained concrete and fiber-reinforced concrete. N. Novgorod: Izd-vo NNGU (in Russ.)
Bragov AM, Konstantinov AY, Kruszka L, Lomunov AK, Filippov AR (2018) Dynamic properties of stainless steel under direct tension loading using a simple gas gun. Eur Phys J Web Conf 183:02035
Bragov AM, Konstantinov AY, Lomunov AK (2018) Experimental and theoretical study of the processes of high-speed deformation and fracture of materials of various physical nature using the Kolsky method and its modifications. Nizhnij Novgorod: Izd-vo Nizhegorodskogo gosuniversiteta im. N.I. Lobachevskogo, 187 s (in Russian)
Bragov AM, Igumnov LA, Konstantinov AY, Lomunov AK, Rusin EE, Eremeyev VA (2020) Experimental analysis of wear resistance of compacts of fine-dispersed iron powder and tungsten monocarbide nanopowder produced by impulse pressing. Wear 456–457. https://doi.org/10.1016/j.wear.2020.203358
Erzar B, Forquin P (2010) An experimental method to determine the tensile strength of concrete at high rates of strain. Exp Mech 50(7):941–955. https://doi.org/10.1007/s11340-009-9284-z
Flores-Johnson EA, Li QM (2017) Structural effects on compressive strength enhancement of concrete-like materials in a split Hopkinson pressure bar test. Int J Impact Eng 109:408–418. https://doi.org/10.1016/j.ijimpeng.2017.08.003
Guo YB, Gao GF, Jing L, Shim VPW (2017) Response of high-strength concrete to dynamic compressive loading. Int J Impact Eng 108:114–135. https://doi.org/10.1016/j.ijimpeng.2017.04.015
Hao Y, Hao H, Jiang GP, Zhou Y (2013) Experimental confirmation of some factors influencing dynamic concrete compressive strengths in high-speed impact tests. Cement Concrete Res 52:63–70. https://doi.org/10.1016/j.cemconres.2013.05.008
Hao Y, Hao H, Li ZX (2013) Influence of end friction confinement on impact tests of concrete material at high strain rate. Int J Impact Eng 60:82–106. https://doi.org/10.1016/j.ijimpeng.2013.04.008
Kettenbeil C, Ravichandran G (2018) Experimental investigation of the dynamic behavior of metaconcrete. Int J Impact Eng 111:199–207. https://doi.org/10.1016/j.ijimpeng.2017.09.017
Kolsky H (1949) An investigation of the mechanical properties of materials at very high rates of loading. Proc Phys Soc London Sect B 62:676–700
Kwak H-G, Gang HG (2015) An improved criterion to minimize FE mesh-dependency in concrete structures under high strain rate conditions. Int J Impact Eng 86:84–95
Lamzin DA (2012) Determination of the shear strength of concrete under dynamic loading. N. Novgorod: NNGASU, S. 31–35 (in Russ.)
Navas P, Yu RC, Li B, Ruiz G (2018) Modeling the dynamic fracture in concrete: an eigensoftening meshfree approach. Int J Impact Eng 113:9–20. https://doi.org/10.1016/j.ijimpeng.2017.11.004
Pedersen RR, Simone A, Sluys LJ (2013) Mesoscopic modeling and simulation of the dynamic tensile behavior of concrete. Cement Concrete Res 50:74–87. https://doi.org/10.1016/j.cemconres.2013.03.021
Pereira LF, Weerheijm J, Sluys LJ (2018) Simulation of compaction and crushing of concrete in ballistic impact with a new damage model. Int J Impact Eng 111:208–221. https://doi.org/10.1016/j.ijimpeng.2017.09.014

Rodriguez T, Navarro C, Sanchez-Galvez V (1994) Splitting tests: an alternative to determine the dynamic tensile strength of ceramic materials. J Phys 04:101–106. https://doi.org/10.1051/jp4:1994815

Xiao J, Li L, Shen L, Poon CS (2015) Compressive behaviour of recycled aggregate concrete under impact loading. Cement Concrete Res 71:46–55. https://doi.org/10.1016/j.cemconres.2015.01.014

# Chapter 12
# Singularities in the Classical Elasticity Theory as Hidden Characteristics of a Non-Euclidean Model of a Continuous Medium

**Mikhail Guzev**

**Abstract** A paper deals with describing a microstructural material on the base of a non-Euclidean model of a continuous medium. The solution for the stress field is presented as a sum of elastic and self-balanced fields, the parameterization of which is given through non-Euclidean characteristics: the scalar curvature of the Ricci tensor. The requirement of the absence of singular contributions in the constructed field leads to the fact that the singularities of the elastic field are compensated by the singularities of the self-balanced stress fields, and a linear transformation relates the coefficients at the singularities of both fields. Thus, the coefficients at the singularities of the classical theory of elasticity are "hidden" parameters of the non-Euclidean model. The compensating role of self-balanced stress fields makes it possible to construct in equilibrium a nonsingular distribution for the stress field of a plane-deformed and spherically symmetric state of a continuous medium. The compensating role of self-balanced stress fields enables to construct in equilibrium a nonsingular distribution for the stress field of a plane-deformed and spherically symmetric state of a continuous medium.

**Keywords** Incompatibility condition · Non-Euclidean model of continuum medium · Singularities · Self-balanced stress field

## 12.1 Introduction

Researchers have analyzed the singular behavior of the stress field in continuous medium models for a long time; see, for example, Sinclair (2004a, b). Although singularities cannot be attributed to physical phenomena of the real world, approaches and their interpretation in a physically meaningful way have been proposed to consider

M. Guzev (✉)
Institute for Applied Mathematics, Far Eastern Branch of the Russian Academy of Sciences, 690041 Vladivostok, Russia
e-mail: guzev@iam.dvo.ru

Department of Oil and Gas Technologies, Perm National Research Polytechnic University, 614990 Perm, Russia

 
H. Altenbach et al. (eds.), *Advanced Materials Modelling for Mechanical, Medical and Biological Applications*, Advanced Structured Materials 155,
https://doi.org/10.1007/978-3-030-81705-3_12

such features. In this case, the researchers' interests are divided, which is associated with the consideration of problems of different quality levels, for the solution of which, in the general case, the use of models of different levels is required. For example, when formulating the criterion for brittle fracture of a material determined by cracks, a stress intensity factor is introduced (Eshelby 1951; Sanders 1960; Cherepanov 1967; Rice 1968) in whose calculation the singular components of the stress and displacement field make the main contribution to it.

On the other hand, the study of various materials by physical methods led to the introduction into the materials science literature of such concepts as dislocations, disclinations, vacancies, etc. The characteristics of crystal structure defects are microscopic parameters of materials, while the stress intensity factor should be considered as a characteristic of the following (macroscopic) scale level of the continuous medium. However, the necessity of introducing singular lines for analyzing the stress field around dislocations was also considered in Burgers (1995).

To interpret such features, the model improvement was proposed on the scale of considering the singularity to exclude it and obtain a physically real result. A review of various approaches to overcoming singularities in the continuum dislocation theory in the twentieth century can be found in Cai et al. (2006). Further studies (Po et al. 2018) are devoted to the development of a nonsingular model for describing the field of elastic stresses and strains for dislocations and disclinations in the framework of Mindlin's gradient theory. In Parisis et al. (2018), nonsingular solutions of gradient elasticity for dislocations and cracks were constructed. The construction of nonsingular solutions for nonlinear problems of the equilibrium of elastic bodies with continuously distributed dislocations was carried out in Goloveshkina and Zubov (2019, 2021).

In continuum mechanics, each type of defect is associated with the state of internal stress of the continuum, for whose description researchers in the 1950s proposed using non-Euclidean geometric objects (Kondo 1952; Bilby 1955). Analysis of this problem for various decades can be found in Stojanovic (1963), Minagawa (1976), Kröner (1985), Edelen (1996), Myasnikov and Guzev (2000), Guzev (2010), Yavari and Goriely (2012). The common approach of all authors is to use geometric non-Euclidean objects as variables characterizing the geometric structure of the incompatibility of the medium material elements.

Although the theoretical approaches in this direction are actively developing, non-Euclidean objects for constructing nonsingular solutions in models of a continuous medium with structural defects have not received a wide representation in the scientific literature. In this regard, we point out Yavari and Goriely (2013), where the authors, using the non-Euclidean model of a continuous medium, constructed an equilibrium solution of the stress field for a system with cylindrical and spherical symmetry. However, they indicate singularities in the stress distribution on the cylinder axis and at the sphere center.

Thus, the studies of many authors obviously indicate the possibility of constructing nonsingular fields of internal stresses based on various physical models of the material's internal structure. However, the classical elasticity theory solution contains singularities (see Sect. 12.2 of this work), whose coefficients are vanished to obtain nonsingular fields. Based on the non-Euclidean continuum model, it is shown in this paper that these coefficients can be interpreted in terms of non-Euclidean characteristics and become nonzero. Thus, the "hidden" coefficients (zero coefficients in the classical elasticity theory) become nontrivial when the classical model is extended, allowing one to construct a solution in this work that does not contain singularities. Let us dwell on the structure of the article.

Section 12.2 considers the plane-deformed and spherically symmetric state of a continuous medium. The solution constructed within the classical elasticity theory has a singularity at $r \to 0$, and nonsingular stress fields are homogeneous, corresponding to the vanishing of the coefficients at singularities. To construct inhomogeneous stress fields, a function determined the self-balanced stress field $\sigma_{ij,\mathrm{SE}}$ is proposed to add to the classical elasticity theory solution.

Here, the general idea of eliminating singularities is also formulated. First, it is necessary to ensure that the type of field singularities $\sigma_{ij,\mathrm{SE}}$ coincides with the type of field singularities classical elasticity theory solution; this will enable to compensate for the singularities at $r \to 0$ and construct a nonsingular total stress field. Second, given the need to introduce non-Euclidean geometric objects, fields $\sigma_{ij,\mathrm{SE}}$ should be represented through such objects. The solution to the second problem was earlier proposed by the author and used in this work.

Section 12.3 describes the construction of a solution for the plane-strained state of a continuous medium based on the non-Euclidean continuum model. The requirement of the absence of singular contributions in the total constructed field leads to the fact that the coefficients at singularities of the classical elasticity theory solution are calculated through the parameters of the stress field $\sigma_{ij,\mathrm{SE}}$, determined through non-Euclidean characteristics. Thus, these singularities are "hidden" characteristics of the non-Euclidean model since they are equal to zero in the classical theory.

Section 12.4 demonstrates that for the spatial problem in the non-Euclidean model of a continuous medium, it is possible to compensate for the singularities of the elastic field due to self-balanced stress fields. The corresponding consideration is carried out for a spherically symmetric stress field. Then, the coefficients at singularities in the classical theory are nonzero and are determined through the amplitude coefficients of non-Euclidean characteristics that parameterize the field $\sigma_{ij,\mathrm{SE}}$ components. Consequently, the expansion of the classical theory of elasticity and the transition to a non-Euclidean model of a continuous medium allows us to consider the singularities of the classical theory as "hidden" characteristics of the non-Euclidean model.

## 12.2 Singularities of the Classical Elasticity Theory and the General Idea of Their Elimination

For a plane-deformed state of a continuous medium, the stress field components, which are principal due to symmetry, will be denoted by $\tau_r$, $\tau_\varphi$, and in the classical elasticity theory, they satisfy the system of equations:

$$\frac{d\tau_r}{dr} + \frac{\tau_r - \tau_\varphi}{r} = 0, \quad \Delta(\tau_r + \tau_\varphi) = 0 \tag{12.2.1}$$

The following formula gives the solution to system (12.2.1):

$$\tau_r = \frac{a_1}{2} + a_2 \ln r - \frac{a_2}{2} + \frac{a_3}{r^2}, \quad \tau_\varphi = \frac{a_1}{2} + a_2 \ln r + \frac{a_2}{2} - \frac{a_3}{r^2} \tag{12.2.2}$$

The absence of a singularity at $r \to 0$ for functions (12.2.2) is ensured under condition $a_2 = 0$, $a_3 = 0$.

The components of a spherically symmetric stress field are also principal due to symmetry, and we denote them by $T_r, T_\varphi = T_\theta$. These fields in the classical elasticity theory satisfy the equations:

$$\frac{dT_r}{dr} + \frac{2(T_r - T_\theta)}{r} = 0, \quad \Delta(T_r + 2T_\theta) = 0 \tag{12.2.3}$$

Solutions for (12.2.3) are easy to write down:

$$T_r = \frac{b_1}{3} - \frac{b_2}{2r} + \frac{b_3}{r^3}, \quad T_\varphi = T_\theta = \frac{b_1}{3} - \frac{b_2}{4r} - \frac{b_3}{2r^3} \tag{12.2.4}$$

The requirement that there is no singularity at $r \to 0$ is satisfied under condition $b_2 = 0$, $b_3 = 0$.

From the above, in the classical elasticity theory for a system with cylindrical and spherical symmetry, the equilibrium nonsingular solutions for the stress field at $r \to 0$ are constant. To construct inhomogeneous distributions of fields, one can use a well-known fact from the mechanics of a continuous medium (Godunov and Romensky 2003). This work indicated that the stress field is determined ambiguously for a given distribution of surface and mass forces.

This result is based on the fact that the Eq. (12.2.5) of mechanical equilibrium for a continuous medium in the absence of external forces can have nontrivial solutions:

$$\frac{\partial \sigma_{ij}}{\partial x_j} = 0, \quad \sigma_{ij} n_j \big|_{\partial V} = 0, \tag{12.2.5}$$

where objects $\sigma_{ij}$ are the components of the stress tensor, and $n_j$ are the coordinates of the normal vector to the boundary $\partial V$; summation is performed over repeated indices. Nevertheless, the ambiguity of the stress field turns out to be useful for correcting the equations of a continuous medium state. In particular, such a correction allowed Godunov and Romensky (2003), without changing the smooth solutions of the elasticity theory, to change the equation of state so that it becomes strictly convex and to bring the system of conservation laws to a symmetric hyperbolic form.

An approach to describing nontrivial solutions (12.2.5) (they are also called self-balanced, residual) can also be found in Kiseliev (2004). It was shown in Myasnikov and Guzev (2001) that solutions $\sigma_{ij}$ of Eq. (12.2.5) could be represented in the form of a classical field of elastic stresses $\sigma_{ij,El}$ and a field of self-balanced stresses $\sigma_{ij,\mathrm{SE}}$:

$$\sigma_{ij} = \sigma_{ij,\mathrm{El}} + \sigma_{ij,\mathrm{SE}} \tag{12.2.6}$$

In this case, the stresses $\sigma_{ij,\mathrm{SE}}$ may not, in the general case, satisfy the conditions for the absence of external forces on the surface of the body: $\sigma_{ij,\mathrm{SE}} n_j|_{\partial V} \neq 0$. Point by point, these conditions are satisfied for total stresses (12.2.6):

$$\sigma_{ij} n_j \big|_{\partial V} = (\sigma_{ij,\mathrm{El}} n_j + \sigma_{ij,\mathrm{SE}} n_j)\big|_{\partial V} = 0 \tag{12.2.7}$$

Thus, the combined action of elastic and self-balanced stresses allows a material object to maintain a given shape.

In our work, the existence of self-balanced stress fields can be used to construct nonsingular solutions of equilibrium fields. Since representation (12.2.6) is valid, the fields of self-balanced stresses must have singularities of the same order as the classical solutions of elasticity (12.2.2), (12.2.4) in order to compensate for their singular behavior at $r \to 0$. This requirement leads to the fact that the total stresses are not singular.

Considering the need to introduce non-Euclidean geometric objects, the fields $\sigma_{ij,\mathrm{SE}}$ should be represented through mathematical objects that do not fit into the framework of the Euclidean geometric description of the deformation properties of an elastic continuous medium. For the model of a continuous medium with an internal structure, non-Euclidean geometric objects were introduced to describe the internal stress state (Guzev 2011).

This made it possible to represent the stress field $\sigma_{ij,\mathrm{SE}}$ as follows:

$$\sigma_{ij,\mathrm{SE}} = \sigma_0 l^4 \left( \delta_{ij} \Delta R - \frac{\partial^2 R}{\partial x^i \partial x^j} \right), \tag{12.2.8}$$

where the constants $\sigma_0$ and $l$ have the dimensions of stress and length, $\Delta$ is the Laplace operator, and the function $R$ is the scalar curvature (the trace of the Ricci tensor, see Novikov and Taimanov 2006) and is unknown in the general case. It is clear that fields (12.2.8) identically satisfy equilibrium Eq. (12.2.5). In three-dimensional space, the symmetric Ricci tensor $R_{jl}$ defines the Riemann tensor $R^q{}_{j,kl}$ (Novikov and Taimanov 2006), which characterizes the difference between the internal geometries of the modeled object from the Euclidean geometry. If we introduce a metric tensor $g_{ij} = \delta_{ij} - 2\varepsilon_{ij}$ and pass to covariant components $R_{ij,kl} = g_{iq} R^q{}_{j,kl}$, $g^{ki} g_{ij} = \delta^k{}_j$, then, for small $\varepsilon_{ij}$, component of the Riemann tensor linear in $\varepsilon_{ij}$ is equal to

$$-R_{ij,kl} = \frac{\partial^2 \varepsilon_{li}}{\partial x^j \partial x^k} - \frac{\partial^2 \varepsilon_{ik}}{\partial x^j \partial x^l} - \frac{\partial^2 \varepsilon_{lj}}{\partial x^i \partial x^k} + \frac{\partial^2 \varepsilon_{jk}}{\partial x^i \partial x^l}. \tag{12.2.9}$$

In continuum mechanics, it is well known (Godunov and Romensky 2003) that the fulfillment of the requirements

$$\frac{\partial^2 \varepsilon_{li}}{\partial x^j \partial x^k} - \frac{\partial^2 \varepsilon_{ik}}{\partial x^j \partial x^l} - \frac{\partial^2 \varepsilon_{lj}}{\partial x^i \partial x^k} + \frac{\partial^2 \varepsilon_{jk}}{\partial x^i \partial x^l} = 0 \qquad (12.2.10)$$

is a necessary and sufficient condition for representing the tensor $\varepsilon_{ij}$ in the form $\varepsilon_{ij} = (\partial u_i/\partial x_j + \partial u_j/\partial x_i)/2$ where $u_i$ are the components of the displacement field of the continuous medium points.

Linear equations (12.2.10) is called Saint–Venant compatibility conditions. The extension of the classical elasticity theory, associated with the rejection of the compatibility condition and the introduction of additional parameters (12.2.9), corresponds to the transition to a non-Euclidean model of a continuous medium. Scalar curvature $R$ was introduced as a thermodynamic parameter in the analysis of the zonal disintegration phenomenon (Guzev and Paroshin 2001) and in models of continuous media with an internal structure (Guzev 2011). Under the assumption that the medium internal energy depends quadratically on the thermodynamic variables, a stationary equation was obtained for $R$ in Guzev (2011), Guzev and Paroshin (2001):

$$\Delta^2 R = \gamma R, \quad \gamma \neq 0. \qquad (12.2.11)$$

The parameter $\gamma$ depends on the spatial scale of the problem, and its dimension is $1/[\text{length}]^4$.

## 12.3 Nonsingular Solutions and Hidden Parameters for a System with Cylindrical Symmetry

The solutions to Eq. (12.2.11) are defined in terms of cylindrical functions with a zero index:

$$R(s) = aJ_0(s) + bN_0(s) + cK_0(s) + dI_0(s), \qquad (12.3.1)$$

where the variables $s = \sqrt[4]{\gamma}r$, $J_0(s)$ and $N_0(s)$ are the Bessel and Neumann functions, respectively, $K_0(s)$ is the MacDonald function, and $I_0(s)$ is the modified Bessel function; *a, b, c, d*—amplitude coefficients, whose dimension is equal to $1/[\text{length}]^2$. In polar coordinates, Formula (12.2.8) is written as

$$\sigma_{r,\text{SE}} = \sigma_0 l^4 \frac{1}{r}\frac{\partial R(s)}{\partial r}, \quad \sigma_{\varphi,\text{SE}} = \sigma_0 l^4 \frac{\partial^2 R(s)}{\partial r^2}. \qquad (12.3.2)$$

Substituting (12.3.1) here, and using the relations between cylindrical functions, we obtain a representation for the principal components of self-balanced stress fields:

$$\sigma_{r,\text{SE}} = \sigma_0 l^4 \sqrt{\gamma}\left(-a\frac{J_1(s)}{s} - b\frac{N_1(s)}{s} - c\frac{K_1(s)}{s} + d\frac{I_1(s)}{s}\right),$$

$$\sigma_{\varphi,\mathrm{SE}} = \sigma_0 l^4 \sqrt{\gamma}\left(a\frac{J_1(s)}{s} + b\frac{N_1(s)}{s} + c\frac{K_1(s)}{s} - d\frac{I_1(s)}{s}\right) \\ - aJ_0(s) - bN_0(s) + cK_0(s) + dI_0(s), \tag{12.3.3}$$

where we renormalized the amplitude coefficients by $\sigma_0 l^2$ keeping their previous notations.

Let us pass to a new variable $s = \sqrt[4]{\gamma} r$ in the representation for the classical field of elastic stresses, then its formula retains the form (12.2.2), but $r \to s$ should be assumed in it. The total stresses $\sigma_r$, $\sigma_\varphi$ can be represented as

$$\sigma_r = \tau_r + \sigma_{r,\mathrm{SE}}, \quad \sigma_\varphi = \tau_\varphi + \sigma_{\varphi,\mathrm{SE}}. \tag{12.3.4}$$

The classical field has a singular behavior at $s \to 0$. On the right-hand side of (12.3.3), individual terms also contain a singularity near zero, since at $s \to 0$ the following asymptotics are valid:

$$N_0(s) \sim \frac{2}{\pi}\ln s,\; K_0(s) \sim -\ln s, \quad N_1(s) \sim \frac{s}{\pi}\ln s - \frac{2}{\pi s},\; K_1(s) \sim \frac{1}{s} + \frac{s}{2}\ln s, \tag{12.3.5}$$

The requirement that there are no singular contributions in the total field (12.3.4) leads to the fact that the coefficients in (12.2.2) and (12.3.3) cannot be arbitrary: They should be chosen so that singularities of the same type compensate each other. The leading singularity in $s$ is of order $1/s^2$. Equating to zero the combination of coefficients at $1/s^2$ in (12.3.3), we obtain

$$a_3 = \frac{c}{\sqrt{\gamma}} - \frac{2b}{\pi\sqrt{\gamma}}. \tag{12.3.6}$$

Note that the requirement that there is no singularity for $\sigma_r$, $\sigma_\phi$ leads to only one condition (12.3.6). The next singular contribution to (12.3.4) contains $\ln s$, and then, the requirement for vanishing the coefficients at $\ln s$ gives

$$a_2 = \frac{b}{\pi\sqrt{\gamma}} + \frac{c}{2\sqrt{\gamma}}. \tag{12.3.7}$$

At singularities in solution (12.2.2), assuming in (12.3.6) and (12.3.7) the coefficients $a_2$, $a_3$ to be known values, one can write the amplitude coefficients $b$, $c$ of the non-Euclidean characteristic (12.3.1) through them. In this way, the "hidden" parameters of the classical theory "reveal" the parameters of the non-Euclidean model.

## 12.4 Nonsingular Solutions and Hidden Parameters for a System with Spherical Symmetry

Since the rheological relationship between the components of the stress and the strain field $\varepsilon_{ij}$ is linear (Hooke's law), we will write the stresses $\sigma_{ij}$ in terms of deformations $\varepsilon_{ij}$:

$$\varepsilon_{ij} = \frac{1}{2\mu}\left[\sigma_{ij} - \delta_{ij}\frac{\lambda\sigma}{3\lambda + 2\mu}\right], \quad \sigma = \sigma_{kk}, \tag{12.4.1}$$

where $\lambda$, $\mu$ are the phenomenological parameters of Lamé. Using (12.2.9), we write the representation for the Ricci tensor's trace as following:

$$\frac{R}{2} = \frac{\partial^2 \varepsilon_{kk}}{\partial x^j \partial x^j} - \frac{\partial^2 \varepsilon_{jk}}{\partial x^j \partial x^k}, \tag{12.4.2}$$

Substituting (12.4.1) into (12.4.2) and using the equilibrium Eq. (12.2.5), we obtain the equation for the first invariant of the stress tensor:

$$\Delta\sigma = \frac{E}{2(1-\nu)}R, \quad E = \mu\frac{3\lambda + 2\mu}{\lambda + \mu}, \quad \nu = \frac{\lambda}{2(\lambda + \mu)}. \tag{12.4.3}$$

Due to (12.2.6), the function $\sigma$ has the following structure:

$$\sigma = \sigma_{\text{El}} + \sigma_{\text{SE}}, \quad \sigma_{\text{El}} = T_r + 2T_\varphi, \quad \sigma_{\text{SE}} = \sigma_{r,\text{SE}} + 2\sigma_{\varphi,\text{SE}}, \tag{12.4.4}$$

where $\sigma_{\text{El}}$, $\sigma_{\text{SE}}$ are the solutions of the equations

$$\Delta\sigma_{\text{El}} = 0, \quad \Delta\sigma_{\text{SE}} = \frac{E}{2(1-\nu)}R. \tag{12.4.5}$$

Since $\Delta^2 - \gamma = (\Delta + \sqrt{\gamma})(\Delta - \sqrt{\gamma})$, the solution of (12.2.11) is given by a formula $R = R_+ + R_-$ in which $R_+$, $R_-$ satisfy the equations:

$$\left(\Delta + \sqrt{\gamma}\right)R_+ = 0, \quad \left(\Delta - \sqrt{\gamma}\right)R_- = 0, \quad \Delta = \frac{1}{r^2}\frac{\partial}{\partial r}r^2\frac{\partial}{\partial r}. \tag{12.4.6}$$

Then,

$$\sigma_{\text{SE}} = \sigma_{+,\text{SE}} + \sigma_{-,\text{SE}}, \quad \Delta\sigma_{+,\text{SE}} = \frac{E}{2(1-\nu)}R_+, \quad \Delta\sigma_{-,\text{SE}} = \frac{E}{2(1-\nu)}R_-. \tag{12.4.7}$$

From (12.4.6) and (12.4.7), we obtain

$$\sigma_{+,\mathrm{SE}} = -\frac{E}{2(1-\nu)\sqrt{\gamma}}R_+, \quad \sigma_{-,\mathrm{SE}} = \frac{E}{2(1-\nu)\sqrt{\gamma}}R_-. \tag{12.4.8}$$

Solutions of Eq. (12.4.6) are written in terms of cylindrical functions with the one-half index. Since these functions are represented in terms of elementary ones, then

$$R_+ = \frac{1}{s}\sqrt{\frac{2}{\pi}}(A\sin s - B\cos s), \quad R_- = \frac{1}{s}\left[C\sqrt{\frac{\pi}{2}}e^{-s} + D\sqrt{\frac{1}{2\pi}}\left(e^{s} - e^{-s}\right)\right]. \tag{12.4.9}$$

with amplitude coefficients *A, B, C, D.* The field $\sigma_{r,\mathrm{SE}}$ satisfies the equilibrium equation:

$$\frac{\mathrm{d}\sigma_{r,SE}}{\mathrm{d}s} + \frac{2(\sigma_{r,\mathrm{SE}} - \sigma_{\varphi,\mathrm{SE}})}{s} = 0. \tag{12.4.10}$$

The solution for $\sigma_{r,\mathrm{SE}}$ is given by quadrature:

$$\sigma_{r,\mathrm{SE}} = \frac{1}{s^3}\int^{s} \tilde{s}^2\sigma_{\mathrm{SE}}\mathrm{d}\tilde{s} = \Sigma_{+,\mathrm{SE}} + \Sigma_{-,\mathrm{SE}},$$

$$\Sigma_{+,\mathrm{SE}} = \frac{1}{s^3}\int^{s} \tilde{s}^2\sigma_{+,\mathrm{SE}}\mathrm{d}\tilde{s}, \quad \Sigma_{-,\mathrm{SE}} = \frac{1}{s^3}\int_{s_0}^{s} \tilde{s}^2\sigma_{-,\mathrm{SE}}\mathrm{d}\tilde{s}. \tag{12.4.11}$$

Using (12.4.8)–(12.4.10), we perform simple calculations, then

$$\Sigma_{+,\mathrm{SE}} = -\frac{E}{2s^3(1-\nu)\sqrt{\gamma}}\sqrt{\frac{2}{\pi}}(A\sin s - As\cos s - Bs\sin s - B\cos s), \tag{12.4.12}$$

$$\Sigma_{-,\mathrm{SE}} = \frac{E}{2s^3(1-\nu)\sqrt{\gamma}}\left[-C\sqrt{\frac{\pi}{2}}\exp(-s)(1+s) + D\frac{1}{\sqrt{2\pi}}(\exp(s)(-1+s) + \exp(-s)(1+s))\right] \tag{12.4.13}$$

Let us turn to the new variable $s = \sqrt[4]{\gamma}r$ in the representation for the classical field of elastic stresses, then the formula for it retains the form (12.2.4), but in it, $r \to s$ should be assumed. The total stresses $\sigma_r$, $\sigma_\varphi$ $\sigma_\theta$ can be represented as

$$\sigma_r = T_r + \sigma_{r,\mathrm{SE}}, \quad \sigma_\theta = \sigma_\varphi = T_\varphi + \sigma_{\varphi,\mathrm{SE}}. \tag{12.4.14}$$

The elastic stress field $T_r$, $\sigma_\varphi$ , $T_\varphi$ (12.2.4) has a singular behavior at $s \to 0$. Functions $\sigma_{\mathrm{SE}}$, $\sigma_{r,\mathrm{SE}}$ also contain a singularity at $s \to 0$. The leading in $s \to 0$ singularity $T_r$, $\sigma_{r,\mathrm{SE}}$ is of order $1/s^3$. The requirement that there are no singular contributions in the total stress fields (12.4.14) leads to the fact that the combination of the coefficients at the singularity is zero:

$$b_3 + \sqrt{\frac{2}{\pi}} B - \sqrt{\frac{\pi}{2}} C = 0. \tag{12.4.15}$$

The next singular contribution contains $1/s$, and then, the requirement for the set of coefficients to vanish at $1/s$ gives

$$b_2 = \sqrt{\frac{2}{\pi}} B + \sqrt{\frac{\pi}{2}} C. \tag{12.4.16}$$

Relations (12.4.15) and (12.4.16) make it possible to recover the amplitude coefficients $B$, $C$ of non-Euclidean characteristics (12.4.9) in terms of the coefficients $b_2$, $b_3$ at singularities of the classical solution (12.4.4). In this way, we "discover" the parameters of the non-Euclidean model through the "hidden" parameters of the classical theory.

## 12.5 Conclusions

The study performed in this work showed that constructing a nonsingular solution based on a non-Euclidean model of a continuous medium leads to further analysis of the solution of the classical elasticity theory. The standard requirement of the absence of a singularity for the classical solution in systems with symmetry gives a uniform distribution of the stress field, corresponding to zeroing the coefficients at singularities. The transition to a non-Euclidean description of a continuous medium, associated with the rejection of the Saint–Venant compatibility condition, allows one to "move" these coefficients, leaving them in the solutions for stresses. Since the latter are formed from the classical stress and the field determined through non-Euclidean geometric characteristics, the amplitude coefficients can be chosen so that the singularities of the fields mutually compensate, and a linear transformation relates the coefficients at the singularities. In other words, information about the singular structure of the solution is an invariant of the system. Moreover, it cannot be analyzed based on the classical elasticity theory since these singularities are absent. Thus, the invariant is "hidden" and manifests itself only when using a non-Euclidean model.

In this work, the Riemannian non-Euclidean model of a continuous medium was used, and the nonsingular solution was used to process the experimental data on measuring residual stresses for systems with cylindrical symmetry (Liu et al. 2021).

In the future, it is of interest to carry out a similar study based on other non-Euclidean models.

**Acknowledgements** This work was financially supported by the Russian Science Foundation (project No. 19-19-00408).

# References

Bilby BA, Bullough R, Smith E (1955) Continuous distributions of dislocations: a new application of the methods of non-Reimannian geometry. Proc R Soc A 231:263–273. https://doi.org/10.1098/rspa.1955.0171

Burgers JM (1995) Physics–some considerations on the fields of stress connected with dislocations in a regular crystal lattice. I, II. In: Nieuwstadt FTM, Steketee JA (eds) Selected papers of J. M. Burgers. Springer, Dordrecht, pp 335–389. https://doi.org/10.1007/978-94-011-0195-0_11

Cai W, Arsenlis A, Weinberger C, Bulatov V (2006) A non-singular continuum theory of dislocations. J Mech Phys Solids 54(3):561–587. https://doi.org/10.1016/j.jmps.2005.09.005

Cherepanov GP (1967) Crack propagation in continuous media. J Appl Math Mech 31:503–512

Edelen DGB (1996) A correct, globally defined solution of the screw dislocation problem in the gauge theory of defects. Int J Eng Sci 34(1):81–86. https://doi.org/10.1016/0020-7225(95)00081-x

Eshelby JD (1951) The force on an elastic singularity. Philos Trans R Soc London. Ser A 244:87–112

Godunov SK, Romensky EI (2003) Elements of continuum mechanics and conservation laws. Kluwer Academic Publisher, Dordrecht

Goloveshkina EV, Zubov LM (2019) Universal spherically symmetric solution of nonlinear dislocation theory for incompressible isotropic elastic medium. Arch Appl Mech 89(3):409–424. https://doi.org/10.1007/s00419-018-1403-9

Goloveshkina EV, Zubov LM (2021) Spherically symmetric tensor fields and their application in nonlinear theory of dislocations. Symmetry 13:830. https://doi.org/10.3390/sym13050830

Guzev MA (2010) Non-Euclidean models of elastoplastic materials with structure defects. Lambert Academic Publishing

Guzev MA (2011) Structure of kinematic and force fields in the Riemannian continuum model. J Appl Mech Tech Phys 52(5):709–716. https://doi.org/10.1134/S002189441105004X

Guzev MA, Paroshin AA (2001) Non-Euclidean model of the zonal disintegration of rocks around an underground working. J Appl Mech Tech Phys 42(1):131–139. https://doi.org/10.1023/A:1018877015940

Kiselev SP (2004) Internal stresses in a solid with dislocations. J Appl Mech Tech Phys 45(4):567–571. https://doi.org/10.1023/B:JAMT.0000030335.96577.6f

Kondo K (1952) On the geometrical and physical foundations of the theory of yielding. Proc Japan Nat Congr Apl Mech 2:41–47

Kröner E (1985) Incompatibility, defects, and stress functions in the mechanics of generalized continua. Int J Solids Struct 21(7):747–756. https://doi.org/10.1016/0020-7683(85)90077-0

Liu W, Guzev MA, Qi C (2021) Non-Euclidean model for description of residual stresses in planar deformations. Appl Math Model 90:615–623. https://doi.org/10.1016/j.apm.2020.09.001

Minagawa S (1976) On the stress functions in elastodynamics. Acta Mechanica 24(3):209–217. https://doi.org/10.1007/BF01190371

Myasnikov VP, Guzev MA (2000) Thermo-mechanical model of elastic-plastic materials with defect structures. Theoret Appl Fract Mech 33(3):165–171. https://doi.org/10.1016/S0167-8442(00)00011-2

Myasnikov VP, Guzev MA (2001) Geometric model of internal self-equilibrated stresses in solids. Dokl Phys 46(10):740–741. https://doi.org/10.1134/1.1415593
Novikov SP, Taimanov IA (2006) Modern geometric structures and fields. Graduate studies in mathematics, vol 71. American Mathematical Society
Parisis K, Konstantopoulos I, Aifantis EC (2018) Non-singular solutions of GradEla models for dislocations: an extension to fractional GradEla. J Micromech Mol Phys. https://doi.org/10.1142/s2424913018400131
Po G, Lazar M, Admal NC, Ghoniem N (2018) A non-singular theory of dislocations in anisotropic crystals. Int J Plast 103:1–22. https://doi.org/10.1016/j.ijplas.2017.10.003
Rice JR (1968) A path independent integral and the approximate analysis of strain concentration by notches and cracks. ASME J Appl Mech 35:379–386
Sanders Jr. JL (1960) On the Griffith-Irwin fracture theory. ASME J Appl Mech 27:352–353
Sinclair GB (2004a) Stress singularities in classical elasticity-I: removal, interpretation, and analysis. Appl Mech Rev 57:251–297
Sinclair GB (2004b) Stress singularities in classical elasticity-II: asymptotic identification. Appl Mech Rev 57:385–439
Stojanovic R (1963) Equilibrium conditions for internal stresses in non-euclidian continua and stress space. Int J Eng Sci 1:323–327. https://doi.org/10.1016/0020-7225(63)90010-7
Yavari A, Goriely A (2013) Nonlinear elastic inclusions in isotropic solids. Proc R Soc a: Math, Phys Eng Sci 469(2160):20130415–20130415. https://doi.org/10.1098/rspa.2013.0415
Yavari A, Goriely A (2012) Riemann–Cartan geometry of nonlinear disclination mechanics. Math Mech Solids 18(1):91–102. https://doi.org/10.1177/1081286511436137

# Chapter 13
# Stress State of a Compound Plane with Interface Absolutely Rigid Inclusion and Crack Having Common Tip

**Vahram Hakobyan, Avetik Sahakyan, and Harutyun Amirjanyan**

**Abstract** The plane stress state of a compound elastic plane, containing a finite crack and an absolutely rigid thin inclusion on the junction line between two heterogeneous half-planes, is considered. The crack and inclusion continue each other and have a common end. Using the discontinuous solutions of the Lame equations for the composite plane with interface defects, the governing system of singular integral equations of the problem is derived with respect to the stress jump function acting on the front sides of the inclusion and the dislocation of displacements of the points of the crack edges. The behavior of unknown functions at the end points of the defects is studied, and the solution of the problem is constructed by the method of mechanical quadratures. The regulations of changes in the contact stresses acting on the front sides of the inclusion, crack opening and the Cherepanov-Rice $J$-integral at the end point of the crack are revealed, depending on the physical, mechanical and geometric parameters of the problem.

**Keywords** Contact problem · Singular integral equation · Elasticity · Interface crack · Thin inclusion

## 13.1 Introduction

The study of the stress–strain state of homogeneous and composite massive bodies with various defects such as cracks and absolutely rigid inclusions, as well as the study of the mutual influence of these defects, has been the subject of many works. Many results in this direction are given in Panasyuk et al. (1976), Berezhnitski et al. (1983), Hakobyan (2014), where, in particular, using Kolosov-Muskhelishvili

V. Hakobyan · A. Sahakyan (✉) · H. Amirjanyan
Institute of Mechanics, National Academy of Sciences of Armenia, Yerevan, Armenia
e-mail: avetik.sahakyan@sci.am

V. Hakobyan
e-mail: vhakobyan@sci.com

H. Amirjanyan
e-mail: amirjanyan@gmail.com

 
H. Altenbach et al. (eds.), *Advanced Materials Modelling for Mechanical, Medical and Biological Applications*, Advanced Structured Materials 155,
https://doi.org/10.1007/978-3-030-81705-3_13

complex potentials and the method of solving the Riemann problem for one and two functions, exact solutions of a number of important problems for a homogeneous isotropic and orthotropic plane with cracks and absolutely rigid thin inclusions are constructed.

The exact solution of the problem for an orthotropic plane with a crack and a collinear absolutely rigid inclusion with one common end was obtained in Hakobyan and Simonyan (2011). We also mention Sahakyan and Dashtoyan (2003), where a solution of an analogous problem for a two-component composite plane with an interface crack and an absolutely rigid thin inclusion located in a stationary temperature field is constructed by numerically analytical method of mechanical quadrature's, when the crack and the inclusion do not have a common end.

In contrast to that, here we consider the case when the interface crack and the inclusion have one common end, which leads to the fact that the unknown functions at the common end of the crack and the inclusions have two different exponents of the singularity. This circumstance mathematically complicates the solution of the problem and requires the improvement of the method of mechanical quadrature's for solving similar problems.

## 13.2 The Statement of the Problem and the Derivation of Governing Equations

Let the composite elastic plane consisting of two homogeneous half-planes with shear moduli $\mu_j$ and Poisson's ratios $\nu_j$ $(j = 1, 2)$ on the segments $(-b, 0)$ and $(0, a)$ of the joining line of the half-planes coinciding with the $O_x$-axis of the rectangular Cartesian coordinate system $O_{xy}$ contain, respectively, a crack and a thin absolutely rigid inclusion. It is assumed that the compound plane is deformed by the distributed normal $q_j(x)$ and tangential $\tau_j(x)$ $(j = 1, 2)$ loads applied to the crack faces, and also the concentrated force $P_0$ applied to the inclusion at a point $x = x_0$ at an angle $\alpha$ relative to the axis (Fig. 13.1). For clarity, tangential loads $\tau_j(x)$ are not shown in the figure.

It is required to determine the contact stresses acting on the long sides of the inclusion, the crack opening and the Cherepanov-Rice $J$-integral at the end point of the crack $x = -b$ and the distance $x_0$ of the point of application of the force $P_0$ from the origin, at which the resultant moment of the forces acting on the inclusion turns to zero, such that, the possibility of rotating the entire system as a whole is excluded.

The stated problem is mathematically formulated in the form of the following boundary value problem: the Lame equations for each half-plane and the boundary conditions on the joining line of the half-planes

$$\begin{cases} \sigma_y^{(j)}(x,0) = -q_j(x) & (-b < x < 0), \\ \tau_{xy}^{(j)}(x,0) = \tau_j(x) & (-b < x < 0), \\ V_1(x,0) = V_2(x,0) = \delta_1 & (0 < x < a), \\ U_1(x,0) = U_2(x,0) = \delta_2 & (0 < x < a), \end{cases} \tag{13.1}$$

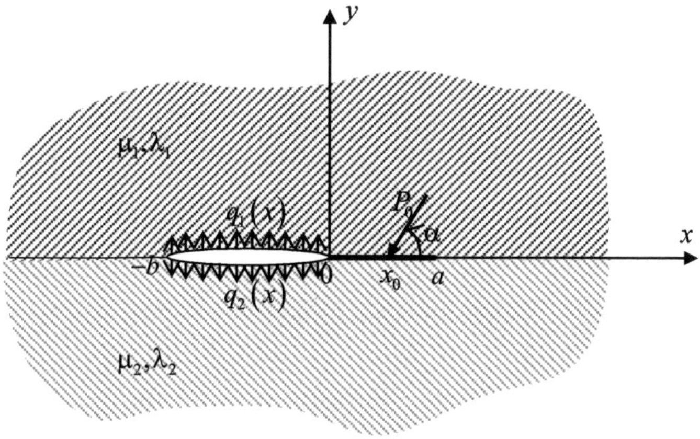


**Fig. 13.1** Schematic representation of the problem

$$\begin{cases} \sigma_y^{(1)}(x,0) = \sigma_y^{(2)}(x,0) \\ \tau_{xy}^{(1)}(x,0) = \tau_{xy}^{(2)}(x,0) \\ U_1(x,0) = U_2(x,0) \\ V_1(x,0) = V_2(x,0) \end{cases} \quad (-\infty < x < -b;\ x > a).$$

Here, $\sigma_y^{(j)}(x, y)$ and $\tau_{xy}^{(j)}(x, y)(j = 1, 2)$ are the normal and tangential components of the stresses, $U_j(x, y)$ and $V_j(x, y)$ are the horizontal and vertical displacements of the points of the corresponding half-plane and $\delta_1, \delta_2$ are the unknown constants expressing a rigid displacement not only of the inclusion, but also of the system as a whole.

In order to write down the governing system of equations for the stated problem, we use discontinuous solutions of the equations of plane elasticity theory for a compound plane with interface defects, obtained in Hakobyan (2014). In this case, they have the following form:

$$U_2'(x,0) = \frac{\mathrm{d}_0}{\Delta_*\mu_2}\sigma(x) - \frac{\mu l_0}{\Delta_*}U'(x) - \frac{\mathrm{d}_1}{\pi\Delta_*\mu_2}\int_{-b}^{a}\frac{\tau(\xi)\mathrm{d}\xi}{\xi - x} + \frac{\mu l_2}{\pi\Delta_*}\int_{-b}^{a}\frac{V'(\xi)\mathrm{d}\xi}{\xi - x},$$

$$V_2'(x,0) = -\frac{\mathrm{d}_0}{\Delta_*\mu_2}\tau(x) - \frac{\mu l_0}{\Delta_*}V'(x) - \frac{\mathrm{d}_1}{\pi\Delta_*\mu_2}\int_{-b}^{a}\frac{\sigma(\xi)\mathrm{d}\xi}{\xi - x} - \frac{\mu l_2}{\pi\Delta_*}\int_{-b}^{a}\frac{U'(\xi)\mathrm{d}\xi}{\xi - x},$$

$$\sigma_y^{(1)}(x,0) = \frac{\mu l_0}{\Delta_*}\sigma(x) + \frac{\mu l_1}{\Delta_*}\mu_2 U'(x) + \frac{\mu l_2}{\pi\Delta_*}\int_{-b}^{a}\frac{\tau(s)\mathrm{d}s}{s - x} - \mu_2\frac{\mu l_3}{\pi\Delta_*}\int_{-b}^{a}\frac{V'(s)\mathrm{d}s}{s - x},$$

$$\tau_{xy}^{(1)}(x,0)=\frac{\mu l_0}{\Delta_*}\tau(x)-\frac{\mu l_1}{\Delta_*}\mu_2 V'(x)-\frac{\mu l_2}{\pi\Delta_*}\int_{-b}^{a}\frac{\sigma(s)\mathrm{d}s}{s-x}-\mu_2\frac{\mu l_3}{\pi\Delta_*}\int_{-b}^{a}\frac{U'(s)\mathrm{d}s}{s-x},$$

$$U_1'(x,0)=U_2'(x,0)+U'(x),\quad V_1'(x,0)=V'(x)+V_2'(x,0),$$
$$\sigma_y^{(2)}(x,0)=-\sigma(x)+\sigma_y^{(1)}(x,0),\quad \tau_{xy}^{(2)}(x,0)=-\tau(x)+\tau_{xy}^{(1)}(x,0). \tag{13.2}$$

Here we use the difference functions

$$U(x)=\begin{cases}U_1(x,0)-U_2(x,0) & (-b\le x\le 0)\\ 0 & (0\le x\le a),\end{cases}$$
$$V(x)=\begin{cases}V_1(x,0)-V_2(x,0) & (-b\le x\le 0),\\ 0 & (0\le x\le a),\end{cases}$$
$$\sigma(x)=\begin{cases}\sigma_y^{(1)}(x,0)-\sigma_y^{(2)}(x,0) & (0\le x\le a),\\ \sigma_0(x)=-q_1(x)+q_2(x) & (-b\le x\le 0),\end{cases}$$
$$\tau(x)=\begin{cases}\tau_{xy}^{(1)}(x,0)-\tau_{xy}^{(2)}(x,0) & (0\le x\le a),\\ \tau_0(x)=\tau_1(x)-\tau_2(x) & (-b\le x\le 0),\end{cases} \tag{13.3}$$

which are finite functions that differ from zero only on the interval $(-b,a)$ occupied by the crack and inclusion, and the following notation:

$$l_0=\vartheta_2^{(1)}(\mu\vartheta_2^{(1)}+\vartheta_2^{(2)})-\vartheta_1^{(1)}(\mu\vartheta_1^{(1)}-\vartheta_1^{(2)}),\quad l_2=\vartheta_1^{(1)}\vartheta_2^{(2)}+\vartheta_2^{(1)}\vartheta_1^{(2)},$$
$$l_1=2(\vartheta_1^{(2)}l_0^*-\vartheta_2^{(2)}l_2^*),$$
$$l_3=2(\vartheta_1^{(2)}l_2^*-\vartheta_2^{(2)}l_0^*),\quad \mathrm{d}_0=\left(\mu\vartheta_1^{(1)}-\vartheta_1^{(2)}\right)/2,\quad \mathrm{d}_1=\left(\mu\vartheta_2^{(1)}+\vartheta_2^{(2)}\right)/2,$$
$$\Delta_*=\left(\mu\vartheta_2^{(1)}+\vartheta_2^{(2)}\right)^2-\left(\mu\vartheta_1^{(1)}-\vartheta_1^{(2)}\right)^2,$$
$$\vartheta_1^{(j)}=\frac{1-2\nu_j}{\kappa_j},\quad \vartheta_2^{(j)}=\frac{2(1-\nu_j)}{\kappa_j},\quad \kappa_j=3-4\nu_j,\quad \mu=\frac{\mu_1}{\mu_2}.$$

As the unknown functions, we choose the complex combinations of jumps of stresses and displacements at the corresponding intervals:

$$\chi(x)=\left[\sigma_y^{(1)}(x,0)-\sigma_y^{(2)}(x,0)\right]-i\left[\tau_{xy}^{(1)}(x,0)-\tau_{xy}^{(2)}(x,0)\right]\quad (0<x<a),$$
$$W(x)=[U_1(x,0)-U_2(x,0)]+i[V_1(x,0)-V_2(x,0)]\quad (-b<x<0). \tag{13.4}$$

Taking into account that in the representations (13.3), the difference of boundary conditions (13.1) has already been used, we write down the sum of these conditions in the form of complex combinations:

$$\begin{cases} \left[\sigma_y^{(1)}(x,0)+\sigma_y^{(2)}(x,0)\right]-i\left[\tau_{xy}^{(1)}(x,0)+\tau_{xy}^{(2)}(x,0)\right]=P_+(x)-iT_+(x) & (-b<x<0), \\ \left[U_1'(x,0)+U_2'(x,0)\right]+i\left[V_1'(x,0)+V_2'(x,0)\right]=0 & (0<x<a) \end{cases} \tag{13.5}$$

$$P_+(x)=-q_1(x)-q_2(x),\quad T_+(x)=\tau_1(x)+\tau_2(x).$$

Substituting in (13.5) the representation (13.2) and using (13.3), to determine the unknown functions (13.4), we obtain the following system of governing equations:

$$\begin{cases} l_1W'(x)+\dfrac{il_3}{\pi}\displaystyle\int_{-b}^{0}\dfrac{W'(\xi)}{\xi-x}d\xi+\dfrac{il_2}{\pi\mu_2}\displaystyle\int_{0}^{a}\dfrac{X(\xi)}{\xi-x}d\xi=\Phi_1^*(x) & (-b<x<0), \\ d_0X(x)-\dfrac{id_1}{\pi}\displaystyle\int_{0}^{a}\dfrac{X(\xi)}{\xi-x}d\xi-\dfrac{i\mu_2l_2}{\pi}\displaystyle\int_{-b}^{0}\dfrac{W'(\xi)}{\xi-x}d\xi=\Phi_2^*(x) & (0<x<a). \end{cases} \tag{13.6}$$

This system must be considered together with the conditions of equilibrium of inclusion and continuity of displacements at the end points of the crack:

$$\int_0^a X(\xi)d\xi=P_0(\sin\alpha-i\cos\alpha),\quad \int_{-b}^{0}W'(\xi)d\xi=0. \tag{13.7}$$

Here

$$\Phi_1^*=-\frac{l_0}{\mu_2}(\sigma_0(x)-i\tau_0(x))-\frac{l_2}{\pi\mu_2}\int_{-b}^{0}\frac{\tau_0(\xi)+i\sigma_0(\xi)}{\xi-x}d\xi$$
$$+\frac{\Delta_*}{2\mu\mu_2}(P_+(x)-iT_+(x)),$$
$$\Phi_2^*=\frac{d_1}{\pi}\int_{-b}^{0}\frac{\tau_0(\xi)+i\sigma_0(\xi)}{\xi-x}d\xi.$$

## 13.3 Algorithm for Solving the Governing System of Equations

We switch to the dimensionless quantities, by dividing the stress components by the shear moduli $\mu_2$, and the linear values by the length of the inclusion $a$, and bringing the integration intervals to the interval $(-1, 1)$:

$$W'(x) = W'\left(-\frac{b}{2}(1-t)\right) = \varphi(t), \quad X(t) = X\left(\frac{a}{2}(1+t)\right) = \mu_2\,\psi(t),$$

$$\Phi_1^*(x) = \Phi_1^*\left(-\frac{b}{2}(1-t)\right) = g_1(t), \quad \Phi_2^*(t) = \Phi_2^*(t)\left(\frac{a}{2}(1+t)\right) = \mu_2\,g_2(t),$$

$$\sigma_0(x) = \mu_2\sigma_0^*\left(1+\frac{2x}{b}\right), \quad \tau_0(x) = \mu_2\tau_0^*\left(1+\frac{2x}{b}\right),$$

$$P_+(x) = \mu_2 P_+^*\left(1+\frac{2x}{b}\right), \quad T_+(x) = \mu_2 T_+^*\left(1+\frac{2x}{b}\right),$$

$$g_1(t) = -l_0\big(\sigma_0^*(t) - i\tau_0^*(t)\big) - \frac{l_2}{\pi}\int_{-1}^{1}\frac{\tau_0^*(\xi)+i\sigma_0^*(\xi)}{\xi - t}\mathrm{d}\xi + \frac{\Delta_*}{2\mu}\big(P_+^*(t) - iT_+^*(t)\big),$$

$$g_2(t) = \frac{\mathrm{d}_1}{\pi}\int_{-1}^{1}\frac{\tau_0^*(\xi)+i\sigma_0^*(\xi)}{\xi - t}\mathrm{d}\xi\,, \quad P_0 = a\mu_2 P_0^*.$$

The system of governing Eq. (13.6) and condition (13.7) takes the form:

$$\begin{cases} l_1\varphi(t) + \frac{il_3}{\pi}\int_{-1}^{1}\frac{\varphi(s)}{s-t}\mathrm{d}s + \frac{il_2}{\pi}\int_{-1}^{1}\frac{\psi(s)}{s+1+\frac{b}{a}(1-t)}\mathrm{d}s = g_1(t), \\ d_0\psi(t) - \frac{id_1}{\pi}\int_{-1}^{1}\frac{\psi(s)}{s-t}\mathrm{d}s - \frac{il_2}{\pi}\int_{-1}^{1}\frac{\varphi(s)}{s-1-\frac{a}{b}(1+t)}\mathrm{d}s = g_2(t), \end{cases} \tag{13.8}$$

$$\int_{-1}^{1}\varphi(s)\mathrm{d}s = 0,$$

$$\int_{-1}^{1}\psi(s)\mathrm{d}s = 2P_0^*(\sin\alpha - i\cos\alpha) = -2iP_0^*e^{i\alpha}. \tag{13.9}$$

The solution of the last system will be constructed using the method of mechanical quadratures (MMQ) Sahakyan and Amirjanyan (2018). The required functions will be sought in the following form:

$$\begin{aligned} &\varphi(t) = \varphi_*(t)(1-t)^{\alpha}(1+t)^{\beta}, \\ &\psi(t) = \psi_*(t)(1-t)^{\gamma}(1+t)^{\delta}, \quad \big(-1 < \mathrm{Re}[\alpha,\beta,\gamma,\delta] < 1\big). \end{aligned} \tag{13.10}$$

According to the standard MMQ procedure, it is required to study the behavior of the equations of the system in the vicinity of the ends of the integration interval using the known results from (Muskhelishvili 1958); in order to determine the exponents in the representations (13.10), we obtain a system of trigonometric equations:

$$l_1 - il_3\mathrm{ctg}\pi\beta = 0, \quad \mathrm{d}_0 - i\mathrm{d}_1\mathrm{ctg}\pi\gamma = 0, \ \alpha = \delta,$$

$$\mathrm{tg}^2\pi\alpha + i\frac{2(\nu_1-\nu_2)}{(1-2\nu_1)(1-2\nu_2)}\mathrm{tg}\pi\alpha + \frac{4(1-\nu_1)(1-\nu_2)}{(1-2\nu_1)(1-2\nu_2)} = 0. \tag{13.11}$$

The exponents $\beta$ and $\gamma$ are uniquely determined, but for $\alpha$, we have two different values:

$$\beta = -\frac{1}{2} - i\theta,\quad \gamma = -\frac{1}{2} + i\left(\theta + \frac{1}{2\pi}\ln\frac{\kappa_1}{\kappa_2}\right),\quad \theta = \frac{1}{2\pi}\ln\frac{1+\mu\kappa_2}{\mu+\kappa_1},$$
$$\alpha_1 = -\frac{1}{2} - \frac{i}{2\pi}\ln\kappa_1,\quad \alpha_2 = -\frac{1}{2} + \frac{i}{2\pi}\ln\kappa_2. \tag{13.12}$$

Therefore, we seek the solution of the defining system (13.8) in the following form:

$$\varphi(t) = \frac{1-2\nu_1}{2\mu\cos\pi\alpha_1}\Phi(t)(1-t)^{\alpha_1}(1+t)^{\beta} + \Psi(t)(1-t)^{\alpha_2}(1+t)^{\beta}$$
$$\psi(t) = \Phi(t)(1-t)^{\gamma}(1+t)^{\alpha_1} + \frac{2\cos\pi\alpha_2}{1-2\nu_2}\Psi(t)(1-t)^{\gamma}(1+t)^{\alpha_2} \tag{13.13}$$

and the new unknown functions will be $\Phi(t)$ and $\Psi(t)$ which satisfy the Hölder continuity condition on the interval $[-1, 1]$. Continuing the MMQ procedure, we can represent the new unknown functions in the form of interpolation polynomials in the roots of the Jacobi polynomial corresponding to the weight function

$$\Phi(t) = \begin{cases} \dfrac{2}{n+\alpha_1+\beta+1}\displaystyle\sum_{j=1}^{n}\frac{X_j^{(\varphi)}P_n^{(\alpha_1,\beta)}(t)}{\left(t-\xi_j^{(\alpha_1,\beta)}\right)P_{n-1}^{(\alpha_1+1,\beta+1)}\left(\xi_j^{(\alpha_1,\beta)}\right)} & \text{for } \varphi(t)\,, \\ \dfrac{2}{n+\gamma+\alpha_1+1}\displaystyle\sum_{j=1}^{n}\frac{X_j^{(\psi)}P_n^{(\gamma,\alpha_1)}(t)}{\left(t-\xi_j^{(\gamma,\alpha_1)}\right)P_{n-1}^{(\gamma+1,\alpha_1+1)}\left(\xi_j^{(\gamma,\alpha_1)}\right)} & \text{for } \psi(t)\,, \end{cases}$$
$$\Psi(t) = \begin{cases} \dfrac{2}{n+\alpha_2+\beta+1}\displaystyle\sum_{j=1}^{n}\frac{Y_j^{(\varphi)}P_n^{(\alpha_2,\beta)}(t)}{\left(t-\xi_j^{(\alpha_2,\beta)}\right)P_{n-1}^{(\alpha_2+1,\beta+1)}\left(\xi_j^{(\alpha_2,\beta)}\right)} & \text{for } \varphi(t)\,, \\ \dfrac{2}{n+\alpha_2+\gamma+1}\displaystyle\sum_{j=1}^{n}\frac{Y_j^{(\psi)}P_n^{(\gamma,\alpha_2)}(t)}{\left(t-\xi_j^{(\gamma,\alpha_2)}\right)P_{n-1}^{(\gamma+1,\alpha_2+1)}\left(\xi_j^{(\gamma,\alpha_2)}\right)} & \text{for } \psi(t)\,. \end{cases} \tag{13.14}$$

Here, $\xi_j^{(\alpha,\beta)}$ are the roots of the Jacobi polynomial $P_n^{(\alpha,\beta)}(t)$, and the coefficients $X_j^{(\varphi)}$, $X_j^{(\psi)}$, $Y_j^{(\varphi)}$, $Y_j^{(\psi)}$ $\left(j=\overline{1,n}\right)$ are to be determined.

We substitute (13.14) into the equations of system (13.8) and use the standard MMQ procedure to replace the integrals by finite sums with respect to each term in (13.13). Taking into account the fact that the real part of the singularity exponent at both ends of the integration interval is equal to $-0.5$, as the collocation points for each equation of the system, it is necessary to choose a system of $n-1$ points.

In the numerical analysis, the roots of the Chebyshev polynomial $U_{n-1}(t)$ were chosen. Adding two conditions (13.9) to $2(n-1)$ equations and equating of different representations of the same functions (13.14) in the roots of the Chebyshev polynomial $T_n(t)$, we obtain a closed system of linear algebraic equations with respect to $4n$ unknown coefficients.

After determining the unknown coefficients, the derivatives of the jump of the crack faces displacements and the jump of the stresses on the edges of the inclusion can be represented as interpolation polynomials, and using the corresponding quadrature formula (Sahakyan 2010), one can write out a representation for the jump of the displacements and construct graphs for crack opening.

It is also possible to determine other values related to the stress state in the vicinity of stress concentrators. For example, from the condition that the resultant moment of the loads acting on the inclusion be zero, we find the point of application of the concentrated force $x_0^* = x_0/a$, when the rotation of the inclusion is excluded

$$x_0^* = \frac{1}{2} - \frac{M_0}{4P_0^* \sin \pi\alpha} . \tag{13.15}$$

Here, $M_0$ is the resultant moment of contact stresses in the case when the force is applied at the center of inclusion, defined by the formula

$$M_0 = \int_{-1}^{1} s\mathrm{Re}[\psi(s)]\mathrm{d}s =$$

$$= \mathrm{Re}\left[\sum_{i=1}^{n}\left\{w_i^{(\gamma,\alpha_1)}\xi_i^{(\gamma,\alpha_1)}X_i^{(\psi)} + \frac{2\cos\pi\alpha_2}{1-2\nu_2}w_i^{(\gamma,\alpha_2)}\xi_i^{(\gamma,\alpha_2)}Y_i^{(\psi)}\right\}\right], \tag{13.16}$$

$$w_j^{(\alpha,\beta)} = \frac{2^{\alpha+\beta+3}}{1-\left(\xi_j^{(\alpha,\beta)}\right)^2}\frac{B(\alpha,\beta+n+1)}{B(\alpha,n+1)}\left[\frac{1}{(\alpha+\beta+n+1)P_{n-1}^{(\alpha+1,\beta+1)}\left(\xi_j^{(\alpha,\beta)}\right)}\right]^2 .$$

In the crack tip disconnected with inclusion, we define such a characteristic for cracks, as the dimensionless Cherepanov-Rice J-integral (Murakami 1987)

$$J_* = \frac{1}{\mu_2}J(-1) = \frac{\mu+1}{4\mu}K(-1)\overline{K}(-1) , \tag{13.17}$$

where $K(-1)$ is the dimensionless complex coefficient of stress concentration at the tip of the crack, defined by the formula

$$K(-1) = i\frac{4\sqrt{2\pi}}{\sin\pi\beta}\frac{1-\nu_1+\mu(1-\nu_2)}{(\mu+\kappa_1)(1+\mu\kappa_2)}\left\{\frac{1-2\nu_1}{2\cos\pi\alpha_1}\Phi(-1)2^{\alpha_1} + \mu\Psi(-1)2^{\alpha_2}\right\},$$

$$\Phi(-1) = \frac{2}{n + \alpha_1 + \beta + 1} \sum_{j=1}^{n} \frac{X_j^{(\varphi)} P_n^{(\alpha_1,\beta)}(-1)}{\left(-1 - \xi_j^{(\alpha_1,\beta)}\right) P_{n-1}^{(\alpha_1+1,\beta+1)}\left(\xi_j^{(\alpha_1,\beta)}\right)},$$

$$\Psi(-1) = \frac{2}{n + \alpha_2 + \beta + 1} \sum_{j=1}^{n} \frac{Y_j^{(\varphi)} P_n^{(\alpha_2,\beta)}(-1)}{\left(-1 - \xi_j^{(\alpha_2,\beta)}\right) P_{n-1}^{(\alpha_2+1,\beta+1)}\left(\xi_j^{(\alpha_2,\beta)}\right)}.$$

## 13.4 Numerical Analysis

For definiteness, suppose that the concentrated load $P_0^* = 1$ acts on the inclusion at an angle $\alpha = \pi/3$, and normal uniformly distributed load $q_0/\mu_2 = 0.01$ acts on the crack faces, hence, $\sigma_0^*(t) = \tau_0^*(t) = 0$; $P_+^* = 0.02$. Initially, for certain values of remaining parameters of the problem: $\mu = 2$; $\nu_1 = 0.25$; $\nu_2 = 0.3$; $b/a = 1$, we will test the convergence of the computational process by calculating some values for different orders of approximation $n$. In Table 13.1, we show the values of $M_0$ and $J_*$ calculated from Formulas (13.16) and (13.17), as well as the absolute values of the root-mean-square deviations $\Delta_1$ and $\Delta_2$ of the left-hand sides of Eq. (13.8) calculated by standard integration programs, from their right-hand parts, calculated on a uniform grid of 20 points.

Judging from the data in the table, which do not change very much with an increase in the order $n$ of approximation from 6 to 20, it can be concluded that, apparently, the accuracy obtained is the maximal for the approach used. However, this accuracy is sufficient for constructing the corresponding graphs and analyzing the dependence of the jump of the contact stresses $\psi(t)$, the opening of the crack $w(t) = \int_{-1}^{t} \varphi(s)\mathrm{d}s$, the $J$-integral and the point of application of the concentrated force, when the rotation of the inclusion is excluded, depended on the mechanical characteristics of the half-planes and the ratio of the lengths of the stress concentrators. Calculations were carried out for $n = 10$. Figures 13.2, 13.3 and 13.4 show graphs of crack opening, jumps of normal and tangential contact stresses for and different values of shear moduli ratio.

**Table 13.1** Convergence of the numerical method

| $n$ | $M_0$ | $J_*$ | $\lvert\Delta_1\rvert$ | $\lvert\Delta_2\rvert$ |
|---|---|---|---|---|
| 6 | −0.0666 | 0.01356 | $3.9 \times 10^{-3}$ | $3.1 \times 10^{-2}$ |
| 8 | −0.0669 | 0.01387 | $2.2 \times 10^{-3}$ | $1.4 \times 10^{-2}$ |
| 10 | −0.0670 | 0.01418 | $1.4 \times 10^{-3}$ | $6.7 \times 10^{-3}$ |
| 12 | −0.0670 | 0.01447 | $1.0 \times 10^{-3}$ | $7.6 \times 10^{-3}$ |
| 14 | −0.0670 | 0.01473 | $8.8 \times 10^{-4}$ | $8.7 \times 10^{-3}$ |
| 20 | −0.0670 | 0.01543 | $4.2 \times 10^{-4}$ | $1.9 \times 10^{-3}$ |

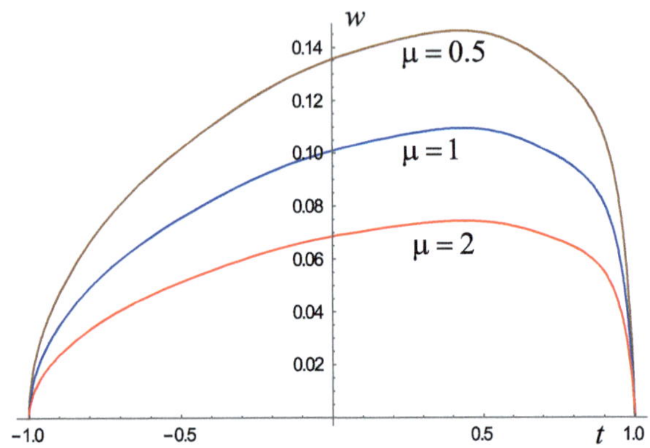


**Fig. 13.2** Crack opening $w(t)$ for $\nu_1 = 0.25,\ \nu_2 = 0.3$

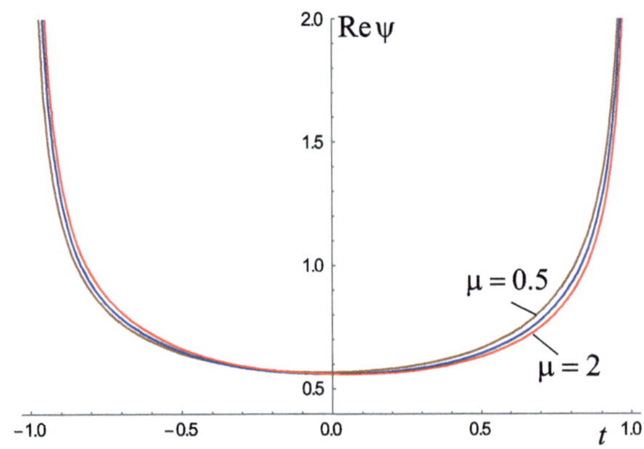


**Fig. 13.3** Distribution of the normal contact pressure

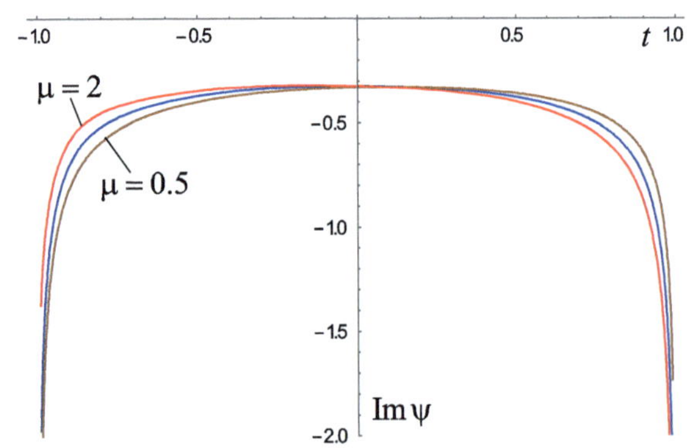


**Fig. 13.4** Distribution of tangential contact stresses

Similar graphs are constructed for different values of the Poisson's ratio of both the upper and the lower half-planes for $\nu_1, \nu_2 = 0.15,\ 0.3,\ 0.45$ at a constant value of the ratio $\mu = 2$. These graphs are not shown, because they qualitatively repeat the curves given in the figures. It should be noted that the order of the curves is similar:

larger values $\mu$ correspond to a larger value $\nu$. Figures 13.5 and 13.6 show curves of the dependence of and on the ratio for different values of the Poisson's ratios.

The graphs in Fig. 13.5 indicate that for any values of the Poisson's coefficients $\nu_1$ and $\nu_2$, $J$-integral takes on a maximum value for values $\mu$ close to 0.5. This means that for these values $\mu$, the probability of crack propagation is much higher.

An interesting result can be seen from comparing the graphs in Fig. 13.7.

Comparing the elastic characteristics for the curves 1 and 5, which are very close to each other, with the characteristics for the curve 2, we note that in both cases, one of the parameters decreases by two times. In the first case, this is the Poisson's ratio $\nu_2$, and in the second case, it is the ratio of the shear moduli $\mu$. The regulation is maintained even for increasing of the last parameters, which corresponds to the curves 3 and 4. The difference between them is slightly larger, since the ratio of the shear moduli $\mu$ is doubled, but the Poisson's ratio $\nu_2$ is increased by one and a half time.

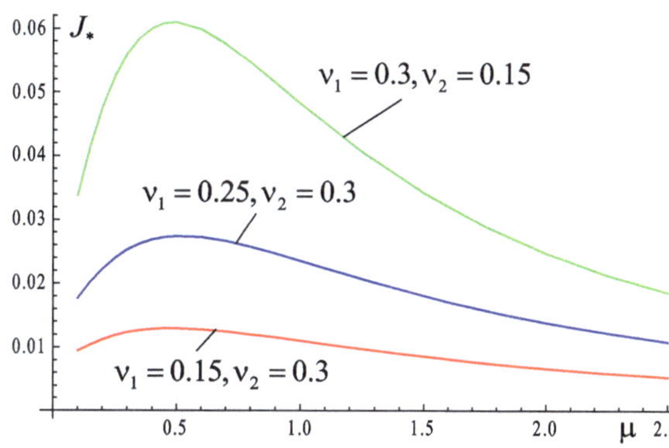


**Fig. 13.5** Dependence $J$-integral of the ratio $\mu$ for different $\nu_1$ and $\nu_2$

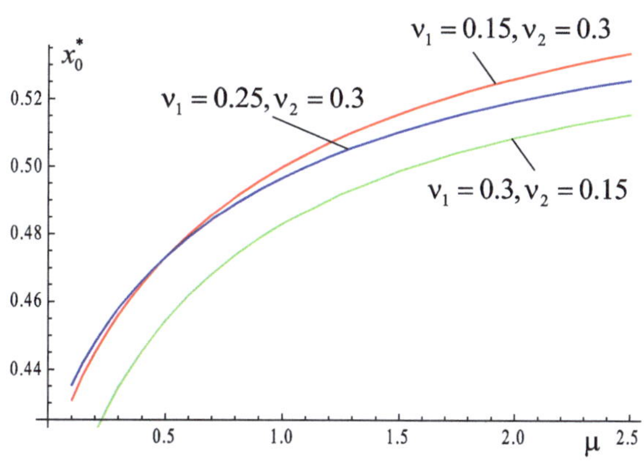


**Fig. 13.6** Dependence of the point of application of force $x_0^*$ on the ratio $\mu$ for different $\nu_1$ and $\nu_2$

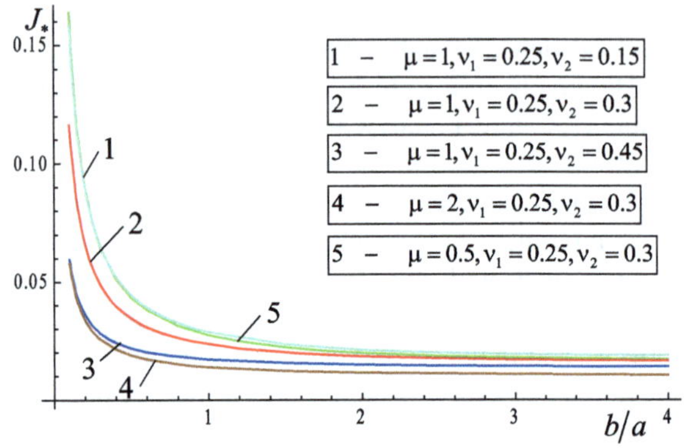


**Fig. 13.7** Dependence $J$-integral of the ratio of the lengths of the crack and the inclusion $b/a$

## 13.5 Conclusions

Unlike the analogous problem for a homogeneous plane that allows closed solution, the heterogeneity of the half-planes and the presence on the junction line of a system of stress concentrators consisting of crack and inclusion that continue each other, significantly complicates the solution of the problem, since it does not allow to obtain an analytical solution by known methods for solution of contact problems of the theory of elasticity. An algorithm for the numerical-analytical solution of the stated problem of the theory of elasticity for a piecewise homogeneous plane by the method of mechanical quadratures is proposed.

New representations for the unknown functions are used, which take into account the presence of singularities of different orders at the junction point between the crack and the inclusion. Numerical analysis of the dependence of contact stresses on the sides of the inclusion, crack opening, the $J$-integral at the end point of the crack, and the coordinate of the point of application $x_0$ of the force at which the resulting moment of the forces acting on the inclusion turns to zero, from the elastic and geometric parameters of the problem, is performed.

## References

Berezhnitski LT, Panasyuk VV, Stashyuk NG (1983) Interaction of hard linear inclusions and cracks in deformed solid. Naukova Dumka, Kiev (in Russ.)

Hakobyan VN (2014) Mixed boundary value problems on interaction of continuum deformable bodies with different types stress concentrators. Gitutyun NAS RA, Yerevan (in Russ.)

Hakobyan VN, Simonyan AR (2011) On plane-deformed state of orthotropic plane with slits. Proc National Acad Sci Armenia Mech 64(2):4–14 (in Russ.)

Murakami Y (Ed) (1987) Stress intensity factors handbook. vol 1. The Society of Materials Science, Japan, Pergamon Press, Oxford

Muskhelishvili NI (1958) Singular integral equations. Springer, Dordrecht. https://doi.org/10.1007/978-94-009-9994-7

Panasyuk VV, Savruk MP, Datsyshyn AA (1976) Distribution of stresses near cracks near plates and shells. Naukova Dumka, Kiev (in Russ.)
Sahakyan AV (2010) Quadrature formulas for calculation of the integral with a variable upper limit. In: Proceeding of II international conference Topical problems of continuum mechanics. Dilijan, Armenia, 4–8 Oct, pp 107–111. http://www.mechins.sci.am/publ/avetik_sahakyan/Dilijan2010.doc (in Russ.)
Sahakyan AV, Amirjanyan HA (2018) Method of mechanical quadrature's for solving singular integral equations of various types. IOP Conf Ser: J Phys: Conf Ser 991:012070. https://doi.org/10.1088/1742-6596/991/1/012070
Sahakyan AV, Dashtoyan LL (2003) On one problem for compound plane with crack and inclusion with the existence of temperature field—THERMAL STRESSES 2003. June 8–11, Blacksburg VA, USA, TM-4–1–1–4–1–4

# Chapter 14
# Finite Inflation of an Inhomogeneous Hyperelastic Cylindrical Membrane

**Alexey M. Kolesnikov and Daria A. Letunova**

**Abstract** We study an inflation of an inhomogeneous thin-walled hyperelastic tube. Either the thickness of the tube or the material properties of tube the change on its cross section. The inhomogeneity leads to a bending of tube subjected by an internal uniform pressure. We analyse the effect of the magnitude of the inhomogeneity, the size of its area and the pressure on the bending of tube.

**Keywords** Hyperelastic membrane · Thin-walled tube · Inhomogeneity · Inflation

## 14.1 Introduction

Inhomogeneities are a common phenomenon in solids and structures. For bodies of natural origin, it is probably more accurate to say that homogeneity in mechanical properties or geometric sizes is the exception. The production of details is always carried out with a certain degree of precision. Also inhomogeneities may appear during the operation of structures. Due to the large number of different inhomogeneities, their investigation is a difficult problem.

In this paper, we consider the problem of inflating a thin-walled tube made of hyperelastic material under finite deformations. For a homogeneous tube, the problem has a known analytical solution both in the framework of non-linear theory of elasticity and in the non-linear theory of shells and membranes. Under the action of pressure the tube changes its length and radius and its stress-strain state is homogeneous. If the tube is inhomogeneous, it may twist and bend when inflated. This effect is used, for example, in the construction of bending actuators (Connolly et al. 2017).

A. M. Kolesnikov (✉) · D. A. Letunova
Southern Federal University, ul. Milchakova 8a, Rostov-na-Donu 344090, Russian Federation
e-mail: Alexey.M.Kolesnikov@gmail.com

D. A. Letunova
e-mail: letunova.daria.a@gmail.com

A. M. Kolesnikov
Don State Technical University, Gagarin square 1, Rostov-na-Donu 344000, Russian Federation

 
H. Altenbach et al. (eds.), *Advanced Materials Modelling for Mechanical, Medical and Biological Applications*, Advanced Structured Materials 155,
https://doi.org/10.1007/978-3-030-81705-3_14

Despite of a variety of variants of the problem statement, there are few studies of elastic tubes with inhomogeneities under large strains. In Hart and Shi (1993), Haughton and McKay (1997) an axisymmetric problem of inflation and tension of two connected thin-walled tubes made of different material or having different dimensions is considered. In Patil et al. (2015) a tube of variable thickness under hydrostatic pressure is investigated.

In this paper, we investigate a cylindrical membrane with variable properties along a cross section. The basic equations of non-linear membrane theory are given in Sect. 14.2. Solving the problem, we use the semi-inverse method presented in Sect. 14.3. The considered deformation corresponds to the pure bending of an inflated cylindrical membrane (Libai and Simmonds 1988). There are a number of studies devoted to the bending of an inflated tube (Haseganu and Steigmann 1994; Haughton and McKay 1996; Kolesnikov and Zubov 2009). In addition, note the closely related bending problem of an inflated curved tube (Kuznetsov and Levyakov 2001; Polyakova et al 2009; Levyakov 2017; Kolesnikov et al. 2019), in which the tube curvature can changes under the action of internal pressure without applying bending moment (Kolesnikov 2015). In Sect. 14.4 the inflation problem of an inhomogeneous tube is presented.

## 14.2 Non-linear Theory of Elastic Membranes

The non-linear shell and membrane theories are presented in Green and Adkins (1960), Zubov (1982), Libai and Simmonds (1988) and among others. We use vectors $\mathbf{r}(q^1, q^2)$ and $\mathbf{R}(q^1, q^2)$ to describe initial and deformed material points on a membrane, respectively. Here $q^1$ and $q^2$ are Gaussian coordinates on the undeformed membrane. The right Cauchy–Green deformation tensor $\mathbf{G}^\times$ is defined as

$$\mathbf{G}^\times = (\nabla_0\mathbf{R}) \cdot (\nabla_0\mathbf{R})^T = G_{\alpha\beta}\mathbf{r}^\alpha \otimes \mathbf{r}^\beta, \quad \alpha, \beta = 1, 2,$$
$$G_{\alpha\beta} = \mathbf{R}_\alpha \cdot \mathbf{R}_\beta, \quad G = G_{11}G_{22} - G_{12}^2,$$
$$g_{\alpha\beta} = \mathbf{r}_\alpha \cdot \mathbf{r}_\beta, \quad g = g_{11}g_{22} - g_{12}^2.$$

Here $g_{\alpha\beta}$ and $G_{\alpha\beta}$ are components of the first metric tensor of initial and deformed membrane surfaces, $\mathbf{r}_\alpha$, $\mathbf{r}^\alpha$, $\mathbf{R}_\beta$ and $\mathbf{R}^\beta$ $(\alpha, \beta = 1, 2)$ are basis vectors on the initial and deformed membrane surfaces defined as

$$\mathbf{R}_\alpha = \frac{\partial \mathbf{R}}{\partial q^\alpha}, \quad \mathbf{R}_\alpha \cdot \mathbf{R}^\beta = \delta_\alpha^\beta, \quad \mathbf{r}_\alpha = \frac{\partial \mathbf{r}}{\partial q^\alpha}, \quad \mathbf{r}_\alpha \cdot \mathbf{r}^\beta = \delta_\alpha^\beta.$$

Here and below we use the nabla operators ($\nabla_0$ and $\nabla$) in the reference and current state

$$\nabla_0 = \mathbf{r}^\alpha \otimes \frac{\partial}{\partial q^\alpha}, \quad \nabla = \mathbf{R}^\alpha \otimes \frac{\partial}{\partial q^\alpha}.$$

We introduce the second metric tensor $\mathbf{B}$ of a deformed surface, the unit normal $\mathbf{N}$ to the deformed surface and the Christoffel symbol of the second kind $\Gamma^{\gamma}_{\alpha\beta}$ which have the form

$$\mathbf{B} = B_{\alpha\beta}\mathbf{R}^{\alpha} \otimes \mathbf{R}^{\beta}, \quad B_{\alpha\beta} = \mathbf{B} \cdot \frac{\partial \mathbf{R}^{\alpha}}{\partial q^{\beta}}, \quad \mathbf{N} = \frac{\mathbf{R}_1 \times \mathbf{R}_2}{|\mathbf{R}_1 \times \mathbf{R}_2|}, \tag{14.1}$$

$$\Gamma^{\gamma}_{\alpha\beta} = \frac{1}{2} G^{\gamma\nu} \left( \frac{\partial G_{\alpha\nu}}{\partial q^{\beta}} + \frac{\partial G_{\beta\nu}}{\partial q^{\alpha}} - \frac{\partial G_{\alpha\beta}}{\partial q^{\nu}} \right), \quad G^{\alpha\beta} = \mathbf{R}^{\alpha} \cdot \mathbf{R}^{\beta}. \tag{14.2}$$

We assume the the material of the membrane is isotropic, incompressible and hyperelastic. We describe mechanical properties of the membrane by a surface strain energy density function $W^{\times}(\mathbf{G}^{\times})$ which is determined from a bulk strain energy density in the form

$$W^{\times}(\mathbf{G}^{\times}) = hW\left[\mathbf{G}^{\times} + (g/G)\mathbf{n} \otimes \mathbf{n}\right],$$

where $h = h(q^1, q^2)$ is an initial thickness of the membrane, $\mathbf{n}$ is a unit normal vector to the undeformed membrane. We can define the surface strain energy function as

$$W^{\times} = k\tilde{W}, \quad k = \mu h, \quad \tilde{W} = \frac{W}{\mu}. \tag{14.3}$$

Here, we introduce a material parameter $\mu = \mu(q^1, q^2)$ and a parameter $k = k(q^1, q^2)$ which we can call the stiffness of the membrane surface. This means that a change in the stiffness of the membrane can result either from a change in its thickness or from a change in the mechanical properties of the material (whilst maintaining its isotropy).

The Cauchy stress resultant tensor is given by

$$\mathbf{L} = 2k\sqrt{\frac{g}{G}} (\nabla_0 \mathbf{R})^T \cdot \frac{\partial \tilde{W}}{\partial \mathbf{G}^{\times}} \cdot \nabla_0 \mathbf{R} = L^{\alpha\beta} \mathbf{R}_{\alpha} \otimes \mathbf{R}_{\beta}. \tag{14.4}$$

The equilibrium equations of non-linear elastic membranes have the form

$$\nabla \cdot \mathbf{L} + \mathbf{p} = 0, \tag{14.5}$$

where $\mathbf{p}$ is a surface load in the deformed configuration.

## 14.3 Pure Bending of Inhomogeneous Tube

We define the undeformed middle surface of the closed tube (Fig. 14.1a) by equations

$$\mathbf{r} = x_1(s)\mathbf{i}_1 + x_2(s)\mathbf{i}_2 + t\mathbf{i}_3, \quad s \in [-S; S], \quad t \in [0; T].$$

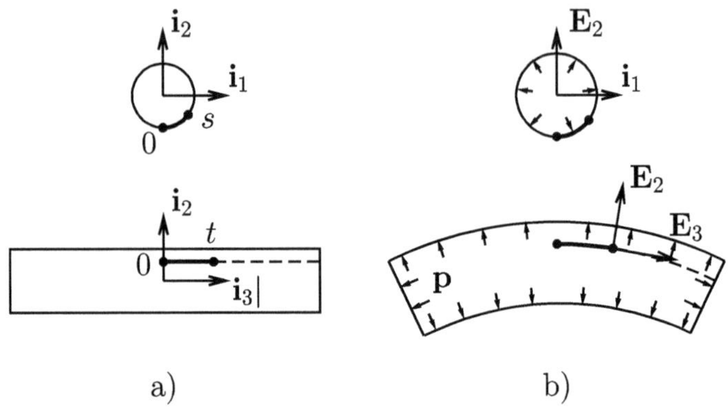


**Fig. 14.1** The middle surface of undeformed (**a**) and inflated (**b**) inhomogeneous tube

Here, $\{\mathbf{i}_1, \mathbf{i}_2, \mathbf{i}_3\}$ is the standard orthonormal basis in a fixed Cartesian coordinate system, $q^1 = s$ and $q^2 = t$ are the Gaussian surface coordinates. The stiffness $k$ of tube is a function of the coordinate $s$ only: $k = k(s)$. We assume that the cross section of the tube is symmetrical relative the $x_2x_3$-plane. So we have

$$x_1(-s) = -x_1(s), \quad x_2(-s) = x_2(s), \quad k(-s) = k(s).$$

We consider the deformation field defined as

$$\mathbf{R} = X_1(s)\mathbf{i}_1 + X_2(s)\mathbf{E}_2, \quad \mathbf{E}_2 = \mathbf{i}_2 \cos Bt + \mathbf{i}_3 \sin Bt. \tag{14.6}$$

That means that the deformed tube is a torus sector (Fig. 14.1b). Its cross section is given by unknown functions $X_1(s)$ and $X_2(s)$. The curvature of the centroidal axis is given by the unknown parameter $B$. This deformation corresponds to the pure bending of the tube (Libai and Simmonds 1988; Zubov 2001).

The natural bases $\mathbf{r}_\alpha$ and $\mathbf{R}_\beta$ in the reference and deformed configurations have the form

$$\begin{aligned} &\mathbf{r}_1 = x_1'\mathbf{i}_1 + x_2'\mathbf{i}_2, \quad \mathbf{r}_2 = \mathbf{i}_3, \\ &\mathbf{R}_1 = X_1'\mathbf{i}_1 + X_2'\mathbf{E}_2, \quad \mathbf{R}_2 = BX_2\mathbf{E}_3, \quad \mathbf{E}_3 = -\mathbf{i}_2 \sin Bt + \mathbf{i}_3 \cos Bt. \end{aligned} \tag{14.7}$$

Here, the prime ( $'$ ) denotes differentiation with respect to $s$. The components of the metric tensors of the initial and deformed surfaces have the form

$$g_{11} = {x_1'}^2 + {x_2'}^2, \quad g_{12} = g_{21} = 0, \quad g_{22} = 1, \tag{14.8}$$

$$G_{11} = {X_1'}^2 + {X_2'}^2, \quad G_{12} = G_{21} = 0, \quad G_{22} = B^2X_2^2. \tag{14.9}$$

We introduce the principal stretches $\lambda_1(s)$, $\lambda_2(s)$ and $\psi(s)$ the angle of inclination of the tangent

$$\lambda_1(s) = \sqrt{\frac{G_{11}}{g_{11}}}, \quad \lambda_2(s) = \sqrt{\frac{G_{22}}{g_{22}}}, \quad \tan\psi(s) = \frac{X_2'}{X_1'}. \tag{14.10}$$

From (14.9), (14.10) the functions $X_1(s)$, $X_2(s)$, $\lambda_1(s)$, $\lambda_2(s)$ and $\psi(s)$ are related to the following differential equations

$$X_1'(s) = \sqrt{g_{11}}\lambda_1 \cos\psi, \quad X_2'(s) = \sqrt{g_{11}}\lambda_1 \sin\psi, \tag{14.11}$$

$$\lambda_2'(s) = B\sqrt{\frac{g_{11}}{g_{22}}}\lambda_1 \sin\psi - \frac{g_{22}'}{2g_{22}}\lambda_2. \tag{14.12}$$

The components of the stress resultant tensor are the functions of the coordinate $s$ only and have the following form

$$L^{11} = \frac{k}{g_{11}\lambda_1^2\lambda_2}\frac{\partial \tilde{W}}{\partial \lambda_1}, \quad L^{22} = \frac{k}{g_{22}\lambda_1\lambda_2^2}\frac{\partial \tilde{W}}{\partial \lambda_2}, \quad L^{12} = L^{21} = 0. \tag{14.13}$$

We assume that the surface load be normal uniform pressure

$$\mathbf{p} = p\mathbf{N}, \quad \mathbf{N} = \mathbf{i}_1 \sin\psi - \mathbf{E}_2 \cos\psi.$$

The equilibrium equation (14.5) can be written in the following form (Kolesnikov and Zubov 2009; Kolesnikov et al. 2019)

$$\frac{dL^{11}}{ds} + L^{11}(2\Gamma_{11}^1 + \Gamma_{21}^2) + L^{22}\Gamma_{22}^1 = 0, \tag{14.14}$$

$$L^{11}B_{11} + L^{22}B_{22} = -p. \tag{14.15}$$

Here,

$$\begin{aligned} \Gamma_{11}^1 &= \frac{\lambda_1'}{\lambda_1} + \frac{g_{11}'}{2g_{11}}, \quad \Gamma_{22}^1 = -\frac{g_{22}\lambda_2\lambda_2'}{g_{11}\lambda_1^2} - \frac{\lambda_2^2 g_{22}'}{2g_{11}\lambda_1^2}, \\ \Gamma_{21}^2 &= \Gamma_{12}^2 = \frac{\lambda_2'}{\lambda_2} + \frac{g_{22}'}{2g_{22}}, \\ B_{11} &= -\sqrt{g_{11}}\lambda_1\psi', \quad B_{22} = B\sqrt{g_{22}}\lambda_2 \cos\psi. \end{aligned} \tag{14.16}$$

Substitution of (14.13) and (14.16) into (14.14) and (14.15) yields to ordinary differential equations for unknowns $\lambda_1(s)$ and $\psi(s)$

$$\frac{\partial^2 \tilde{W}}{\partial \lambda_1^2}\lambda_1' = \left(\frac{\partial \tilde{W}}{\partial \lambda_2} - \lambda_1 \frac{\partial^2 \tilde{W}}{\partial \lambda_1 \lambda_2}\right) B \sqrt{\frac{g_{11}}{g_{22}}} \sin\psi \\ - \left(\frac{\partial \tilde{W}}{\partial \lambda_1} - \lambda_2 \frac{\partial^2 \tilde{W}}{\partial \lambda_1 \lambda_2}\right) \frac{g_{22}'}{2g_{22}} - \frac{k'}{k}\frac{\partial \tilde{W}}{\partial \lambda_1}, \tag{14.17}$$

$$\frac{\partial \tilde{W}}{\partial \lambda_1}\psi' = B\sqrt{\frac{g_{11}}{g_{22}}}\frac{\partial \tilde{W}}{\partial \lambda_2}\cos\psi + \frac{p}{k}\sqrt{g_{11}}\lambda_1\lambda_2. \tag{14.18}$$

Thus, we reduce the equilibrium equation (14.5) to the system of the ordinary differential equations (14.11), (14.12), (14.17) and (14.18). We consider the tube with a closed cross-section. The boundary conditions are periodical conditions to the unknown functions

$$\psi(-S) = \psi(S), \quad \lambda_1(-S) = \lambda_1(S), \quad \lambda_2(-S) = \lambda_2(S), \\ X_1(-S) = X_1(S), \quad X_2(-S) = X_2(S). \tag{14.19}$$

For fixed values of the data $(p, B)$ we have boundary–value problem which we can solve by a numerical method.

The end conditions at $t = 0, T$ are satisfied in the integral sense. The deformation (14.6) is the pure bending (Libai and Simmonds 1988). So the stress-strain state of each cross-section is the same. For closed and pressurized tube the sum of the pressure resultant and the net force of the stress resultants at any cross-section is equal to zero (Libai and Simmonds 1988). The total moment is the sum of the net moment of stress resultants and the couple of pressure. Under the in-plane bending only the moment about $\mathbf{i}_1$ is not zero (Libai and Simmonds 1988). We compute the total bending moment relative to the centre of mass of a deformed cross section with coordinates $\{X_{1C}, X_{2C}\}$. Then the resultant bending moment of the pressure is equal to zero. The net moment of the stress resultants is given by

$$M = \int_{-S}^{S} \sqrt{G_{11}} G_{22} L^{22} (X_2 - X_{2C}) \mathrm{d}s. \tag{14.20}$$

Thus, the closed pressurized curved tube is subjected by the following external edge moments

$$\mathbf{M}_T = -\mathbf{M}_0 = -M\mathbf{i}_1.$$

## 14.4 Inflation

Here, we demonstrate results for a tube composed of the neo-Hookean material. In this case, we have

$$\tilde{W} = \frac{1}{2}(I_1 - 3), \quad I_1 = \lambda_1^2 + \lambda_2^2 + \frac{1}{\lambda_1^2 \lambda_2^2}.$$

We consider a circular cross section with a radius $r_0$

$$x_1(s) = r_0 \sin s, \quad x_2(s) = -r_0 \cos(s), \quad s \in [-\pi; \pi].$$

Symmetry of the problem allows us to change the boundary conditions (14.19). We use the following conditions at the point $s = 0$

$$\psi(0) = 0, \quad X_1(0) = 0, \quad \lambda_1(0) = A_1, \quad \lambda_2(0) = A_2.$$

Here, $A_1$ and $A_2$ are unknown constants, which we define using the symmetry conditions at the point $s = \pi$

$$\psi(\pi) = 0, \quad X_1(\pi) = 0.$$

The curvature $B$ is the parameter of the problem. In this paper we assume that the bending moment is absent. So we determine $B$ from $M = 0$.

And we assume that the stiffness of the tube has a form

$$k(s) = k_1 - \frac{k_1 - k_2}{1 + e^{\varkappa(S_0 - s)}}, \quad s \in [0, \pi]. \tag{14.21}$$

Here, $k_1$ and $k_2$ are the maximum and minimum stiffness, respectively, $\varkappa$ is the transition coefficient, $S_0$ is the coordinate of average stiffness. We note that the function $k(s)$ is defined on $s \in [0, \pi]$. For the segment $s \in [-\pi, 0]$ we have to take the function symmetric with respect to $s = 0$. In this case, at the points $s = 0$ and $s = \pm\pi$ the stiffness function is continuous, but its derivative has the jump discontinuity. For considered parameters of the problem, the magnitude of this discontinuity has same order to the error of the used numerical methods.

To illustrate the stiffness function (14.21) let us introduce the relative stiffness difference $\Delta$, the relative $k^*$ and the average $k_0$ stiffness as

$$\Delta = \frac{k_1 - k_2}{k_0}, \quad k^* = \frac{k(s)}{k_0}, \quad k_0 = k(S_0) = \frac{k_1 + k_2}{2}.$$

Fig. 14.2 shows plots of the function $k^*(s)$ and the shape of the undeformed section. For clarity, we assume that the stiffness changes due to changes in membrane thickness. Figure 14.2b shows cross sections of a tube with the thickness increased 10 times.

Let us introduce the dimensionless curvature, pressure and radius

$$B^* = r_0 B, \quad p^* = \frac{r_0 p}{k_0}, \quad R^* = \frac{R_{av}}{r_0} \tag{14.22}$$

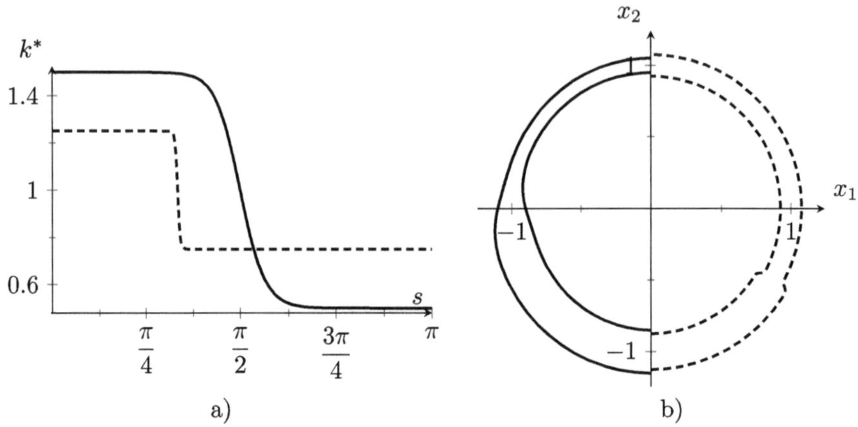


**Fig. 14.2** The stiffness of the membrane: **a** the relative stiffness $k^*(s)$; **b** the initial cross section with thickness increased 10 times. The dashed line corresponds to $k_0 = 0.01$, $\Delta = 0.5$, $\varkappa = 100$, $S_0 = \pi/3$. The solid line corresponds to $k_0 = 0.01$, $\Delta = 1$, $\varkappa = 10$, $S_0 = \pi/2$

Here, $R_{av}$ is an average radius of a deformed cross-section. Figure 14.3 shows the pressure dependence of the curvature and radius of the inflated tube at the following parameters: $S_0 = \pi/2$, $\varkappa = 100$ (solid lines), $\varkappa = 10$ (dashed lines), $\Delta = 0.1, 0.3, 0.5$ and 1. We note that there is a maximum pressure for the neo–Hookean model, after which the curvature and radius of the tube increase with decreasing pressure. Similar behaviour is observed in many problems of structures made of neo–Hookean material under large strains. For example, in the problem of inflation of a homogeneous tube there is a maximum internal pressure after which the tube radius increases as the pressure decreases. This behaviour is usually related to the instability of the neo–Hookean material. Therefore, in the following we will analyse the behaviour of the structure until the maximum pressure is reached.

We obtain that the deformed cross section of the inflated inhomogeneous tube is close to a circle. For the considered cases, the maximum deviation is less than 2% at $\Delta = 1$. It can be seen from Fig. 14.3a that the influence of $\varkappa$ is small for the considered range [10; 100]. And the change of membrane stiffness $\Delta$ significantly influences the "curvature-pressure" dependences. The growth $\Delta$ leads to increasing the bending of the inhomogeneous tube. For the considered parameters the curvature radius of the inflated tube reaches about 10 of the cross-sectional radius. For the curves in Fig. 14.3 the average thickness is constant ($k_0 = 0.01$). Then, an increasing $\Delta$ leads to an increasing $k_1$ and a decreasing $k_2$. Reducing the stiffness $k_2$ for part of the tube leads to a lower maximum pressure.

The effect of parameter $S_0$ on the "curvature-pressure" dependence is shown in Figs. 14.4 and 14.5 for $\varkappa = 100$, $\Delta = 0.1$ and $\Delta = 1$, respectively. The different curves correspond to different parameters $S_0$. For a small change in membrane stiffness ($\Delta = 0.1$) the maximum bending occurs at $S_0 = \pi/2$ (Fig. 14.4). Changing $S_0$ (from $\pi/2$) decreases the curvature of the tube, whilst maintaining the pressure value.

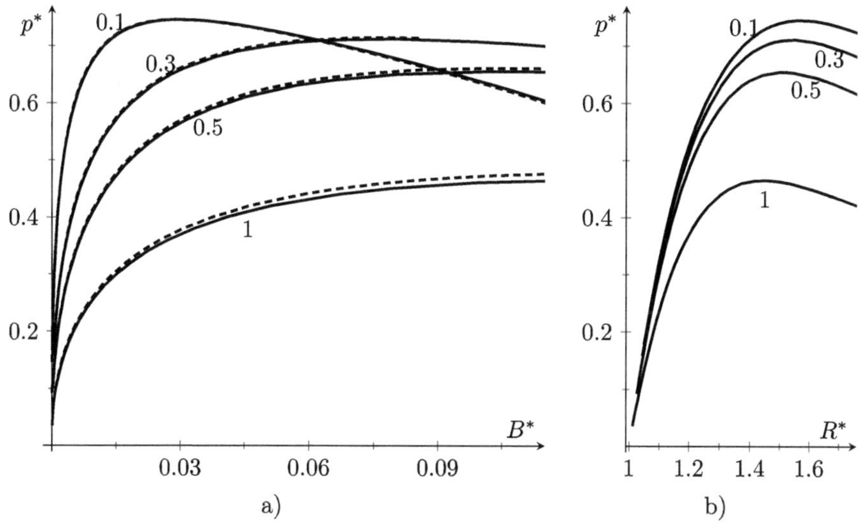


**Fig. 14.3** The dependence of tube curvature and radius on pressure. **a** Curvature versus pressure. **b** Average radius versus pressure. ($S_0 = \pi/2$, the solid lines correspond to $\varkappa = 100$, the dashed lines correspond to $\varkappa = 10$, $\Delta$ is presented in the figure)

We note that with the same decrease and increase in $S_0$ the behaviour of the tubes is close, especially at not very high pressures.

For $\Delta = 1$ the "curvature-pressure" curves for $S_0 \geq \pi/2$ are close (Fig. 14.5) We note that the maximum bending effect occurs at $S_0 < \pi/2$. As $S_0$ increases above $\pi/2$ the curvature of the tube decreases significantly.

Thus, for small change of stiffness, a maximum bending occurs when the small and large stiffness areas have the same length ($S_0 = \pi/2$) (Fig. 14.4). If the higher stiffness area is larger than the lower stiffness area, the maximum bending deflection decreases as the stiffness difference increases. If the higher stiffness region is smaller than the lower stiffness region, it has little effect on the "curvature-pressure" relation (Fig. 14.5).

## 14.5 Conclusions

This paper investigates the inflation of a hyperelastic cylindrical membrane with a non-uniform cross section. The inhomogeneous stiffness can result from changes in the thickness of the membrane or from changes in the mechanical properties of the material. The law of its change is symmetrical in relation to the central axis of the cross section and is described by a step function with smooth transition between minimum and maximum values. It leads to bending of the tube under the action of internal pressure.

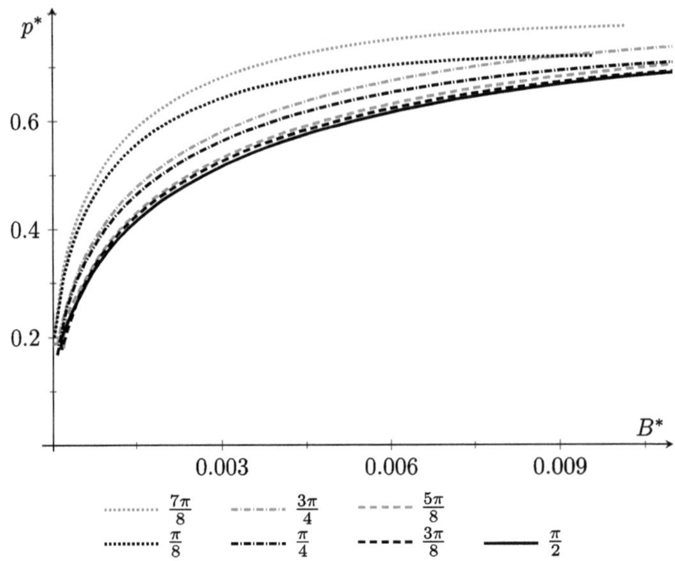


**Fig. 14.4** Curvature v/s pressure. $\Delta = 0.1$, $\varkappa = 100$. The different lines correspond to different $S_0$, which are presented in figure

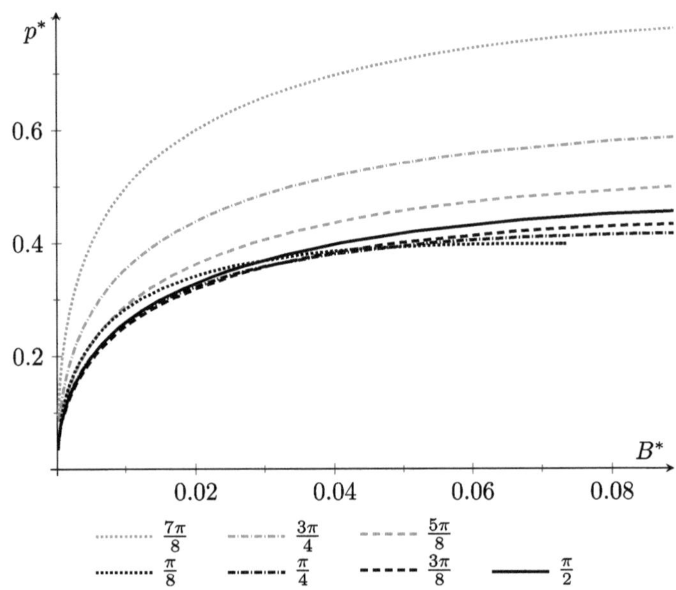


**Fig. 14.5** Curvature vs. pressure. $\Delta = 1$, $\varkappa = 100$, $S_0$ is presented in figure

The analysis of influence on "curvature-pressure" relation of following parameters is carried out: magnitude, smoothness and coordinate of stiffness change. It is found that the smoothness of stiffness change has little effect on the dependence between curvature and pressure. It is significantly affected by the magnitude and coordinate of the stiffness change.

**Acknowledgements** This research was supported by the Government of Russian Federation grant no. 14.Z50.31.0046.

## References

Connolly F, Walsh C, Bertoldi K (2017) Automatic design of fiber-reinforced soft actuators for trajectory matching. Proc Natl Acad Sci 114(1):51–56

Green A, Adkins J (1960) Large elastic deformation and non-linear continuum mechanics. Clarendon Press, Oxford

Hart V, Shi J (1993) Joined dissimilar orthotropic elastic cylindrical membranes under internal pressure and longitudinal tension. J Aust Math Soc Ser B App Math 34(03):296–317

Haseganu E, Steigmann D (1994) Theoretical flexure response of a pressurized cylindrical membrane. Int J Solids Struct 31(1):27–50

Haughton D, McKay B (1996) Wrinkling of inflated elastic cylindrical membranes under flexure. Int J Eng Sci 34(13):1531–1550

Haughton D, McKay B (1997) Inflation of joined dissimilar elastic tubes. Q J Mech Appl Math 50(1):69–83

Kolesnikov A (2015) The finite inflation of a curved elastic tube. Math Mech Solids 20:823–835. https://doi.org/10.1177/1081286514553372

Kolesnikov A, Zubov L (2009) Large bending deformations of a cylindrical membrane with internal pressure. ZAMM 89:288–305

Kolesnikov A, Popov A, Shubchinskaya N (2019) Bending of inflated curved hyperelastic tubes. ZAMM-J Appl Math Mech/Zeitschrift für Angewandte Mathematik und Mechanik 99(7):e201800,093

Kuznetsov V, Levyakov S (2001) Nonlinear pure bending of toroidal shells of arbitrary cross section. In J Solids Struct 38:7343–7354

Levyakov S (2017) The effect of inflating pressure on the finite pure bending of hyperelastic tubes. ZAMM J Appl Math Mech/Zeitschrift für Angewandte Mathematik und Mechanik 97(5):561–575

Libai A, Simmonds J (1988) Nonlin Theory Elastic Shells. Academic Press, San Diego

Patil A, Nordmark A, Eriksson A (2015) Wrinkling of cylindrical membranes with non-uniform thickness. Eur J Mech A/Solids 54:1–10

Polyakova E, Tovstik P, Philippov S, Chaykin V (2009) Axisymmetric deformation of the toroidal shell under internal pressure. Vestnik StPetersburg University Ser 1(4):98–113 (in Russian)

Zubov L (1982) The methods of nonlinear elastisity in the Shell theory. Rostov. Univ. Press, Rostov-on-Don (in Russian)

Zubov L (2001) Semi-inverse solution in non-linear theory of elastic shells. Arch Mech 53(4–5):599–610

# Chapter 15
# Finite Element Analysis of Foam Models Based on Regular and Irregular Arrays of Cubic Open Cells Having Uniform or Normal Distributions

**Alexandr S. Kornievsky and Andrey V. Nasedkin**

**Abstract** The paper considers models of foam materials in the form of Gibson–Ashby cell arrays. The method of effective moduli, based on the equality of the potential energies of the porous and homogeneous material, is described. Six boundary value problems of the linear static elasticity theory for a representative volume are given. These problems together allow determining all the coefficients of the effective stiffness matrix for any anisotropy class of the frame material and geometric asymmetry. The finite element package ANSYS and the capabilities of its command language APDL are used to construct representative volumes and to solve homogenization problems numerically. The procedures for creating solid and finite element models of arrays composed of open Gibson–Ashby cells with regular and irregular structure are described in detail. Two different algorithms for regular lattices of low and high porosity are offered. For an irregular lattice, the sizes of the cube frames are randomly generated with a uniform and normal distribution. The results of numerical calculations for stainless steel lattices in a wide range of porosity are presented. The dependencies of the effective elastic moduli on porosity for a single Gibson–Ashby cell and for regular and irregular lattices with uniform and normal distribution are analyzed. It is shown that the applied Gibson–Ashby model predicts the elastic properties of highly porous materials quite well. But the prediction for lattices with porosity less than 75% gives a sufficiently large error. It is noted that regular and irregular lattices with a large number of cells give similar results for effective elastic stiffness moduli. Meanwhile, individual irregular structures with strongly different sizes of cubic cell frames can take extreme values of effective moduli with pronounced anisotropy in different directions. These effects depend on the geometric asymmetry of the irregular lattices and on the stress concentrations. Examples of the stress–strain state in strongly irregular Gibson–Ashby lattices are given. The analysis

A. S. Kornievsky · A. V. Nasedkin (✉)
Institute of Mathematics, Mechanics and Computer Science, Southern Federal University, Milchakova Street 8a, Rostov-on-Don 344090, Russia
e-mail: nasedkin@math.sfedu.ru



H. Altenbach et al. (eds.), *Advanced Materials Modelling for Mechanical, Medical and Biological Applications*, Advanced Structured Materials 155,
https://doi.org/10.1007/978-3-030-81705-3_15

of the value scatter of various effective elastic moduli, demonstrating the anisotropy of strongly irregular Gibson–Ashby lattices, is given.

**Keywords** Highly porous material · Foam · Open porosity · Homogenization · Effective modulus · Gibson–Ashby model · Geometric anisotropy · Finite element method

## 15.1 Introduction

Highly porous materials have been actively studied recently. Such materials include foam-like, cellular and honeycomb structures. Their main advantages are relatively high rigidity, low density and low-thermal conductivity, see Scheffler and Colombo (2005), Gibson and Ashby (1997). The increased interest in cellular materials is due to the fact that they can be effectively produced artificially in various ways, including using 3D printing (Lv et al. 2021; Maconachie et al. 2019; Maheo et al. 2013; Pabst et al. 2016, 2018; Xiao et al. 2018). Therefore, the improvement of analytical and numerical methods for determining the effective material properties and the optimal geometric structure of foam materials has become the subject of numerous studies. The results obtained in this area have been presented in the recent reviews (Hössinger-Kalteis et al. 2020; Lv et al. 2021; Pan et al. 2020; Srivastava and Srivastava 2014).

Among the models of foam-like and honeycomb structures, the most popular seems to be the Gibson–Ashby cell, see Gibson and Ashby (1997). This model and its modifications were described in detail in Ashby (1983, 2006), Gibson (2005), Gibson and Ashby (1982, 1997). A large number of natural and computer experiments have shown that the Gibson–Ashby models satisfactorily describe the elastic properties of highly porous foams with a regular structure (Andresen et al. 2020; Avalle and Scattina 2014; Maheo et al. 2013; Pabst et al. 2018; Uhlířová and Pabst 2019; Xiao et al. 2018; Zhou et al. 2005, etc.).

Irregular highly porous structures with cells other than Gibson–Ashby cells have been studied by theoretical, experimental and numerical methods in Andresen et al. (2020), Jang et al. (2008), Kaoua et al. (2009), Marvi-Mashhadi et al. (2018a, b), Mukhopadhyay and Adhikari (2016, 2017), Pabst et al. (2018), Roberts and Garboczi (2000, 2001) and in others. Irregular lattices based on Gibson–Ashby cells were investigated in Dillard et al. (2005), Koudelka et al. (2011), Singh et al. (2010). In these papers, the lattice was modeled from identical Gibson-Ashby cells, but the cells were asymmetric along the coordinate axes. Thus, such structures can be considered as partially regular with irregularity only in the basic cell. When calculating these structures in Dillard et al. (2005), Koudelka et al. (2011), anisotropy of the effective properties was observed. The anisotropy is caused by the asymmetry of the lattices along the axes.

Note that for the theoretical determination of the effective properties of foam and cellular structures, in addition to the Gibson–Ashby models, we can use many well-known averaging methods used for composites of various types (Scheffler and Colombo 2005; Hössinger-Kalteis et al. 2020; Kachanov and Sevostianov 2018; Mills 2006), as well as computational homogenization methods that take into account the features of foam-like and cellular materials (Hössinger-Kalteis et al. 2020). In the finite element solution of homogenization problems for various foam, cellular and honeycomb structures, various models of representative volumes were used in Dillard et al. (2005), Hössinger-Kalteis et al. (2020), Kaoua et al. (2009), Koudelka et al. (2011), Maheo et al. (2013), Marvi-Mashhadi et al. (2018a, b), Mukhopadhyay and Adhikari (2016), Ortona et al. (2014), Pabst et al. (2018), Roberts and Garboczi (2001), Singh et al. (2010), Uhlířová and Pabst (2019), Zhu et al. (2000, and others). Special software has also been developed for cellular structures, such as Digimat FE (MSC Software Corporation, Part of Hexagon); GeoDict (Math2Market); MAVI—Modular Algorithms for Volume Images (Fraunhofer Institute for Industrial Mathematics, ITWM); Rhinoceros (Robert McNeel and Associates); and others. However, irregular lattices composed of Gibson–Ashby cells of a random structure have apparently not been studied earlier.

In this paper, arrays of Gibson–Ashby cells of both regular and irregular structures with different types of random distribution of cell frame sizes are investigated. First, for the general case of an anisotropic two-phase material, the homogenization problem by the method of effective moduli is presented. The next section describes in detail the algorithms for creating solid models of regular and irregular Gibson–Ashby cell arrays using the ANSYS finite element package. Two algorithm variants are designed to create regular structures of foam materials both with high porosity and with relatively low porosity. The constructed solid models allow further generation of high-quality hexahedral finite element meshes with controlled characteristics by means of APDL ANSYS.

Regular and irregular lattices of a foam steel material are considered as a numerical example. The results of calculations for a single Gibson–Ashby cell, for regular arrays, and for irregular arrays of Gibson–Ashby cells, depending on the porosity, are compared. At the same time, for irregular lattices, two more variants with uniform and normal random distributions of the values of Gibson–Ashby cell sizes were considered. The dependencies of the effective Young's moduli, stiffness moduli, and Poisson's coefficients on porosity are analyzed. The effect of the lattice geometry irregularity on the anisotropy of the materials is studied.

## 15.2 Theoretical Aspects of the Homogenization Method

Let $\Omega$ be a representative volume of a porous composite material, which we will compare with a volume of the same size filled with a homogeneous material. This volume includes both the material itself inside the volume $\Omega^s$ and the pore volume $\Omega^p$, i.e. $\Omega = \Omega^s \cup \Omega^p$. The volume $\Omega^s$ is a connected structure $\Omega^s = \cup_j \Omega_j^c$ consisting of interconnected three-dimensional Gibson–Ashby cells $\Omega_j^c$. We will assume that the second region is filled with a hypothetical elastic material with very low-elastic stiffnesses. This region $\Omega^p$ simulates voids in the porous material and is included in the model to size–align the volumes $\Omega$ of the composite and homogeneous medium.

Let us assume that in the region $\Omega^s$ the material is linearly elastic, in the general case is anisotropic, and is characterized by density $\rho^s$ and by the coefficients of elastic stiffness $c^s_{\alpha\beta}$, $\alpha$, $\beta = 1, 2, \ldots, 6$. In the area $\Omega^p$ for the fictitious material, we assume: $\rho^p = \kappa\rho^s$, $c^p_{\alpha\beta} = \kappa c^s_{\alpha\beta}$, where $\kappa \ll 1$. Thus, the region $\Omega$ is considered as a two-phase elastic composite with piecewise inhomogeneous density $\rho$ and elastic moduli $c_{\alpha\beta}$: $\rho = \rho^m$, $c_{\alpha\beta} = c^m_{\alpha\beta}$, $m = e, p$. Thus, here, in contrast to a conventional composite, for a porous material, the contrast parameter $\kappa$ between the moduli of different phases is simply extremely small.

We will consider in the volume $\Omega$ the static problem of the linear elasticity theory for the displacement vector $\mathbf{u}(\mathbf{x})$ with special boundary conditions on the outer boundary $\Gamma = \partial\Omega$. We present the equations of the elastic theory in some non-standard vector-matrix form

$$\mathbf{L}^{\mathrm{T}}(\nabla) \cdot \mathbf{T} = 0, \quad \mathbf{T} = \mathbf{c} \cdot \mathbf{S}, \quad \mathbf{S} = \mathbf{L}(\nabla) \cdot \mathbf{u}, \tag{15.1}$$

$$\mathbf{L}^{\mathrm{T}}(\nabla) = \begin{bmatrix} \partial_1 & 0 & 0 & 0 & \partial_3 & \partial_2 \\ 0 & \partial_2 & 0 & \partial_3 & 0 & \partial_1 \\ 0 & 0 & \partial_3 & \partial_2 & \partial_1 & 0 \end{bmatrix}, \quad \nabla = \begin{Bmatrix} \partial_1 \\ \partial_2 \\ \partial_3 \end{Bmatrix}, \tag{15.2}$$

where $\mathbf{T} = \{\sigma_{11}, \sigma_{22}, \sigma_{33}, \sigma_{23}, \sigma_{13}, \sigma_{12}\}$ is the array of mechanical stress components $\sigma_{ij}$; $\mathbf{S} = \{\varepsilon_{11}, \varepsilon_{22}, \varepsilon_{33}, 2\varepsilon_{23}, 2\varepsilon_{13}, 2\varepsilon_{12}\}$ is the array of the strain components $\varepsilon_{kl}$; $\mathbf{c}$ is the $6 \times 6$ matrix of elastic stiffness moduli $c_{\alpha\beta}$; $\mathbf{L}(\nabla)$ and $\nabla$ are matrix and vector operators, respectively; $\mathbf{x} = \{x_1, x_2, x_3\} = \{x, y, z\}$ is the vector of spatial coordinates; $(\ldots)^{\mathrm{T}}$ denotes the transpose operation; and $(\ldots) \cdot (\ldots)$ is the scalar product operation for matrices or vectors.

Hereinafter, Latin indices vary from 1 to 3, and Greek indices vary from 1 to 6. In the transition from tensor notation $\sigma_{ij}$, $\varepsilon_{kl}$, $c_{ijkl}$ to vector-matrix notation $T_\alpha$, $S_\beta$, $c_{\alpha\beta}$ the following correspondence rules between indices are used: $(ij) \Leftrightarrow \alpha$, : $(kl) \Leftrightarrow \beta$, : $(ii) \Leftrightarrow i$, $i = 1, 2, 3$, $(23) \sim (32) \Leftrightarrow 4$, $(13) \sim (31) \Leftrightarrow 5$, $(12) \sim (21) \Leftrightarrow 6$.

As is often accepted in the homogenization problems of composites (Kachanov and Sevostianov 2018; Milton 2002), on the external boundary $\Gamma$ we accept linear by spatial coordinates essential boundary conditions

$$\mathbf{u} = \mathbf{L}^{\mathrm{T}}(\mathbf{x}) \cdot \mathbf{S}_0, \quad \mathbf{x} \in \Gamma, \tag{15.3}$$

where $\mathbf{L}(\mathbf{x})$ is the matrix of the same structure as (15.2), when the vector argument $\nabla$ is replaced by $\mathbf{x}$; $\mathbf{S}_0$ is the six-dimensional array made up of constant components $S_{0\beta} = \varepsilon_{0kl}$.

Then, we will suppose in (15.1) only one nonzero components of the array $\mathbf{S}_0$

$$S_{0\beta} = \delta_{\beta\zeta} S_0, \tag{15.4}$$

where $\delta_{\beta\zeta}$ is the Kronecker symbol, the index $\zeta$ takes some integer from 1 to 6, and $S_0$ is some constant value.

Then, in accordance with the general theory, after solving problem (15.1)–(15.4) and finding the stress fields $\sigma_{ij}$, it is possible to determine the column of the matrix of effective moduli $\mathbf{c}^{\text{eff}}$ with the number $\zeta$:

$$c_{\alpha\zeta}^{\text{eff}} = \langle\sigma_{ij}\rangle / S_0, \tag{15.5}$$

where the angle brackets $\langle(\bullet)\rangle$ denote the averaged integral volume

$$\langle(\bullet)\rangle = \frac{1}{|\Omega|}\int_{\Omega}(\bullet)\mathrm{d}\Omega. \tag{15.6}$$

If the initial fractions of the composite are anisotropic and/or the geometry of the composite or its phases are strongly asymmetric, then the homogeneous medium will also be anisotropic. Then, for the general case of anisotropy, it is required to solve six problems (15.1)–(15.6), which together will allow finding all the effective moduli of the matrix $\mathbf{c}^{\text{eff}}$. From a mechanical point of view, the first three problems with $\zeta = 1, 2, 3$ are the problems of s tension the volume $\Omega$ along the axes $x_1$, $x_2$, $x_3$, respectively, and the problems with $\zeta = 4, 5, 6$ are the problems of the shears of the representative volume in different faces.

The method of effective moduli presented above is based on the fact that for problems (15.1)–(15.4) for a composite body with elastic stiffness moduli $c_{\alpha\beta}$ and for a homogeneous body with moduli $c_{\alpha\beta}^{\text{eff}}$ under the same external influences (15.3), (15.4) in geometrically the same volume $\Omega$ turn out to be the same average strains $\langle\mathbf{S}\rangle = \langle\mathbf{S}_0\rangle = \mathbf{S}_0$, and Hill's identity $\langle\mathbf{S}\cdot\mathbf{T}\rangle = \langle\mathbf{S}\rangle\cdot\langle\mathbf{T}\rangle$ holds (Kachanov and Sevostianov 2018; Milton 2002). Therefore, acceptance of the condition of equality of potential energies $W = W_0$, $W = \langle\mathbf{S}\cdot\mathbf{T}\rangle/2$, $W_0 = \langle\mathbf{S}_0\cdot\mathbf{T}_0\rangle/2 = \mathbf{S}_0\cdot\mathbf{T}_0/2$, entails equality of average stresses $\langle\mathbf{T}\rangle = \langle\mathbf{T}_0\rangle = \mathbf{T}_0$, which is the justification of Eq. (15.5).

In the case of the material of the orthorhombic system, the elastic stiffness matrix has nine independent coefficients. For example, the compliance matrix $\mathbf{s} = \mathbf{c}^{-1}$ can be represented as

$$\mathbf{s} = \begin{bmatrix} 1/E_1 & -\nu_{12}/E_1 & -\nu_{12}/E_1 & 0 & 0 & 0 \\ -\nu_{21}/E_2 & 1/E_2 & -\nu_{23}/E_2 & 0 & 0 & 0 \\ -\nu_{31}/E_3 & -\nu_{32}/E_3 & 1/E_3 & 0 & 0 & 0 \\ 0 & 0 & 0 & 1/G_{23} & 0 & 0 \\ 0 & 0 & 0 & 0 & 1/G_{13} & 0 \\ 0 & 0 & 0 & 0 & 0 & 1/G_{12} \end{bmatrix}, \tag{15.7}$$

and by virtue of the symmetry of the stiffness and compliance matrices, the equalities $\nu_{ij}/E_i = \nu_{ji}/E_j, i \neq j$, hold. Here, $E_i$ are the Young's moduli, $\nu_{ij}$ are the Poisson's ratios, and $G_{ij}$ are the shear moduli.

For an isotropic material, there are only two independent moduli, since $E_i = E$, $\nu_{kl} = \nu_{lk} = \nu$, $G_{kl} = G$, $k \neq l$, $G = E/[2(1+\nu)]$.

Note that if the initial moduli of the two composite phases had some type of anisotropy, then its effective moduli may have a different type of anisotropy, since the representative volume may not have the same symmetry properties which have materials of separate phases. Therefore, solving problems (15.1)–(15.4) for different and determining the effective moduli from (15.5), (15.6), it is necessary to check what type of anisotropy the homogeneous material will have. In the numerical solution of problems (15.1)–(15.4) by the finite element method, the equalities of some effective moduli can be fulfilled not exactly, but with a certain error. Naturally, to estimate the type of anisotropy of effective moduli, this computational error should be neglected.

In the numerical example below, an isotropic material was considered. Therefore, for a regular representative volume with voids, an equivalent homogeneous material is also obtained isotropic, and to find its effective constants, we can solve only one problem (15.1)–(15.6), for example, with $\zeta = 1$. If the representative volume is geometrically strongly asymmetric, then it is necessary to solve all six problems (15.1)–(15.4) and evaluate the degree of difference between the effective moduli obtained from the solutions of different problems. In this case, as follows from the proof for the more general case of a piezoelectric composite (Nasedkin et al. 2020), the matrix of effective elastic moduli should be symmetric; i.e., equalities $c^{\text{eff}}_{\alpha\zeta} = c^{\text{eff}}_{\zeta\alpha}$ must hold.

The presented homogenization methodology is applicable to two-phase elastic composites with full contact conditions at the interface $\Gamma^m = \partial\Omega^s \cap \Omega^p$

$$[\mathbf{u}] = 0, \quad [\mathbf{L}^{\mathrm{T}}(\mathbf{n}) \cdot \mathbf{T}] = 0, \quad \mathbf{x} \in \Gamma^m, \tag{15.8}$$

where $[(\bullet)]$ is the jump of the function $(\bullet)$ through the interface boundary, $\mathbf{n}$ is the unit normal vector to $\Gamma^m$, external, for example, with respect to the volume $\Omega^s$.

Equation (15.8) expresses the continuity of the displacement vector and the stress vector at the interface $\Gamma^m$. For the case under consideration, when the region $\Omega^p$ is filled with fictitious material with very small stiffness moduli, it is possible to solve problem (15.1), (15.2) only for the region $\Omega^s$ with the boundary $\Gamma^s = \partial\Omega^s$, setting condition (15.3) only on the boundary $\Gamma^e = \Gamma \cap \Gamma^s$, and assuming that the boundary $\Gamma^m$ is free from stresses

$$\mathbf{u} = \mathbf{L}^{\mathrm{T}}(\mathbf{x}) \cdot \mathbf{S}_0, \quad \mathbf{x} \in \Gamma^e, \tag{15.9}$$

$$\mathbf{L}^{\mathrm{T}}(\mathbf{n}) \cdot \mathbf{T} = 0, \quad \mathbf{x} \in \Gamma^m. \tag{15.10}$$

For small stiffness moduli $c^p_{\alpha\beta} = \kappa c^s_{\alpha\beta}$, $\kappa \ll 1$, the effective moduli calculated by (15.5), (15.6) from the solution of the problem (15.1)–(15.4), (15.8) for the volume $\Omega$, and from the solution of the problem (15.1), (15.2), (15.9), (15.4), (15.10) for the volume $\Omega^s$ without pores will practically coincide. In this case, the averaging over (15.5) in (15.6) for the second problem should be replaced by averaging over the volume $\Omega$ of functions $(\bullet)$ given only on $\Omega^s$:

$$\langle(\bullet)\rangle \equiv \langle(\bullet)\rangle^s = \frac{1}{|\Omega|}\int\limits_{\Omega^s}(\bullet)\mathrm{d}\Omega. \tag{15.11}$$

For highly porous materials, the first version of the homogenization problem is more convenient for the theoretical substantiation of the effective moduli method, and the second version is more convenient for a practical solution. In this regard, below we will use the second variant with the solution of the homogenization problem (15.1), (15.2), (15.9), (15.4), (15.10), (15.5), (15.11) only in $\Omega^s$.

## 15.3 Description of the Algorithm for Constructing a Representative Volume

According to the Gibson–Ashby (Ashby 1983, 2006; Gibson 2005; Gibson and Ashby 1982, 1997) at the simplest level, a three-dimensional highly porous material or foam can be modeled as an array of cubic elements (cells) consisting of thin beams. Let us assume that the individual cells have length $L$ along all the axes of the Cartesian coordinate system. All the beams have the same square cross section of size $h$. The size of the cube-shaped inner skeleton, measured as the distance between the midpoints of its opposite beams, is $a$. Figure 15.1a shows one of the variants of the solid-state Gibson–Ashby cell model generated in the ANSYS package, and Fig. 15.1b shows the same model in the $xz$-plane.

While constructing a Gibson–Ashby cell, we first create its main part, consisting of an inner cubic skeleton, and then we build the connecting beams. In Fig. 15.3a, an example of constructing the lower beams of the main cubic part of the cell is shown. Note that this lower part consists of 16 bricks of two types: cubes and elongated parallelepipeds. The volumes were divided into two types in order to avoid further problems in generating regular finite element meshes consisting only of hexahedral

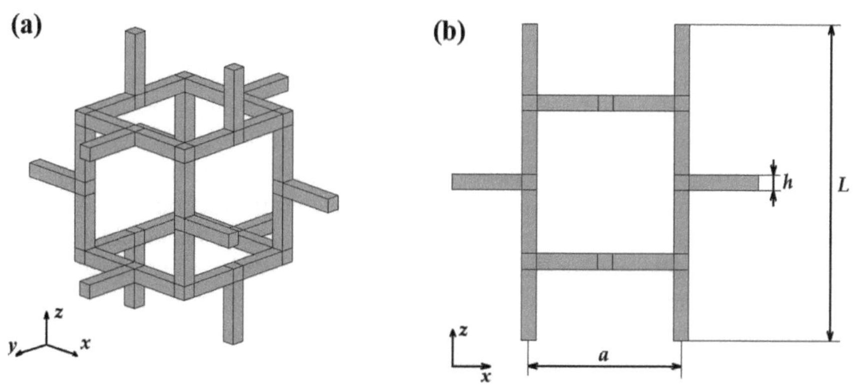


**Fig. 15.1** Example of Gibson–Ashby cell: **a** general view in 3D space, **b** view in $xz$-plane

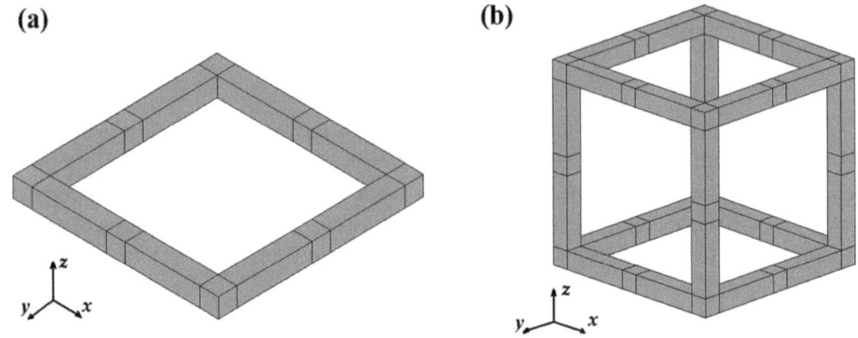


**Fig. 15.2** Generating of cubic skeleton for cell: **a** bottom part, **b** full skeleton

finite elements. Therefore, the special cubic volumes were placed at the points of contact with the connecting beams. Then, the main inner cell part was completed in the same way (Fig. 15.2b). At the last stage of the cell building, the connecting beams were added, and as a result, the Gibson–Ashby cell was obtained (Fig. 15.1).

Note that the porosity of the whole structure changes when the side size $h$ of the beam cross-section varies. If you gradually increase the thickness of the beams, you can come to the case when the geometric volumes that make up the cell will intersect with each other. This is bad for the subsequent generation of a finite element grid. Therefore, for cases of large thicknesses $h$, we have developed another model. Assume that $h > a/2$. In this case, the cubes which are located in the corners and in the middle of the beams of the main cell part will be so large that they will intersect with each other. In order to avoid such overlaps, we additionally distinguished the lines of the supposed intersections. Then, we built non-intersecting volumes in the form of parallelepipeds along all the boundary lines. Finally, we completed the cell skeleton and added connecting beams. In Fig. 15.3a an example of constructing the lower beams of the main cell part for the case $h > a/2$ is shown. In this example, the planes of the initial intersection of the large corner cubes are clearly traced. In Fig. 15.3b, one of the variants of the generated full cell for the described case is shown.

It is known (Alabort et al. 2019; Avalle and Scattina 2014; Andresen et al. 2020; Maheo et al. 2013), artificially produced foams, as well as natural highly porous materials, can have pores of various sizes. The size of individual pores can vary quite a lot. Therefore, we have developed algorithms for constructing two types of lattices: regular and irregular. A regular lattice is an array consisting of Gibson–Ashby cells of the same internal structure. An irregular lattice is constructed from cells in which the sizes of the inner cubic frames are chosen at random. Moreover, in this study, the random variable determining the size $a$ of the internal cubic skeleton obeyed both uniform and normal distribution.

In the case of a regular lattice, the model assembling from individual Gibson–Ashby cells is quite simple. Since all the cells are identical, they are perfectly joined together with the help of connecting beams.

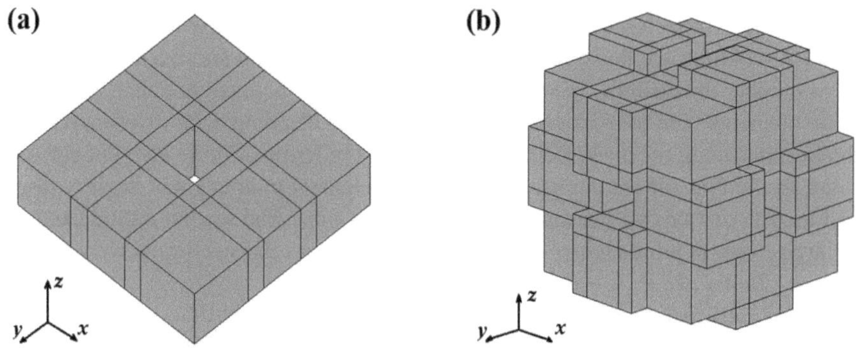


**Fig. 15.3** Generating of a cell with large thicknesses of beams: **a** bottom part of cubic skeleton, **b** full cell

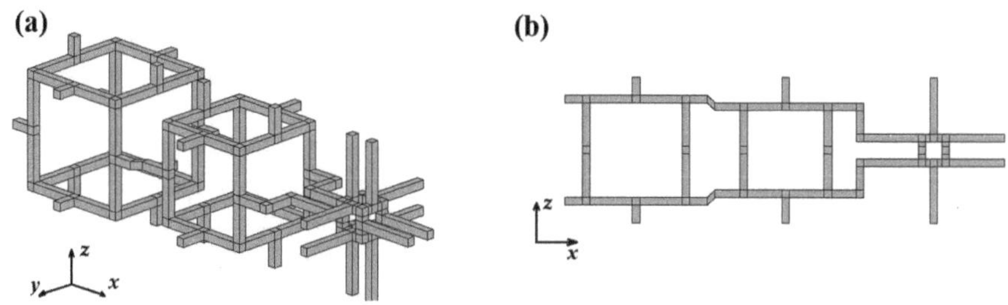


**Fig. 15.4** Irregular chainlet with three Gibson–Ashby cells: **a** view in 3D space, **b** view in $xz$-plane

In the case of an irregular lattice, the algorithm becomes more complex. Let us assume that the thickness $h$ of the beams is the same for all cells, which ensures the simplest contact between different cells. Note that at large thicknesses $h$ the dimensions of the main inner cell part can vary within smaller limits than at small $h$. Therefore, with thick beams, the lattice becomes more regular. In this regard, for an irregular lattice, we consider only the case when $h < a/2$, which corresponds to a porosity of more than 60%.

Note that in the presented model all cells have the same size $L$. But for each cell, the size $a$ of the inner part is chosen at random. While the size $a$ decreases, the cell's inner frame decreases too, but the length of the connecting beams increases. In contrast, with increasing $a$, the inner part of the cell increases, but the lengths of the beams decrease. In Fig. 15.4, a sequence of cells with different inner part sizes is shown.

Before starting the geometric construction of the whole structure using a random number generator, two arrays are created. These arrays contain the values $a$ for all the lattice cells. The values $a$ of the first array are subject to a uniform distribution, while these values of the second array are subject to a normal distribution. Of course, there are restrictions for the values $a$. Obviously, $a \in (2h, L - 2h)$. But given these limitations, we still have a computational problem. For example, if the size $a$ is close to the extreme values, then problem volumes will appear when cells are built. Problem volumes are those in which sizes in one dimension are much smaller than sizes in other dimensions. In such a situation, we have to generate either very many finite elements with proportional dimensions or use elements with disproportionate dimensions along the axes. Both options are bad in terms of finite element calculations. In the first variant, the sizes of the finite element matrices will be very large, and the calculations will be carried out for a long time. The second option can lead to large errors in finite element solutions. In this regard, the size limits were chosen as: $a \in (2h + \delta, L - 2h - \delta)$, $\delta = 0.01L$.

When constructing irregular lattices with a normal distribution, we additionally conducted the following steps. Taking into account the constraints for the values $a$, let us set the parameters for the random variable generator which is subject to the Gaussian distribution. Let the mean be equal to $\mu = 0.25L$ (note that for a regular lattice $a = 0.25L$). Then, the difference between the mean value and the boundary value is taken as $3\sigma$. There is a small probability that $a$ will take on values outside of the boundaries. For this case, the algorithm provides an error handler, and $a$ will be equal to either the leftmost or rightmost possible value.

After the arrays with the values $a$ are generated, the cells are built along the coordinate axes by analogy with a regular lattice. Since, the sizes of the cell skeletons are different, and they must be connected with additional beams. We propose two types of these additional beams. If the distance between the beams of the two cells is less than $h$, then the additional beams are built at an angle, as shown in Fig. 15.4 between the left and center cells. If the distance between the beams is greater than the beam thickness, then additional beams are built perpendicularly, as shown in Fig. 15.4 between the center and right cells.

In Fig. 15.5 for clarity, examples of a regular lattice (a) and an irregular lattice with a uniform distribution (b) are given. These structures contain three cells along each axis. As you can see, in the irregular case, cells with large frames can be generated at random locations in the lattice. Obviously, in this case, cells with small frames are

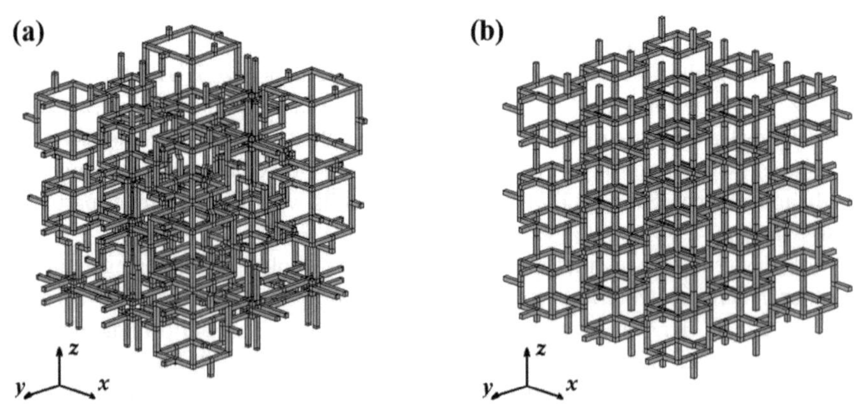


**Fig. 15.5** Examples of regular (**a**) and irregular (**b**) lattices

concentrated in other places. This arrangement can lead to geometric anisotropy of the cell frame, which will entail the anisotropy of the effective medium. This will be further confirmed by numerical experiments. At the same time, it is obvious that for an irregular lattice with a normal distribution sizes of the skeletons are also different. But in most cases sizes are close to the mean. Hence, for this case, we should expect a weaker anisotropy of the equivalent medium than for the case of an irregular lattice with a uniform distribution.

Since the homogenization problems were solved numerically in ANSYS, the solid-state representative volume had to be divided into finite elements. For these purposes, a finite element SOLID185 from the ANSYS library was selected. This element is defined by eight nodes having three degrees of freedom at each node. The average size of the finite elements was taken equal to $L/20$. As a result, finite element meshes were generated from solid models in ANSYS, including only hexahedral elements.

Note that due to the linearity of homogenization problems, the values of effective moduli calculated by (15.5), (15.6), or (15.11) will be the same for geometric structures of representative volumes with similar values of $L$, $a$, $h$ and $\delta$. Therefore, we can take, for example, $L = 1$ (m). So, $a$, $h$ and $\delta$ can be defined as relative sizes from $L$.

## 15.4 Numerical Results

In computational experiments, we considered 316L stainless steel, which a linear isotropic elastic material is. The following material constants were used for calculations (Xiao et al. 2018): Young's modulus $E^s = 187$ GPa, Poisson's ratio $\nu^s = 0.3$.

Representative volumes for both regular and irregular lattices were composed of $3 \times 3 \times 3$ Gibson–Ashby cells (i.e., three cells along each axis). The calculations were carried out at different porosity $p$, which depends on the specified thickness of the beams $h$. Numerical experiments were carried out with porosity from 5 to 99% for regular lattices, and from 62 to 99% for irregular lattices. All the results are given below in relative terms: $r(c_{\alpha\beta}) = c^{\text{eff}}_{\alpha\beta}/c^s_{\alpha\beta}$ is the ratio of the components of the elastic stiffness matrices, $r(E_i) = E^{\text{eff}}_i/E^s_i$ is the Young's modulus ratio, $r(\nu_{ij}) = \nu^{\text{eff}}_{ij}/\nu^s_{ij}$ is the relative Poisson's ratio. Here $c^{\text{eff}}_{\alpha\beta}$, $E^{\text{eff}}_i$, $\nu^{\text{eff}}_{ij}$ are the values of the effective moduli of the porous body, and $c^s_{\alpha\beta}$, $E^s_i$, $\nu^s_{ij}$ are the corresponding values of the moduli of the solid material (i.e., a body with zero porosity).

According to the formulas of the Gibson–Ashby model for an isotropic material, the relative value of the Young's modulus is equal to the square of the relative density multiplied by a certain coefficient. For open Gibson–Ashby cells with prevailing tensile-compression strains, this coefficient is equal to one, and the formula for the relative Young's modulus can be represented as (Ashby 1983, 2006; Gibson 2005; Gibson and Ashby 1982, 1997; Uhlířová and Pabst 2019)

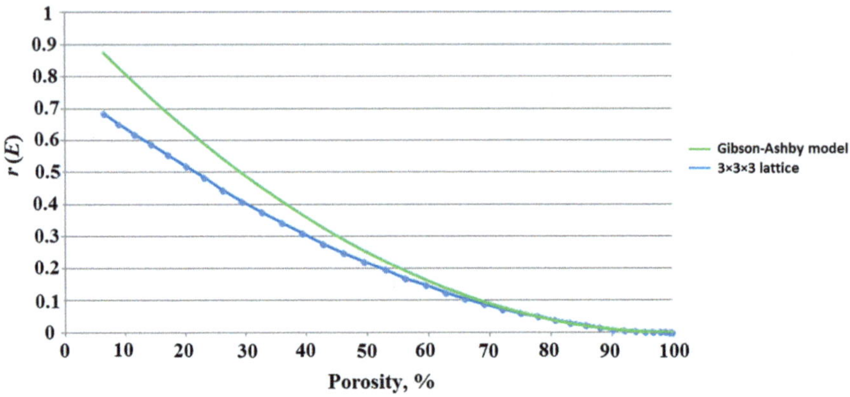


**Fig. 15.6** Dependencies of the relative effective Young's modulus versus porosity for the Gibson–Ashby model and for a regular lattice

$$r(E^{GA}) = E^{GA}/E^s = (\rho^{GA}/\rho^s)^2 = (1-p)^2. \tag{15.12}$$

In Fig. 15.6, the plots of the dependencies on the porosity of the relative Young's modulus for Gibson–Ashby formula and for a regular lattice of 27 Gibson–Ashby cells are shown. The green line corresponds to the values of the relative Young's moduli for the Gibson–Ashby model calculated by Eq. (15.12). The blue line corresponds to the values obtained by numerical experiments in ANSYS from the solutions of homogenization problems (15.1), (15.2), (15.9), (15.4), (15.10), (15.5), (15.11) in $\Omega^s$. From the plots in Fig. 15.6 we can note that formula (15.12) gives values close to values obtained numerically for lattices of Gibson–Ashby cells only for highly porous foams.

In Figs. 15.7, 15.8, 15.9 and 15.10, the porosity dependencies of the relative Young's modulus $r(E)$, Poisson's ratio $r(\nu)$, and stiffness moduli $r(c_{11})$ and $r(c_{12})$ for both regular and irregular Gibson–Ashby cell lattices are shown.

Moreover, in Fig. 15.7 also there is the green curve corresponding to the values of the relative Young's moduli obtained by the formula (15.12). The blue curves in Figs. 15.7, 15.8, 15.9 and 15.10 are plotted according to the values obtained for regular lattices; orange areas contain values obtained for irregular lattices with a uniform distribution, and red areas with normal distribution. It should be noted that for irregular lattices, the cell sizes were chosen randomly, and therefore the final results varied from experiment to experiment. To reduce the influence of randomness, ten calculations were performed for each fixed beam thickness and envelope curves were plotted from the obtained values. At the same time, to compare the results for various models of Gibson–Ashby cell lattices in the case of irregular lattices, an additional averaging of the coefficients of effective elastic moduli (15.7) was carried out to the moduli of an isotropic material.

It can be seen from the plots above that the values obtained for irregular lattices can have a quite large scatter, especially with a uniform distribution of the random

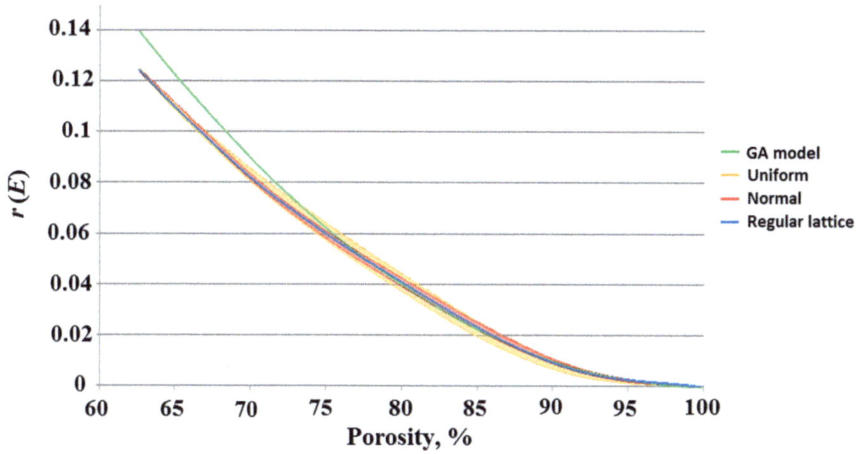


**Fig. 15.7** Dependencies of the relative effective Young's modulus versus porosity for the Gibson–Ashby model and for regular and irregular lattices

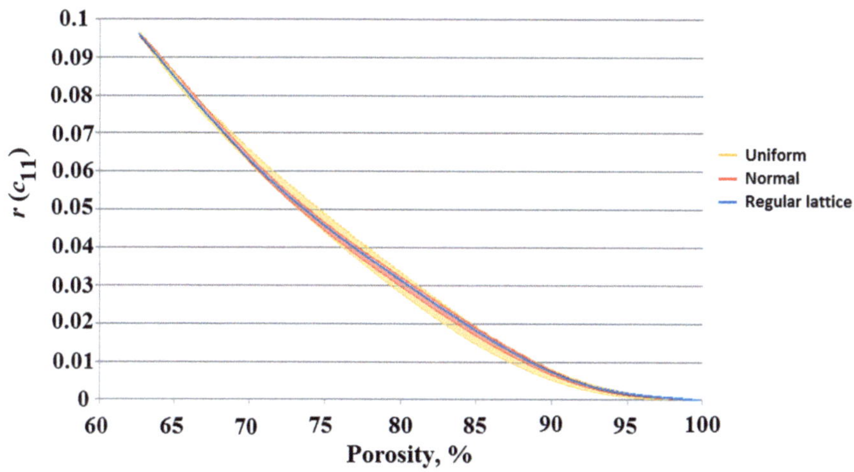


**Fig. 15.8** Dependencies of the relative effective stiffness modulus $r(c_{11})$ versus porosity for regular and irregular lattices

variable. However, they have a similar tendency and, except for the Poisson's ratio values, are close to the values obtained for regular lattices. But, as it is known (Roberts and Garboczi 2000, 2001), the effective Poisson's ratios of porous materials depend on the microstructure to the greatest extent. So, the tendencies can depend not only on porosity but also on the Poisson's ratios of the skeleton material and the internal structure of representative volumes.

As noted above, an irregular lattice may not retain the isotropic properties of solid material. To analyze this phenomenon, we calculated the relative deviations of

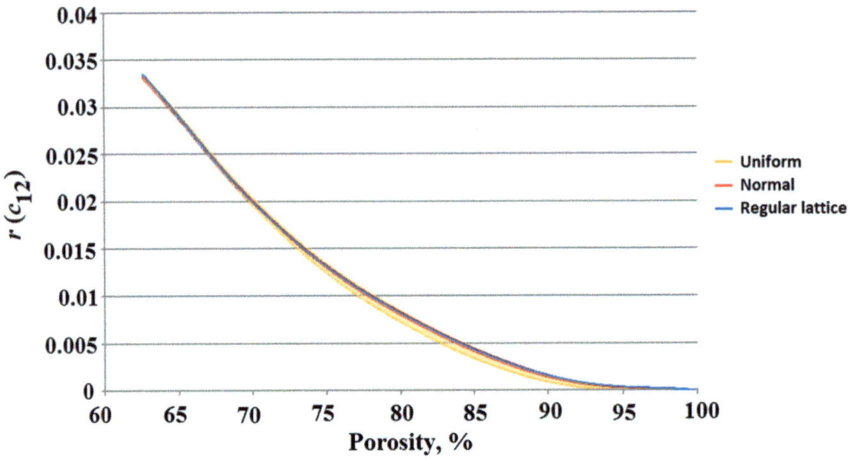


**Fig. 15.9** Dependencies of the relative effective stiffness modulus $r(c_{12})$ versus porosity for regular and irregular lattices

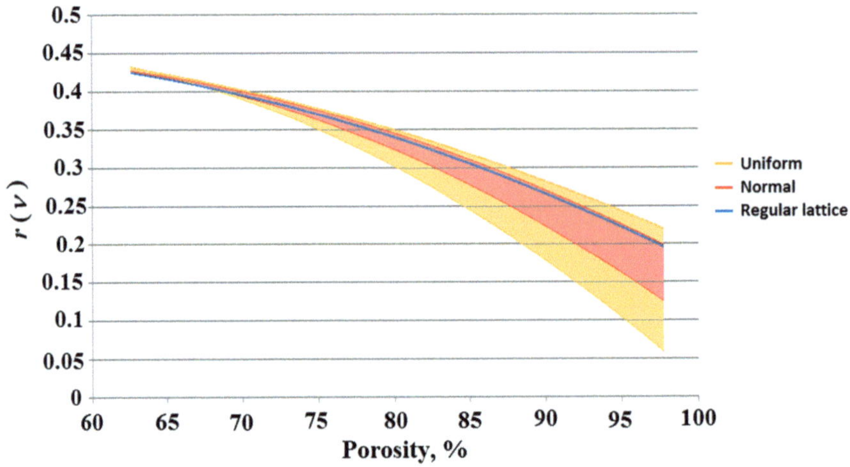


**Fig. 15.10** Dependencies of the relative effective Poisson's ratio versus porosity for regular and irregular lattices

the effective moduli for composites with irregular lattices. So, the following values were calculated: $\delta(E) = (|E_1^{\text{eff}} - E_3^{\text{eff}}|/E_1^{\text{eff}}) \cdot 100\ \%$ are the relative deviations of Young's moduli, $\delta(c_{\alpha\beta}, c_{\gamma\zeta}) = (|c_{\alpha\beta}^{\text{eff}} - c_{\gamma\zeta}^{\text{eff}}|/c_{\alpha\beta}^{\text{eff}}) \cdot 100\%$ are the relative deviations of the stiffness moduli $c_{\alpha\beta}$ and $c_{\gamma\zeta}$; $\delta(\nu) = (|\nu_{12}^{\text{eff}} - \nu_{13}^{\text{eff}}|/\nu_{12}^{\text{eff}}) \cdot 100\%$ are the relative deviations of Poisson's ratios.

In Fig. 15.11 plots of the average relative deviations for stiffness moduli $c_{11}$ and $c_{33}$ (blue curve), moduli $c_{12}$ and $c_{13}$ (green curve), Young's moduli (red curve) and Poisson's ratios (yellow curve) are shown. The solid lines correspond to the normal

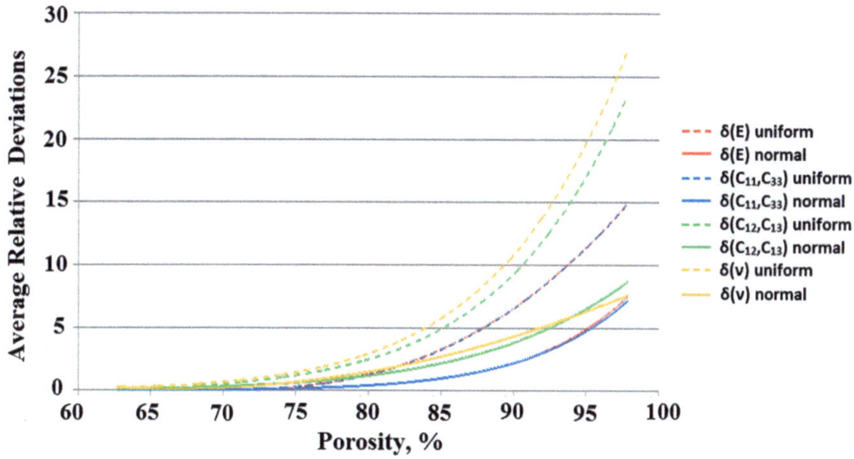


**Fig. 15.11** Dispersion of the relative effective moduli versus porosity, characterizing anisotropy of irregular lattices

distribution of the random variable. The dashed lines correspond to a uniform distribution. Note that these curves are an approximation of the calculated deviations for numerical experiments. It is also interesting to note that the dependencies on the porosity of the relative deviations $\delta(c_{11}, c_{33})$ and $\delta(E)$ are very close, and therefore, the blue curve practically coincides with the red one.

From Fig. 15.11 we can note that the average relative deviations do not exceed 30%, however, their non-averaged values in some cases can be much larger, especially for very highly porous structures. The presented results confirm the statement that porous foams with an irregular lattice can have significant anisotropic properties, even if the material of their skeleton is isotropic. This anisotropy is caused by the asymmetry of geometries and high stresses in irregular lattices.

In Fig. 15.12 there are two examples with irregular lattices of various structures, in which high-stress concentrations are observed in cells along one axis. Note that the skeletons of these cells are small with respect to skeletons of other cells. Obviously, the smaller the dimensions of the frames, the longer their connecting beams, and therefore there are high stresses in the long beams and near small cells (Kaoua et al. 2009; Koudelka et al. 2011).

Thus, Fig. 15.12a shows the distribution of the $x$-component of the tensile stress $\sigma_{xx} = \sigma_{11}$ along the $Ox$-axis. It can be seen from this figure that the highest stress is observed directly in the part of the representative volume, where the long connecting beams are located along the $Ox$-axis. A similar trend is seen in Fig. 15.12b, but in contrast to Fig. 15.12a, it shows the distribution of the $y$-component of the tensile stress $\sigma_{yy} = \sigma_{22}$ along the $Oy$-axis of another irregular lattice.

Note that this paper presents the results for regular and irregular lattices composed of three Gibson–Ashby cells along each axis. With a large number of cells in regular and irregular lattices, more similar results are obtained for effective elastic moduli,

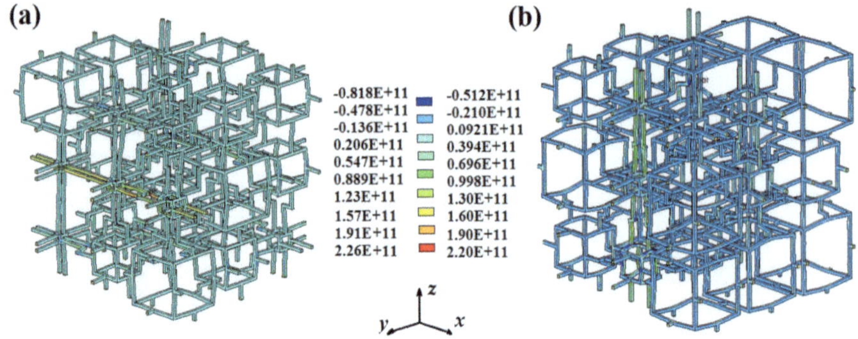


**Fig. 15.12** Stress concentration on the connecting beams during extension along the $x$-axis (**a**) and $y$-axis (**b**)

and the effects of stress concentration are leveled. However, for some variants of highly irregular lattices, the effective moduli can still take on extreme values. The developed models of irregular structures from Gibson–Ashby cells can be used in specialized software for the computer design of foam materials, both within the approximations of the linear theory of elasticity and for studying their nonlinear behavior.

## 15.5 Conclusions

So, in this paper, we studied the effective elastic properties of a porous foam material modelled by arrays of Gibson–Ashby cells of regular and irregular structures. A set of programs in the command language APDL ANSYS has been developed, which allows us to create such arrays of cells, generate finite element models with partially controllable characteristics, solve numerically boundary-value homogenization problems and determine the full set of effective stiffness moduli for elastic materials of an arbitrary class of anisotropy. For arrays of irregular structure, two algorithms were considered for choosing the thicknesses of the beams in the Gibson–Ashby cells: based on a constant distribution and based on the normal distribution of a random variable. Computational experiments for steel foams have shown that for irregular lattices, a fairly significant scatter in the values of elastic moduli can be observed, especially for effective Poisson's ratios. It is noted that irregular lattices are characterized by the appearance of geometric anisotropy of the effective composite properties, which can be quite appeared strong enough for highly porous structures. These effects are leveled for structures of higher sizes and when choosing the normal distribution of a random variable characterizing the irregularity of the Gibson–Ashby cells.

Using the developed finite element technologies, it is planned in future to investigate the properties of representative volumes from Gibson–Ashby cells of a very large dimensions, as well as to investigate nanostructured foams (Hou et al. 2019; Pan and Delogu 2015) according to the Gurtin–Murdoch theory of surface stresses by adding surface finite elements to the developed models similarly to Gao et al. (2006), Javili et al. (2015), Nasedkin and Kornievsky (2017a, b, 2018), Riaz and Ashraf (2012).

**Acknowledgements** This research was supported by the Government of the Russian Federation, contract No. 075-15-2019-1928, and by Russian Foundation for Basic Research, project No. 20-31-90057.

## References

Alabort E, Barba D, Reed RC (2019) Design of metallic bone by additive manufacturing. Scr Mater 164:110–114
Andresen S, Bäger A, Hamm C (2020) Eigenfrequency maximisation by using irregular lattice structures. J Sound Vib 465:115027
Ashby MF (1983) The mechanical properties of cellular solids. Metall Mater Trans A 14(9):1755–1769
Ashby MF (2006) Philos Trans R Soc A 364(1838):15–30
Avalle M, Scattina A (2014) Mechanical properties and impact behavior of a microcellular structural foam. Lat Am J Solids Struct 11(2):200–222
Dillard T, N'guyen F, Maire E, Salvo L, Forest S, Bienvenu Y et al (2005) 3-D quantitative image analysis of open-cell nickel foams under tension and compression loading using X-ray microtomography. Philos Mag 85(19):2147–2175
Gao W, Yu SW, Huang GY (2006) Finite element characterization of the size-dependent mechanical behaviour in nanosystem. Nanotechnology 17:1118–1122
Gibson LJ (2005) Biomechanics of cellular solids. J Biomech 38:377–399
Gibson LJ, Ashby MF (1982) The mechanics of three-dimensional cellular materials. Proc R Soc Lond A 382(1782):43–59
Gibson LJ, Ashby MF (1997) Cellular solids: structure and properties. Cambridge University Press, Cambridge
Hössinger-Kalteis A, Reiter M, Jerabek M, Major Z (2020) Overview and comparison of modelling methods for foams. J Cell Plast 15:1–51
Hou Y, Xu Z, Yuan Y, Liu L, Ma S, Wang W, Hu Y, Hu W, Gui Z (2019) Nanosized bimetal-organic frameworks as robust coating for multi-functional flexible polyurethane foam: rapid oil-absorption and excellent fire safety. Compos Sci Technol 177:66–72
Jang W-Y, Kraynik AM, Kyriakides S (2008) On the microstructure of open-cell foams and its effect on elastic properties. Int J Solids Struct 45:1845–1875
Javili A, Chatzigeorgiou G, McBride AT, Steinmann P, Linder C (2015) Computational homogenization of nano-materials accounting for size effects via surface elasticity. GAMM-Mitt 38(2):285–312
Kachanov M, Sevostianov I (2018) Micromechanics of materials, with applications. Series: Solid mechanics and its applications, vol 249. Springer Int. Publ. AG, Switzerland
Kaoua SA, Dahmoun D, Belhadj AE, Azzaz M (2009) Finite element simulation of mechanical behaviour of nickel-based metallic foam structures. J Alloys Compd 471(1–2):147–152
Koudelka P, Jiroušek O, Valach J (2011) Determination of mechanical properties of materials with complex inner structure using microstructural models. Mach Technol Mater 1(3):39–42

Lv Y, Wang B, Liu G, Tang Y, Lu E, Xie K, Lan C, Liu J, Qin Z, Wang L (2021) Metal material, properties and design methods of porous biomedical scaffolds for additive manufacturing: a review. Front Bioeng Biotechnol 9:641130
Maconachie T, Leary M, Lozanovski B, Zhang X, Qian M, Faruque O, Brandt M (2019) SLM lattice structures: properties, performance, applications and challenges. Mater Des 183:108137
Maheo L, Viot P, Bernard D, Chirazi A, Ceglia G, Schmitt V, Mondain-Monval O (2013) Elastic behavior of multi-scale, open-cell foams. Compos Part B Eng 44(1):172–183
Marvi-Mashhadi M, Lopes CS, LLorca J (2018a) Modelling of the mechanical behavior of polyurethane foams by means of micromechanical characterization and computational homogenization. Int J Solids Struct 146:154–166
Marvi-Mashhadi M, Lopes CS, LLorca J (2018b) Effect of anisotropy on the mechanical properties of polyurethane foams: an experimental and numerical study. Mech Mater 124:143–154
Mills NJ (2006) Finite element models for the viscoelasticity of open-cell polyurethane foam. Cell Polym 25(5):293–316
Milton GW (2002) The theory of composites. Cambridge University Press, Cambridge
Mukhopadhyay T, Adhikari S (2016) Equivalent in-plane elastic properties of irregular honeycombs: an analytical approach. Int J Solids Struct 91:169–184
Mukhopadhyay T, Adhikari S (2017) Effective in-plane elastic moduli of quasi-random spatially irregular hexagonal lattices. Int J Eng Sci 119:142–179
Nasedkin AV, Kornievsky AS (2017a) Finite element modeling and computer design of anisotropic elastic porous composites with surface stresses. In: Sumbatyan MA (ed) Wave dynamics and composite mechanics for microstructured materials and metamaterials. Series: Advanced structured materials, vol 59. Springer, Singapore, pp 107–122
Nasedkin AV, Kornievsky AS (2017b) Finite element modeling of effective properties of elastic materials with random nanosized porosities. Vycisl. meh. splos. sred.—Comput Continuum Mech 10(4):375–387
Nasedkin AV, Kornievsky AS (2018) Finite element homogenization of elastic materials with open porosity at different scale levels. AIP Conf Proc 2046:020064
Nasedkin AV, Nasedkina AA, Nassar ME (2020) Homogenization of porous piezocomposites with extreme properties at pore boundaries by effective moduli method. Mech Solids 55(6):827–836
Ortona A, Rezaei E (2014) Modeling the properties of cellular ceramics: from foams to lattices and back to foams. Adv Sci Technol 91:70–78
Pabst W, Gregorová E, Uhlířová T (2016) Processing, microstructure, properties, applications and curvature-based classification schemes of porous ceramics. In: Newton A (ed) Advances in porous ceramics. Nova Science Publ, New York, pp 1–52
Pabst W, Uhlířová T, Gregorová E, Wiegmann A (2018) Young's modulus and thermal conductivity of closed-cell, open-cell and inverse ceramic foams-model-based predictions, cross-property predictions and numerical calculations. J Eur Ceram Soc 38:2570–2578
Pan C, Han V, Lu J (2020) Design and optimization of lattice structures: a review. Appl Sci 10:6374
Pia G, Delogu F (2015) Mechanical properties of nanoporous Au: from empirical evidence to phenomenological modeling. Metals 5(3):1665–1694
Riaz U, Ashraf SM (2012) Application of finite element method for the design of nanocomposites. In: Musa SM (ed) Computational finite element methods in nanotechnology. CRC Press, Boca Raton, pp 241–290
Roberts AP, Garboczi EJ (2000) Elastic properties of model porous ceramics. J Am Ceram Soc 83(12):3044–3048
Roberts AP, Garboczi EJ (2001) Elastic moduli of model random three-dimensional closed-cell cellular solids. Acta Mater 49(2):189–197
Scheffler M, Colombo P (eds) (2005) Cellular ceramics: structure, manufacturing, properties and applications. Wiley, Hoboken
Singh R, Lee PD, Lindley TC, Kohlhauser C, Hellmich C, Bram M, Imwinkelried T, Dashwood RJ (2010) Characterization of the deformation behavior of intermediate porosity interconnected

Ti foams using micro-computed tomography and direct finite element modeling. Acta Biomater 6(6):2342–2351

Srivastava V, Srivastava R (2014) On the polymeric foams: modeling and properties. J Mater Sci 49:2681–2692

Uhlířová T, Pabst W (2019) Conductivity and Young's modulus of porous metamaterials based on Gibson-Ashby cells. Scr Mater 159:1–4

Xiao Z, Yang Y, Xiao R, Bai Y, Song C, Wang D (2018) Evaluation of topology-optimized lattice structures manufactured via selective laser melting. Mater Des 143:27–37

Zhou J, Gao Z, Allameh S, Akpan E, Cuitino AM, Soboyejo WO (2005) Multiscale deformation of open cell aluminum foams. Mech Adv Mater Struct 12(3):201–216

Zhu HX, Hobdell JR, Windle AH (2000) Effects of cell irregularity on the elastic properties of open-cell foams. Acta Mater 48(20):4893–4900

# Chapter 16
# Effect of Atmosphere During Deposition on the Morphology, Mechanical Properties and Microfriction of Zr-Based Coatings

**Tatyana A. Kuznetsova, Vasilina A. Lapitskaya, Sergei A. Chizhik, Bogdan Warcholinski, and Adam Gilewicz**

**Abstract** Zr-based coatings are multifunctional materials and very perspective in macro- and microtechnics as protective and tribological coatings. ZrN and ZrCN coatings were deposited on steel substrates by reactive magnetron sputtering at the constant magnetron power of 700 W and at the substrate bias voltage of − 10 V with different nitrogen or acetylene flow rate. The surface investigations were carried out using scanning electron microscopy, atomic force microscopy, nanoindentation and glancing angle X-ray diffraction. Tribological tests in the condition of microcontact were carried out using atomic force microscopy under ambient temperature and humidity controlled conditions. It was found that the mechanical and tribological properties and the size of grains in the polycrystalline ZrN and ZrCN coatings can be controlled using different gas flows.

**Keywords** Coatings · ZrN · ZrCN · ZrC · Atomic force microscopy · Nanoindentation · Roughness · Elasticity modulus · Microhardness · Friction · Wear · Grain

---

T. A. Kuznetsova (✉) · V. A. Lapitskaya · S. A. Chizhik
Nanoprocesses and Technology Laboratory, A.V. Luikov Heat and Mass Transfer Institute of the National Academy of Science of Belarus, 15 P. Brovki str, 220072 Minsk, Belarus
e-mail: kuzn06@mail.ru

V. A. Lapitskaya
e-mail: vasilinka.92@mail.ru

S. A. Chizhik
e-mail: chizhik_sa@tut.by

B. Warcholinski · A. Gilewicz
Faculty of Mechanical Engineering, Koszalin University of Technology, 2 Sniadeckich, 75-453 Koszalin, Poland
e-mail: bogdan.warcholinski@tu.koszalin.pl

A. Gilewicz
e-mail: adam.gilewicz@tu.koszalin.pl

 
H. Altenbach et al. (eds.), *Advanced Materials Modelling for Mechanical, Medical and Biological Applications*, Advanced Structured Materials 155,
https://doi.org/10.1007/978-3-030-81705-3_16

## 16.1 Introdution

Zirconium and its compounds (nitrides, carbonitrides and carbides) are multifunctional materials. Therefore the coatings from them are promising in different areas of application. Zr has many valuable properties: high hardness and strength, increased fracture toughness, good wear resistance, corrosion resistance, biostability and resistance to the destructive effects of radiation (Ul-Hamid, 2020a). Zirconium significantly affects the alloys properties as an ligation addition (Yang et al. 2021). The tribological properties of zirconium nitrides and carbonitrides are significantly improved compared to pure Zr (Singh et al. 2012; Silva et al. 2010). The addition of carbon to the ZrN coatings significantly improves its tribological characteristics (Bai et al. 2017).

ZrN and ZrC have a cubic crystal structure with an FCC lattice and covalent bonds (Shirvani et al. 2021; Pang et al. 2021). ZrCN is a quasi-binary solid solution as a result of carbon and nitrogen atoms occupation of octahedral interstitial locations (Harrison et al. 2016). Non-oxide ceramics based on Zr contain a mixture of ionic, covalent and metallic bonds. Their combination gives the materials a special combination of useful properties, such as the high hardness over 20 GPa, melting point over 3000 K, good thermal and electrical conductivity and a small neutron capture cross section.

The coatings of these compounds are used as functional layers on dental implants and endoimplants (Patil 2015; Puente Reyna et al. 2018; Rizzi et al. 2014), as protective layers on cutting and medical instruments (Oganyan et al. 2021; Geyao et al. 2020), as friction parts (Clavería et al. 2019; Zhang et al. 2009; Ruden et al. 2013; Auger et al. 2007), on surfaces of gas turbine engines (Bhattacharya et al. 2020; Muboyadzhyan 2009), as a barrier material of integrated circuits (Zou et al. 2018) in parts exposed to high temperatures (Maksakova et al. 2019), erosion (Zhang et al. 2020; Chen et al. 2020) and aggressive media (Shen et al. 2021; Mejía et al. 2021), in optical components and electronics (Baqir et al. 2021; Martin et al. 2017; Chandiramouli et al. 2015).

The ZrN, ZrCN and ZrC coatings are remarkable by an adjustable color in the entire color palette and have decorative properties (Budke et al. 1999). ZrN has a specific light golden color (Niyomsoan et al. 2002). At high pressure and without the substrate heating, the color can be very dark, up to dark brown (Valaiauksornlikit et al. 2020). When the additional elements are added, the light yellow color of ZrN darkens and acquires the other shades (da Silva Oliveira et al. 2018). So, ZrCN has a muted orange color of copper (Ul-Hamid, 2020b). Amorphous ZrN with thickness till 100 nm can have a blue or dark blue color (Klumdoung et al. 2012). ZrC has a metallic gray color (Jarman et al. 2021). A wide color palette allows them to be used as aesthetic decorative coatings for industry, household appliances and jewelry.

Having such an attractive properties combination, the Zr-based coatings are multifunctional materials and are expected to be very promising in macro- and microtechnology as protective and tribological coatings with their good resistance to extreme loads. The main feature of micro-engineering parts is the small size of the contact, which turns even an insignificant normal load into serious mechanical stresses on the surface. Improved mechanical properties, high electrical conductivity and chemical inertness allow considering the possible use of the ZrN, ZrC and ZrCN coatings in the sliding electrical contacts (Yin et al. 2019; Nguyen et al. 2020). Multifunctional properties of the Zr-based coatings can be used in microelectromechanical systems (MEMS) and devices (González-Carmona et al. 2020; Craciun et al. 2014).

The ZrN, ZrCN and ZrC coatings can be obtained by various methods such as chemical vapor deposition (CVD) (Ren et al. 2021; Wei et al. 2013), physical vapor deposition (PVD) (Meng et al. 2011; Larijani et al. 2006; Rizzo et al. 2018) and among them cathode-arc evaporation (Arias et al., 2006a, b; Johnson et al. 2016; Kuznetsova et al. 2005; Krysina et al. 2020; Braic et al. 2011), magnetron sputtering (Wu et al. 2016; Lin et al. 2018; Singh et al. 2013; Sandu et al. 2009), pulsed laser deposition (Craciun et al. 2011; Gu et al. 2018), etc. The phase composition of the coatings can be changed by controlling of reactive gas pressure or substrate bias voltage. The structure of the coating affects its mechanical and chemical properties (Kuznetsova et al., 2020a; Qi et al. 2011).

The magnetron sputtering produces the smallest macroparticles amount among the all deposition methods. The size of the macroparticles does not exceed 1–3 μm (Lin et al. 2018). Surface roughness is an important parameter in the microcontact conditions. For comparison, the particle size in ion plating reaches 10 μm (Du et al. 2011; Dearnley 2017). In addition, the ion bombardment prior to the reactive magnetron sputtering alters the substrate structure and improves the coatings properties (Dearnley 2017). Therefore, the magnetron sputtering is preferable to other methods for parts of micro-engineering operating under the conditions of the tribological contact.

The Zr-based coatings can be obtained both in a crystalline or amorphous form depending on the deposition conditions and subsequent processing. To enhance the mechanical properties due to the material crystallinity, the ZrN, ZrCN and ZrC coatings need a high energy treatment. It can be annealing, heating of the substrate or negative bias on the substrate during the coating deposition (Ul-Hamid, 2020a; Valaiauksornlikit et al. 2020; Meng et al. 2011). The coating with a nanocrystalline structure is characterized by higher mechanical properties than amorphous one of the same composition. The basis for the guaranteed obtaining of the crystalline form is the heating of the substrate during the deposition (Ul-Hamid, 2020a). An increase in the pressure of reactive and neutral gases also leads to a change in the surface parameters of the Zr-based coatings deposited by magnetron sputtering (Silva et al. 2010). An increase in the nitrogen pressure decreases the crystallites size of the ZrN coating and simultaneously rises the roughness (Lin et al. 2018; Singh et al. 2013). Here we are talking about nanoscale roughness detected by atomic force microscopy (AFM) in areas of $1 \times 1$–$2 \times 2$ μm$^2$. Thus, an increase the nitrogen pressure from 0.3 to 1.1 Pa during the ZrN coatings synthesis leads to an increase in the roughness from about 7 nm to about 30 nm (Lin et al. 2018).

Many applications of the ZrN, ZrCN and ZrC coatings as biomaterials, wear protection surfaces or optical materials are related precisely to their surfaces microstructure and roughness (Brunello et al. 2018; Ramoul et al. 2019; Usmani et al. 2017). The high hardness of the coatings also promotes the surface topography specified in the synthesis process.

The growing demand for the new materials with improved mechanical and tribological properties is driving research toward controlled surface engineering for the coatings (Sadyrin et al. 2019; Aizikovich et al. 2011). Coatings that not only have the desired composition but also the desired surface microstructure are relevant (Liu et al. 2021). For a material intended to be used in a variety of harsh operating conditions, it is important to understand the effect of deposition conditions on the mechanical properties, coating microstructure and surface topography in order to optimize the relationship between the structure and properties. It is important to know in advance which topography will be preferable in terms of tribological properties under the microcontact conditions.

High hardness and toughness of the coating do not guarantee its good wear resistance (Zelentsov et al. 2016). The wear process is very specific for the each contact pair and is sensitive to the sliding conditions (Kragelsky et al. 1982). Wear is a complex interaction of the surface forces, contact mechanics, deformation behavior, contact geometry and environmental parameters (Khonsari et al. 2021). Therefore, in order to select the optimal tribological coating that is promising for working under the microcontact conditions, it is necessary to conduct the appropriate tests with high resolution in the topography and forces (Vasiliev et al. 2017). For MEMS and microtechnology, there is a need to evaluate the tribological characteristics directly on the part surface and not on a reference sample, where the complete surface identity can be violated due to the size effect (Arias et al., 2006b; Grzywacz et al. 2021). The interaction of the contacting surfaces of submicron and nanometer sizes has its own peculiarities in comparison with details of macro sizes (Carpick et al. 1997; Wen et al. 2017). The influence of the adhesion forces, capillary and electrostatic effects in the microcontact should not be neglected (Bhushan 2005). For such micro-sized parts, AFM is a suitable experimental technique (Chizhik et al. 2007). The AFM method makes it possible to take into account the microgeometry of the interaction surface on a nanometer scale, conduct the tests to determine the friction coefficient *Cfr* and wear and accurately to visualize their result (Kuznetsova et al. 2017; Anishchik et al. 2005). In addition, the nano- and submicrocontact of the AFM probe and the surface makes it possible to experimentally simulate a single contact between submicro- and nanoasperities in macro-parts (Jacobs et al. 2019; Lapitskaya et al. 2018).

The main feature in the wear of hard coatings under the microcontact conditions is the ability to create an increased level of mechanical stresses (Kudish et al. 2018). For comparison, the level of mechanical stresses during the scratch test of the hard coatings with an elastic modulus of 270 GPa and a normal load of 100 N is 20–30 GPa (Warcholinski et al. 2018). And the stresses are about 50 GPa in the contact of the same coating with a diamond probe of 40 nm radius even at the load of 3 μN (Kuznetsova et al. 2020b). In addition to the high level of the mechanical stresses, a distinctive feature of contact interaction at submicron and nanometer sizes of the contact is the small depth of maximum stresses, which is limited to nanometers, and not micrometers as in the macrocontact (Gilewicz et al. 2019). This feature makes it possible to implement a special mode of wear—the removal of material from the surface by atomic layers. Such a feature can be in demand in the modern technology of tip-based nanomanufacturing (TBN) (Yan et al. 2015), as it will ensure the accuracy of the contour, unattainable by classical technologies.

The literature contains the data on *Cfr* and materials wear by the AFM method (Wen et al. 2017), and the most common results are the wear on soft polymer materials (Grzywacz et al. 2021). The materials for TBN processing are mostly polymers (Yan et al. 2015). In this work, the wear of hard wear-resistant coatings is carried out using AFM. The effectiveness and accuracy of AFM processing are shown, which expand the basis of materials that can be treated by TBN. In addition to the formation of nanostructures, surface wear by the AFM method is successfully used to test the materials properties and their thin layers (Jarzabek et al. 2014). The ability of AFM to preliminarily study the surface morphology, to determine the roughness, grain size and the distribution of various phases by the contrasts of lateral forces or adhesion forces provides complete information about the microcontact with the surface and allows explaining many phenomena of microtribology and micromechanics (Gosvami et al. 2020).

This article presents the experimental studies by AFM and nanoindentation (NI) of the microstructure, mechanical and tribological properties of the ZrN, ZrC and ZrCN nanocrystalline hard coatings synthesized using the magnetron sputtering at different working atmosphere. In the experiment, a real nanocontact of a single asperity with a radius of 1.5–40 nm (the tip of the AFM probe) with a surface consisting of nanosized grains with a diameter of 50–200 nm is realized.

The aim of this work is to provide an understanding of the complex relationship between the microstructure and mechanical properties of the hard Zr-based coatings (ZrN, ZrC and ZrCN) with a thickness of about 3 μm, deposited by the magnetron sputtering methods and their tribological properties under the microcontact conditions.

## 16.2 Coatings Formation and Characterization

### *16.2.1 Coatings Formation*

The ZrN and ZrCN coatings were formed using magnetron sputtering system equipped with magnetron sources of 100 mm in diameter zirconium cathodes 99.2% purity (Zr grade R60702). Cylindrical HS6-5-2 steel substrates with a diameter of 32 mm and a thickness of 3 mm were polished to *Ra* of about 0.02 μm. They were separately cleaned in an alkaline ultrasonic bath, rinsed with deionized water and dried with warm air. The substrates were placed on a rotating holder parallel to the surface of the cathode in the distance of about 80 mm in the working chamber, Fig. 16.1.

After evacuating the chamber to the pressure of $1 \cdot 10^{-3}$ Pa, the substrates were heated to the temperature of about 400 °C. In order to remove the oxide layer from

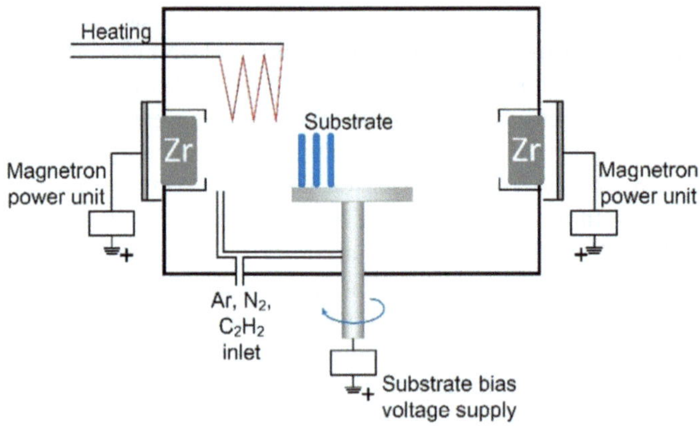


**Fig. 16.1** Schematic diagram of the magnetron sputtering system

the substrate surface, as well as the adsorbed oxygen, the substrates were etched with argon ions. The parameters of this process were as follows: argon pressure 0.5 Pa, bias − 600 V and time 5 min.

To improve the adhesion of the coatings, zirconium layer was deposited on the substrate as a layer between substrate and proper coating. The adhesion underlayer thickness was about 0.25 μm and constant for all coatings.

To deposit the coatings, unbalanced sputtering sources were used in a pulsed mode at the frequency of 1 kHz with modulation of 100 kHz.

The ZrN coatings were formed in argon–nitrogen gas mixture. The ZrCN coatings were formed in argon–nitrogen–acetylene gas mixture. The composition of the technological atmosphere was controlled in the first case by the nitrogen flow rate and in the second by the acetylene flow rate. The thickness of all coatings was 3 μm.

The ZrC coating was deposited in argon–acetylene gas mixture at the acetylene flow rate of 6 sccm.

The details of the two-step deposition process of coating formation in the working chamber, i.e., adhesion layer deposition and the proper coating are gathered in Table 16.1.

### *16.2.2 Coatings Characterization*

The roughness of the substrates (*Ra*) before the coatings deposition was controlled using Hommel Tester T8000 contact profilometer (Germany).

The surface microstructure was investigated using the scanning electron microscope JEOL JSM 5500LV produced by JEOL Ltd. (Japan).

The structure of the coatings was analyzed using X-ray diffraction (XRD). To avoid the diffraction lines from the substrate and from the Zr sublayer, a grazing angle X-ray diffraction was used with a Cu-Kα radiation beam incident at an angle of 3°.

**Table 16.1** Parameters of the two-step formation of the coatings

| | Parameter/coating | ZrN | ZrCN | ZrC |
|---|---|---|---|---|
| | Set of samples | I | II | III |
| Adhesion layer | Sublayer | Zr | | |
| | Power [W] | 700 | | |
| | Argon pressure [Pa] | 0.5 | | |
| | Substrate bias voltage [V] | − 10 | | |
| | Rotation frequency [Hz] | 0.3 | | |
| | Sublayer thickness [μm] | 0.25 | | |
| Proper layer | Power [W] | 700 | | |
| | Total pressure [Pa] | 0.5 | | |
| | Argon flow rate [sccm] | 46 | | |
| | Nitrogen flow rate [sccm] | 3, 4, 5, 6 | 4 | – |
| | Acetylene flow rate [sccm] | – | 0, 1, 1.5, 2, 3 | 6 |
| | Substrate bias voltage [V] | − 10 | | |
| | Rotation frequency [Hz] | 0.3 | | |
| | Thickness [μm] | 3 | | |

An X'Pert Panalytical powder diffractometer was applied. Angular reproducibility of the X-ray diffraction unit is < 0.0002°. For ZrN coatings, the texture coefficient of the cubic phase of ZrN with orientation (111) was determined, and for coatings of ZrCN and ZrC, the texture coefficient of the cubic phase of ZrC with orientation (111). The texture coefficient of (111) plane was calculated using the following equation (Qi et al. 2011):

$$C_{\text{text of}(111)} = \frac{I(111)}{\sum I}, \tag{16.1}$$

where $I(111)$ is the intensity of ZrN (111) for ZrN coatings with 3–6 sccm of nitrogen, of ZrC (111) for ZrCN coatings with 1–6 sccm of acetylene,

$\sum I$ is the sum of the intensities of the peaks $I(200)$, $I(220)$, $I(311)$, $I(222)$, $I(400)$, $I(331)$, $I(420)$, $I(422)$ and $I(511)$.

The coating surface was examined using Dimension FastScan atomic force microscope (Bruker, USA). To study the morphology and adhesive forces, the PeakForce Tapping QNM (Quantitative Nanoscale Mechanical Mapping) mode was used. QNM used standard MPP-12120-10 type silicone cantilevers (Bruker, USA) with a console stiffness of 6.2 N/m, NSC-11 (Micromash, Estonia) with the stiffness of 7.4 N/m and CSG10 (TipsNano, Russia) with the stiffness of 0.3 N/m. The initial radii of the tip were 1.5 nm for the CSG10 probe, 10.0 nm for the MPP-12120-10 probe and 20.1 nm for the NSC11 probe. Determination of the stiffness of the probe console was carried out on the basis of a standard calibration sample of sapphire, which always allows to clarify the passport values. The radius of the probe tip was controlled using a titanium calibration sample RS-12 M (manufactured by Bruker). The error in measuring the surface relief is ± 0.2 nm in XY and ± 0.03 nm in Z. In the PeakForce QNM mode, the probe makes an "approach-retraction" to the sample surface at the each point of the image, recording the force curves.

Using the PeakForce QNM AFM mode, one can measure the adhesion force between the probe tip and the surface and estimate the specific surface energy of the coatings (Park et al. 2020). The strength of adhesion was determined by the maximum force of separation of the cantilever tip from the surface of the samples from the force curves of the "approach-withdrawal". Specific surface energy (work of adhesion) was determined according to the Deryagin–Muller–Toporov theory for the contact of rigid materials by the formula (Wang et al. 2013):

$$\gamma = \frac{F_{\text{ad}}}{2\pi R} \tag{16.2}$$

where $F_{\text{ad}}$ is the force of adhesive interaction between the AFM probe tip and the surface, N; $R$ is the radius of the probe tip, *m*. The AFM matrix of images for topography and adhesion forces was 256 × 256 points (curves).

The roughness parameters of the coatings (*Ra*, *Rq* and *Rz*) in the QNM mode were assessed on the three different areas of 2 × 2 $\mu m^2$ for each sample. On this scale, the contribution of the relief of the individual crystallites of a smooth surface to the roughness value increases, and on the other hand, the number of crystallites is still sufficient for statistics and amounted to about several hundred crystallites. The adhesion forces were measured in areas of 2 × 2, 1 × 1 and 0.5 × 0.5 $\mu m^2$. They have the highest values of adhesion forces in comparison with the areas of 5 × 5–90 × 90 $\mu m^2$, which is associated with the influence of the speed of the "approach-disapproach" of the force curves on the adhesion values for different sizes of areas.

In this work, the maximum diameter was taken as the grain size (of an individual crystallite). Convex areas in AFM images bounded by a concave border were considered grains.

To study the tribological characteristics of the coating surface, the lateral force mode (LFM) AFM was applied using probes with a diamond tip on a D300 silicon console (manufactured by SCDprobes, Estonia) with the stiffness of 38.2 N/m. The precise experimental assessment of the cantilever stiffness is very important before the friction experiment and is especially important for probes with the high stiffness, since it can differ significantly from the passport values (Jarzabek 2015). During operation, the radius of the probe was changed from 20 to 40 nm.

The AFM allows to measure the friction force at the nanoscale with a high resolution. *Cfr* was calculated from the ratio of the friction force between two bodies (tip and surface) and the normal force acting on them, and their measurement is described in (Bhushan 2005). To measure the frictional force, the AFM probe scans the surface using the so-called two-pass method (Kuznetsova et al. 2017). During the first pass, the probe moves along the surface. On the second pass, it moves along the first line, which corresponds to the relief. During forward and backward moving, frictional forces from the surface act on the tip, which causes the cantilever to twist, in addition to deflecting it in the normal direction. The angle of this twist depends on the value of the friction force. This mode of operation is called LFM or frictional force microscopy (FFM) (Bhushan 2005).

Tribological tests of all coatings were carried out for 40 cycles of reciprocating motion. A multi-cycle AFM-tribo-test differs significantly from a single-cycle test, where forces and coefficient of friction are determined on the surface in one scan without any running-in. The multi-cycle mode makes it possible to assess the change in *Cfr* over time and to carry out the wear of the hard coatings. The worn area of $1 \times 1\ \mu m^2$ was scanned for $128 \times 128$ points (lines) at the speed of 1.99–8.00 μm/s. The main load on the probe during the wear test of the ZrN coatings was of 3.6 μN and during the ZrCN coatings testing was 3.1 μN.

Using the example of the ZrN coating obtained at 4 sccm nitrogen in the chamber, the influence of the scanning frequency, the number of cycles and the load on the wear value was determined. In this case, two parameters were kept constant. The scanning frequency varied from 1 to 4 Hz, the number of cycles from 10 to 50 and the load from 8 to 27 μN. The control of the tip radius was carried out before and after the testing each sample against the reference calibration titanium RS-12M sample. The finished wear result was visualized in the PeakForce QNM mode. The amount of the worn out material was determined using the "Bearing Analysis" method in the "NanoScope Analysis" processing program. "Bearing Analysis" allows to determine the total volume of the material under the examined surface according to all pixels in the image for a given height. To determine the volume of the worn material, we subtracted the volume of the segment bounded by the surface obtained after 40 friction cycles from the volume of the segment bounded by the surface obtained after the first cycle of friction. The given height was determined as the difference between the heights of the highest and lowest point in the image. Specific volumetric wear $\omega$ was determined as the ratio of the volume of the material $V$ removed during the friction test of the AFM to the value of the normal load $L$ and the sliding distance $s$ (Kragelsky et al. 1982):

$$\omega = \frac{V}{L \cdot s} \tag{16.3}$$

The microhardness $H$ and the elastic modulus $E$ of the coatings were determined using Hysitron 750 Ubi nanoindenter (Bruker, USA) equipped with a Berkovich diamond tip with a radius of curvature of 60 nm. The tip radius was calibrated by indenting into a calibration sample of fused silica. Additionally, the radius of the indenter was checked using the AFM image of the top of its imprint in aluminum. During the indentation, the values of the test load $F$ and the corresponding indentation depth $h_{\mathrm{Ber}}$ were recorded. As a result, the data are obtained in the form of the dependence $F = f(h_{\mathrm{Ber}})$. The average values of H and E were calculated using the Oliver–Pharr formula using experimental curves of continuous recording of the applied load–indentation depth at the load of 5 and 10 mN. The indentation depth was about 50–100 nm and did not exceed one tenth of the coating thickness so that the substrate did not affect the measurement results (Vasiliev et al. 2018).

NI was performed not only on the coatings surface, but also for the ZrN sample over the cross section of the coating with the substrate. For this, a sample of the coated substrate was cut perpendicular to the surface, pressed into a plastic washer and mechanically polished using a standard metallographic technique. The cross section of the coating was previously scanned by a nanoindenter, and then, 48 prints were marked on the coatings image at the load of 1 mN. Based on the results, maps of the E and H values distribution over the cross section of the coating were built. The reduced elastic modulus was used in the work.

## 16.3 Results

### *16.3.1 ZrN Coatings*

According to the XRD data, only one phase was defined in the ZrN coatings system obtained at the different nitrogen rates in the chamber (Fig. 16.2a). It was the ZrN phase with a cubic crystal lattice. The diffraction lines (111), (200), (220), (311), (222), (331), (420), (422) and (551) were revealed in the coatings. There are no diffraction peaks from the substrate and Zr sublayer. The lines (111), (200) and (220) have the highest intensity. Textures (111) and (200) are the basic in the cubic phase of ZrN, which were described in detail in many works (Maksakova et al. 2019; Arias et al., 2006a; Krysina et al. 2020; Kuznetsova et al. 2020a, b; Qi et al. 2011; Du et al. 2011). The difference between the (111) and (200) textures in the same ZrN phase lies in the packing density of atoms on the atomic plane (Meng et al. 2011). The (111) orientation has a denser arrangement of atoms than the (200) orientation, and the coatings with this orientation have a lower roughness (Ul-Hamid, 2020a). There are a number of works in which the difference in the properties of the ZrN coatings is explained by the preferred orientation (111) or (200) (Maksakova et al. 2019; Gu et al. 2018; Kuznetsova et al. 2020a). In this work, we will be able to trace the effect of texture on the tribological properties of the ZrN coatings, fixing a preferred orientation for each of the samples, according to the XRD data: (111) is the main peak for the coatings obtained at 3 and 5 sccm (Fig. 16.2b), (200) is the main peak for 6 sccm (Fig. 16.2c), and (220) peak is present in all coatings (Fig. 16.2d).

The relationship between the peaks (111) and (200) is influenced by all parameters of the deposition: bias voltage, nitrogen pressure, substrate orientation, addition of chemical elements, etc. (Johnson et al. 2016; Singh et al. 2013; Sandu et al. 2009; Qi et al. 2011). Therefore, it is impossible to say in advance which peak will be dominant in the ZrN coatings without an experiments conducting, not taking into account the mutual influence of the parameters. In this experiment, the ratio between the peaks intensity changes twice with the increase in the nitrogen rate

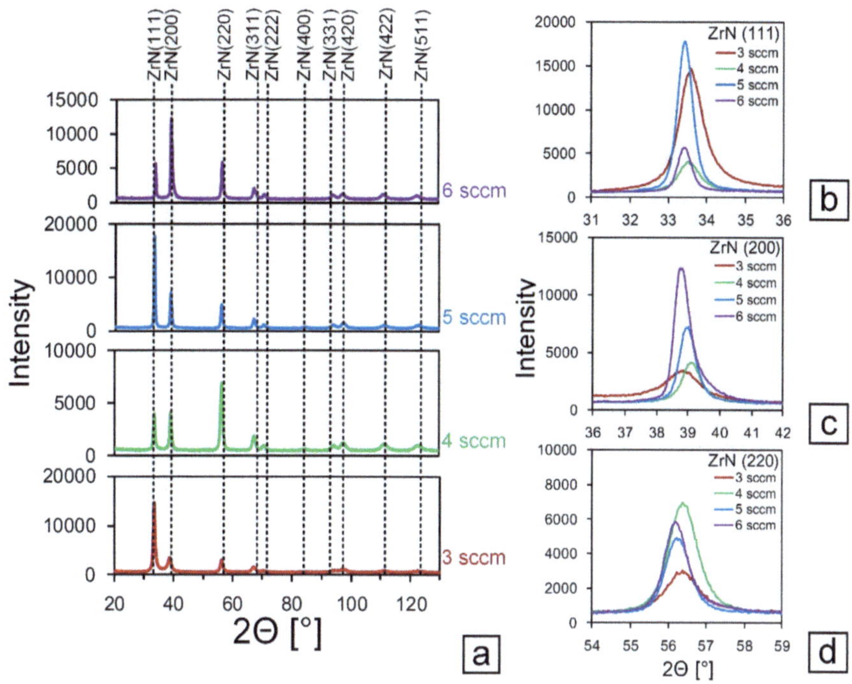


**Fig. 16.2** Glancing angle X-ray diffractograms of the ZrN coatings obtained at different nitrogen flow rate: **a** diffractograms of the ZrN coatings obtained at 3, 4, 5 and 6 sccm; **b** ZrN (111) peaks; **c** ZrN (200) peaks; **d** ZrN (220) peaks

during deposition (Fig. 16.2a). The dependence of the orientation (111) intensity ratio to the other phases on the nitrogen flow rate is shown in Fig. 16.3.

It was found (Meng et al. 2011) that the substrate temperature also affects the ratio of the basic textures, among other parameters. Orientation (111) is preferred

**Fig. 16.3** Dependences of texture coefficient of $I(111)$ of the ZrN coatings on the nitrogen flow rate

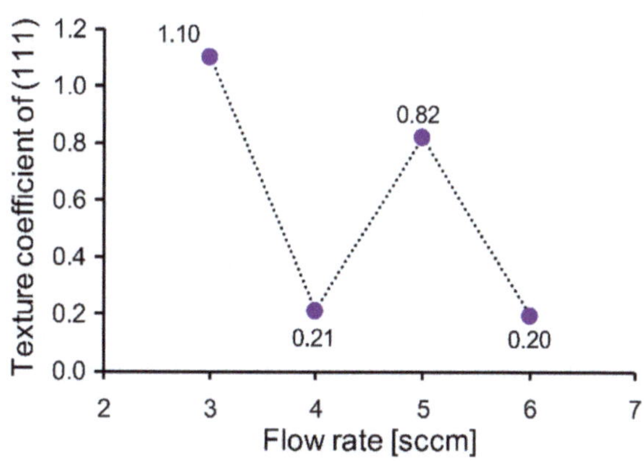

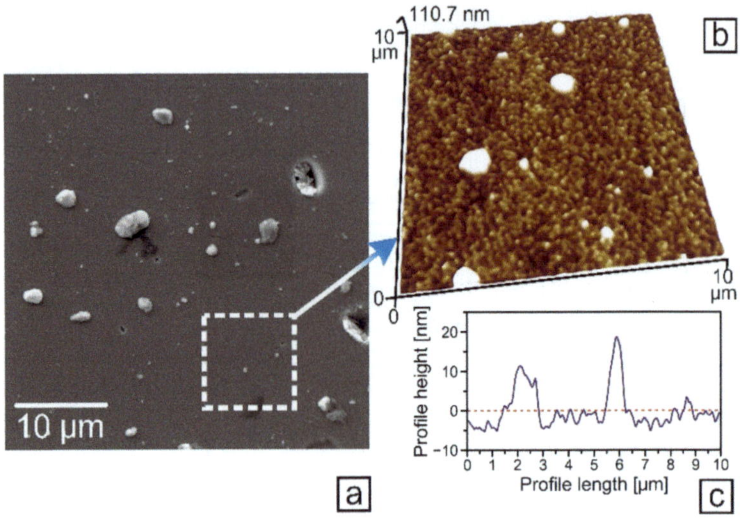


**Fig. 16.4** Surface morphology of the ZrN coating obtained at 4 sccm of $N_2$: **a** SEM, magnification 5000×; **b** AFM topography, scanning area 10 × 10 $\mu m^2$, **c** AFM profile

for forming up to 300 °C. The amount of texture (200) increases at the substrate temperature of 500 °C (Meng et al. 2011). In this work, a "boundary" temperature of 400 °C was used. It led to a "mixed" results: The coatings obtained at 3 and 5 sccm nitrogen flow rates have the advantage of (111) orientation, and the coatings obtained at 4 and 6 sccm nitrogen flow rates have the advantage of (200) and (220) orientation (Fig. 16.3).

The structure of the ZrN coatings is nanocrystalline (Figs. 16.4 and 16.5). According to the ratio $T_h = T_s/T_m$, called the homological temperature, where $T_s$ is the substrate temperature and $T_m$ is the melting point of the coating material, expressed in Kelvin, which for ZrN in this work is about 0.21 (for the substrate temperature of 673 K and the ZrN melting point of 3253 K), the obtained coatings are referred to the T-zone according to the Thornton diagram (Dearnley 2017). Zone T has no pores and no trapped gas. In the magnetron sputtering, the coatings are collected on surfaces from adatoms, which migrate under the magnetic field influence. On the polished substrate surface, pretreated by the ion bombardment and which has almost no small defects, the adatoms of the coating can form nuclei uniformly. At the low Th level of 0.21, adatoms subsequently crystallize in the form of hemispherical shapes (Dearnley 2017). However, on irregularities, scratches and pits found on the polished surface, the adatoms can fix earlier. The preferential formation of crystallization nuclei can here take place and the adatoms have time to gather in the form of larger nodule (or macroparticles). SEM images of the ZrN coatings show a smooth surface with the small amount of the macroparticles ranging in size from 200 to 3 µm and rare craters on the surface from removed macroparticles (Figs. 16.3a and 16.4a). Their amount is no more than 2% of the surface area. Round shapes of

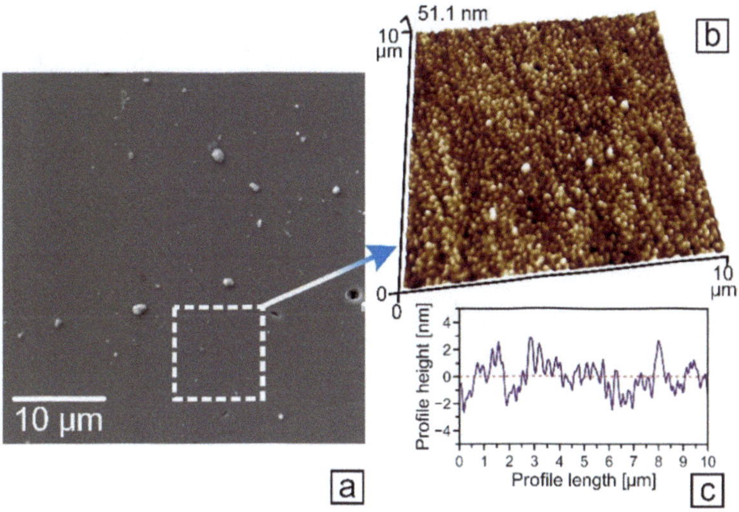


**Fig. 16.5** Surface morphology of the ZrN coating obtained at 5 sccm of $N_2$: **a** SEM, magnification 5000×; **b** AFM topography, scanning area 10 × 10 μm$^2$, **c** AFM profile

individual crystallites (grains) of the approximately same size on the coating surface are found only by AFM with the scanning size of 10 × 10 μm$^2$ (Figs. 16.4b and 16.5b).

The size and shape of the individual nanocrystallites are important for the mechanical and tribological properties of the Zr-based coatings. In work (Wu et al. 2016), the evolution of the microstructure of the ZrN coating after the deformation under nanoindentation condition at the load of 22 μN was studied using TEM. It is shown that the plastic deformation of the coating occurs due to the gliding dislocations that form and propagate along the grain boundaries. This result refutes the statement that the grain boundaries are effective barriers to dislocation motion. Hence the importance of the grain boundaries for the mechanical properties of the ZrN coatings. The grain size characterizes the boundaries density. The finer the grains, the greater the boundaries length per area unit.

The increase in the nitrogen flow rate during the ZrN coating formation leads to an increase in the pressure of the gas mixture in the system and should lead to the decrease in the crystallite size if the other formation conditions were completely equal. However, analyzing the AFM images of the surface (Fig. 16.6), it can be noted that the average size of the individual crystallites increases with an increase in the nitrogen supply rate to the chamber. Obviously, the crystallite size is influenced by both the nuclei number on the surface and the distribution over the surface of the phases with (111) and (200) orientations.

Crystallites formed by the planes of (200) orientation have the lower atomic density but the larger volume. Crystallites of the preferred orientation (111) are denser and finer. Both types are available in the ZrN coatings. It can be concluded

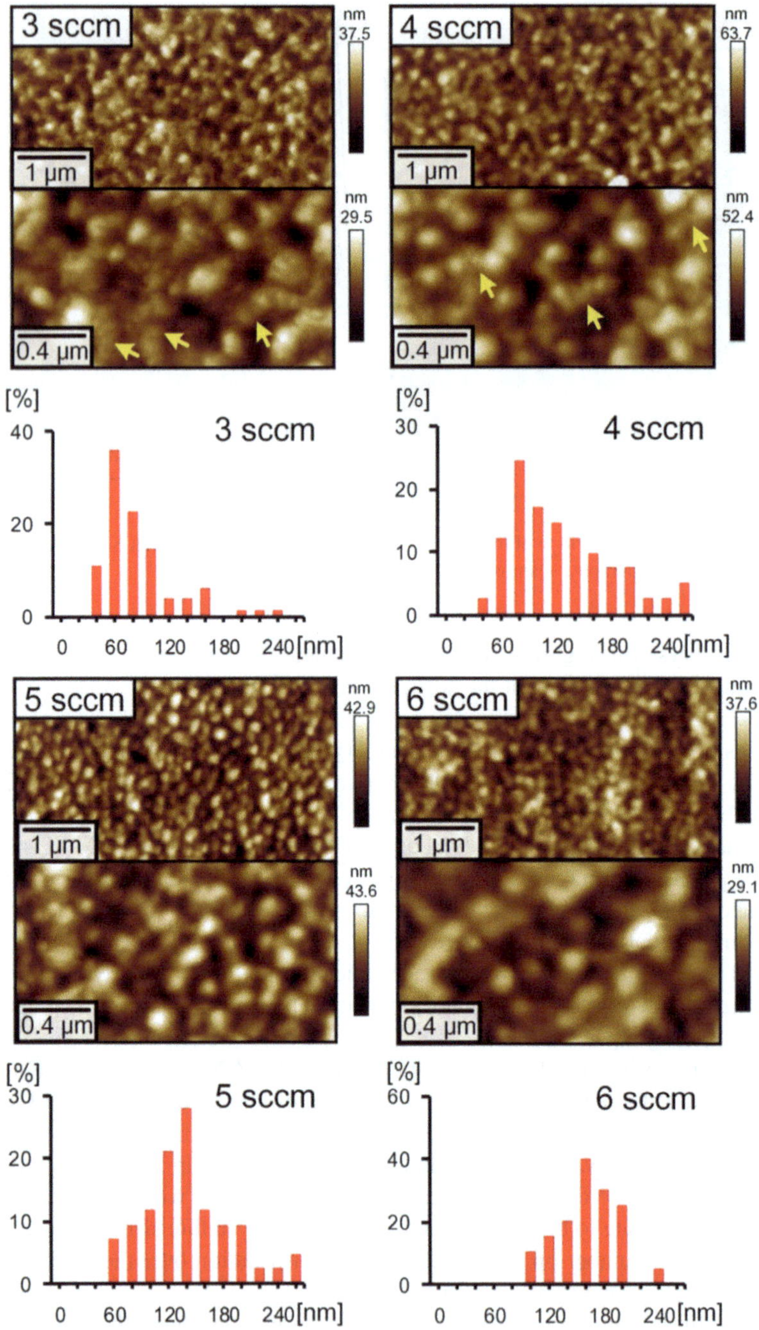


**Fig. 16.6** Influence of nitrogen flow ratio on the surface morphology and grain distribution in the ZrN coatings with marked grain chains

with an increase in the nitrogen flow rate, the number of adatoms entering the surface increases, but they are collected in large and small fractions (grains). As is shown in (Kuznetsova et al. 2020a), crystallites of different sizes are simultaneously present on the ZrN coating surface. The largest number of the small crystallites is in the ZrN coatings obtained at 3 sccm, and the large ones prevail in the coatings obtained at 4 and 6 sccm. At 3 and 4 sccm, the individual crystallites are assembled into chains (Fig. 16.6, arrows). This is probably due to the various conditions for the formation of different orientations (111), (200) and (220). The orientation (111) is formed on already existing nuclei, the phase of (200) orientation can itself form nuclei and be evenly distributed over the entire surface. As the nitrogen supply increases, the preferred orientation of phases changes several times between (111) and (200) (Singh et al. 2013).

The smallest roughness *Ra* is observed in the ZrN coating obtained at 3 sccm of nitrogen, which has the largest amount of the (111) orientation phase. The maximum average grain size is found for the nitrogen flow rate of 5 sccm, and the maximum roughness is discovered for the nitrogen flow rate of 4 sccm (Fig. 16.7). This result agrees with the statement (Meng et al. 2011) that the ZrN phase with the preferred orientation (111) has the highest atomic density in the plane and the lowest roughness.

The roughness of the ZrN coatings usually corresponds to the ratio of the intensities of the (200) peaks to (111) peaks. The (111) planes densely filled with the atoms are smoother, and the coatings with a preferred (111) orientation have a lower roughness (Ul-Hamid, 2020a). The texture (200) with a lower density of atoms provides a larger volume of crystals. In this work, this dependence is fulfilled not for all coatings,

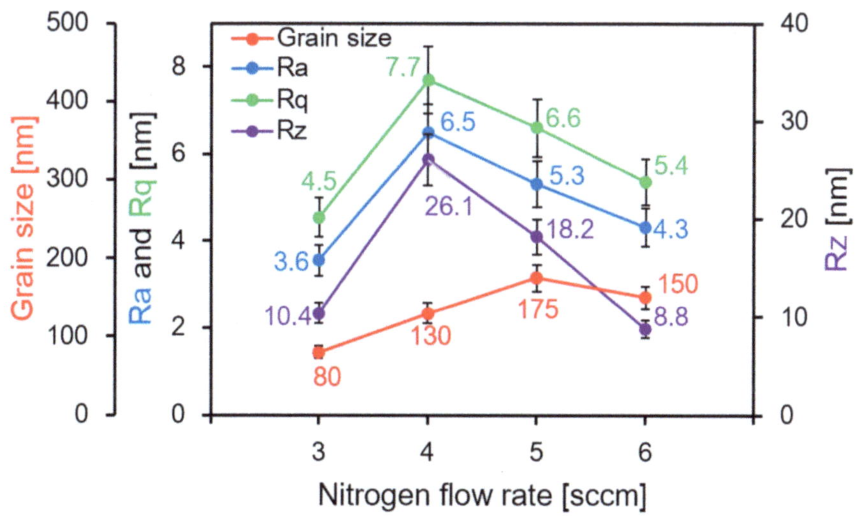


**Fig. 16.7** Dependences of the roughness parameters (*Ra*, *Rq*, *Rz*) on the area $2 \times 2\ \mu m^2$ and grain size of the ZrN coatings on the nitrogen flow rate

but for those with similar lines in the diffractograms and a similar texture coefficient $I(111)$.

If we estimate the grain size and compare the $I(200)/I(111)$ ratio for the 3 and 5 sccm coatings with a similar texture coefficient (111) (1.10 and 0.82), then the maximum grain size of 175 nm will be for 5 sccm with the $I(200)/I(111)$ of 0.412, and the minimum is 80 nm for 3 sccm with a ratio of 0.167. In this case, this ratio works—the grain sizes differ by approximately twice (2.46), as well as the ratio of intensities (2.19). The 4 and 6 sccm coatings, containing a significant line (220), except for (111) and (200), should also be considered separately and compared with each other. The grain size of 6 sccm coating is 150 nm, and the $I(200)/I(111)$ ratio is 2.4. The grain size of the 4 sccm coatings is 130 nm, and the $I(200)/I(111)$ ratio is 1. The presence of planes of other orientations violates the numerical ratio, but the trend persists.

The results of $H$ and $E$ measurements of the ZrN coatings under a constant load obtained at the different nitrogen flow rates are shown in Fig. 16.8. The E values are in the range from 267 to 329 GPa, and the $H$ values are from 29.3 to 30.8 GPa. $H/E$ are 0.089–0.115, and $H^3/E^2$ amount 0.233–0.410 GPa. Comparing these results with other works, it can be noted that in (Singh et al. 2013) $H/E$ is significantly lower and does not exceed 0.07, and $H^3/E^2$ does not exceed 0.12 GPa. $H$ and $E$ of obtained in this work are in good agreement with the data of (Rizzo et al. 2018), where $H$ varies from 22 to 32 GPa, and $E$ from 320 to 380 GPa. In work (Qi et al. 2011), the values of $H$ and $E$ for the ZrN nanocrystalline coatings are 20–34 GPa and 404–421 GPa, respectively. In this case, $H/E$ does not exceed 0.08. Relatively low $E$ value at the high $H$ guarantees the high $H/E$ relation of the coating. In work (Fragiel et al. 2008), the $H$ is 20–26 GPa, and $E$ is 380 GPa. The values of $H/E$ obtained in this work are in good agreement with the data of (Musil et al. 2016).

The inversion of the Hall–Petch law found for the nanocrystalline ZrN coatings in (Qi et al. 2011) was not confirmed in this work. For the grain size of nanocrystals, *Ra*

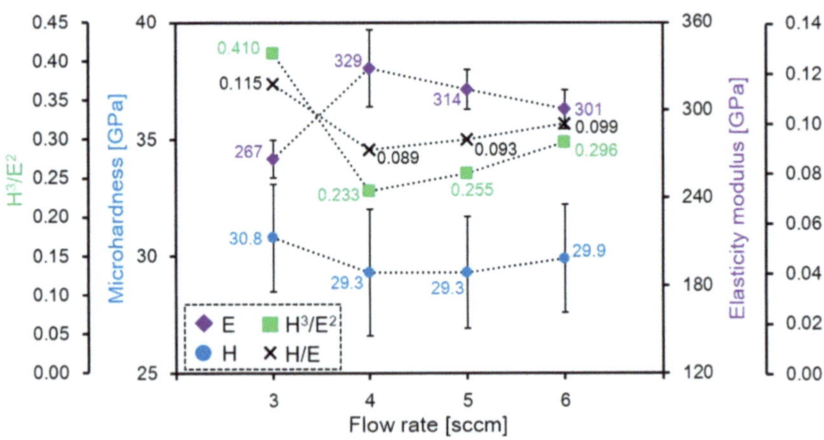


**Fig. 16.8** Dependences of $E$, $H$, $H/E$ and $H^3/E^2$ of the ZrN coatings on the nitrogen flow rate

and $C_{\text{texture of (111)}}$, high correlations (correlation coefficient $r > 0.7$) with mechanical properties were found. The correlation of the grain size with $H$ was negative (−0.8) as in a classical relation between $H$ and grain size and with $E$ the correlation was positive (0.7). For *Ra,* the correlation coefficient of the ZrN coatings with $E$ and $H$ was especially high (0.9 and − 0.9). The correlation of $C_{\text{texture of (111)}}$ with $E$ was negative (− 0.7) and with $H$ was positive (0.6). The higher correlation of *Ra* with $E$ and $H$ than the grain size shows that *Ra* of the areas of $1 \times 1\ \mu m^2$ can be a more accurate indicator of the atoms density on planes and due to it the index of mechanical properties. The $C_{\text{texture of (111)}}$ for all samples determined in this work was significantly higher than in work (Qi et al. 2011). The discrepancy may also be due to the fact that in work (Qi et al. 2011) the crystallite size was estimated from XRD data as coherent regions and in this work—by the AFM method of direct measurements on the surface. Comparing the mechanical properties of the nanocrystalline ZrN coatings, not only the grain size but the texture of the coating should also be taken into account. As was found in (Kuznetsova et al. 2020a), the fraction of the phase with a higher density of atoms is very important for mechanical properties.

Traditionally, the hardening of the coating according to the Hall law and Petch is achieved by the reducing the grain size. However, in these coatings, this is not the main hardening mechanism. The strengthening of the ZrN coatings is explained by the complex effects of grain refinement and structure compaction due to the evolution of the preferred orientation (111).

The $E$ and $H$ values obtained on the coating surface (Fig. 16.8) at the constant load and the indentation depth of 80 nm are not constant throughout the whole coating thickness. This is clearly shown in Fig. 16.9 in the form of $E$ and $H$ maps on the cross section of the coating with the substrate. The maximum values correspond to the middle of the coating thickness. Closer to the surface, the values decrease, as well as toward the substrate. The presence of the Zr sublayer reduces $E$ from 283 to 184 GPa, but after the sublayer, the steel substrate is strengthened by the Zr doping to 221 GPa after 1 μm, decreasing to 173 GPa of the substrate value. Thus, the ZrN coating has a double gradient of properties: decreasing from the middle of the coating to the surface and decreasing from the middle of the coating to the substrate (Aizikovich et al. 2015).

In addition to the heterogeneity of $E$ and $H$ in coating thickness, the ZrN coating surface cannot be considered as a flat and monolithic at the nanoscale level. The coating surface has a polycrystalline island structure. When the macroindenter comes into the contact with such a surface, the contact area is divided into many subcontacts with the diameter of 80–200 nm. During the friction and wear by the AFM tip, not the plane is worn out, but the bumpy polycrystalline surface of a complex shape (Fig. 16.10), consisting of quasi-asperities (Vasiliev et al. 2017). Wear is expressed in smoothing and gradual displacement of the protrusions in the direction of the slow AFM scanning (Fig. 16.10).

The wear of the ZrN coating under these conditions has a boundary state between the elastic and plastic deformation. Based on the $h/R_{\text{tip}}$ ratio where $R_{\text{tip}}$ is a single irregularity radius introduced into the coating (AFM diamond tip with $R_{\text{tip}} = 20$ nm) and $h$ is the thickness of the material removed in one pass, the friction

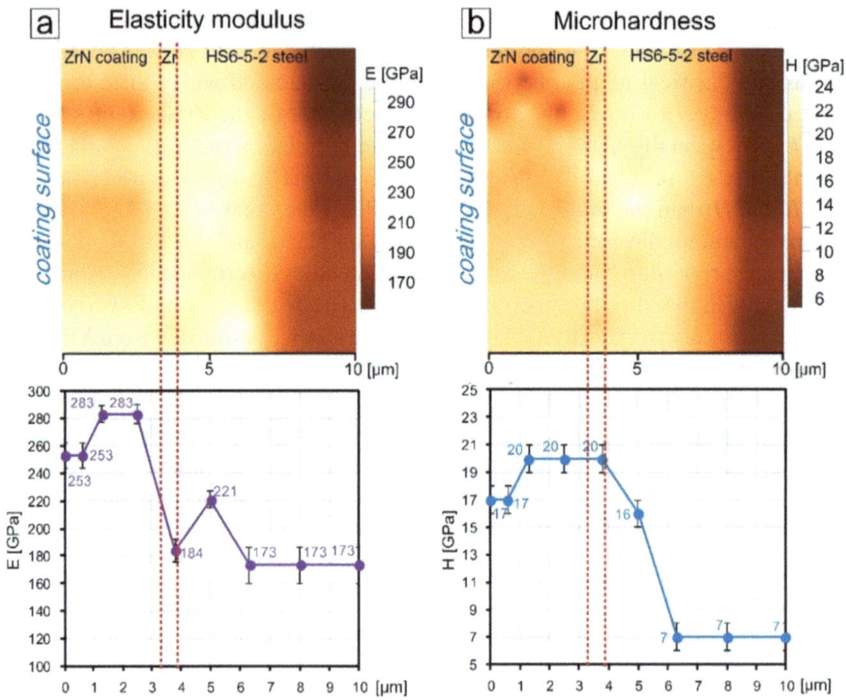


**Fig. 16.9** Cross section elasticity modulus (**a**) and microhardness (**b**) maps (area of $10 \times 10\ \mu m^2$) of the ZrN coating and profiles through the maps

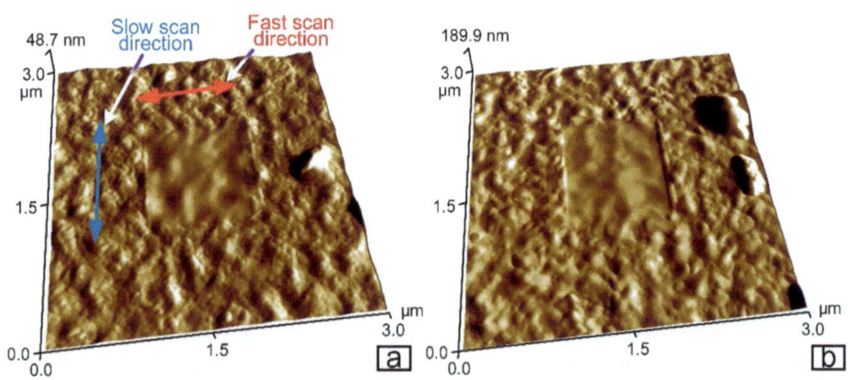


**Fig. 16.10** AFM images of wearing results at the load of 3.6 μN, frequency of 1 Hz and 40 cycles on the ZrN coatings surfaces obtained at different nitrogen flow rates: **a** 3 sccm; **b** 4 sccm with marked slow (blue arrow) and fast (red arrow) scan direction

processes in which $h/R_{tip} < 0.01$, according to Kragelsky, are attributed to the elastic displacement of the material, and when $h/R_{tip} > 0.01$, the plastic displacement takes place (Kragelsky et al. 1982). The wear of the ZrN coatings, according to Kragelsky, should be attributed to elastic because the thickness of the displaced material is commensurate with the atoms layer. The thickness of the material removed from the ZrN coatings surface in 40 cycles and determined by a step at the wear trace boundary was 4 nm for 3 sccm, 40 nm for 4 sccm, 4 nm for 5 sccm and 2 nm for 6 sccm. The average thickness of 0.1 nm was removed for 3 sccm, 1 nm for 4 sccm, 0.1 nm for 5 sccm and 0.05 nm for 6 sccm in one pass. The $h/R_{tip}$ ratios were 0.005 for the coating obtained at 3 sccm, 0.05 at 4 sccm, 0.005 at 5 sccm and 0.0025 at 6 sccm. For all ZrN coatings, the ratio $h/R_{tip} < 0.01$ makes it possible to attribute the wear process to the elastic displacement of the material. There are no pile-up around the wear trace (Fig. 16.10). The absence of the worn out material by the AFM visualizing of the wear trace indicates that material is fully displaced and uniformly distributed over the surface. At the same time, the complete recovery of the material removed in 40 cycles in its previous forms in the wear trace does not occur either. It was established in (Bhushan 2005) that wear ceases to be elastic if one wear cycle can be determined using AFM. It is not yet clear how to more accurately qualify wear if one wear cycle is not detected by AFM at all, but several tens of cycles can be determined.

The dependences of the worn out volume and the specific volumetric wear of the ZrN coating on the probe movement speed are decreasing as is shown in Fig. 16.11. The worn out volume reduced from $11.33 \times 10^{-21}$ m$^3$ to $3.50 \times 10^{-21}$ m$^3$, and the specific volumetric wear decreased by more than 3 times with an increase in the scanning speed from 1 to 8 μm/s.

This result shows that the material removing by this mechanism requires a certain contact time. A sufficient number of dislocations and other defects which “prepare” the material for displacement and help to carry it out have time to arise. When the speed increases, the contact time of the probe and the coating at each local point decrease, and the less amount of dislocation arise and smaller material can be displaced.

The volume of wear with the cycles number grows predictably, but the increase is according to the logarithmic dependence (Fig. 16.11). So, the volume of wear is $2.69 \times 10^{-21}$ m$^3$ after 20 cycles, and after 40 cycles is $16.42 \times 10^{-21}$ m$^3$. This disproportionality can be explained by the energy which must be spent in the uppermost atoms layer destruction (displacement) due to the special state of the surface layer, which is additionally acted upon by surface tension forces.

With the load increase from 9 to 26 μN, the wear volume also increases according to the logarithmic dependence from $1.34 \times 10^{-21}$ m$^3$ to $5.87 \times 10^{-21}$ m$^3$, but the specific volumetric wear increases insignificantly from $1.5 \times 10^{-14}$ m$^3$ N/m up to $2.1 \times 10^{-14}$ m$^3$ N/m, taking into account the load (Fig. 16.11). The flow of the ZrN material during wear occurs at a certain level of mechanical stresses in the contact reaching the yield point, and upon achieving this level, a further increase in the load does not lead to a significant increase in the wear rate. The explanation of the logarithmic dependence of *Cfr* on the load according to the logarithmic dependences

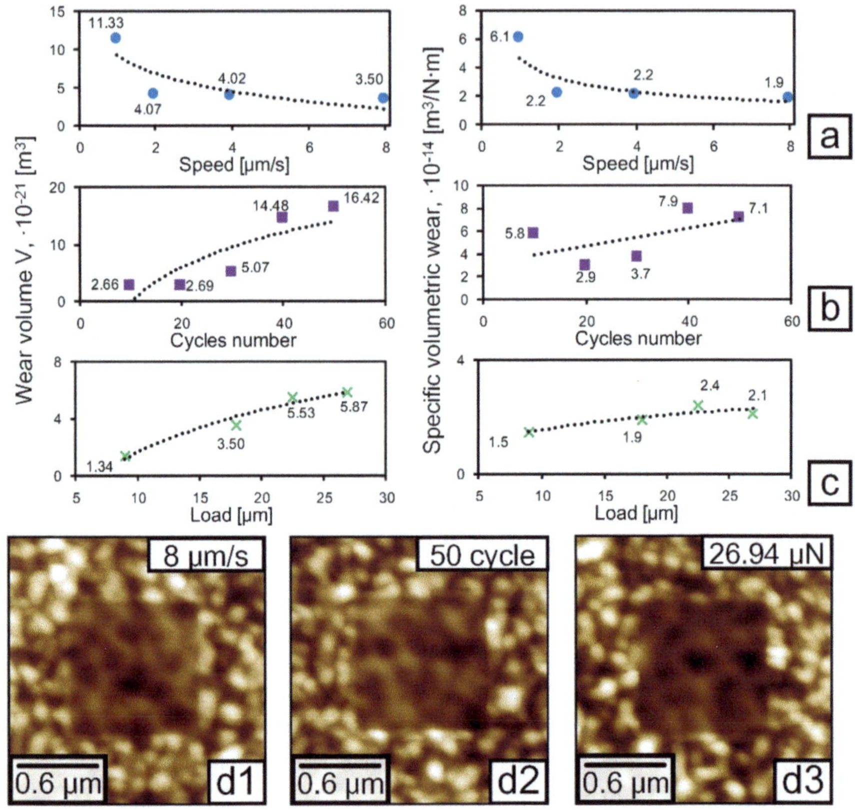


**Fig. 16.11** Dependences of wear volume and specific volumetric wear of the ZrN coatings obtained at 5 sccm on the scanning speed (**a**) cycles number (**b**) and normal load (**c**) and worn surface images (**d**): **a** at 40 cycles and 17.96 μN; **b** at 4 Hz (8 μm/s) and 17.96 μN; **c** at 40 cycles and 4 Hz; **d1** at 40 cycles and 17.96 μN; **d2** at 50 cycles and 17.96 μN; **d3** at 40 cycles and 26.94 μN

of the contact area on the load in all used models JKR, DMT and Herzt (Carpick et al. 1997) is not suitable in this case, since the contact cannot be recognized as elastic when the material displacement is set. This result is closer to the data of (Bhushan 2005), where the loads from 3 to 40 μN are used on the surfaces of Si(111), $SiO_2$ and natural diamond in the contact with $Si_3N_4$ tip. It is founded that for the low loads up to 25 μN, the coefficient of friction remains constant and independent of the load and then increases after reaching a certain value of the critical load. The values of critical loads for Si and $SiO_2$ correspond to stresses equal to their values of hardness, which suggests that the transition to plasticity plays a key role in this effect.

Another experiment from the same work (Bhushan 2005) showed that when an increasing load is applied in a scratch on the Si(100) surface, *Cfr* increases after reaching the critical load, which corresponds to the plastic flow of the material. In this work, an increase in the wear volume (and an increase in *Cfr*) begins immediately, from the minimum load value of 8 μN, which is associated with the attainment of stresses in the contact (at 8 μN, $R_{tip} = 23.9$ nm and the contact stresses of 60.5 GPa, or at 26.9 μN, $R_{tip} = 40.19$ nm and the stresses of 61.7 GPa) exceeding the yield point of ZrN, which is about 10 GPa. Comparing the convex dependence of the wear volume *V* on the load with the concave dependence of the Si(100) wear depth per cycle in the range of 20–80 μN (Bhushan 2005), it should be noted that our convex shape is associated with the wear of the probe tip, due to which the stress in the contact practically did not increase. The small values of wear at the load of less than 20 μN in (Bhushan 2005) are associated with the failure to reach the level of the critical stresses. Elastic deformation is primarily responsible for the low wear at the loads below 20 μN, as indicated in (Bhushan 2005). Consequently, the wear of ZrN after one cycle can be formally considered elastic, since it cannot be determined even by AFM. However, although the wear of one cycle on the ZrN surface is not detected and does not lead to material shear, it creates defects—dislocations and their accumulations—and thus "prepares" the material for displacement. The question of whether the result of the action on the surface can be considered elastic in this case is debatable.

The results of the AFM friction tests of the ZrN coating obtained at 5 sccm of nitrogen flow with the varying load, speed and the cycles number show that with an increase in the probe speed from 0.5 to 4 Hz, the wear volume *V* decreases three times (Fig. 16.11). With an increase in the scanning cycles number from 10 to 50, *V* increases sixfold, and with an increase in the load from 6.5 to 26.5 μN, *V* increases more than 4 times. All dependencies are nonlinear and clearly show that the probe speed is a parameter that determines the wear result to the same extent as the load and the cycles number. Slowing down the probe speed from 4 to 1 Hz increases the depth of wear by 1.5 times at the same load. Probe tip control allows the accurate determination of mechanical stress in the contact, which is the main parameter of the AFM friction.

Figure 16.12 shows the relationship between the depth of wear and the load for a comparable number of cycles and scan cycle speed. The contrast in the height of the worn area in the topography image becomes darker. The roughness histograms of the worn areas show how the material is "smoothed". The maximum roughness height decreases from 29 to 20 nm, and the average (distribution center) from 15 to 10 nm (Fig. 16.12). It can then be assumed that approximately 5 nm of material was uniformly removed from the surface. The "Bearing Analysis" method allows an accurate determination of the worn surface segment volume *V*. The value of *V* is shown in Fig. 16.11 as a function of load. The method for assessing wear along the step of the profile at the boundary of the wear area, used in (Kuznetsova et al. 2020b), was less accurate in relation to the nanometer thicknesses of the removed material against the background of the grain topography profiles of the surface with differences in the grains of tens nanometers.

Tests carried out with the different speeds and cycles number on the surface of the coating obtained with 5 sccm of nitrogen showed a slight increase in *Cfr* with the increasing probe speed (Fig. 16.13a, b). If we make a "cross section" of the dependences of *Cfr* (*Ffr*) on the number of scanning cycles for one, 10, 20 and 40 cycles, the dependences of *Cfr* (*Ffr*) on the speed (frequency) will be increasing

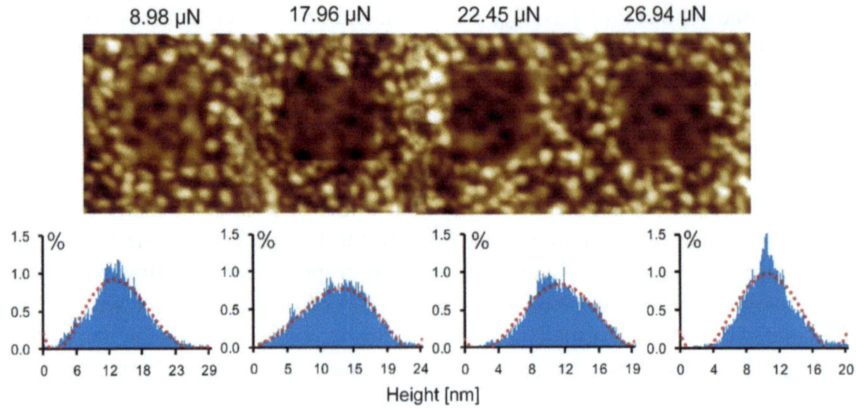


**Fig. 16.12** Changes of surface morphology in wear track of the ZrN coatings obtained at 5 sccm nitrogen flow during the wear at different load, 40 cycles and $v = 8$ μm/s: **a** AFM images; **b** histograms

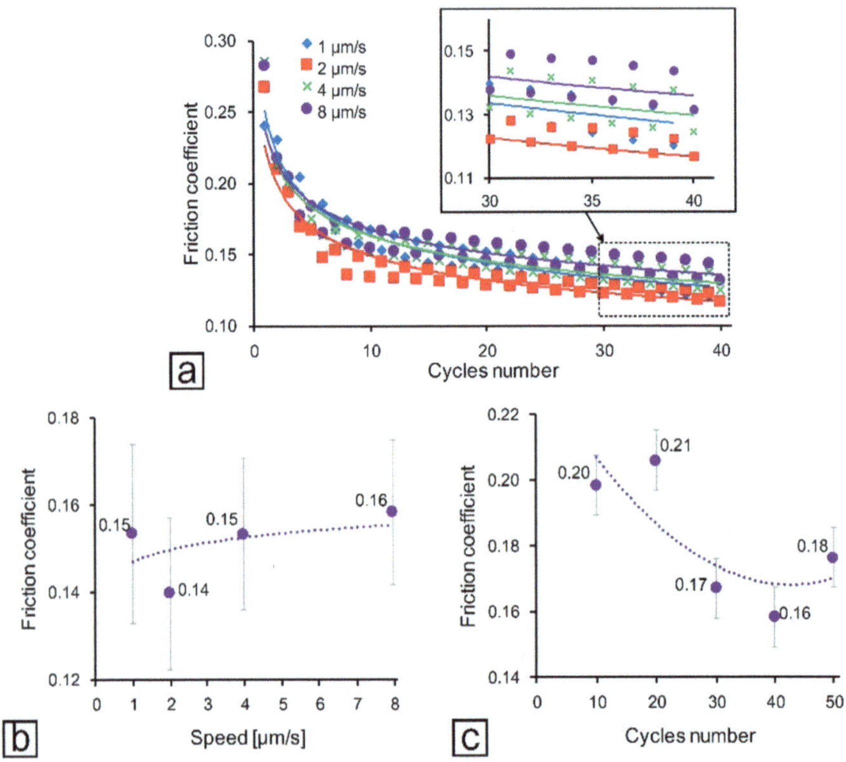


**Fig. 16.13** Dependences of *Cfr* of the ZrN coatings obtained at 5 sccm nitrogen flow rate on the scanning cycles number and on the sliding speed: **a** the dependences of *Cfr* on the cycles obtained at the different sliding speed; **b** the dependence of *Cfr* on speed; **c** the dependence of *Cfr* on cycles number

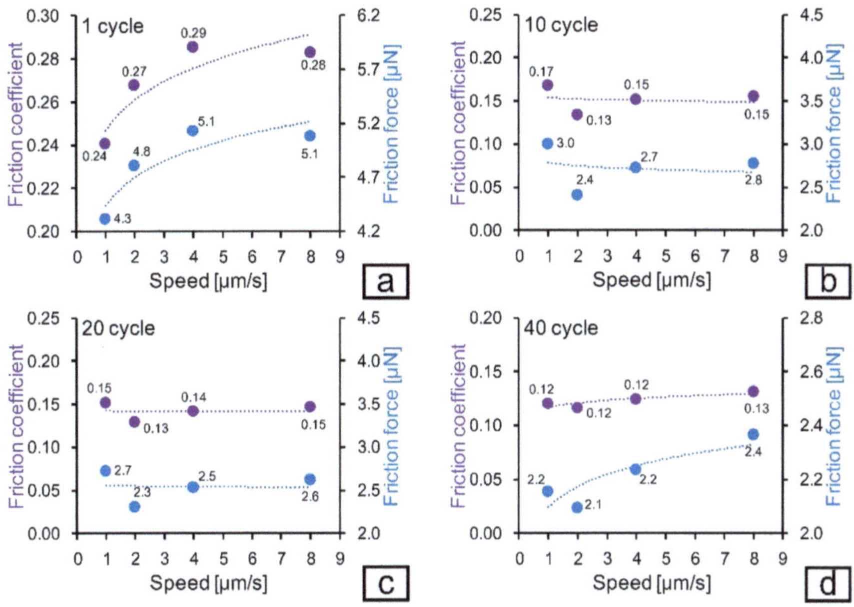


**Fig. 16.14** Dependences of *Cfr* of the ZrN coatings obtained at 5 sccm nitrogen flow rate on the sliding speed: **a** at the first cycle; **b** at the 10th cycle; **c** at the 20th cycle; **d** at the 40th cycle

and will also be logarithmic. The increase in *Cfr* is affected by a shorter contact time and, therefore, a smaller amount of defects. With an increase in the number of cycles, *Cfr* decreases, since in the previous cycles, a greater number of defects were formed, and the material moves more easily (Fig. 16.13c). The effect of scan speed is not the same for different scan cycles. Its greatest influence was established for the first cycle (Fig. 16.14a), during which the upper atomic layer of the coating is moved. In the first scan cycle, *Cfr* increases with the increasing speed (Fig. 16.14a). The scanning speed has practically no effect on the *Cfr* for the second and subsequent cycles (Fig. 16.14b–d). An increase in *Cfr* with an increase in the scanning speed, in addition to the wearening of defect formation process, can also be explained by the extrusion of nanometer water film from the contact.

The application of the new technique of the total volume determination by the “Bearing Analysis” method to the wear traces on the ZrN coatings showed that the greatest wear was found in the coating obtained at 4 sccm, and the lowest at 6 sccm (Fig. 16.15). The correction made it possible to refine the values of specific volumetric wear obtained in (Kuznetsova et al. 2020b) and decreased the wear for the coatings formed at 3, 5, 6 sccm and increased for coating formed at 4 sccm. The ranking of the coatings in terms of wear is practically preserved, but it made it possible to more accurately separate close values and find more accurate correlations with the mechanical properties and deposition regimes.

Comparison of the worn surfaces images after the first and fortieth wear cycles shows their evolution (Fig. 16.15). Applying the multi-cycle wear and successively

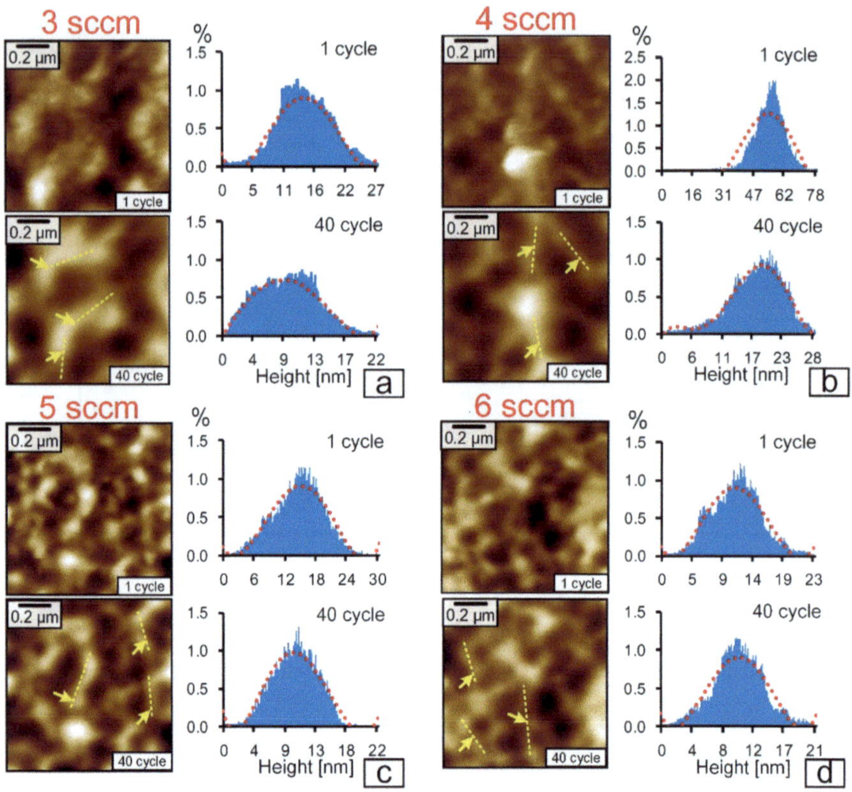


**Fig. 16.15** AFM images of wear results at the load of 3.6 μN, frequency of 1 Hz and 40 cycles on the ZrN coatings surfaces obtained at different nitrogen flow rates: **a** 3 sccm; **b** 4 sccm; **c** 5 sccm; **d** 6 sccm with marked linear structures

obtaining images of the worn surface, we are able to observe a gradual surface "flow" of a solid material according to its grain microstructure. The grains gradually unfold with their axes (maximum diameters) along the "slow scanning" direction, merge into linear structures (Fig. 16.15, arrows). "Slow scan" is the direction where the probe gradually moves along the Y-axis. It is perpendicular to the "fast scan" along the *X*-axis, along which the probe moves back and forth. The crystallite boundaries have retained their direction, while the main material is moved by the probe. The displacement (slippage) of the material does not occur randomly, but along the grain axes. Thus, stronger boundaries guide the flow of the material.

This wear mechanism differs from established by Thomas et al. (Thomas et al. 1993) for gold film. They found that when the *W* tip of 250 nm radius comes into the contact with a protruding grain in the passivated Au film, an initial elastic response, as described by the Hertz model, occurs. When the load increased, the entire nanometer-sized gold grains were plastically displaced, pushing downward relative to neighboring grains, filling some subsurface pores (free volume). Thus, the grain boundary slip is defined as the initial mechanism of plastic deformation. At even higher loads, the subsurface free volume no longer remains, and the grains begin to deform plastically with a measured shear stress threshold for plastic deformation of 2.6 GPa, which is much higher than the value for bulk gold of 200 MPa. The ideal theoretical yield point for Au in the absence of dislocations is 1.5 GPa. Thus, the wear mechanism of relatively soft low-density materials with porosity under elastic action differs from the wear mechanism of high-density hard material at stresses exceeding the plastic flow threshold. The load and the magnitude of mechanical stresses have a decisive influence on the wear mechanism.

The dependence of *Cfr* on the number of wear cycles of the ZrN coatings obtained at the different nitrogen flow rates is shown in Fig. 16.16. In almost all dependences, elevated *Cfr* are observed in the first cycles, which then decrease (Fig. 16.16a, b). The elevation of *Cfr* in the first scan cycle during AFM tests can be associated with both a more difficultly movement of the material during the initial defect formation and the capillary effect. More water can be found on the rough surfaces. This assumption is also confirmed for these coatings—the coatings obtained with 4 and 5 sccm have the highest roughness and the highest *Cfr* in the first cycles (Fig. 16.16b).

In the work of Capric et al. (Carpick et al. 1997), a decrease in the friction force from scan to scan was also noted during tests in high vacuum with friction of the Pt tip on mica and was associated with a decrease in the work of adhesion (Carpick et al. 1997). The decrease in friction and adhesion was caused by chemical and structural modification of the tip from the mica surface, which reduced the surface free energy of Pt. The decrease in *Cfr* with an increase in the wear cycles number can also be explained by the accumulation of defects due to the effect of mechanical stresses and the subsequent easier displacement of the material.

LFM studies of various materials [Si(100), Au, $SiO_2$, AgBr, $C_{60}$, mica] have shown that the observed wear (which is subsequently visualized by AFM) can occur on the surface by the tip sliding only at a certain threshold load, below which no damage is observed. Some of these studies tracked the friction forces with the load increase and typically observed a sudden increase in friction when intense wear began, indicating that shear processes are causing wear. Hu et al. found a sudden appearance of the significant wear after repeated scans at the low load (Hu et al. 1995). This suggests that point defects accumulated during scanning even at low load.

The lowest roughness and the lowest *Cfr* in the first cycles were determined in the ZrN coatings obtained at 3 and 6 sccm (Fig. 16.16b). The *Cfr* for all coatings decrease with the cycles number (Fig. 16.16c–e). On the 40th cycle, the highest Cfr is in the ZrN coating obtained with 5 sccm (Fig. 16.16e). The coating obtained at 3 sccm of nitrogen for all cycles had the lowest *Cfr*, several times lower than the other coatings. The low *Cfr* values can be explained by the small grain size and the high content of the ZrN phase with the (111) orientation, which facilitate the defect formation ability. The coating has the highest density of grain boundaries allowing material slippage.

The dependences of *Cfr*, friction forces, adhesion forces, specific surface energy, wear volume *V* and specific volumetric wear of the ZrN coatings on the nitrogen flow rate are shown in Fig. 16.17. The widespread approach to the *Cfr* and wear

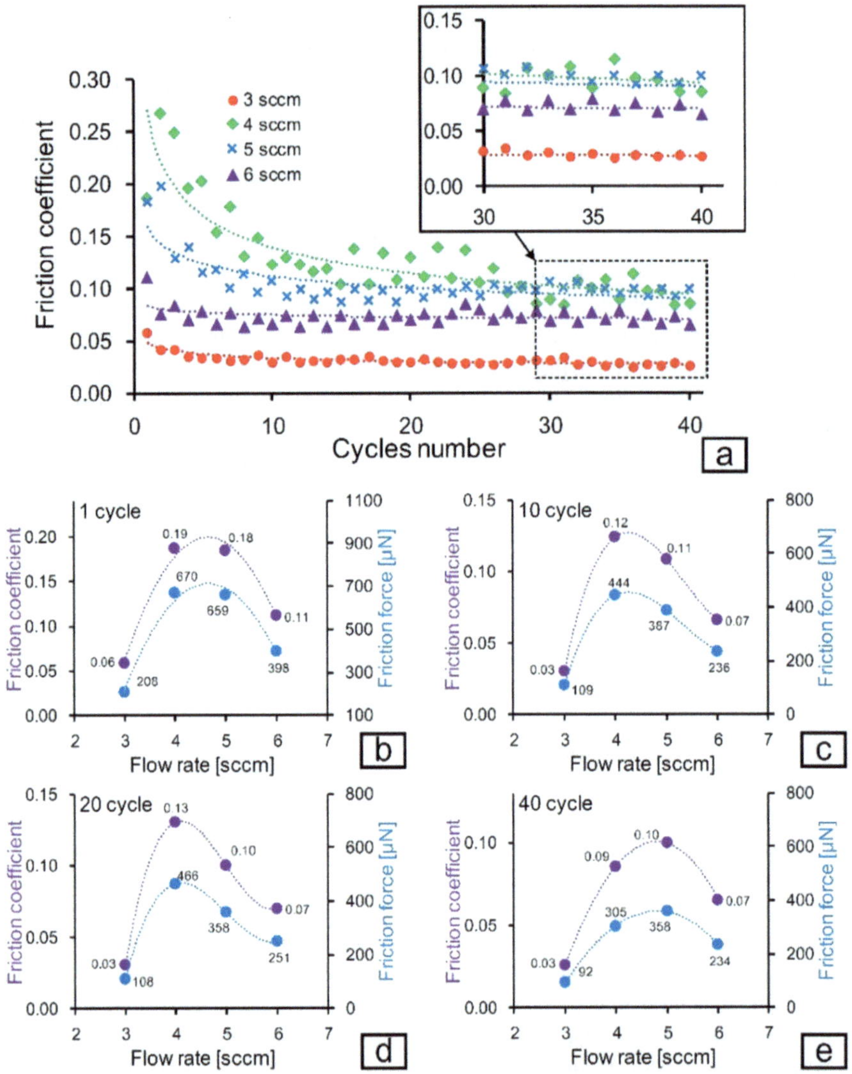


**Fig. 16.16** Dependences of friction coefficient and friction forces on the scanning cycles number of the ZrN coatings obtained at the different nitrogen flow rate of 3–6 sccm (loading 3.6 μN, speed 1 Hz): **a** the dependences of *Cfr* for all nitrogen flow rate; **b** for the first cycle; **c** for the tenth cycle; **d** for the 20th cycle; **e** for the 40th cycle

assessing, which assumes lower values of these characteristics for harder materials, does not allow predicting the tribological characteristics of the ZrN coatings under the microcontact conditions. *H* values of all ZrN coatings, determined by NI, are close and range from 29.3 to 30.8 GPa. They cannot fully explain the difference in *Cfr* or wear of the ZrN coatings. The ZrN coatings obtained at 4 and 5 sccm have the

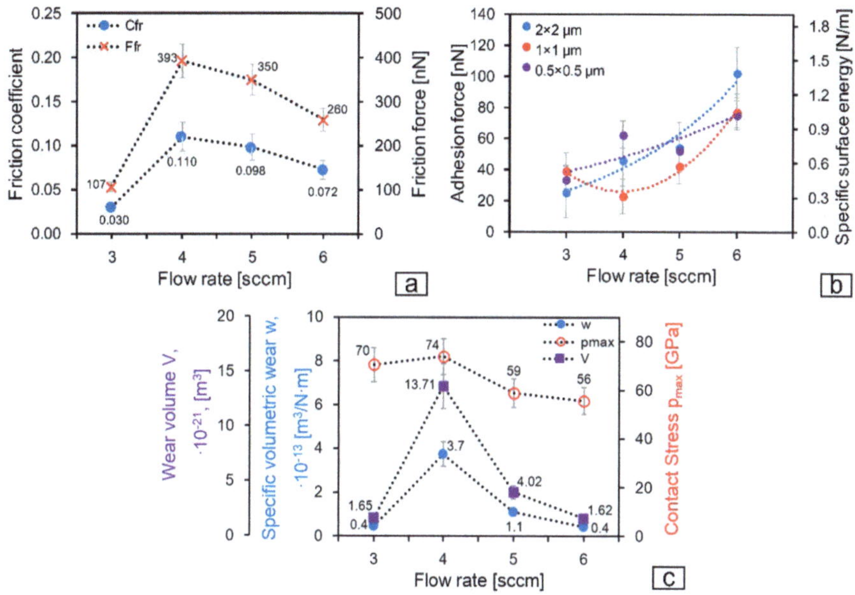


**Fig. 16.17** Dependences of friction coefficient and friction force (**a**), adhesion force and specific surface energy (**b**) wear volume and specific volumetric wear (**c**) of the ZrN coatings on the nitrogen flow rates during deposition

same microhardness of 29.3 GPa, close *Cfr* (0.11 and 0.10) and almost in 1.8 times different specific volumetric wear values – $3.42 \times 10^{-13}$ and $1.90 \times 10^{-13}$ $m^3/(Nm)$ (Fig. 16.17). The elastic modulus determined by NI better explains the difference in the tribology characteristics. A material with a higher *E* value has stronger bonds in the crystal lattice, and when its layer is displaced by the probe, more force is required, which leads to a larger twist of the AFM probe and a higher *Cfr*. The coatings are arranged according to decreasing wear volume (Fig. 16.17c) practically in the same order as *Cfr* and *E*, which corresponds to the direct dependence of the wear rate on the elastic modulus of the material (Kragelsky et al. 1982). In general, *Cfr* has the strong positive correlations with *Ra* (0.9) and with *E* (1.0) and the strong negative correlation with *H* (– 1.0). *V* has the strong positive correlations with *Ra* (0.9) and with *E* (0.8) and the large negative correlation with *H* (– 0.6) and with *H*/*E* (– 0.7). The correlation of *Cfr* with the grain size was large positive (0.7), and the correlation of *V* with the grain size was week (0.1). The correlations of *Cfr* and *V* with $C_{\text{texture of (111)}}$ were large negative (– 0.6) and (– 0.5), respectively. The increased wear of the ZrN coating obtained at 4 sccm nitrogen can be explained by the elevated mechanical stresses in the contact up to 74 GPa at the lowest *H*/*E* ratio among all ZrN coatings (Kuznetsova et al. 2020b).

The obtained values of the specific volumetric wear of the ZrN coatings are somewhat overestimated in comparison with the literature data. This phenomenon can be explained by the special conditions at the nanocontact—the increased mechanical stresses localized in the thin surface layer several nanometers thick. At the same time, the wear of the counterbody (AFM probe) was precisely controlled. The diamond tip radius increased from sample to sample during scanning of the coatings obtained at 3, 4, 5 and 6 sccm nitrogen flow rate as 21, 23, 31 and 32 nm. Taking into account the radius of the AFM probe and *E* of each coating determined by NI, the mechanical stress in the contact of the ZrN coatings obtained at 3 and 4 sccm (70.5 and 73.9 GPa) decreased for the ZrN coatings obtained at 5 and 6 sccm, up to 58.9 and 55.7 GPa. The contact stress and stress depth were calculated using Advanced Mechanical Engineering Solutions (Amesweb) calculator (Advanced mechanical engineering solutions/hertzian contact stress calculator. Calculator. http://www.amesweb.info/HertzianContact/HertzianContact.aspx). A decrease in mechanical stresses of the ZrN coatings obtained at 5 and 6 sccm changes the contact conditions and can introduce some error in the determination of wear. The dependence type of *Cfr* on the nitrogen flow rate completely coincides with the dependence of *Ra* on the nitrogen flow rate.

Considering the friction forces under the AFM probe microcontact conditions, the adhesion forces cannot be ignored. The determination of the adhesion forces for the ZrN coatings was carried out in the QNM mode at the loads of 3–5 nN, and they can be compared with other works where the AFM adhesion forces were determinated. In (Bhushan 2005), the level of adhesion forces on the Si(100) surface for approximately the same conditions of humidity (30%) and the tip radius (about 20 nm) is about 50 nN. In (Kaynak et al. 2017), the average values of adhesive forces, depending on the modification of the hydrophobilized steel surface, varied from 20 to 60 nN. In this work, for the ZrN coatings, the adhesive forces range from 20 to 80 nN (Fig. 16.17b). The dependences of the specific surface energy on the nitrogen flow rate are shown using the same curves and an additional axis (Fig. 16.17b). Adhesion forces have the large negative correlations with *Ra* (− 0.5). The adhesive forces influence on the AFM friction test in air by the capillary forces that always exist in the microcontact due to the thin film of water. It was established that the water is squeezed out of the contact and its effect on *Cfr* decreases at the increased loads as in the wear condition. The influence of the capillary forces during the tribological microcontact of the AFM probe with the ZrN surface cannot be completely excluded, but the degree of this influence is not decisive, since the level of the adhesive forces is comparable to the level on a hydrophobic substrate (Musil et al. 2016). The presence of water on the surface can affect the *Cfr* by two mechanisms: The formation of nanomenisks increases *Cfr*, and the presence of water in the form of a thin film reduces the *Cfr* (Bhushan 2005). The increased load reduces the impact of both mechanisms. Water is pushed back to the edge of the meniscus, which increases the radius of the meniscus and therefore reduces the force of interaction and reduces the force of friction, or a thin film of water is squeezed out of the contact, which increases the force of friction.

The maps of the adhesive forces of the ZrN coatings obtained at the different nitrogen flow rates demonstrate the fact that the distribution of adhesive forces contrast does not coincide with the topography (Fig. 16.18). Considering the capillary forces as an important component of the adhesion forces, it can be assumed that a thin film of water present in the contact between the AFM probe and the surface

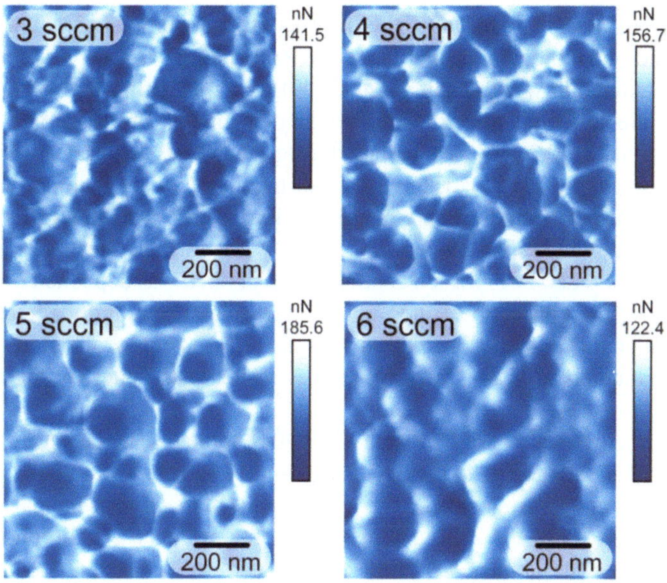


**Fig. 16.18** Influence of nitrogen flow rate at the deposition on the adhesion forces maps in the ZrN coatings, area size is 1 × 1 $\mu m^2$

during tests in air will accumulate at deeper grain boundaries. The map of the adhesive forces distribution on which the maximum values correspond to the light blue color confirms this statement. But not all the grain boundaries correspond to maxima, and there are still smooth regions, which also correspond to maximum values. The minima corresponding to the protrusions and the maxima corresponding to the cavities on the map of adhesive forces can be explained by the size of the contact of the probe with the different parts of the surface: It decreases on the protrusions and increases on the cavities. However, there are some deviations from this principle, when the maxima of the adhesive forces are distributed differently in the cavities of the same size, allow us to assume the influence of the islands of a water film on the adhesion forces, the distribution of which, in addition to cavities, is also sensitive to the nanograined structure. An increased adhesion contrast selectively surrounds the protrusions on the surface of the ZrN coating obtained at 3 sccm of nitrogen, here there are linear areas and surface fragments filled with the light blue color. In the ZrN coating obtained at 4 sccm of nitrogen, an increased adhesive contrast is distributed on some of the boundaries and on fragments of the surface. In the ZrN coating obtained at 5 sccm of nitrogen, an increased adhesive contrast surrounds all the protrusions evenly. The ZrN coating obtained at 6 sccm nitrogen has an increased adhesion contrast surrounding some of the protrusions, but the width of the elements is much wider than in all previous coatings. The smooth surface also has a light contrast, higher than that in other coatings.

The *Cfr* values obtained under the macrocontact conditions can be a guideline for the trend of tribological properties of the ZrN coatings in the AFM tests. *Cfr* of 0.6 were obtained in work (Zhang et al. 2009) for ZrN coating with a counterbody of WC in the form of ball with a diameter of 6 mm at the load of 2 N with a sliding velocity of 0.2 m/s. *Cfr* values during macrotests with an $Al_2O_3$ ball are also 0.6 (Ruden et al. 2013). The material of the counterbody is important for *Cfr*, but its size, surface roughness are also important. The *Cfr* values of the coatings are influenced by the substrate material. A fairly low for macrotests *Cfr* of 0.2 was obtained in (Bai et al. 2017) when coating of 2 μm thick was deposited on Ti6Al4V substrates. At the condition of the microcontact, the substrate does not affect the tribological characteristics, since the maximum stresses are localized at a depth of several nanometers, and not micrometers, as in the tests under the macrocontact conditions.

In addition to the standard Cfr value of 0.5–0.6, some reseachers have *Cfr* of 0.1–0.2 in the ZrN coatings due to the content of amorphous phases in them or previous annealing of samples (Auger et al. 2007). The results of macrotests indicate a fivefold decrease in the *Cfr* of the ZrN coatings due to the presence of amorphous phases in the contact. Such amorphous phases can also form in the microcontact on the ZrN coatings surface during the accumulation of defects in thin surface layers and displacement of the material by subnanometer layers.

### *16.3.2 ZrCN Coatings*

The results of X-ray phase analysis of the ZrCN coatings are shown in Fig. 16.19. Several phases are identified in the composition of the ZrCN coatings: ZrN, ZrC and Zr. The diffraction lines originating from planes (111), (200), (220), (311), (222), (400), (331), (422) and (511) are determined for the ZrN phase. The same lines listed above and the line from plane (420) are found for the ZrC phase. In contrast to the ZrN coatings series (paragraph 16.3.1), the lines of Zr from plane (103) and of Fe from plane (110) are found. In each coating, the intensity of the characteristic diffraction lines is different. The ZrC phase lines are absent in the ZrN coating obtained at 0 sccm acetylene, as well as the ZrN phase lines are not detected in the ZrC coatings obtained at acetylene 6 sccm. The intensities of the Zr and Fe lines in all coatings, except of ZrC, are practically the same. There are no Zr and Fe lines in the ZrC coating at all. It can be assumed that Zr and Fe are present in the ZrCN coating diffractograms. The influence of the substrate and the sublayer on the diffractograms in the glancing angle XRD method with a coating thickness of 3 μm is unlikely. The peaks from the substrate and sublayer are not found in the ZrN coatings series (paragraph 16.3.1). The possible presence of Zr and Fe in the coatings makes their phase composition even more diverse and can affect tribological properties (Fig. 16.19).

The main orientation of the ZrN phase in the ZrN coating obtained at 0 sccm of acetylene is (200) and not (111) as it was in the ZrN coatings series. The intensity of the orientation (200) decreases with a further increase in the acetylene flow rate: first

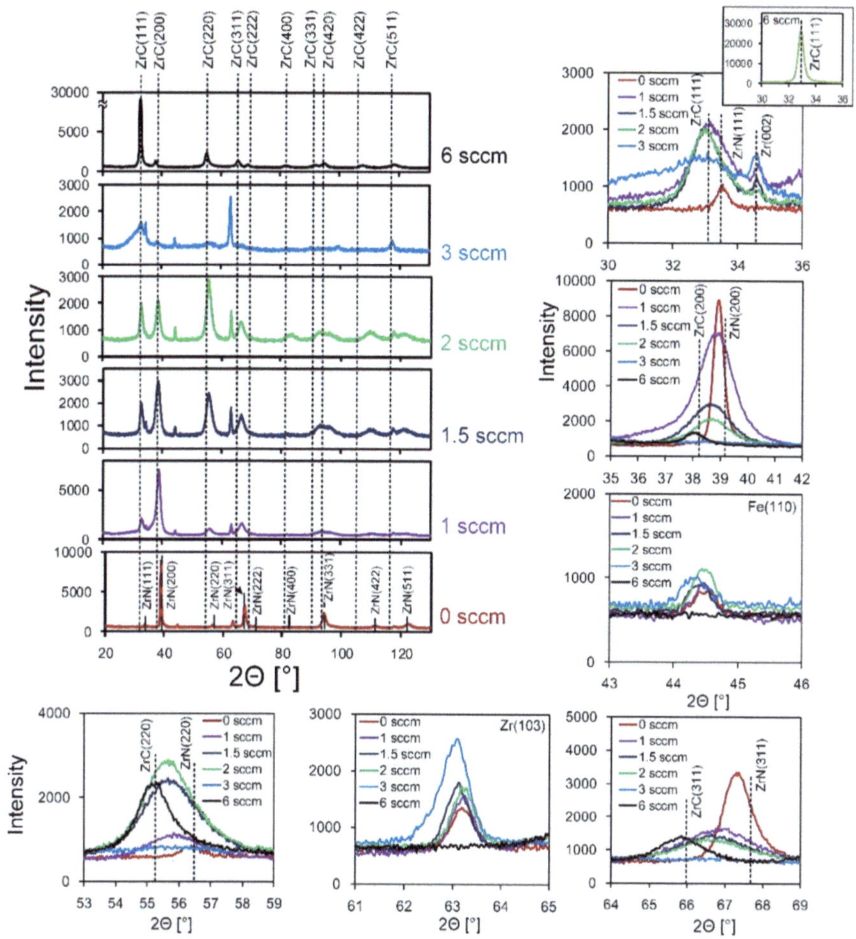


**Fig. 16.19** Glancing angle X-ray diffractograms of the ZrCN coatings obtained at different acetylen flow rate

at 1 sccm by a quarter of the initial intensity, then at 1.5 and 2 sccm—significant, and at 3 sccm it disappears completely. The ZrC phase of (111) orientation appears in the ZrCN coatings already at 1 sccm acetylene, and the intensity of its peak remains unchanged at 1.5 and 2 sccm acetylene. The intensity of the ZrC (111) peak decreases by half at 3 sccm of acetylene, and the peak broadens significantly, as possibly happens during the phase amorphization. The ZrC phase of (111) orientation becomes the main one in the coating at 6 sccm acetylene.

The texture coefficient of the phases with the densest packing of atoms in the (111) orientation was determined for the ZrCN coatings by formula (16.1). The dependence of the (111) texture on the acetylene flow rate is shown in Fig. 16.20. The phase variety of the ZrCN coatings makes it possible to assume that they can

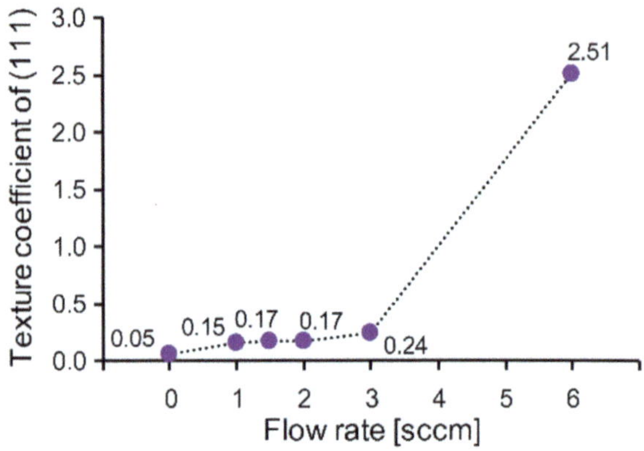


**Fig. 16.20** Dependence of texture coefficient of $I(111)$ ZrN and ZrC phases in the ZrCN coatings on the acetylene flow rate

have more attractive tribological properties in comparison with the monophase ZrN coatings due to the simultaneous presence of cubic (ZrN, ZrC) and hexagonal (Zr) phases and different packing densities of atoms in the cubic phases.

SEM microstructure of the ZrCN coatings is typical for the magnetron sputtering with a smooth surface and small amount of macroparticles (Figs. 16.21, 16.22, 16.23 and 16.24). The amount of microparticles is even less than in the ZrN coatings series and decreases with an increase in the acetylene flow rate. An exception is the ZrC coating, the smooth surface of which has a more developed relief in comparison with other coatings (Fig. 16.24). The smooth surface of the ZrC coatings is covered with "islands" of phases of different morphology: There are the flat

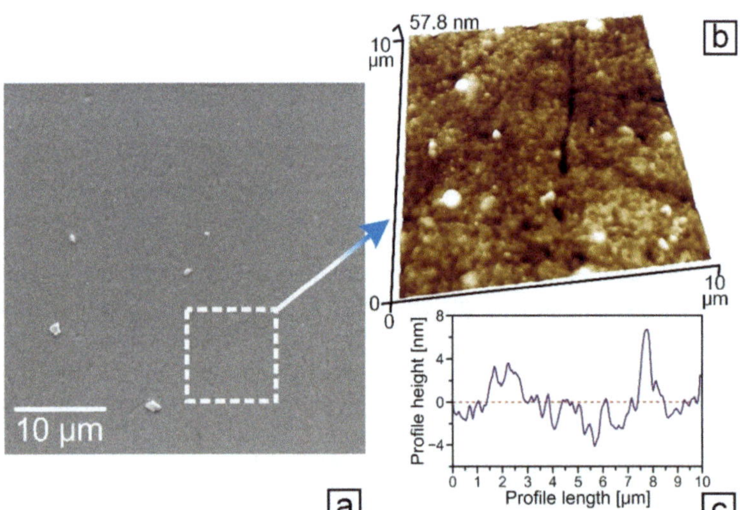


**Fig. 16.21** Surface morphology of the ZrN coating (ZrCN obtained at 0 sccm of $C_2H_2$): **a** SEM, magnification 5000×; **b** AFM topography, scanning area 10 × 10 $\mu m^2$, **c** AFM profile

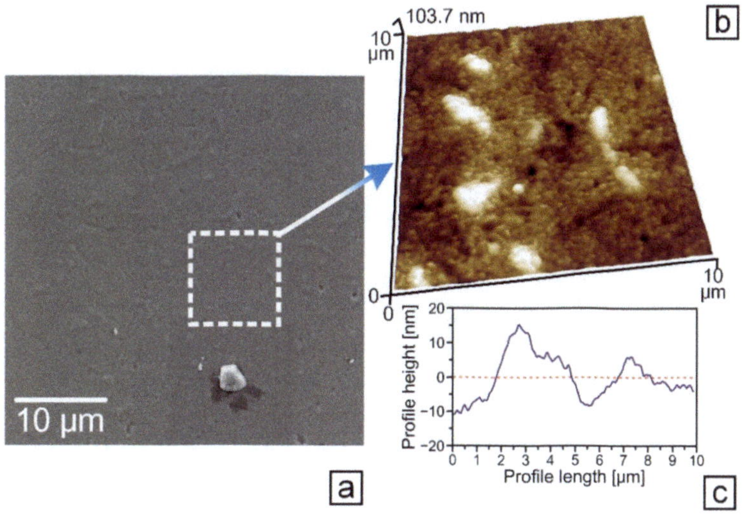


**Fig. 16.22** Surface morphology of the ZrCN coating obtained at 1 sccm of $C_2H_2$: **a** SEM, magnification 5000×; **b** AFM topography, scanning area 10 × 10 $\mu m^2$, **c** AFM profile

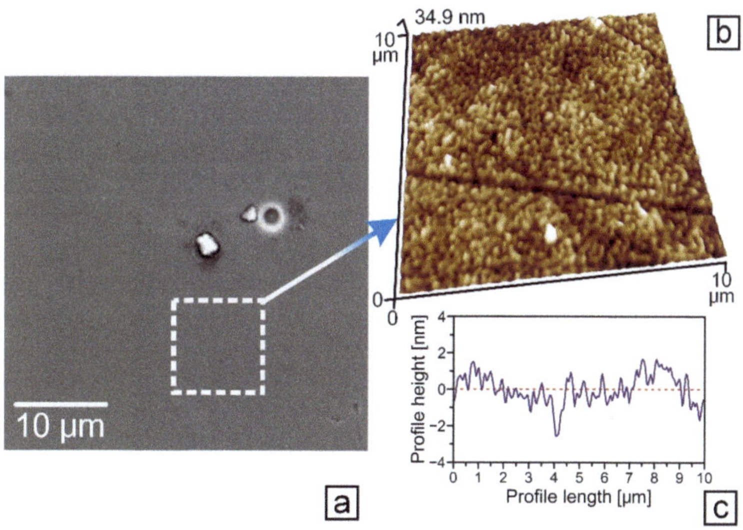


**Fig. 16.23** Surface morphology of the ZrCN coating obtained at 3 sccm of $C_2H_2$: **a** SEM, magnification 5000×; **b** AFM topography, scanning area 10 × 10 $\mu m^2$, **c** AFM profile

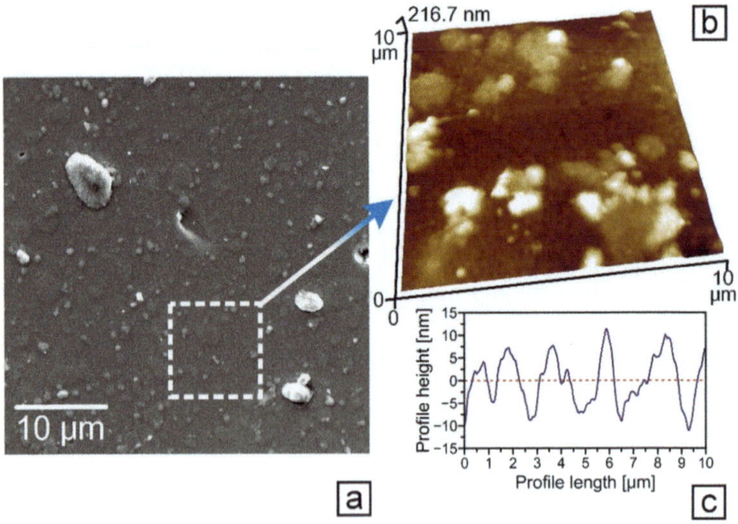


**Fig. 16.24** Surface morphology of the ZrC coating obtained at 6 sccm of $C_2H_2$ and 0 sccm of $N_2$: **a** SEM, magnification 5000×; **b** AFM topography, scanning area 10 × 10 $\mu m^2$, **c** AFM profile

islands of different heights, the spherical microparticles and elongated microparticles (Fig. 16.24). Their size is tens times smaller than the traditional for the magnetron sputtering macroparticles up to 10 μm in size and ranges from 100 nm to 2 μm.

The AFM microstructure of the ZrN coating surface contains nanocrystallites, some of which are 2–3 times larger than the average size (Fig. 16.24b). On an area of 10 × 10 $\mu m^2$, up to 10 pieces such large crystallites are observed (Fig. 16.24b).

The addition of acetylene to the chamber for the the ZrCN coating synthesis changes the shape of large crystallites into flat and elongated ones, and their number on an area of 10 × 10 $\mu m^2$ decreases (Figs. 16.22b and 16.24b). It can be assumed that the planar formations are the layers of the ZrC phase, which, according to the X-ray diffraction data, appears in the coatings already upon the addition of 1 sccm of acetylene. The greatest variety of the planar formations types on the surface consisting of nanocrystallites is noted for the ZrC coating (Fig. 16.24b).

The surface morphology of the ZrCN coatings (Fig. 16.25) using AFM is much more diverse than that of the ZrN coatings (Fig. 16.6). Nanocrystallites form not only the grain chains in the process of crystallization as in the ZrN coating (Fig. 16.25). When acetylene is supplied, nanocrystallites can form globules (Fig. 16.25, 1 sccm). Globules are the rounded formations of several crystallites. Probably, the conditions for their formation are related to the fact that the number of nucleation centers remains the same, while the inflow of the substance for the coating formation increases as the reaction gas flows faster. In the ZrCN coating obtained at 1 sccm of acetylene, the individual crystallites have almost the same size of 50–60 nm. At 1.5 sccm of acetylene, the scatter in the size of the single crystallites increases—a fraction of small (10–30 nm) and a fraction of large (100–150 nm) crystallites appear, the

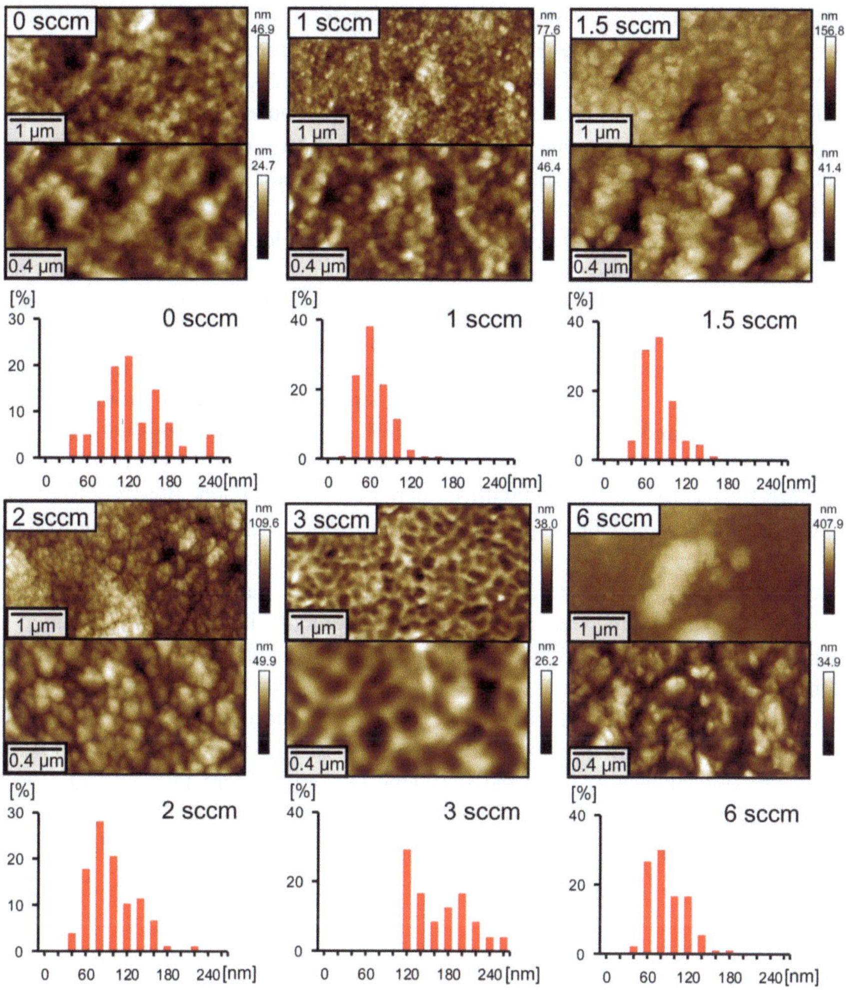


**Fig. 16.25** Influence of acetylene flow ratio on the surface morphology and grain distribution in the ZrCN coatings

average size increases to 70 nm. The size of the globules is 200–400 nm (Fig. 16.25, 1.5 sccm). The size of the globules decreased to 100–150 nm with an increase in the acetylene flow to 2 sccm, which can be explained by an increase in the pressure in the chamber. In this case, the individual crystallites did not decrease, and their average size was 88 nm. The surface morphology changed significantly at 3 sccm of acetylene (Fig. 16.25, 3 sccm). The surface appearance has changed from "grainy" to "cellular". In the case of the "cellular" structure, crystallite boundaries protrude, and the internal volume is concave. Probably, during the coating synthesis, the processes of knocking out the coating atoms by the plasma atoms took place, and the surface was "sprayed", but the harder boundaries remained. In this case, the grain size was taken to be the diameter of the protruding crystallites when three cell boundaries were connected. The average grain size is 155 nm. The greatest variety of crystallite forms morphology is found in the ZrC coating (Fig. 16.25, 6 sccm).

The crystallites that make up the chains are not round, many have faces, and many are elongated along one direction (Fig. 16.25, 6 sccm). The ZrC coating is characterized by layer-by-layer growth and many islands of the next layer on the surface. The average grain size is 79 nm. Since in the ZrCN coatings, supercrystalline formations of the individual crystallites are more developed—globules, islands of additional layers, the dependence of roughness on the acetylene flow rate and the dependence of the grain size on the acetylene flow rate differ greatly from each other, stronger than in the ZrN coating (Fig. 16.26). The ZrCN coatings are multiphase, and the rates of phase crystallization under the same conditions may not coincide. The rule about the relationship between a higher packing density of atoms with the (111) orientation and a lower roughness is fulfilled among the coatings with a similar phase composition. The texture coefficient of the ZrC phase of the (111) orientation increases from 0.15–0.17 to 0.24 with an increase in the acetylene flow rate from 1 to 3 sccm (Fig. 16.20), and the roughness decreases from 6–7 nm to 3.1 nm (Fig. 16.26). The grain size of the ZrCN coatings grows almost twofold with an increase in the acetylene flow rate, which is a consequence of the accelerated supply of adatoms to the substrate. The obtained results are in good agreement with the results presented in (Braic et al. 2011), where it was reported that the addition of C reduces the roughness of the coatings based on Zr.

The average *H* and *E* values of the ZrCN coatings, determined at the load of 5 mN and an indentation depth of about 80 nm, are 32.8–40.2 GPa for *H* and 195–300 GPa for *E*. The dependence of *H* on the acetylene flow rate completely coincides with the dependence of *Ra* on the acetylene flow rate (Figs. 16.26 and 16.27). *H* has the large positive correlations with *Ra* (0.6) in the ZrCN coatings. The correlation of *H*

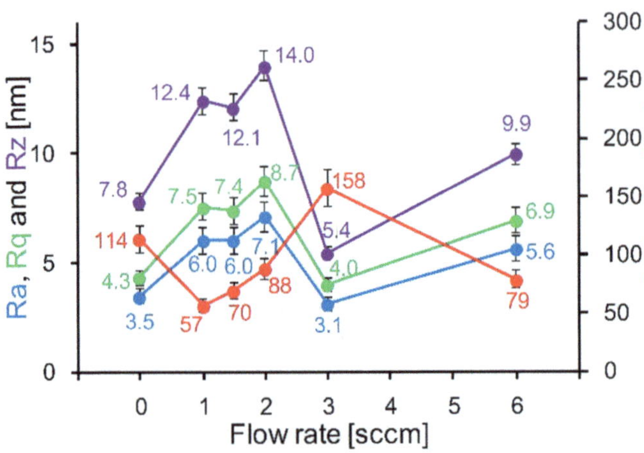


**Fig. 16.26** Dependences of the roughness parameters (*Ra*, *Rq*, *Rz*) on area 2 × 2 $\mu m^2$ and grain size of the ZrCN coatings on the acetylene flow rate

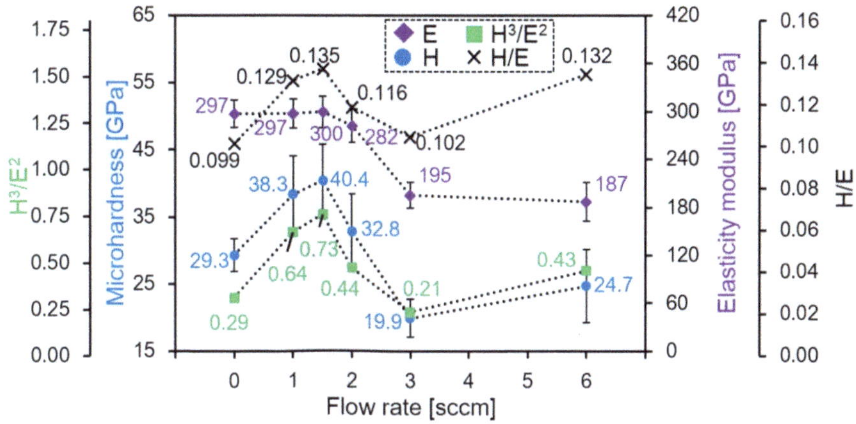


**Fig. 16.27** Dependences of $E$, $H$, $H/E$ and $H^3/E^2$ of the ZrCN coatings on the acetylene flow rate

with *Ra* of the ZrN coatings was strong negative (– 0.9). The correlation of $H$ with the grain size in the ZrCN coatings is strong negative (– 0.8). The ZrN coatings have also the strong negative correlation with the grain size. In general, the direct Hall–Petch law is fulfilled for the ZrCN coatings.

The $E$ and $H$ values obtained in this work are in good agreement with the data obtained in (Silva et al. 2010; Ul-Hamid 2020b). The $H/E$ ratios obtained for the ZrCN coatings are 0.102–0.135 and suggest that the tribological properties of these coatings will be slightly higher than those of the ZrN coatings series.

The microgeometrical parameters and mechanical properties of the ZrCN coatings demonstrate the complex behavior with the change of the acetylene flow rate (Figs. 16.26 and 16.27). The reason for the sudden drop of most graphs from 2 to 3 sccm can be explained by significant phase changes in the Zr–C–N system at 3 sccm of acetylene. They are the most radical in the entire group of the ZrCN coatings and are described in Sect. 16.3.2. In the ZrCN coating obtained at 3 sccm, ZrN phase completely disappears, there is the largest Zr peak among all ZrCN coatings, and the ZrC peak of (111) orientation has the broadest profile, which may indicate the amorphization.

Figure 16.28 shows the surface morphology evolution of the ZrCN coatings during friction. After 40 cycles of friction, linear "guides" were formed in the surface microstructure. These are coalesced grains, the axes of which are oriented along the slow scan direction ($Y$-axis). The softer material removed from the surface of the grains was displaced, the grid from the grain boundaries remained, and its elements are oriented mainly along the direction of slow scanning. Having in the initial coating grain chains oriented in all directions, we obtain a texture in the friction zone adapted to the direction of material displacement after high-cycle wear. The grains of the ZrCN coatings are displaced along the $Y$-axis. The chains of grains directed almost parallel to the $Y$-axis retain their direction, merge into single formations, become thinner and represent fragments of boundaries. The chains of grains oriented in the

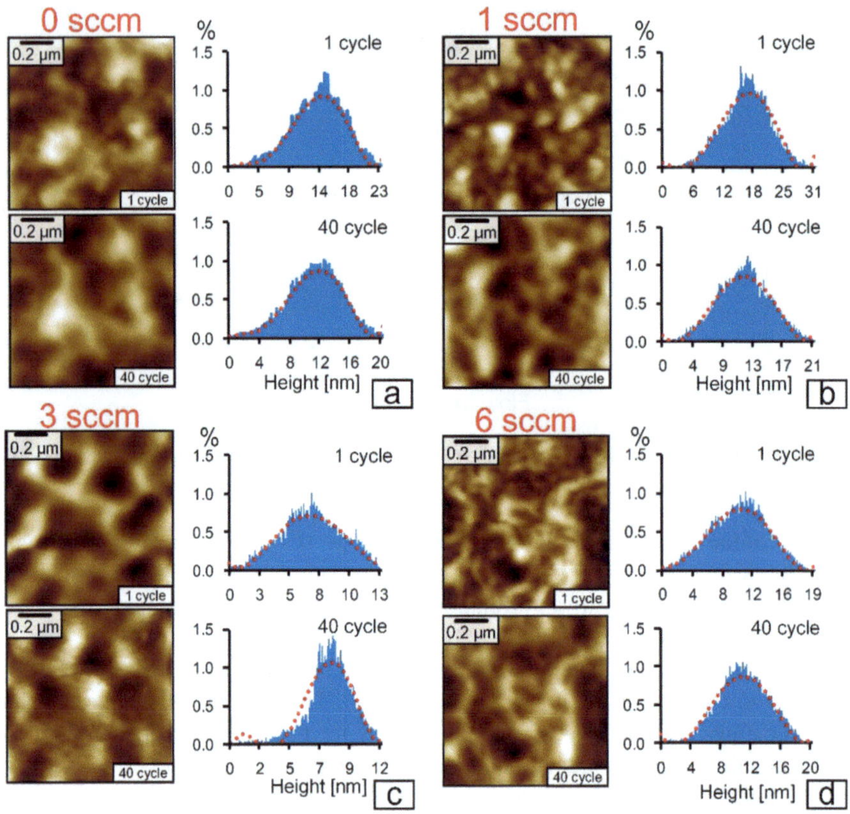


**Fig. 16.28** AFM images of wear results at the load of 3.1 μN, frequency of 1 Hz and 40 cycles on the ZrCN coatings surfaces obtained at different acetylene flow rates: **a** 0 sccm; **b** 1 sccm; **c** 3 sccm; **d** 6 sccm

initial structure along the *X*-axis are shortened, and some of them gradually unfold along the *Y*-axis. According to the roughness histograms accompanying the AFM images of the worn surfaces in Fig. 16.28, the largest wear volume was established for the ZrCN coating obtained at 3 sccm of acetylene.

After the wear tests, no pile-ups were formed on the ZrN and ZrCN coatings. The volume of the worn out material is so small that this amount is evenly distributed in atomic layers over the surface of the wear track. Since the initial surface has protruding grains and deepenings between the grains, the displaced material partially fills the deepenings, contributing to smoothing the wear track and is partially redistributed over the surface.

The next most durable ZrCN coating is obtained at 1 sccm of acetylene. The step at the edge of the wear track for the ZrCN coating obtained at 3 sccm of acetylene was 7.6 nm after 40 cycles, i.e., for one pass $h = 0.19$ nm. $R_{tip} = 22$ nm, the ratio $h/R_{tip} = 0.0086 < 0.01$ (Kragelsky et al. 1982). In other coatings, the wear depths for

40 cycles were even lower. Consequently, according to Kragelsky, the displacement of the material can be considered elastic. However, the wear track determined by AFM after the 40th cycle does not allow to consider the wear completely elastic.

The dependences of *Cfr* on the number of wear cycles for the ZrCN coatings obtained at the different acetylene flow are shown in Fig. 16.29. Overestimated *Cfr* in the first cycle, which decreases in subsequent cycles, is established for the ZrC and ZrCN coatings obtained at 1, 1.5 and 2 sccm (Fig. 16.29a, b). For the ZrCN coating obtained at 3 sccm of acetylene, the *Cfr* in the first cycle is practically equal to the values in the subsequent cycles. The dependence of *Cfr* on the acetylene flow rate tends to be similar to the dependence of *Ra* on the acetylene flow (Figs. 16.26, 16.30a). The capillary effect is more pronounced on the surfaces with a higher roughness.

*Cfr* by cycle 40 for the ZrCN coatings decreases, which is associated with a decrease in surface roughness during wear and the formation of an easily displaced defective layer (Fig. 16.29c, d). At cycle 40, the *Cfr* values were 0.055 for the ZrN coating, 0.081 for the ZrC coating and 0.051–0.088 for the ZrCN coatings (Fig. 16.29). In general, with comparable roughness and similar test conditions, the *Cfr* for the ZrCN coatings is slightly lower than for ZrN obtained at different nitrogen rates (Fig. 16.31a). *Cfr* of the ZrCN coatings correlates with *Ra*, *H* and $H/E$. The correlation of *Cfr* with *Ra* is strong positive (0.8), with *H* is large positive (0.7), with $H/E$ and $H^3/E^2$ is strong positive (0.8 and 0.9).

The adhesion forces on the ZrCN coatings surfaces are even lower than in the ZrN coatings series and range from 10 to 50 nN, which is probably associated with a lower roughness (Fig. 16.30b). Adhesion forces have the strong negative correlations with *Ra* (− 0.9). Perhaps due to the lower roughness, the ZrCN coatings have a higher correlation coefficient with Ra than ZrN (− 0.5). *Cfr* has the strong negative correlations with $F_{ad}$ (− 1.0).

The distribution of the maxima and minima of the adhesion forces is influenced by the surface morphology but does not match it. The distribution of the adhesion forces contrast in the ZrN coating from the ZrCN series (Fig. 16.31, 0 sccm) is similar to the ZrN obtained at 3 sccm of nitrogen in the ZrN series (Fig. 16.18). Regions of high contrast also occupy part of the cavities, and the minima corresponding to protruding nanosized grains are evenly distributed (Fig. 16.31, 0 sccm). An increased adhesion contrast surrounds all the protrusions evenly, forming a connected network in the ZrCN coating obtained at 1 sccm of acetylene (Fig. 16.31, 1 sccm). The diameters of the protrusions are the smallest among all coatings, which corresponds to the smallest grain size of 57 nm. The adhesion contrast of the ZrCN coating obtained at 3 sccm of acetylene (Fig. 16.31, 3 sccm) is similar to the the adhesion map of the ZrN coating obtained at 6 sccm of nitrogen (Fig. 16.31, 6 sccm).

The similarity lies in the equally large width of the areas of increased contrast, their uneven distribution and approximately the same number over an area of $1 \times 1\ \mu m^2$. Assuming the width of light blue elements to be the width of local water islands at deepened grain boundaries, it can be noted that in these coatings they are the widest. It can be assumed that this coating is the most hydrophilic in the studied coatings group. A special place in the contrast of adhesion forces is occupied by the ZrC coating (Fig. 16.31, 6 sccm). The elements of increased contrast are associated

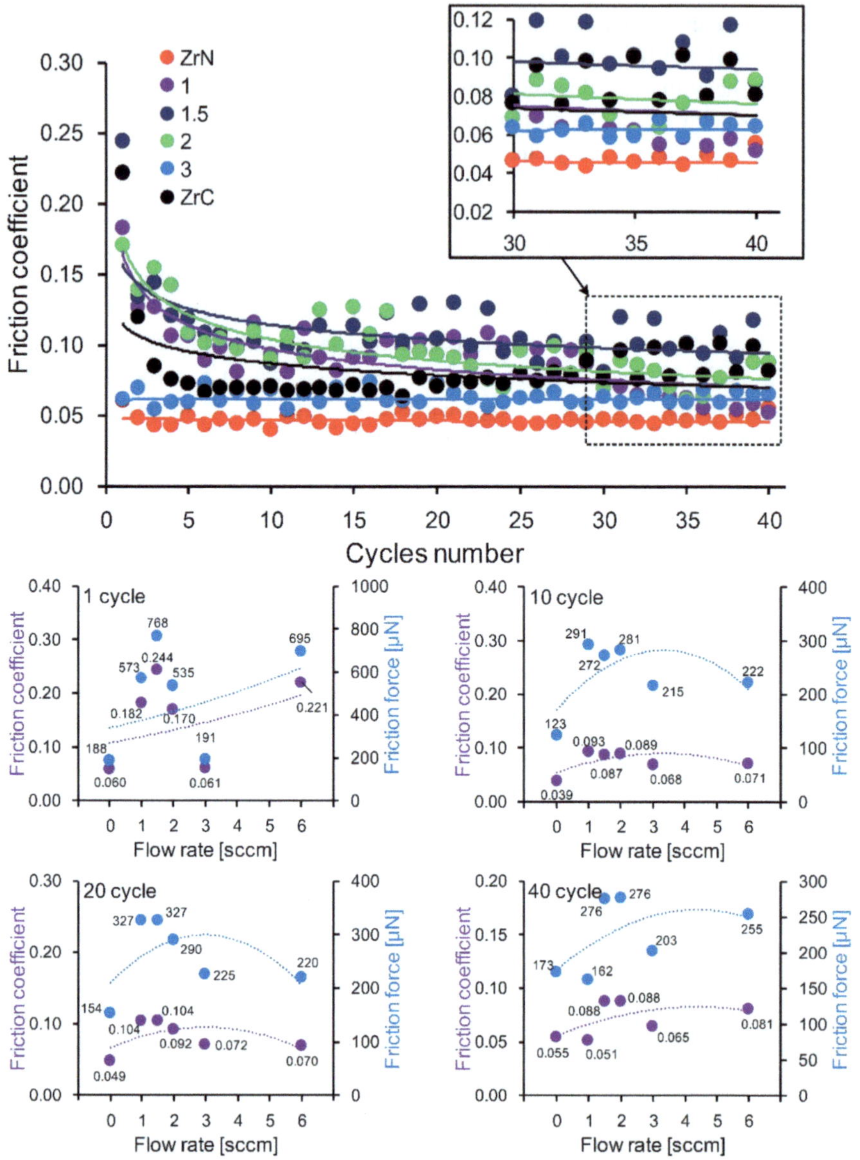


**Fig. 16.29** Dependences of friction coefficient on the scanning cycles number of the ZrCN coatings obtained at the different acetylene flow rate of 0–6 sccm (loading 3.1 μN, speed 1 Hz)

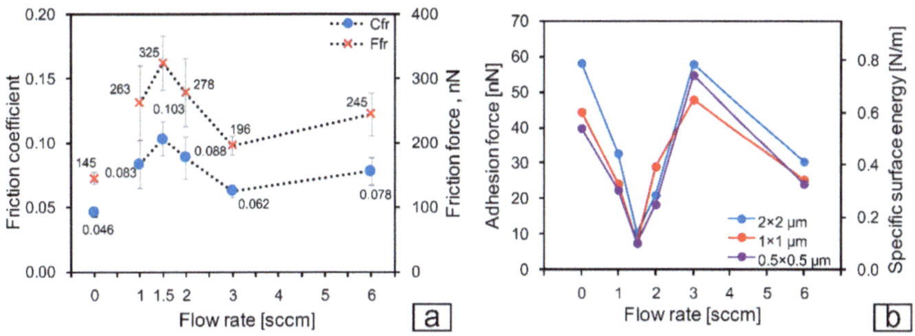


**Fig. 16.30** Dependences of friction coefficient, friction force (**a**) and adhesion forces, specific surface energy (**b**) of the ZrCN coatings on acetylene flow rates

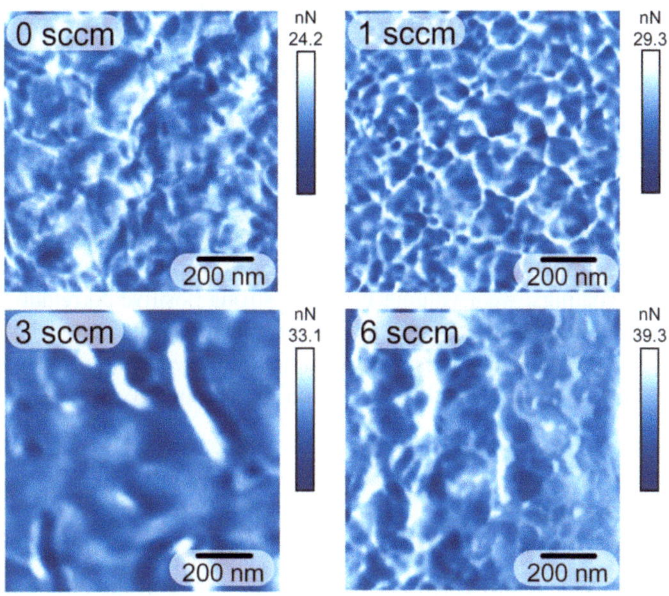


**Fig. 16.31** Adhesion force maps of the ZrCN coatings obtained at different acetylene flow rate

not only with deepened boundaries, but probably also with the distribution on the surface of thin layers of phases with different filling densities of atomic planes in it (Fig. 16.31, 6 sccm). These layers occupy at least 30–50% of the entire surface area.

The dependence of the specific volumetric wear on the acetylene flow rate is close to the dependence of *Cfr* on the acetylene flow rate and has a maximum of 1 sccm (Fig. 16.32). The greater wear of the coating obtained at 1 sccm than that obtained at 3 sccm at close values of *E* and *H*, and *Ra* is explained by the highest contact stress in this series of the coatings (69 GPa). In general, if the *Cfr* of the ZrCN coatings is

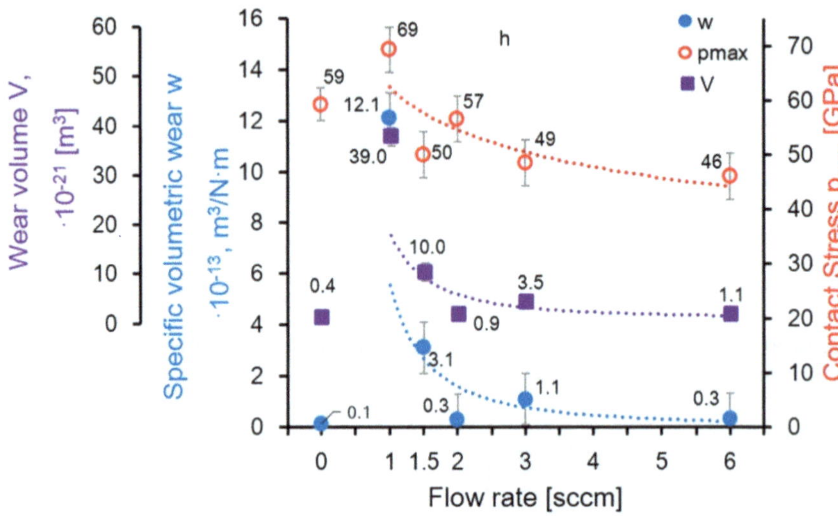


**Fig. 16.32** Dependences of the contact stress, worn out volume, specific volumetric wear of the ZrCN coatings on acetylene flow rates

comparable to the *Cfr* of the ZrN coatings, the wear of the ZrCN coatings obtained at 2 sccm is lower than that of ZrN.

In the ZrN coatings, wear volume *V* has the strong positive correlations with *Ra* (0.9) and with *E* (0.8) and the large negative correlation with *H* (− 0.6) and with *H*/*E* (− 0.7). In the ZrCN coatings, wear volume *V* has the large positive correlations with *H* (0.6) and the large negative correlation with grain size (− 0.5).

## 16.4 Conclusion

In this article, we have presented results for friction tests of the zirconium nitride, carbonitride and zirconium carbide coatings under the microcontact conditions. The coatings were formed by the magnetron sputtering on the steel substrates at the different flow rates of nitrogen and acetylene, which affected their morphology, phase composition and mechanical properties.

We found in many cases that the texture of the ZrN and ZrC phases affects the grain size of the coatings, their roughness and mechanical properties in accordance with the classical Hall–Petch law. The texture, grain size and mechanical properties of ZrN and ZrCN coatings can be managed using the different flow rates of nitrogen or acetylene.

Nonlinear dependences of the coefficient of friction and specific volumetric wear on the reaction gases flow rate were obtained. The maximum in the dependences corresponds to 4 sccm of nitrogen for ZrN and 1.5 sccm of acetylene for ZrCN. The maximum correlation coefficients of the coefficient of friction were obtained with mechanical properties (1.0 with $E$ and − 1.0 with $H$) in the ZrN coating and with the adhesion force (− 0.9) in the ZrCN coating.

The dependences of the coefficient of friction and the specific volumetric wear on the probe movement speed, the load and the number of scanning cycles are obtained. It was shown that a decrease in the sliding speed affects the wear equivalent to an increase in the load, since it contributes to the accumulation of defects in the surface layer.

It was found that the stress in the contact has a decisive influence on the wear result of the ZrN and ZrCN coatings under the microcontact conditions.

The mechanism of the coating material moving during the wear under the microcontact conditions is revealed, in which the defects formation in the nanometer surface layer and its successive displacement by atomic layers is the determining factor. Under the mechanical stresses influence, a wear texture is formed with a gradual reorientation of the grain boundaries perpendicular to the sliding direction of the probe.

We propose to consider as a criterion of elastic friction during its experimental observation not the impossibility of detecting a single trace using ACM, but the absence of any deepening after several dozen cycles. The friction of coatings of nitride, carbonitride and zirconium carbide under the microcontact conditions, carried out in this work, cannot be considered elastic even if a single track is indefinable. It was shown that the ZrCN coating obtained at 2 sccm of acetylene possesses the best tribological properties.

We believe that these studies give a solid base for further analysis of materials with coatings including mathematical and numerical modeling.

**Acknowledgements** This research was supported by the grant of Belarusian Republican Foundation for Fundamental Research No. T18PLDG-002 and No. T21MS-029. Publication partly financed by the National Centre for Research and Development, Poland, BIOSTRATEG3/344303/14/NCBR/2018.

## References

Aizikovich S, Krenev L, Sevostianov I, Trubchik I, Evich L (2011) Evaluation of the elastic properties of a functionally-graded coating from the indentation measurements. Z Angew Math Mech 91(6):493–515. https://doi.org/10.1002/zamm.201000179

Aizikovich SM, Vasil'ev AS, Volkov SS (2015) The axisymmetric contact problem of the indentation of a conical punch into a half-space with a coating inhomogeneous in depth. J Appl Math Mech 79:500–505. http://dx.doi.org/10.1016/j.jappmathmech.2016.03.011

Anishchik VM, Uglov VV, Kuleshov AK, Filipp AR, Rusalsky DP, Astashynskaya MV, Samtsov MP, Kuznetsova TA, Thiery F, Pauleau Y (2005) Electron field emission and surface morphology of a-C and a-C: H thin films. Thin Solid Films 482:248–252. https://doi.org/10.1016/j.tsf.2004.11.153

Arias DF, Arango YC, Devia A (2006a) Study of TiN and ZrN thin films grown by cathodic arc technique. Appl Surf Sci 253(4):1683–1690. https://doi.org/10.1016/j.apsusc.2006.03.017

Arias DF, Marulanda DM, Baena AM, Devia A (2006b) Determination of friction coefficient on ZrN and TiN using lateral force microscopy (LFM). Wear 261:1232–1236. https://doi.org/10.1016/j.wear.2006.03.007
Auger MA, Araiza JJ, Falcony C, Sánchez O, Albella JM (2007) Hardness and tribology measurements on ZrN coatings deposited by reactive sputtering technique. Vacuum 81:1462–1465. https://doi.org/10.1016/j.vacuum.2007.04.009
Bai WQ, Li LL, Li RL, Gu CD, Wang XL, Jin G, Liu DG, Tu JP (2017) Deposition and characterization of a ZrN/Zr/a-C multilayer: Implication on bio-tribological and corrosion behaviors. Surf Coatings Technol 324:509–517. https://doi.org/10.1016/j.surfcoat.2017.05.058
Baqir, MA, Choudhury PK, Akhtar MN (2021) ZrN fractal-graphene-based metamaterial absorber in the visible and near-IR regimes. Optik (Stuttg) 237:166769. https://doi.org/10.1016/j.ijleo.2021.166769
Bhattacharya S, Mo K, Mei Z, Seidman D, Stepnik B, Pellin MJ, Yacout AM (2020) Improving stability of ALD ZrN thin film coatings over U-Mo dispersion fuel. Appl Surf Sci 533:147378. https://doi.org/10.1016/j.apsusc.2020.147378
Bhushan B (2005) Micro/nanotribology and materials characterization studies using scanning probe microscopy. In: Bhushan B (eds) Nanotribology and nanomechanics. Springer, Berlin, Heidelberg. https://doi.org/10.1007/3-540-28248-3_8
Braic M, Braic V, Balaceanu M, Zoita CN, Kiss A, Vladescu A, Popescu A, Ripeanu R (2011) Structure and properties of Zr/ZrCN coatings deposited by cathodic arc method. Mater Chem Phys 126:818–825. https://doi.org/10.1016/j.matchemphys.2010.12.036
Brunello G, Brun P, Gardin C, Ferroni L, Bressan E, Meneghello R, Zavan B, Sivolella S (2018) Biocompatibility and antibacterial properties of zirconium nitride coating on titanium abutments: an in vitro study. PLoS ONE 13:e0199591. https://doi.org/10.1371/journal.pone.0199591
Budke E, Krempel-Hesse J, Maidhof H, Schüssler H (1999) Decorative hard coatings with improved corrosion resistance. Surf Coatings Technol 112:108–113. https://doi.org/10.1016/S0257-8972(98)00791-9
Carpick RW, Salmeron M (1997) Scratching the surface: fundamental investigations of tribology with atomic force microscopy. Chem Rev 97:1163–1194. https://doi.org/10.1021/cr960068q
Chandiramouli R, Nagarajan V (2015) Tuning band structure and electronic transport properties of ZrN nanotube—a first-principles investigation. Spectrochim. Acta Part A Mol Biomol Spectrosc 136:1018–1026. https://doi.org/10.1016/j.saa.2014.09.125
Chen J, Zhang Z, Yang G, Fang Z, Yang Z, Li Z, He G (2020) Performance and damage mechanism of TiN/ZrN nano-multilayer coatings based on different erosion angles. Appl Surf Sci 513:145457. https://doi.org/10.1016/j.apsusc.2020.145457
Chizhik SA, Rymuza Z, Chikunov VV, Kuznetsova TA, Jarzabek D (2007) Micro-and nanoscale testing of tribomechanical properties of surfaces. In: Jabłoński R, Turkowski M, Szewczyk R (eds) Recent advances in mechatronics. Springer, Berlin, Heidelberg. https://doi.org/10.1007/978-3-540-73956-2_106
Clavería I, Lostalé A, Fernández Á, Castell P, Elduque D, Mendoza G, Zubizarreta C (2019) Enhancement of tribological behavior of rolling bearings by applying a multilayer ZrN/ZrCN. Coatings 9(7):434. https://doi.org/10.3390/coatings9070434
Craciun D, Bourne G, Socol G, Stefan N, Dorcioman G, Lambers E, Craciun V (2011) Characteristics of ZrC/ZrN and ZrC/TiN multilayers grown by pulsed laser deposition. Appl Surf Sci 257(12):5332–5336. https://doi.org/10.1016/j.apsusc.2010.11.106
Craciun D, Socol G, Stefan N, Dorcioman G, Hanna M, Taylor CR, Lambers E, Craciun V (2014) The effect of deposition atmosphere on the chemical composition of TiN and ZrN thin films grown by pulsed laser deposition. Appl Surf Sci 302:124–128. https://doi.org/10.1016/j.apsusc.2013.10.095
da Silva Oliveira CI, Martinez-Martinez D, Cunha L, Rodrigues MS, Borges J, Lopes C, Alves E, Barradas NP, Apreutesei M (2018) Zr–O–N coatings for decorative purposes: study of the system stability by exploration of the deposition parameter space. Surf Coatings Technol 343:30–37. https://doi.org/10.1016/j.surfcoat.2017.11.056

Dearnley PA (2017) Surface engineering with deposition technologies. In: Introduction to surface engineering. Cambridge University Press, Cambridge, pp 116–230. https://doi.org/10.1017/9781139031509.005

Du J, Zhang P, Cai Z, Zhao J (2011) The influence of substrate-target orientation on the properties of ZrN films deposited by arc ion plating. Phys Proc 18:181–186. https://doi.org/10.1016/j.phpro.2011.06.078

Fragiel A, Machorro R, Muñoz-Saldaña J, Salinas J, Cota L (2008) Correlation between optical characterization of the plasma in reactive magnetron sputtering deposition of Zr–N on SS 316L and surface and mechanical properties of the deposited films. Appl Surf Sci 254:4632–4637. https://doi.org/10.1016/j.apsusc.2008.01.104

Geyao L, Yang D, Wanglin C, Chengyong W (2020) Development and application of physical vapor deposited coatings for medical devices: a review. Proc CIRP 89:250–262. https://doi.org/10.1016/j.procir.2020.05.149

Gilewicz A, Jedrzejewski R, Myslinski P, Warcholinski B (2019) Influence of substrate bias voltage on structure, morphology and mechanical properties of AlCrN coatings synthesized using cathodic arc evaporation. Tribol Ind 41:484–497. https://doi.org/10.24874/ti.2019.41.04.03

González-Carmona JM, Triviño JD, Gómez-Ovalle Ortega C, Alvarado-Orozco JM, Sánchez-Sthepa H, Avila A (2020) Wear mechanisms identification using Kelvin probe force microscopy in TiN, ZrN and TiN/ZrN hard ceramic multilayers coatings. Ceram Int 46:24592–24604. https://doi.org/10.1016/j.ceramint.2020.06.248

Gosvami NN, Lahouij I, Ma J, Carpick RW (2020) Nanoscale in situ study of ZDDP tribofilm growth at aluminum-based interfaces using atomic force microscopy. Tribol Int 143:106075. https://doi.org/10.1016/j.triboint.2019.106075

Grzywacz H, Milczarek M, Jenczyk P, Dera W, Michałowski M, Jarząbek DM (2021) Quantitative measurement of nanofriction between PMMA thin films and various AFM probes. Meas J Int Meas Confed 168:108267. https://doi.org/10.1016/j.measurement.2020.108267

Gu C, Sui Z, Li Y, Chu H, Ding S, Zhao Y, Jiang C (2018) The growth of the metallic ZrN x thin films on P-GaN substrate by pulsed laser deposition. Appl Surf Sci 433:306–311. https://doi.org/10.1016/j.apsusc.2017.09.262

Harrison RW, Lee WE (2016) Processing and properties of ZrC, ZrN and ZrCN ceramics: a review. Adv Appl Ceram 115:294–307. https://doi.org/10.1179/1743676115Y.0000000061

Hu J, Xiao XD, Ogletree DF, Salmeron M (1995) Imaging the condensation and evaporation of molecularly thin films of water with nanometer resolution. Science 268:267–269. https://doi.org/10.1126/science.268.5208.267

Jacobs TDB, Greiner C, Wahl KJ, Carpick RW (2019) Insights into tribology from in situ nanoscale experiments. MRS Bull 44:478–486. https://doi.org/10.1557/mrs.2019.122

Jarman JD, Fahrenholtz WG, Hilmas GE, Watts JL, King DS (2021) Characterization of fusion welded ceramics in the SiC–ZrB2–ZrC system. J Eur Ceram Soc 41:2255–2262. https://doi.org/10.1016/j.jeurceramsoc.2020.10.067

Jarzabek DM (2015) Precise and direct method for the measurement of the torsion spring constant of the atomic force microscopy cantilevers. Rev Sci Instrum 86:013701. https://doi.org/10.1063/1.4904866

Jarzabek DM, Kaufmann AN, Schift H, Rymuza Z, Jung TA (2014) Elastic modulus and fracture strength evaluation on the nanoscale by scanning force microscope experiments. Nanotechnology 25. https://doi.org/10.1088/0957-4484/25/21/215701

Johnson LJS, Ghafoor N, Engberg D, Thuvander M, Stiller K, Odén M, Hultman L (2016) Self-organized nanostructuring in $Zr_{0.69}Al_{0.31}N$ thin films studied by atom probe tomography. Thin Solid Films 615:233–238. https://doi.org/10.1016/j.tsf.2016.07.034

Kaynak B, Alpan C, Kratzer M, Ganser C, Teichert C, Kern W (2017) Anti-adhesive layers on stainless steel using thermally stable dipodal perfluoroalkyl silanes. Appl Surf Sci 416:824–833. https://doi.org/10.1016/j.apsusc.2017.04.197

Khonsari MM, Ghatrehsamani S, Akbarzadeh S (2021) On the running-in nature of metallic tribo-components: a review. Wear 474–475:203871. https://doi.org/10.1016/j.wear.2021.203871

Klumdoung P, Buranawong A, Chaiyakun S, Limsuwan P (2012) Variation of color in Zirconium nitride thin films prepared at high Ar flow rates with reactive dc magnetron sputtering. Proc Eng 32:916–921. https://doi.org/10.1016/j.proeng.2012.02.032

Kragelsky IV, Dobychin MN, Kombalov VS (1982) The wear of solids. Frict Wear:237–296. https://doi.org/10.1016/b978-0-08-025461-6.50013-5

Krysina OV, Ivanov YF, Prokopenko NA, Shugurov VV, Petrikova EA, Denisova YA, Tolkachev OS (2020) Influence of Nb addition on the structure, composition and properties of single-layered ZrN-based coatings obtained by vacuum-arc deposition method. Surf Coatings Technol 387:125555. https://doi.org/10.1016/j.surfcoat.2020.125555

Kudish II, Volkov SS, Vasiliev AS, Aizikovich SM (2018) Some criteria for coating effectiveness in heavily loaded line elastohydrodynamically lubricated contacts-part I: dry contacts. J Tribol 23(7):1061–1080. https://doi.org/10.1177/1081286517704689

Kuznetsova TA, Andreev MA, Markova LV (2005) Research of wear resistance of the combined vacuum electroarc coatings on the basis of ZrHf. Trenie Iznos 26:521–529

Kuznetsova T, Lapitskaya V, Khabarava A, Chizhik S, Warcholinski B, Gilewicz A (2020a) The influence of nitrogen on the morphology of ZrN coatings deposited by magnetron sputtering. Appl Surf Sci 522:146508. https://doi.org/10.1016/j.apsusc.2020.146508

Kuznetsova T, Lapitskaya V, Warcholinski B, Gilewicz A, Chizhik S (2020b) Friction and wear of ZrN coatings under conditions of microcontact using atomic-force microscopy. J Frict Wear 41:287–294. https://doi.org/10.3103/S1068366620040078

Kuznetsova TA, Zubar TI, Lapitskaya VA, Sudzilouskaya KA, Chizhik SA, Didenko AL, Svetlichnyi VM, Vylegzhanina ME, Kudryavtsev VV, Sukhanova TE (2017) Tribological properties investigation of the thermoplastic elastomers surface with the AFM lateral forces mode. IOP Conf Ser Mater Sci Eng 256:012022.. https://doi.org/10.1088/1757-899X/256/1/012022

Lapitskaya VA, Kuznetsova TA, Chizhik SA, Sudzilouskaya KA, Kotov DA, Nikitiuk SA, Zaparozhchanka YV (2018) Lateral force microscopy as a method of surface control after low-temperature plasma treatment. IOP Conf Ser Mater Sci Eng 443:012019. https://doi.org/10.1088/1757-899X/443/1/012019

Larijani MM, Tabrizi N, Norouzian S, Jafari A, Lahouti S, Haj Hosseini H, Afshari N (2006) Structural and mechanical properties of ZrN films prepared by ion beam sputtering with varying $N_2$/Ar ratio and substrate temperature. Vacuum 81:550–555. https://doi.org/10.1016/j.vacuum.2006.07.018

Lin S, Zhang J, Zhu R, Fu S, Yun D (2018) Effects of sputtering pressure on microstructure and mechanical properties of ZrN films deposited by magnetron sputtering. Mater Res Bull 105:231–236. https://doi.org/10.1016/j.materresbull.2018.04.054

Liu Q, Liu Y, Li X, Dong G (2021) Pulse laser-induced cell-like texture on surface of titanium alloy for tribological properties improvement. Wear 477:203784. https://doi.org/10.1016/j.wear.2021.203784

Maksakova OV, Simoẽs S, Pogrebnjak AD, Bondar OV, Kravchenko YO, Koltunowicz TN, Shaimardanov ZK (2019) Multilayered ZrN/CrN coatings with enhanced thermal and mechanical properties. J Alloys Compd 776:679–690. https://doi.org/10.1016/j.jallcom.2018.10.342

Martin C, Miller KH, Makino H, Craciun D, Simeone D, Craciun V (2017) Optical properties of Ar ions irradiated nanocrystalline ZrC and ZrN thin films. J Nucl Mater 488:16–21. https://doi.org/10.1016/j.jnucmat.2017.02.041

Mejía V, Claudia P, Chellali MR, Garzón CM, Olaya JJ, Hahn H, Velasco L (2021) Effect of discharge current on the corrosion resistance and microstructure of ZrTiSiN coatings deposited by magnetron co-sputtering. Mater Today Commun 26:102151. https://doi.org/10.1016/j.mtcomm.2021.102151

Meng QN, Wen M, Qu CQ, Hu CQ, Zheng WT (2011) Preferred orientation, phase transition and hardness for sputtered zirconium nitride films grown at different substrate biases. Surf Coatings Technol 205:2865–2870. https://doi.org/10.1016/j.surfcoat.2010.10.060

Muboyadzhyan SA (2009) Erosion-resistant coatings for gas turbine compressor blades. Russ Metall 2009:183–196. https://doi.org/10.1134/S003602950903001X

Musil J, Zenkin S, Kos ČR, Haviar S (2016) Flexible hydrophobic ZrN nitride films. Vacuum 131:34–38.https://doi.org/10.1016/j.vacuum.2016.05.020
Nguyen TP, Pazhouhanfar Y, Delbari SA, Sabahi Namini A, Babapoor A, Mohammadpourderakhshi Y, Shaddel S, Le QV, Shokouhimehr M, Shahedi Asl M (2020) Physical, mechanical and microstructural characterization of TiC–ZrN ceramics. Ceram Int 46:22154–22163. https://doi.org/10.1016/j.ceramint.2020.05.292
Niyomsoan S, Grant W, Olson DL, Mishra B (2002) Variation of color in titanium and zirconium nitride decorative thin films. Thin Solid Films 415:187–194. https://doi.org/10.1016/S0040-6090(02)00530-8
Oganyan M, Vereschaka A, Volosova M, Gurin V (2021) Influence of the application of wear-resistant coatings on force parameters of the cutting process and the tool life during end milling of titanium alloys. Mater Today Proc 38(4):1428–1432. https://doi.org/10.1016/j.matpr.2020.08.119
Pang X, Yang X, Yang J, Zhao Y, Pang M (2021) Investigation on the interface characteristic between ZrC (111) and diamond (111) surfaces by first-principles calculation. Diam Relat Mater 113:108297. https://doi.org/10.1016/j.diamond.2021.108297
Park W, Müller S, Baumann RP, Becker S, Hwang B (2020) Surface energy characterization of nanoscale metal using quantitative nanomechanical characterization of atomic force microscopy. Appl Surf Sci 507:145041. https://doi.org/10.1016/j.apsusc.2019.145041
Patil R (2015) Zirconia versus titanium dental implants: a systematic review. J Dent Implant 5:39. https://doi.org/10.4103/0974-6781.154430
Puente Reyna AL, Fritz B, Schwiesau J, Schilling C, Summer B, Thomas P, Grupp TM (2018) Metal ion release barrier function and biotribological evaluation of a zirconium nitride multilayer coated knee implant under highly demanding activities wear simulation. J Biomech 79:88–96. https://doi.org/10.1016/j.jbiomech.2018.07.043
Qi ZB, Sun P, Zhu FP, Wang ZC, Peng DL, Wu CH (2011) The inverse Hall-Petch effect in nanocrystalline ZrN coatings. Surf Coatings Technol 205:3692–3697. https://doi.org/10.1016/j.surfcoat.2011.01.021
Ramoul C, Beliardouh NE, Bahi R, Nouveau C, Djahoudi A, Walock MJ (2019) Surface performances of PVD ZrN coatings in biological environments. Tribol Mater Surfaces Interf 13:12–19. https://doi.org/10.1080/17515831.2018.1553820
Ren J, Feng E, Zhang Y, Zhang J, Ding D, Li L (2021) Influences of deposition temperature, gas flow rate and ZrC content on the microstructure and anti-ablation performance of CVD-HfC-ZrC coating. Ceram Int 47:556–566. https://doi.org/10.1016/j.ceramint.2020.08.163
Rizzi M, Gatti G, Migliario M, Marchese L, Rocchetti V, Renò F (2014) Effect of zirconium nitride physical vapor deposition coating on preosteoblast cell adhesion and proliferation onto titanium screws. J Prosthet Dent 112:1103–1110. https://doi.org/10.1016/j.prosdent.2014.04.010
Rizzo A, Valerini D, Capodieci L, Mirenghi L, Di Benedetto F, Protopapa ML (2018) Reactive bipolar pulsed dual magnetron sputtering of ZrN films: The effect of duty cycle. Appl Surf Sci 427:994–1002. https://doi.org/10.1016/j.apsusc.2017.08.032
Ruden A, González JM, Restrepo JS, Cano MF, Sequeda F (2013) Tribology of ZrN, CrN and TiAlN thin films deposited by reactive magnetron sputtering. Dyna 80(178):95–100
Sadyrin EV, Vasiliev AS, Volkov SS, Mitrin BI, Aizikovich SM (2019) Simplified analytical solution of the contact problem on indentation of a coated half-space by a spherical punch. WIT Trans Eng Sci 122:209–221. https://doi.org/10.2495/BE410191
Sandu CS, Cusnir N, Oezer D, Sanjinés R, Patscheider J (2009) Influence of bias voltage on the microstructure and physical properties of magnetron sputtered Zr–Si–N nanocomposite thin films. Surf Coatings Technol 204:969–972. https://doi.org/10.1016/j.surfcoat.2009.06.042
Shen X, Gao N, Shi Z, Wang X, Zhang L, Huang J, Li K (2021) New insight into the ablation behavior of C/C-ZrC composites in a nitrogen plasma torch with a high heat flux of ∼25 MW/m2. Corros Sci 185:109409. https://doi.org/10.1016/j.corsci.2021.109409

Shirvani F, Shokri A, Ravan BA (2021) An ab-initio study of structure and mechanical properties of rocksalt ZrN and its bilayers. Solid State Commun 328:114218. https://doi.org/10.1016/j.ssc.2021.114218
Singh A, Kumar N, Kuppusami P, Prasanthi TN, Chandramohan P, Dash S, Srinivasan MP, Mohandas E, Tyagi AK (2012) Tribological properties of sputter deposited ZrN coatings on titanium modified austenitic stainless steel. Wear 280–281:22–27. https://doi.org/10.1016/j.wear.2012.01.013
Singh A, Kuppusami P, Khan S, Sudha C, Thirumurugesan R, Ramaseshan R, Divakar R, Mohandas E, Dash S (2013) Influence of nitrogen flow rate on microstructural and nanomechanical properties of Zr-N thin films prepared by pulsed DC magnetron sputtering. Appl Surf Sci 280:117–123. https://doi.org/10.1016/j.apsusc.2013.04.107
Advanced mechanical engineering solutions/hertzian contact stress calculator. Calculator. http://www.amesweb.info/HertzianContact/HertzianContact.aspx
Silva E, Rebelo de Figueiredo M, Franz R, Escobar Galindo R, Palacio C, Espinosa A, Calderon VS, Mitterer C, Carvalho S (2010) Structure-property relations in ZrCN coatings for tribological applications. Surf Coatings Technol 205(7):2134–2141.https://doi.org/10.1016/j.surfcoat.2010.08.126
Thomas RC, Houston JE, Michalske TA, Crooks RM (1993) The mechanical response of gold substrates passivated by self-assembling monolayer films. Science 259:1883–1885. https://doi.org/10.1126/science.259.5103.1883
Ul-Hamid A (2020a) Microstructure, properties and applications of Zr-carbide, Zr-nitride and Zr-carbonitride coatings: a review. Mater Adv 1:1012–1037. https://doi.org/10.1039/d0ma00233j
Ul-Hamid A (2020b) Synthesis, microstructural characterization and nanoindentation of Zr, Zr-nitride and Zr-carbonitride coatings deposited using magnetron sputtering. J Adv Res 29:107–119. https://doi.org/10.1016/j.jare.2020.11.010
Usmani B, Vijay V, Chhibber R, Dixit A (2017) Optimization of sputtered zirconium thin films as an infrared reflector for use in spectrally-selective solar absorbers. Thin Solid Films 627:17–25. https://doi.org/10.1016/j.tsf.2017.02.055
Valaiauksornlikit M, Rakreungdet W, Chananonnawathorn C, Limwichean S, Eiamchai P, Lertvanithphol T, Nuntawong N, Patthanasettakul V, Horprathum M (2020) Sputtering of crystalline ZrN thin films on unheated substrates for decorative coating applications. Thai J Nanosci Nanotechnol 5(1):1–9
Vasiliev AS, Volkov SS, Aizikovich SM, Mitrin BI (2017) Plane contact problem on indentation of a flat punch into a transversely-isotropic half-plane with functionally graded transversely-isotropic coating. Z Angew Math Phys 68:4. https://doi.org/10.1007/s00033-016-0746-8
Vasiliev AS, Sadyrin EV, Mitrin BI, Aizikovich SM, Nikolaev AL (2018) Nanoindentation of ZrN coatings on silicon and copper substrates. Russ Eng Res 38:735–737. https://doi.org/10.3103/S1068798X18090289
Wang QJ, Chung Y (2013) Encyclopedia of tribology. Springer US.https://doi.org/10.1007/978-0-387-92897-5.
Warcholinski B, Gilewicz A, Kuprin AS, Tolmachova GN, Ovcharenko VD, Kuznetsova TA, Zubar TI, Khudoley AL, Chizhik SA (2018) Mechanical properties of Cr–O–N coatings deposited by cathodic arc evaporation. Vacuum 156:97–107. https://doi.org/10.1016/j.vacuum.2018.07.017
Wei S, Hua HZ, Xiang X (2013) Thermodynamic analysis and growth of zirconium carbide by chemical vapor deposition. Phys Proc 46:88–101. https://doi.org/10.1016/j.phpro.2013.07.049
Wen S, Huang P (2017) Tribology of micro electromechanical system. In: Wen S, Huang P (eds) Principles of tribology. https://doi.org/10.1002/9781119214908.ch20
Wu ZT, Qi ZB, Zhang DF, Wang ZC (2016) Nanoindentation induced plastic deformation in nanocrystalline ZrN coating. Mater Lett 164:120–123. https://doi.org/10.1016/j.matlet.2015.10.091
Yan Y, Geng Y, Hu Z (2015) Recent advances in AFM tip-based nanomechanical machining. Int J Mach Tools Manuf 99:1–18. https://doi.org/10.1016/j.ijmachtools.2015.09.004

Yang Y, Tan P, Sui Y, Jiang Y, Zhou R (2021) Influence of Zr content on microstructure and mechanical properties of As-cast Al–Zn–Mg–Cu alloy. J Alloys Compd 867:158920. https://doi.org/10.1016/j.jallcom.2021.158920

Yin L, Gao W, Jones MI (2019) Wear behaviour and electrical conductivity of β-Sialon-ZrN composites fabricated by reaction bonding and gas pressure sintering process. Ceram Int 45:2266–2274. https://doi.org/10.1016/j.ceramint.2018.10.140

Zelentsov VB, Mitrin BI, Aizikovich SM (2016) Dynamic and quasi-static instability of sliding thermoelastic frictional contact. J Frict Wear 37:213–220. https://doi.org/10.3103/S1068366616030181

Zhang ZG, Rapaud O, Allain N, Mercs D, Baraket M, Dong C, Coddet C (2009) Microstructures and tribological properties of CrN/ZrN nanoscale multilayer coatings. Appl Surf Sci 255:4020–4026. https://doi.org/10.1016/j.apsusc.2008.10.075

Zhang Y, Xu J, Cheng L, Zhao Y, Peng S, Jiang S (2020) Exploring cavitation erosion resistance of ZrN nanocrystalline coating prepared by double-cathode glow discharge plasma technique. Vacuum 182:109697. https://doi.org/10.1016/j.vacuum.2020.109697

Zou J, Liu B, Lin L, Lu Y, Dong Y, Jiao G, Ma F, Li Q (2018) Investigation of microstructure and properties of ultrathin graded ZrN x self-assembled diffusion barrier in deep nano-vias prepared by plasma ion immersion implantation. Appl Surf Sci 427:950–955. https://doi.org/10.1016/j.apsusc.2017.08.114

# Chapter 17
# Flexomagneticity in Functionally Graded Nanostructures

**Mohammad Malikan, Tomasz Wiczenbach, and Victor A. Eremeyev**

**Abstract** Functionally graded structures have shown the perspective of materials in a higher efficient and consistent manner. This study reports a short investigation by concentrating on the flexomagnetic response of a functionally graded piezomagnetic nano-actuator, keeping in mind that the converse magnetic effect is only taken into evaluation. The rule of mixture assuming exponential composition of properties along with the thickness is developed for the ferromagnetic bulk. Nonlocal effects are assigned to the model, respecting Eringen's hypothesis. The derived equations deserve to be analytically solved. Therefore, numerical results are generated for fully fixed ends. It is denoted that the functionality grading feature of a magnetic nanobeam can magnify the flexomagnetic effect leading to high-performance nanosensors/actuators.

**Keywords** Flexomagneticity · Nonlocal elasticity · FGM · Nanostructures

## Nomenclature

| | |
|---|---|
| $\sigma_{xx}$ | Stress component |
| $\tau_{xz}$ | Shear stress |
| $\xi_{xxz}$ | Hyperstress |

M. Malikan · T. Wiczenbach · V. A. Eremeyev (✉)
Department of Mechanics of Materials and Structures, Faculty of Civil and Environmental Engineering, Gdansk University of Technology, 80-233, Gdansk, Poland
e-mail: victor.eremeev@pg.edu.pl

M. Malikan
e-mail: mohammad.malikan@pg.edu.pl

T. Wiczenbach
e-mail: tomasz.wiczenbach@pg.edu.pl

V. A. Eremeyev
Don State Technical University, Gagarina sq., 1, Rostov on Don 344000, Russia

DICAAR, Università degli Studi di Cagliari, Via Marengo, 2, 09123 Cagliari, Italy



H. Altenbach et al. (eds.), *Advanced Materials Modelling for Mechanical, Medical and Biological Applications*, Advanced Structured Materials 155,
https://doi.org/10.1007/978-3-030-81705-3_17

| | |
|---|---|
| $\eta_{xxz}$ | Hyperstrain |
| $\varepsilon_{xx}$ | Strain component |
| $\gamma_{xz}$ | Shear strain |
| $E$ | Elasticity modulus |
| $G$ | Shear modulus |
| $u_1$ | Displacement along $x$ |
| $u_3$ | Displacement along $z$ |
| $\nu$ | Poisson's ratio |
| $L$ | Length of the beam |
| $b$ | Width of the beam |
| $z$ | Thickness coordinate |
| $h$ | Thickness of the beam |
| $k_s$ | Shear correction factor |
| $k$ | Material property variation |
| $I_z$ | Area moment of inertia |
| $u$ | Axial displacement of the mid-plane |
| $w$ | Transverse displacement of the mid-plane |
| $\phi$ | Rotation of beam nodes around the $y$-axis |
| $q_{31}$ | Component of the third-order piezomagnetic tensor |
| $g_{31}$ | Component of the sixth-order gradient elasticity tensor |
| $f_{31}$ | Component of fourth-order flexomagnetic tensor |
| $a_{33}$ | Component of the second-order magnetic permeability tensor |
| $A$ | Area of the cross section of the beam |
| $N_x$ | Axial stress resultant |
| $M_x$ | Moment stress resultant |
| $Q_x$ | Shear stress resultant |
| $T_{xxz}$ | Hyperstress resultant |
| $\psi$ | Magnetic potential. |

## 17.1 Introduction

The development of technology leads to the discovery of new, more complex properties of materials. One of the recently discovered ones is flexomagnetism (FM). This phenomenon is currently being investigated by researchers under the influence of static and dynamic states. This effect is based on a strain gradient, which can be called the inverse flexomagnetic effect. The flexo-effect can be observed in direct impact by the presence of an exterior magnetic field gradient. FM effect can occur in crystalline structures and all types of materials, see Fahrner (2005), Lukashev and Sabirianov (2010), Pereira et al. (2012), Zhang et al. (2012), Zhou et al. (2014), Moosavi et al. (2017), Eliseev et al. (2019), Kabychenkov and Lisovskii (2019), Eliseev et al. (2009).

A special case of the composite material is functionally graded materials (FGMs). A characteristic feature of FG materials is a continuous and smooth transition between the various components, see Mostefa et al. (2018), Loh et al. (2018), Mahamood et al. (2012), Vasiliev et al. (2017), Malikan and Eremeyev (2020a), Liu et al. (2021), Volkov et al. (2019). For instance, between a ceramic material with a very high thermal resistance (low thermal conductivity and expansion coefficient) and a metal with good heat conduction (high thermal conductivity and expansion coefficient), there is a transition phase (interface), in which a smooth, continuous transition, eliminating the formation of microdamage or delamination of the material at the interface is obtained. The smooth transition is achieved by changing the volumetric fraction of the individual components of the composite, causing the effective thermomechanical properties to change from one piece (e.g., metal or metal alloy) to the other part (e.g., ceramic material). In the case of thermal barrier coating (TBC), the thickness of the FGM interface is small compared to the entire volume of the material.

Afterward, flexomagneticity effect being discovered, several research has been conducted on small-scale sensors and actuators and their static or dynamic response subjected to the effect. Within these publications, Zhang et al. (2019) concentrated on a nano-actuator beam exposed to bending and subjected to the flexomagneticity effect. The static bending equations have been developed following the Euler–Bernoulli beam theory. What is more, the surface elasticity was considered. Diverse boundary conditions were chosen following direct and converse magnetization. According to the obtained results, it could be obtained that the material property, which is the flexomagneticity effect, is size-dependent. On the other hand, Sidhardh and Ray (2018) investigated the Euler–Bernoulli nanosize beam subjected to a bending load containing piezo-flexomagnetic effect with clamped-free boundary conditions. Direct and inverse magnetization effects were considered. With the elasticity surface, the examination of the size dependency was performed for the small beam. The following results present scale-dependent flexomagneticity behavior. Studies of nanostructures have shown that this effect is very significant for the obtained results, despite the omission of the piezomagneticity.

A comprehensive examination of functionally graded materials with electro-magneto-elastic coupling may be found considering a detailed literature review. Nonetheless, the flexomagnetic phenomena in FGM materials cannot be found in the literature. According to this, the need to investigate the flexomagnetic effect in FGM materials is crucial. Moreover, the impact of flexomagneticity on micro-and nanostructures can be found in recent publications, see Zhang et al. (2019), Sidhardh and Ray (2018), Malikan and Eremeyev (2020b, c), Malikan et al. (2020a, b), Malikan et al. (2021), Malikan and Eremeyev (2021). Although research on flexomagnetism has been conducted in recent years, this phenomenon still requires much additional research and raises many questions.

Best of our knowledge, no studies are performed about examining functionally graded materials (FGMs) composed as a nano-actuator beam with the flexomagnetic effect. In this study, we intend to investigate the nanostructured FGM beams containing the flexomagnetic effect. We Perform a computation to evaluate the small scale effect respecting the nonlocal model. The computed numerical results refer to the semi-analytical method. According to the variations in crucial and significant criteria, the illustrating graphs present a magneto-mechanical model.

## 17.2 Mathematical Modeling

Let us attach the square nanobeam containing $L$ and $h$ as length and thickness dimensions to a rectangular coordinate system as manifested by Fig. 17.1. The ferromagnetic functionally graded structure with apparent piezomagnetic effect and obscure flexomagnetic influence should concatenate shear deformations to the modeling. Due to this, the Timoshenko beam is preferred in the theoretical model as in Thanh et al. (2021)

$$u_1(x, z) = u(x) + (z - z_0)\phi(x) \tag{1a}$$

$$u_3(x, z) = w(x) \tag{1b}$$

In a FGM structure with the physical neutral plane, the neutral surface is not matched to the mid-plane and deviates a bit; however, it could be valid for the case of a geometrical model of FGMs. Physically, the neutral plane does not include any strain and stress during a pure bending deformation. One can describe the position of the neutral surface by, see Thanh et al. (2021), Chu et al. (2018), Ahmed Hassan et al. (2020),

$$z_0 = \frac{\int_{-h/2}^{h/2} zE(z)\mathrm{d}z}{\int_{-h/2}^{h/2} E(z)\mathrm{d}z}, \tag{17.2}$$

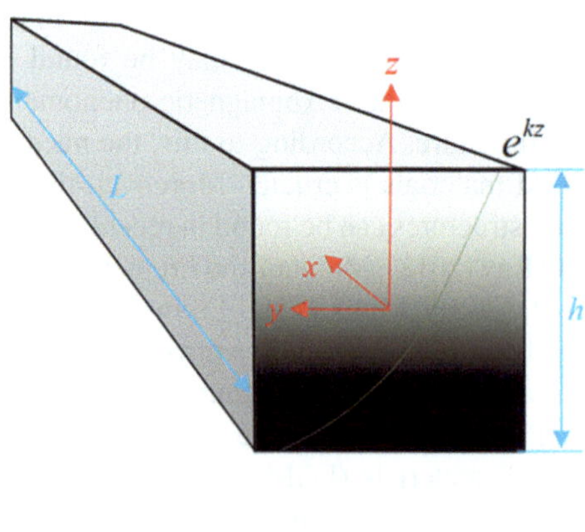


**Fig. 17.1** Hypothetical perspective image of a smart FGM nanobeam in the Cartesian coordinate

where the shifting between the geometric mid-plane and the physical neutral surface is denoted by $z_0$.

The use of Voigt's theorem can define functionally graded properties, also called the rule of mixture, a micromechanics model specifying functionality at any point that generally falls within three functions, namely sigmoid exponential and power ones. These functions can model the functionality grading property through the thickness or another dimension for a beam/plate. In between those functions, the exponential one is applied in this study as in Atmane et al. (2011),

$$P(z) = P_0 e^{kz} \tag{17.3}$$

where the property $P(z)$ can be any effective property (elastic properties etc.,) at any point of $z$, $P_0$ depicts that property at mid-plane ($z = 0$), $k$ shows the index of material property variation in line with the thickness direction. The homogeneous isotropic nanobeam can be described by $k = 0$. In this study, in light of the narrow range of Poisson's ratio for various materials, it is assumed that the value of Poisson's ratio is constant for the FGM.

In the skeleton of Lagrangian strain and based on constituting linearization relevance to stability phenomenon, one writes

$$\varepsilon_{xx} = \frac{\mathrm{d}u}{\mathrm{d}x} + (z - z_0)\frac{\mathrm{d}\phi}{\mathrm{d}x}, \tag{17.4}$$

$$\gamma_{xz} = \phi + \frac{\mathrm{d}w}{\mathrm{d}x}, \tag{17.5}$$

$$\eta_{xxz} = \frac{\mathrm{d}\varepsilon_{xx}}{\mathrm{d}z} = \frac{\mathrm{d}\phi}{\mathrm{d}x}. \tag{17.6}$$

Lagrange's principle expresses energy formula on the time-independent theme as

$$\delta \int (W + U) = 0, \tag{17.7}$$

in which $W$ and $U$, respectively, state thermodynamic work accomplished by external forces and the internal strain energy. The expanded and variated form of strain energy for a beam is written by

$$\delta U = \int_V \left(\sigma_{xx}\delta\varepsilon_{xx} + \tau_{xz}\delta\gamma_{xz} + \xi_{xxz}\delta\eta_{xxz} - B_z\delta H_z\right)\mathrm{d}V, \tag{17.8}$$

subsequent to Eq. (17.8), there will be risen a few one- or two-dimensional relations which can be arranged as follows (one-dimensional relates to non-classical boundary conditions, and two-dimensional correlates to the equilibrium equations),

$$\delta U^{Mech} = -\int_0^L \left( \frac{\mathrm{d}N_x}{\mathrm{d}x}\delta u + \frac{\mathrm{d}Q_x}{\mathrm{d}x}\delta w - Q_x\delta\phi + \frac{\mathrm{d}M_x}{\mathrm{d}x}\delta\phi + \frac{\mathrm{d}T_{xxz}}{\mathrm{d}x}\delta\phi \right)\mathrm{d}x, \quad (9a)$$

$$\delta U^{\mathrm{Mag}} = -\int_0^L \int_{-h/2}^{h/2} \frac{\mathrm{d}B_z}{\mathrm{d}z}\delta\Psi \mathrm{d}z\mathrm{d}x, \quad (9b)$$

$$\delta U^{\mathrm{Mech}} = \left( N_x\delta u + Q_x\delta w + M_x\delta\phi + T_{xxz}\delta\phi \right)\Big|_0^L, \quad (10a)$$

$$\delta U^{\mathrm{Mag}} = \int_0^L (B_z\delta\Psi)\big|_{-h/2}^{h/2}\mathrm{d}x, \quad (10b)$$

in which

$$N_x = \int_{-h/2}^{h/2} \sigma_{xx}\mathrm{d}z, \quad (17.11)$$

$$M_x = \int_{-h/2}^{h/2} \sigma_{xx}z\mathrm{d}z, \quad (17.12)$$

$$Q_x = k_s \times \int_{-h/2}^{h/2} \tau_{xz}\mathrm{d}z, \quad (17.13)$$

$$T_{xxz} = \int_{-h/2}^{h/2} \xi_{xxz}\mathrm{d}z \quad (17.14)$$

Axial compressive load acting on the ends of the nanobeam works on the domain as in Malikan et al. (2020c); Malikan and Eremeyev (2020d).

$$W = \frac{1}{2}\int_0^L N_x^0 \left( \frac{\mathrm{d}w}{\mathrm{d}x} \right)^2 \mathrm{d}x. \quad (17.15)$$

After that, Eq. (17.15) can be developed by variational technique as

$$\delta W = \int_0^L \left( N_x^0 \frac{\mathrm{d}\delta w}{\mathrm{d}x}\frac{\mathrm{d}w}{\mathrm{d}x} \right)\mathrm{d}x, \quad (17.16)$$

in which the $N_x^0$ plays the role of buckling force.

It is considered that the magnetic field acts laterally for which one can formulate the following component along the transverse axis,

$$H_z + \frac{\mathrm{d}\Psi}{\mathrm{d}z} = 0. \tag{17.17}$$

It is deemed that the smart beam is exposed to the reverse flexomagnetic effect. This is mathematically possible concerning the pursuant relation by which one defines a closed-circuit magnetic potential with the utmost potential on the top surface and the least one on the undermost surface of the thickness,

$$\Psi\left(+\frac{h}{2}\right) = \psi,\ \Psi\left(-\frac{h}{2}\right) = 0. \tag{17.18a,b}$$

Let us mingle Eqs. (17.6, 17.9b, 17.10b, 17.17, and 17.18) to each other alongside with a few mathematical attempts, then one can obtain the relations of magnetic potential in the thickness direction and the vertical magnetic field,

$$\Psi = \frac{q_{31}(z)}{2a_{33}(z)}\left((z - z_0)^2 - \frac{h^2}{4}\right)\frac{\mathrm{d}\phi}{\mathrm{d}x} + \frac{\psi}{h}\left((z - z_0) + \frac{h}{2}\right), \tag{17.19}$$

$$H_z = -(z - z_0)\frac{q_{31}(z)}{a_{33}(z)}\frac{\mathrm{d}\phi}{\mathrm{d}x} - \frac{\psi}{h}, \tag{17.20}$$

There are various ways to study nanostructures. Experimental observations are one of the methods of modeling small-scale materials, which due to its high cost, other methods such as atomic modeling and mechanics of continuous environments are proposed. Modeling by continuous media is much less expensive than the previous two methods and includes Eringen's theory of nonlocal elasticity as Dastjerdi et al. (2021), nonlocal strain gradient elasticity theory as Malikan et al. (2020d), and modified coupling stress theory as Malikan (2017), Skrzat and Eremeyev (2020), etc., as Romano and Barretta (2017) for nanostructures. Therefore, continuous environment modeling can be used to analyze nanostructures such as buckling of these scaled-down structures. Meanwhile, Eringen's theory of nonlocal elasticity has relatively more straightforward governing relations and calculates the nonlocal effect of small scale to nanoscale structures which is given by the expression as in Eringen (1983),

$$\left(1 - \mu\frac{\mathrm{d}^2}{\mathrm{d}x^2}\right)\sigma_{ij} = C_{ijkl}\varepsilon_{ij}, \tag{17.21}$$

where an additional length scale parameter is exhibited that is called nonlocal parameter, and its value has been obtained for some nanostructures in the range of zero up to four square nanometers as in Ansari et al. (2010). The parameter $\mu$ equals $(e_0 a)^2$ including $e_0$ as a nonlocal quantity and $a$ that is a characteristic length of materials.

Afterward, Eq. (17.21) is imposed on Eqs. (17.11–17.14) leading to the following relations,

$$\left(1-\mu\frac{\mathrm{d}^2}{\mathrm{d}x^2}\right)\xi_{xxz}=\left(g_{31}(z)+\frac{q_{31}(z)f_{31}(z)(z-z_0)}{a_{33}(z)}\right)\frac{\mathrm{d}\phi}{\mathrm{d}x}+\frac{f_{31}(z)\psi}{h},\quad(17.22)$$

$$\left(1-\mu\frac{\mathrm{d}^2}{\mathrm{d}x^2}\right)\sigma_{xx}=E(z)\frac{\mathrm{d}u}{\mathrm{d}x}+(z-z_0)\left(E(z)+\frac{q_{31}^2(z)}{a_{33}(z)}\right)\frac{\mathrm{d}\phi}{\mathrm{d}x}+\frac{q_{31}(z)\psi}{h},\quad(17.23)$$

$$\left(1-\mu\frac{\mathrm{d}^2}{\mathrm{d}x^2}\right)\tau_{xz}=G(z)A\left(\phi+\frac{\mathrm{d}w}{\mathrm{d}x}\right).\quad(17.24)$$

Substituting Eq. (17.21) in Eqs. (17.11–17.14) eventuates the nonlocal stress resultants as

$$\left(1-\mu\frac{\mathrm{d}^2}{\mathrm{d}x^2}\right)N_x=I_1\frac{\mathrm{d}u}{\mathrm{d}x}+\left(I_2+I_3\right)\frac{\mathrm{d}\phi}{\mathrm{d}x}+I_4,\quad(17.25)$$

$$\left(1-\mu\frac{\mathrm{d}^2}{\mathrm{d}x^2}\right)M_x=I_5\frac{\mathrm{d}u}{\mathrm{d}x}+\left(I_6+I_7\right)\frac{\mathrm{d}\phi}{\mathrm{d}x}+I_8,\quad(17.26)$$

$$\left(1-\mu\frac{\mathrm{d}^2}{\mathrm{d}x^2}\right)Q_x=H_{44}\left(\phi+\frac{\mathrm{d}w}{\mathrm{d}x}\right),\quad(17.27)$$

$$\left(1-\mu\frac{\mathrm{d}^2}{\mathrm{d}x^2}\right)T_{xxz}=\left(I_9+I_{10}\right)\frac{\mathrm{d}\phi}{\mathrm{d}x}+I_{11},\quad(17.28)$$

where the determined parameters can be organized as

$$\left\{I_1,I_2\right\}=\int_{-h/2}^{h/2}E(z)\{1,(z-z_0)\}\mathrm{d}z,$$

$$I_3=\int_{-h/2}^{h/2}(z-z_0)\frac{q_{31}^2(z)}{a_{33}(z)}\mathrm{d}z,\ I_4=\int_{-h/2}^{h/2}\frac{\psi q_{31}(z)}{h}\mathrm{d}z,$$

$$\left\{I_5,I_6\right\}=\int_{-h/2}^{h/2}E(z)\left\{(z-z_0),(z-z_0)^2\right\}\mathrm{d}z,$$

$$I_7=\int_{-h/2}^{h/2}(z-z_0)^2\frac{q_{31}^2(z)}{a_{33}(z)}\mathrm{d}z,\ I_8=\int_{-h/2}^{h/2}(z-z_0)\frac{\psi q_{31}(z)}{h}\mathrm{d}z,$$

$$I_9 = \int\limits_{-h/2}^{h/2} g_{31}(z)\mathrm{d}z,\ I_{10} = \int\limits_{-h/2}^{h/2} (z - z_0)\frac{q_{31}(z) f_{31}(z)}{a_{33}(z)}\mathrm{d}z,$$

$$I_{11} = \int\limits_{-h/2}^{h/2} \frac{\psi f_{31}(z)}{h}\mathrm{d}z,\ H_{44} = k_s \int\limits_{-h/2}^{h/2} G(z)A\mathrm{d}z \tag{17.29}$$

This is the time to pull the governing equations out of the Eqs. (17.9a, b, 17.10a, b, and 17.16) and sort them as

$$\frac{\mathrm{d}N_x}{\mathrm{d}x} = 0, \tag{17.30}$$

$$\frac{\mathrm{d}Q_x}{\mathrm{d}x} + N_x^0\frac{\mathrm{d}^2 w}{\mathrm{d}x^2} = 0, \tag{17.31}$$

$$\frac{\mathrm{d}M_x}{\mathrm{d}x} + \frac{\mathrm{d}T_{xxz}}{\mathrm{d}x} - Q_x = 0. \tag{17.32}$$

Thereupon, Eqs. (17.25–17.28) with the aid of Eqs. (17.30–17.32) can be uncomplicated as

$$N_x = I_1\frac{\mathrm{d}u}{\mathrm{d}x} + (I_2 + I_3)\frac{\mathrm{d}\phi}{\mathrm{d}x} + I_4, \tag{17.33}$$

$$M_x = -\mu\left[(I_9 + I_{10})\frac{\mathrm{d}^3\phi}{\mathrm{d}x^3} + N_x^0\frac{\mathrm{d}^2 w}{\mathrm{d}x^2}\right] + I_5\frac{\mathrm{d}u}{\mathrm{d}x} + (I_6 + I_7)\frac{\mathrm{d}\phi}{\mathrm{d}x} + I_8, \tag{17.34}$$

$$Q_x = -\mu N_x^0\frac{\mathrm{d}^3 w}{\mathrm{d}x^3} + H_{44}\left(\phi + \frac{\mathrm{d}w}{\mathrm{d}x}\right), \tag{17.35}$$

$$T_{xxz} = \mu\frac{\mathrm{d}^2 T_{xxz}}{\mathrm{d}x^2} + (I_9 + I_{10})\frac{\mathrm{d}\phi}{\mathrm{d}x} + I_{11}. \tag{17.36}$$

Let us recast Eqs. (17.30–17.32) based on Eqs. (17.30–17.36) as

$$I_1\frac{\mathrm{d}^2 u}{\mathrm{d}x^2} + (I_2 + I_3)\frac{\mathrm{d}^2\phi}{\mathrm{d}x^2} = 0, \tag{17.37}$$

$$\left(1 - \mu\frac{\mathrm{d}^2}{\mathrm{d}x^2}\right)\left\{N_x^0\frac{\mathrm{d}^2 w}{\mathrm{d}x^2}\right\} + H_{44}\left(\frac{\mathrm{d}\phi}{\mathrm{d}x} + \frac{\mathrm{d}^2 w}{\mathrm{d}x^2}\right) = 0, \tag{17.38}$$

$$\left(1-\mu\frac{d^2}{dx^2}\right)\left\{\left(I_9+I_{10}\right)\frac{d^2\phi}{dx^2}\right\}+I_5\frac{d^2u}{dx^2}$$
$$+\left(I_6+I_7\right)\frac{d^2\phi}{dx^2}-H_{44}\left(\phi+\frac{dw}{dx}\right)=0. \quad (17.39)$$

To compute the buckling load, Eqs. (17.37–17.39) which are coupled together shall be solved. The combination of longitudinal magnetic force together with the axial mechanical compressive load results in the total axial load as

$$N_x^0=P_{cr}+q_{31}\psi. \quad (17.40)$$

## 17.3 Solving Procedure

Mathematically, we can implement all boundary conditions for a FGM system. However, inspecting the physics of a FGM involving the transition in the neutral plane, Karamanli and Aydogdu (2020) approved that while considering a FGM model consisting of $z_0$, there is no buckling mode for simple boundary conditions since the beam/plate will bend before bifurcation buckling. It means the buckling will not occur at this boundary condition for the physical model of FGMs. Thus, in what follows, we urge to present clamped–clamped boundary conditions resulted from an analytical process as

$$u(x)=\sum_{m=1}^{\infty}\frac{d}{dx}\left[\sin^2\left(\frac{m\pi}{L}x\right)\right]\exp(i\omega_n t), \quad (17.41)$$

$$w(x)=\sum_{m=1}^{\infty}\sin^2\left(\frac{m\pi}{L}x\right)\exp(i\omega_n t), \quad (17.42)$$

$$\phi(x)=\sum_{m=1}^{\infty}\cos^2\left(\frac{m\pi}{L}x\right)\exp(i\omega_n t), \quad (17.43)$$

where $t$ is the identification of time in time-dependent problems.

The foregoing series will satisfy the clamped condition ($u=w=0$, $M_x\neq 0$) at both ends for present numerical calculations for axial buckling loads of a square ferromagnetic nanobeam made of a functionally graded compound continued by obtaining residual of Eqs. (17.30–17.32),

$$\int_0^L R_1(x)u(x)dx=0, \quad (17.44)$$

$$\int_0^L R_2(x)w(x)\mathrm{d}x = 0, \tag{17.45}$$

$$\int_0^L R_3(x)\phi(x)\mathrm{d}x = 0, \tag{17.46}$$

where $R_i(x)(i = 1, .., 3)$ are exhibitions of equations' residuals. The next equation computes the critical buckling load.

$$\begin{bmatrix} K_{11} & K_{12} & K_{13} \\ K_{21} & K_{22} & K_{23} \\ K_{31} & K_{32} & K_{33} \end{bmatrix} \begin{Bmatrix} u \\ w \\ \phi \end{Bmatrix} = 0. \tag{17.47}$$

Then,

$$\det\left[K_{ij}\right] = 0. \tag{17.48}$$

Consequently, Eq. (17.48) confronts us with a polynomial relation to which the solution will keep going $P_{\mathrm{cr}}$.

## 17.4 Results and Discussions

Let us here illustrate the influence of functionally graded fabrication on the flexomagnetic effect detailedly upon the conduction of a parametric study. The postulation of size dependency for functionality composition is also demonstrated. First of all, Table 17.1 incorporates the required material properties, which will be employed in the analysis. Second, by presenting a scientific interpretation, one observes the importance of FGMs in embossing the flexomagnetic response. Pursuing this, there are two kinds of structures, piezomagnetic FGM shown by FG-PM and piezomagnetic FGM in conjunction with flexomagneticity displayed by FG-PFM. The mechanical and structural properties are used in the sense of Lu et al. (2016), Balsing Rajput et al. (2013), Senthil et al. (2018)

**Table 17.1** Employed structural properties

| $CoFe_2O_4$ |
|---|
| $E_0 = 286$ GPa |
| $f_{31} = 10^{-9}$ N/A |
| $q_{31} = 580.3$ N/Am |
| $a_{33} = 1.57 \times 10^{-4}$ N/A$^2$ |

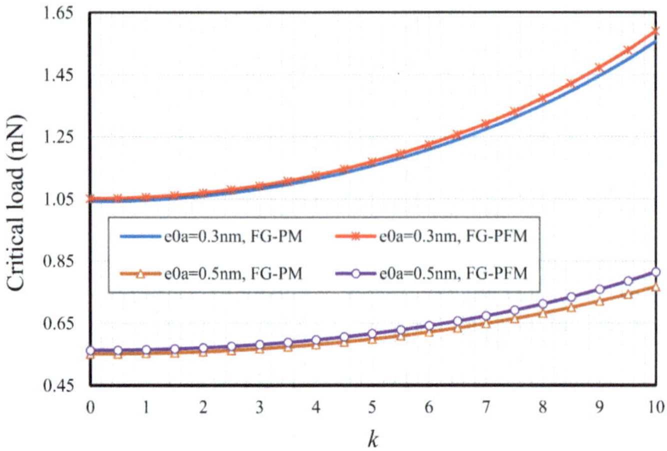


**Fig. 17.2** FGM variation index vs. critical buckling load for various cases ($\Psi = 1$ mA, $m = 1$, $L/h = 10$)

The purpose of this short study is to investigate whether functional grading can lead to changes in the flexomagnetic response of materials or not. In fact, it will be more difficult to construct matter in a functional state. However, if it gives us the privilege of having a more significant flexomagnetic effect and greater polarization overall, it will have unique advantages that can persuade designers of smart magnet nanosensors and actuators to turn to functionally graded magnetic nanomaterials. In this section, with the help of Fig. 17.2, this vital issue related to the property of flexomagnetism will be investigated. What can be seen is that the functional grading property has a direct effect on the polarization and will lead to an increase in the flexomagnetic property. As shown, with increasing the value of the *k-index*, one sees an increase in the distance between the results of PFM and PM, especially for larger values of the nonlocal parameter. Increasing the value of the nonlocal parameter increases this difference to prove that the functionally graded property, in turn, can exhibit size-dependent behavior. Therefore, the fabrication of ferromagnetic nanomaterials as a functional scale, regardless of fabrication complexity, will provide more efficient nanosensors and magnetic nano-actuators with greater efficiency and a more substantial flexomagnetic effect.

## 17.5 Conclusions

Galerkin weighted residual analytical approach has been implemented to compute critical buckling load for a ferroic functionally graded shear deformable structure inclusive of flexomagneticity. Besides, the composition of the material has been presumed as the rule of mixture corresponding to the exponential distribution. The physical neutral surface has been taken into the investigation. The size-dependent behavior has been replaced in the derived linear stability equations using Eringen's nonlocal elasticity differential model to provide nanosize effects. Further, in the framework of performing a short analysis, some new achievements have been established. It was indicated that producing piezomagnetic functionally graded nanostructure creates a more noticeable flexomagnetic effect. Notwithstanding this, the current research is short, and the results will open novel potential studies on ferric functionally graded structures and their application in small-scale sensors and actuators.

**Acknowledgements** V. A. Eremeyev acknowledges the support of the Government of the Russian Federation (contract No. 14.Z50.31.0046).

## References

Ahmed Hassan AH, Kurgan N (2020) Bending analysis of thin FGM skew plate resting on Winkler elastic foundation using multi-term extended Kantorovich method. Eng Sci Technol Int J 23:788–800

Ait Atmane H, Tounsi A, Ahmed Meftah S, Abdesselem Belhadj H (2011) Free vibration behavior of exponential functionally graded beams with varying cross-section. J Vib Control 17:311

Ansari R, Sahmani S, Arash B (2010) Nonlocal plate model for free vibrations of single-layered graphene sheets. Phys Lett A 375(1):53–62

Balsing Rajput A, Hazra S, Nath Ghosh N (2013) Synthesis and characterisation of pure single-phase CoFe2O4 nanopowder via a simple aqueous solution-based EDTA-precursor route. J Exp Nanosci 8:629–639

Chu L, Dui G, Ju Ch (2018) Flexoelectric effect on the bending and vibration responses of functionally graded piezoelectric nanobeams based on general modified strain gradient theory. Compos Struct 186:39–49

Dastjerdi S, Malikan M, Dimitri R, Tornabene F (2021) Nonlocal elasticity analysis of moderately thick porous functionally graded plates in a hygro-thermal environment. Composite Struct 255:112925

Eliseev EA, Morozovska AN, Glinchuk MD, Blinc R (2009) Spontaneous flexoelectric/flexomagnetic effect in nanoferroics. Phys Rev B 79:165433

Eliseev EA, Morozovska AN, Khist VV, Polinger V (2019) Effective flexoelectric and flexomagnetic response of ferroics. Elsevier, Netherlands

Eringen AC (1983) On differential equations of nonlocal elasticity and solutions of screw dislocation and surface waves. J Appl Phys 54:4703–4710

Fahrner W (2005) Nanotechnology and nanoelectronics. Springer, Germany

Hadj Mostefa A, Merdaci S, Mahmoudi N (2018) An overview of functionally graded materials "FGM". Springer, Cham

Kabychenkov AF, Lisovskii FV (2019) Flexomagnetic and flexoantiferromagnetic effects in centrosymmetric antiferromagnetic materials. Tech Phys 64:980–983

Karamanli A, Aydogdu M (2020) Bifurcation buckling conditions of FGM plates with different boundaries. Composite Struct 245:112325

Liu TJ, Yang F, Yu H, Aizikovich SM (2021) Axisymmetric adhesive contact problem for functionally graded materials coating based on the linear multi-layered model. Mech Based Des Struct Mach 49:41–58

Loh GH, Pei E, Harrison D, Monzón MD (2018) An overview of functionally graded additive manufacturing. Addit Manuf 23:34–44
Lu Z-l, Gao P-z, Ma R-x, Xu J, Wang Z-h, Rebrov EV (2016) Structural, magnetic and thermal properties of one-dimensional $CoFe_2O_4$ microtubes. J Alloy Compd 665:428–434
Lukashev P, Sabirianov RF (2010) Flexomagnetic effect in frustrated triangular magnetic structures. Phys Rev B 82:094417
Mahamood RM, Akinlabi ET, Shukla, M & Pityana, SL 2012, *Functionally Graded Material: An overview*, London, UK.
Malikan M (2017) Electro-mechanical shear buckling of piezoelectric nanoplate using modified couple stress theory based on simplified first order shear deformation theory. Appl Math Model 48:196–207
Malikan M, Eremeyev VA (2020a) A new hyperbolic-polynomial higher-order elasticity theory for mechanics of thick FGM beams with imperfection in the material composition. Composite Struct 249:112486
Malikan M, Eremeyev VA (2020b) Free vibration of flexomagnetic nanostructured tubes based on stress-driven nonlocal elasticity. Springer Nature, Switzerland
Malikan M, Eremeyev VA (2020c) On the geometrically nonlinear vibration of a piezo-flexomagnetic nanotube. Math Methods Appl Sci. https://doi.org/10.1002/mma.6758
Malikan M, Eremeyev VA (2020d) On the dynamics of a visco–piezo–flexoelectric nanobeam. Symmetry 12(4):643
Malikan M, Eremeyev VA (2021) Flexomagnetic response of buckled piezomagnetic composite nanoplates. Composite Struct 267:113932
Malikan M, Uglov NS, Eremeyev VA (2020a) On instabilities and post-buckling of piezomagnetic and flexomagnetic nanostructures. Int J Eng Sci 157:103395
Malikan M, Eremeyev VA, Żur KK (2020b) Effect of axial porosities on flexomagnetic response of in-plane compressed piezomagnetic nanobeams. Symmetry 12:1935
Malikan M, Eremeyev VA, Sedighi HM (2020c) Buckling analysis of a non-concentric double-walled carbon nanotube. Acta Mech 231:5007–5020
Malikan M, Krasheninnikov M, Eremeyev VA (2020d) Torsional stability capacity of a nano-composite shell based on a nonlocal strain gradient shell model under a three-dimensional magnetic field. Int J Eng Sci 148:103210
Malikan M, Wiczenbach T, Eremeyev VA (2021) On thermal stability of piezo-flexomagnetic microbeams considering different temperature distributions. Continuum Mech Thermodyn. https://doi.org/10.1007/s00161-021-00971-y
Moosavi S, Zakaria S, Chia CH, Gan S, Azahari NA, Kaco H (2017) Hydrothermal synthesis, magnetic properties and characterization of $CoFe_2O_4$ nanocrystals. Ceram Int 43:7889–7894
Pereira C, Pereira AM, Fernandes C, Rocha M, Mendes R, Fernández-García MP, Guedes A, Tavares PB, Grenèche J-M, Araújo JP, Freire C (2012) Superparamagnetic $MFe_2O_4$ (M = Fe Co, Mn) nanoparticles: tuning the particle size and magnetic properties through a novel one-step coprecipitation route. Chem Mater 24:1496–1504
Romano G, Barretta R (2017) Nonlocal elasticity in nanobeams: the stress-driven integral model. Int J Eng Sci 115:14–27
Senthil VP, Gajendiran J, Gokul Raj S, Shanmugavel T, Ramesh Kumar G, Parthasaradhi Reddy C (2018) Study of structural and magnetic properties of cobalt ferrite ($CoFe_2O_4$) nanostructures. Chem Phys Lett 695:19–23
Skrzat A, Eremeyev VA (2020) On the effective properties of foams in the framework of the couple stress theory. Continuum Mech Thermodyn 32:1779–1801
Sidhardh S, Ray MC (2018) Flexomagnetic response of nanostructures. J Appl Phys 124:244101
Thanh Tran T, Nguyen P-C, Pham Q-H (2021) Vibration analysis of FGM plates in thermal environment resting on elastic foundation using ES-MITC3 element and prediction of ANN. Case Stud Thermal Eng 24:100852
Vasiliev AS, Volkov SS, Belov AA, Litvinchuk SY, Aizikovich SM (2017) Indentation of a hard transversely isotropic functionally graded coating by a conical indenter. Int J Eng Sci 112:63–75

Volkov SS, Vasiliev AS, Aizikovich SM, Mitrin BI (2019) Axisymmetric indentation of an electroelastic piezoelectric half-space with functionally graded piezoelectric coating by a circular punch. Acta Mech 230:1289–1302
Zhang N, Zheng S, Chen D (2019) Size-dependent static bending of flexomagnetic nanobeams. J Appl Phys 126:223901
Zhang JX, Zeches RJ, He Q, Chu YH, Ramesh R (2012) Nanoscale phase boundaries: a new twist to novel functionalities. Nanoscale 4:6196–6204
Zhou H, Pei Y, Fang D (2014) Magnetic field tunable small-scale mechanical properties of nickel single crystals measured by nanoindentation technique. Sci Rep 4:1–6

# Chapter 18
# An Iterative Technique for Rational Approximation of Laplace Domain Vector-Valued Functions

**Ivan P. Markov, Andrey N. Petrov, and Aleksandr V. Boev**

**Abstract** In this paper, iterative procedure for partial fraction approximation of vector functions in Laplace domain is presented. First, the essentials of the original formulation of vector fitting methodology for strictly proper rational approximations are outlined. Then a modification is presented for more efficient application to vector functions. Proposed modification reduces the problem of fitting all elements of vector function to fitting only some of them and sum of the rest. After the set of common poles is obtained, the residues are calculated for all functions as usual. Next, the algorithm for adaptive rational approximation of vector functions is thoroughly presented. Proposed technique consists of successive determination of partial fraction approximations with increasing orders until the specified convergence criterion is satisfied. Detailed results of the numerical example are provided to assess accuracy and efficiency of the proposed procedure.

**Keywords** Laplace transform · Vector fitting · Adaptive rational approximation · Vector function

## 18.1 Introduction

The Laplace integral transform is powerful and versatile tool that is extensively used in numerical and analytical solutions of different problems in various fields of mechanics and mathematics. Development of efficient and accurate numerical procedures for inversion of Laplace transforms is an important issue, and ample number of different approaches have been proposed (e.g., Schapery 1962; Weeks

I. P. Markov (✉) · A. N. Petrov · A. V. Boev
Research Institute for Mechanics, National Research Lobachevsky State University of Nizhny Novgorod, 23 Gagarin Avenue, building 6, Nizhny Novgorod 603950, Russian Federation
e-mail: teanku@gmail.com

A. N. Petrov
e-mail: andrey.petrov@mech.unn.ru

A. V. Boev
e-mail: alexboev11@mail.ru

 
H. Altenbach et al. (eds.), *Advanced Materials Modelling for Mechanical, Medical and Biological Applications*, Advanced Structured Materials 155,
https://doi.org/10.1007/978-3-030-81705-3_18

1966; Dubner and Abate 1968; Stehfest 1970; Durbin 1974; Talbot 1979; Zhao 2004; Xiao et al. 2013; Schanz 2010; Schanz et al. 2016). In all approaches, the values of the complex frequencies in which the function in Laplace domain is calculated are determined by the specific parameters of a particular method, without taking into account the nature of the function. This is not much of a problem if the closed form of the function is known in Laplace domain. If the function itself is a solution of some other problem and its values can only be obtained using numerical methods, then it is preferable to use inversion method which requires fewer function samples. The situation becomes even more complicated when several functions are needed to be simultaneously inverted to the original domain.

Vector fitting (VF) is a method for partial fraction approximation of functions in Laplace domain by pole relocation (Gustavsen and Semlyen 1999; Gustavsen 2006; Deschrijver et al. 2008). In this paper, we present a modification of VF method for more efficient application to vector functions. Using modified VF formulation, we developed an iterative technique for rational approximation of vector-valued functions in Laplace domain. The main idea of the algorithm is to successively determine rational approximations with increasing orders until the specified convergence criterion is satisfied. The obtained partial fraction functions can be analytically inverted back to the original domain. Detailed numerical example is provided. Accuracy of the obtained approximations is confirmed by comparison with analytical solutions in the Laplace and time domains.

## 18.2 Vector Fitting Technique

Laplace transform of a function $f(t)\colon \mathbb{R} \to \mathbb{R}$, with $f(t) = 0$ for $t < 0$, is defined by

$$\bar{f}(s) = L\{f(t)\} = \int_{0}^{+\infty} e^{-\mathrm{st}} f(t)\mathrm{d}t, \tag{18.1}$$

where $s$ is complex frequency.

Consider a rational function approximation of $\bar{f}(s)$

$$\bar{f}(s) \approx \frac{a_0 + a_1 s + \cdots + a_{N_q-1} s^{N_q-1} + a_{N_q} s^{N_q}}{b_0 + b_1 s + \cdots + b_{N_p-1} s^{N_p-1} + b_{N_p} s^{N_p}} = \frac{Q(s)}{P(s)}, \tag{18.2}$$

where $a_i, b_j \in \mathbb{R}$, $i = \overline{0, N_q}$, $j = \overline{0, N_p}$.

We consider partial fraction form of a strictly proper $\left(\text{i.e., } N_q < N_p\right)$ rational function approximation (18.2) with simple poles

$$\bar{f}(s) \approx \frac{Q(s)}{P(s)} = F(s) = \sum_{k=1}^{N_p} \frac{r_k}{s - p_k}, \tag{18.3}$$

where $r_k$ are residues corresponding to the poles $p_k$ and $N_p$ essentially represents the order of approximation. Since coefficients $a_i$ and $b_j$ in (18.2) are real, $r_k$ and $p_k$ are either real or complex conjugate pairs: $\{r_k \in \mathbb{R},\ p_k \in \mathbb{R}\}$, or $\{r_k, r_{k+1} = \alpha_k \pm i\beta_k,\ p_k, p_{k+1} = \gamma_k \pm i\xi_k\}$, $\alpha_k, \beta_k, \gamma_k, \xi_k \in \mathbb{R}$.

Now we briefly recall an original vector fitting methodology for vector functions (Gustavsen and Semlyen 1999). Consider a Laplace transform $\bar{\mathbf{f}}(s)$ of a vector-valued function $\mathbf{f}(t)$ consisting of $N_f$ elements:

$$\bar{\mathbf{f}}(s) = L\{\mathbf{f}(t)\} = \begin{bmatrix} \bar{f}_1(s) \\ \bar{f}_2(s) \\ \vdots \\ \bar{f}_{N_f}(s) \end{bmatrix}. \tag{18.4}$$

We seek a strictly proper rational approximation for $\bar{\mathbf{f}}(s)$ in partial fraction form (18.3) assuming that all elements in the vector have a common set of unknown poles $p_1, p_2 \ldots, p_{N_p}$ :

$$\bar{\mathbf{f}}(s) \approx \mathbf{F}(s) = \begin{bmatrix} F_1(s) \\ F_2(s) \\ \vdots \\ F_{N_f}(s) \end{bmatrix} = \begin{bmatrix} \sum_{k=1}^{N_p} \frac{r_k^1}{s-p_k} \\ \sum_{k=1}^{N_p} \frac{r_k^2}{s-p_k} \\ \vdots \\ \sum_{k=1}^{N_p} \frac{r_k^{N_f}}{s-p_k} \end{bmatrix}, \tag{18.5}$$

where $r_k^j$ are the unknown residues corresponding to the function $F_j(s)$, $j = \overline{1, N_f}$.

Let us specify a set of initial poles $a_1, a_2 \ldots, a_{N_p}$ with *known* values, introduce an auxiliary *unknown* rational function $\sigma(s)$, and consider a partial fraction approximation of $\sigma(s)\bar{\mathbf{f}}(s)$

$$\sigma(s) = \sum_{k=1}^{N_p} \left( \frac{b_k}{s - a_k} \right) + 1, \tag{18.6}$$

$$\sigma(s)\bar{\mathbf{f}}(s) \approx \begin{bmatrix} \sum_{k=1}^{N_p} \frac{c_k^1}{s-a_k} \\ \vdots \\ \sum_{k=1}^{N_p} \frac{c_k^{N_f}}{s-a_k} \end{bmatrix}, \tag{18.7}$$

Multiplying the right-hand side of (18.6) by $\bar{\mathbf{f}}(s)$ gives

$$\begin{bmatrix} \sum_{k=1}^{N_p} \frac{c_k^1}{s-a_k} \\ \vdots \\ \sum_{k=1}^{N_p} \frac{c_k^{N_f}}{s-a_k} \end{bmatrix} - \begin{bmatrix} \bar{f}_1(s) \sum_{k=1}^{N_p} \frac{b_k}{s-a_k} \\ \vdots \\ \bar{f}_{N_f}(s) \sum_{k=1}^{N_p} \frac{b_k}{s-a_k} \end{bmatrix} = \begin{bmatrix} \bar{f}_1(s) \\ \vdots \\ \bar{f}_{N_f}(s) \end{bmatrix}. \tag{18.8}$$

Reordering (18.8) to group equations related to each function $\bar{f}_1(s), \bar{f}_2(s), \dots \bar{f}_{N_f}(s)$ for a given set of complex frequencies $s_1, s_2 \dots, s_{N_s}$ yields a linear system that is *overdetermined* in general case

$$\mathbf{Mx} = \mathbf{y}, \tag{18.9}$$

where

$$\mathbf{M} = \underset{(N_s N_f \times N_p(N_f+1))}{\begin{bmatrix} \mathbf{D} & 0 & \cdots & 0 & \mathbf{V}_1\mathbf{D} \\ 0 & \mathbf{D} & \cdots & 0 & \mathbf{V}_2\mathbf{D} \\ \vdots & \vdots & \ddots & \vdots & \vdots \\ 0 & 0 & \cdots & \mathbf{D} & \mathbf{V}_{N_f}\mathbf{D} \end{bmatrix}}, \tag{18.10}$$

$$\mathbf{D} = \underset{(N_s \times N_p)}{\begin{bmatrix} \frac{1}{s_1-a_1} & \cdots & \frac{1}{s_1-a_{N_p}} \\ \vdots & \ddots & \vdots \\ \frac{1}{s_{N_s}-a_1} & \cdots & \frac{1}{s_{N_s}-a_{N_p}} \end{bmatrix}}, \mathbf{V}_j = \underset{(N_s \times N_s)}{\begin{bmatrix} -\bar{f}_j(s_1) & & 0 \\ & \ddots & \\ 0 & & -\bar{f}_j(s_{N_s}) \end{bmatrix}}, \tag{18.11}$$

$$\mathbf{x} = \underset{(1 \times N_p(N_f+1))}{\left[ c_1^1 \cdots c_{N_p}^1 \; c_1^2 \cdots c_{N_p}^2 \; \cdots \; c_1^{N_f} \cdots c_{N_p}^{N_f} \; b_1 \cdots b_{N_p} \right]}^{\mathrm{T}}, \tag{18.12}$$

$$\mathbf{y} = \underset{(1 \times N_s N_f)}{\left[ \bar{f}_1(s_1) \cdots \bar{f}_1(s_{N_s}) \; \bar{f}_2(s_1) \cdots \bar{f}_2(s_{N_s}) \cdots \bar{f}_{N_f}(s_1) \cdots \bar{f}_{N_f}(s_{N_s}) \right]}^{\mathrm{T}}. \tag{18.13}$$

If the specified initial poles $a_k \in \mathbb{C}$ and $a_{k+1} \in \mathbb{C}$ are a complex conjugate pair, i.e., $a_{k+1} = a_k^*$, then unknown residues $c_k^j$, $c_{k+1}^j$ and $b_k$, $b_{k+1}$ should also constitute a complex conjugate pairs:

$$c_k^j = c_k^{j'} + ic_k^{j''}, c_{k+1}^j = c_k^* = c_k^{j'} - ic_k^{j''}, c_k^{j'}, c_k^{j''} \in \mathbb{R}, j = \overline{1, N_f}, \tag{18.14}$$

$$b_k = b_k' + ib_k'', b_{k+1} = b_k^* = b_k' - ib_k'', b_k', b_k'' \in \mathbb{R}. \tag{18.15}$$

The elements in each row of **D** in columns $k$ and $k+1$ are modified as follows

$$\mathbf{D}_{l,k} = \frac{1}{s_l - a_k} + \frac{1}{s_l - a_k^*}, l = \overline{1, N_s}, \tag{18.16}$$

$$\mathbf{D}_{l,k+1} = \frac{i}{s_l - a_k} - \frac{i}{s_l - a_k^*}. \tag{18.17}$$

This makes corresponding unknown residues $c_k^j$, $c_{k+1}^j$, $j = \overline{1, N_f}$, and $b_k$, $b_{k+1}$ in vector **x** become equal to $c_k^{j'}$, $c_k^{j''}$ and $b_k'$, $b_k''$, respectively. Performing transformations (18.16) and (18.17) for all pairs of complex conjugate initial poles makes all unknowns in **x** become real. In order to maintain this property, linear system (18.9) is reformulated in terms of real quantities

$$\underset{(2N_sN_f \times N_p(N_f+1))}{\begin{bmatrix} \mathrm{Re}(\mathbf{M}) \\ \mathrm{Im}(\mathbf{M}) \end{bmatrix}} \underset{(N_p(N_f+1)\times 1)}{\mathbf{x}} = \underset{(2N_sN_f \times 1)}{\begin{bmatrix} \mathrm{Re}(\mathbf{y}) \\ \mathrm{Im}(\mathbf{y}) \end{bmatrix}}, \tag{18.18}$$

Unknown values of residues $c_1^j \ldots c_{N_p}^j$, $j = \overline{1, N_f}$, and $b_1 \ldots b_{N_p}$ are obtained after solving (18.18) as a least squares problem. Now we rewrite Eqs. (18.6) and (18.7) as

$$\sigma(s) = \sum_{k=1}^{N_p} \left( \frac{b_k}{s - a_k} \right) + 1 = \frac{\prod_{k=1}^{N_p} (s - z_k)}{\prod_{k=1}^{N_p} (s - a_k)}, \tag{18.19}$$

$$\sigma(s)\bar{\mathbf{f}}(s) \approx \begin{bmatrix} \sum_{k=1}^{N_p} \frac{c_k^1}{s-a_k} \\ \vdots \\ \sum_{k=1}^{N_p} \frac{c_k^{N_f}}{s-a_k} \end{bmatrix} = \frac{1}{\prod_{k=1}^{N_p} (s - a_k)} \begin{bmatrix} \prod_{k=1}^{N_p-1} \left(s - \tilde{z}_k^1\right) \\ \vdots \\ \prod_{k=1}^{N_p-1} \left(s - \tilde{z}_k^{N_f}\right) \end{bmatrix}, \tag{18.20}$$

where $z_k$ are the zeros of $\sigma(s)$ and $\tilde{z}_k^j$ are the zeros of partial fraction approximation of function $\sigma(s) f_j(s)$, $j = \overline{1, N_f}$.

Multiplying the right-hand side of (18.19) by $\bar{\mathbf{f}}(s)$ and equating it to the right-hand side of (18.20) gives

$$\bar{\mathbf{f}}(s)\frac{\prod_{k=1}^{N_p}(s-z_k)}{\prod_{k=1}^{N_p}(s-a_k)} \approx \frac{1}{\prod_{k=1}^{N_p}(s-a_k)}\begin{bmatrix}\prod_{k=1}^{N_p-1}\left(s-\tilde{z}_k^1\right)\\ \vdots \\ \prod_{k=1}^{N_p-1}\left(s-\tilde{z}_k^{N_f}\right)\end{bmatrix}, \tag{18.21}$$

$$\bar{\mathbf{f}}(s) \approx \mathbf{F}(s) = \begin{bmatrix} F_1(s) \\ F_2(s) \\ \vdots \\ F_{N_f}(s)\end{bmatrix} = \frac{1}{\prod_{k=1}^{N_p}(s-z_k)}\begin{bmatrix}\prod_{k=1}^{N_p-1}\left(s-\tilde{z}_k^1\right)\\ \vdots \\ \prod_{k=1}^{N_p-1}\left(s-\tilde{z}_k^{N_f}\right)\end{bmatrix}, \tag{18.22}$$

meaning that the zeros $z_k$ of $\sigma(s)$ are the poles of $\mathbf{F}(s)$ which is the rational function approximation of $\bar{\mathbf{f}}(s)$.

The zeros $z_k$ of $\sigma(s)$ can be determined as eigenvalues of the matrix (Gustavsen and Semlyen 1998)

$$\mathbf{H} = \mathbf{A} - \mathbf{q}\mathbf{b}^{\mathrm{T}}, \tag{18.23}$$

$$\mathbf{A} = \underset{(N_p \times N_p)}{\begin{bmatrix} a_1 & & 0 \\ & \ddots & \\ 0 & & a_{N_p}\end{bmatrix}}, \mathbf{q} = \underset{(N_p \times 1)}{\begin{bmatrix} 1 \\ \vdots \\ 1 \end{bmatrix}}, \mathbf{b}^{\mathrm{T}} = \underset{(1 \times N_p)}{\begin{bmatrix} b_1 & \cdots & b_{N_p}\end{bmatrix}}, \tag{18.24}$$

i.e.,

$$\mathbf{H} = \begin{bmatrix} a_1 & & & 0 \\ & a_2 & & \\ & & \ddots & \\ 0 & & & a_{N_p}\end{bmatrix} - \begin{bmatrix} b_1 & b_2 & \cdots & b_{N_p} \\ b_1 & b_2 & \cdots & b_{N_p} \\ \vdots & \vdots & \ddots & \vdots \\ b_1 & b_2 & \cdots & b_{N_p}\end{bmatrix} = \begin{bmatrix} a_1-b_1 & -b_2 & \cdots & -b_{N_p} \\ -b_1 & a_2-b_2 & \cdots & -b_{N_p} \\ \vdots & \vdots & \ddots & \vdots \\ -b_1 & -b_2 & \cdots & a_{N_p}-b_{N_p}\end{bmatrix}. \tag{18.25}$$

For the complex conjugate pairs of initial poles, e.g., $a_k = a'_k + ia''_k$, $a_{k+1} = a^*_k = a'_k - ia''_k$, $a'_k, a''_k \in \mathbb{R}$, and the respective complex conjugate residues $b_k$, $b_{k+1} = b^*_k$ defined in (18.15), the corresponding submatrices in (18.24) are modified using the similarity transformation (Gustavsen and Semlyen 1999) as follows:

$$\hat{\mathbf{A}} = \begin{bmatrix} a_k & 0 \\ 0 & a_{k+1} \end{bmatrix} = \begin{bmatrix} a'_k & a''_k \\ -a''_k & a'_k \end{bmatrix}, \hat{\mathbf{q}} = \begin{bmatrix} 2 \\ 0 \end{bmatrix}, \hat{\mathbf{b}}^{\mathrm{T}} = \begin{bmatrix} b'_k & b''_k \end{bmatrix}. \tag{18.26}$$

These transformations make matrix **H** become real, and its eigenvalues are either real or come in complex conjugate pairs. Obtaining zeros $z_k$ of $\sigma(s)$ concludes pole identification part of vector fitting algorithm. In practice, this part is often repeated with calculated zeros $z_k$ as new initial poles $a_k$ in each iteration until rational approximation converges. If some of new poles are unstable then signs of their real parts are inverted. In our work, we do not perform iterations of pole identification part as will be explained later.

Now we recall Eq. (18.5)

$$\begin{bmatrix} \bar{f}_1(s) \\ \bar{f}_2(s) \\ \vdots \\ \bar{f}_{N_f}(s) \end{bmatrix} \approx \begin{bmatrix} F_1(s) \\ F_2(s) \\ \vdots \\ F_{N_f}(s) \end{bmatrix} = \begin{bmatrix} \sum\limits_{k=1}^{N_p} \frac{r_k^1}{s-p_k} \\ \sum\limits_{k=1}^{N_p} \frac{r_k^2}{s-p_k} \\ \vdots \\ \sum\limits_{k=1}^{N_p} \frac{r_k^{N_f}}{s-p_k} \end{bmatrix}, \tag{18.27}$$

To obtain unknown residues $r_1^j \ldots r_{N_p}^j$, $j = \overline{1, N_f}$, $N_f$ linear problems with zeros $z_k$ as poles $p_k$ are solved

$$\bar{f}_j(s) \approx \sum_{k=1}^{N_p} \frac{r_k^j}{s - p_k}, \; j = \overline{1, N_f}. \tag{18.28}$$

For already used set of complex frequencies $s_1, s_2 \ldots, s_{N_s}$, a linear overdetermined system with $N_f$ different right-hand sides is setup

$$\sum_{k=1}^{N_p} \frac{r_k^j}{s - p_k} = \bar{f}_j(s_l), \, l = \overline{1, N_s}, \tag{18.29}$$

or

$$\mathbf{D}\mathbf{x}_j = \mathbf{y}_j, \tag{18.30}$$

here **D** is defined in Eq. (18.11) and

$$\mathbf{x}_j = \underset{(N_p \times 1)}{\begin{bmatrix} r_1^j \\ \vdots \\ r_{N_p}^j \end{bmatrix}}, \mathbf{y}_j = \underset{(N_s \times 1)}{\begin{bmatrix} \bar{f}_j(s_1) \\ \vdots \\ \bar{f}_j(s_{N_s}) \end{bmatrix}}, \tag{18.31}$$

To ensure that unknown complex residues come in conjugate pairs, transformations (18.14)–(18.17) are applied to $\mathbf{D}$ and $\mathbf{x}_j$ and (18.30) is once again formulated in terms of real quantities

$$\underset{(2N_s \times N_p)}{\begin{bmatrix} \mathrm{Re}(\mathbf{D}) \\ \mathrm{Im}(\mathbf{D}) \end{bmatrix}} \underset{(N_p \times 1)}{\mathbf{x}_j} = \underset{(2N_s \times 1)}{\begin{bmatrix} \mathrm{Re}(\mathbf{y}_j) \\ \mathrm{Im}(\mathbf{y}_j) \end{bmatrix}}. \tag{18.32}$$

This concludes the residue identification part of vector fitting algorithm.

## 18.3 Iterative Algorithm for Vector Functions Approximation

Now consider a problem when it is needed to obtain strictly proper rational approximation $\mathbf{F}(s)$ defined in (18.5) of vector-valued function $\bar{\mathbf{f}}(s)$ over the predefined frequency range. In general, values of $\bar{\mathbf{f}}(s)$ can come from numerical solution of some computationally demanding problem. Hence, it is preferable to construct approximation $\mathbf{F}(s)$ of $\bar{\mathbf{f}}(s)$ in adaptive way using as little samples $\bar{\mathbf{f}}(s_l)$ as possible while maintaining desirable accuracy of approximation. In present work, we consider complex frequencies with constant real part $s = \alpha + i\omega$, $\alpha, \omega \in \mathbb{R}$, $\alpha = \mathrm{const}$, $\omega \in [0, \omega_{\max}]$, $\omega_{\max}$ is a predefined value. Obtained rational approximations $\mathbf{F}(s)$ of $\bar{\mathbf{f}}(s)$ is then used for analytical inversion of the Laplace transform

$$f_j(t) \approx f_{j;fit}(t) = L^{-1}\left[F_j(s)\right] = \sum_{k=1}^{N_p} r_k^j \exp(p_k t), t \in [0, t_{\max}], j = \overline{1, N_f}. \tag{18.33}$$

The real part of complex frequencies $\mathrm{Re}(s) = \alpha$ is determined from the relation (Crump 1976)

$$\mathrm{Re}(s) = \alpha = \frac{\xi \ln 10}{t_{\max}}, \tag{18.34}$$

where $\xi > 0$. In practical applications, value of parameter $\xi$ is often chosen in the range $2.0 \le \xi \le 3.0$.

To present an algorithm for adaptive rational approximation of vector functions in Laplace domain over the given frequency range, first we propose a modification to previously outlined vector fitting method.

Dimensions of matrix in linear system (18.18) are $\left(2N_sN_f \times N_p\left(N_f+1\right)\right)$. With increasing numbers of functions to be fitted, number of poles and frequencies, size of this matrix increases rapidly. Solving system (18.18) becomes prohibiting even for sparse linear least squares solvers. Expanding for convenience Eq. (18.9) we obtain

$$\mathbf{Mx} = \mathbf{y}, \tag{18.35}$$

$$\begin{bmatrix} \mathbf{D} & 0 & \cdots & 0 & \mathbf{V}_1\mathbf{D} \\ 0 & \mathbf{D} & \cdots & 0 & \mathbf{V}_2\mathbf{D} \\ \vdots & \vdots & \ddots & \vdots & \vdots \\ 0 & 0 & \cdots & \mathbf{D} & \mathbf{V}_{N_f}\mathbf{D} \end{bmatrix} \begin{bmatrix} \mathbf{c}^1 \\ \mathbf{c}^2 \\ \vdots \\ \mathbf{b} \end{bmatrix} = \begin{bmatrix} \bar{f}_1(\mathbf{s}) \\ \bar{f}_2(\mathbf{s}) \\ \vdots \\ \bar{f}_{N_f}(\mathbf{s}) \end{bmatrix}, \tag{18.36}$$

$$\mathbf{c}^j = \left[ c_1^j \cdots c_{N_p}^j \right]^{\mathrm{T}}, \mathbf{b} = \left[ b_1 \cdots b_{N_p} \right]^{\mathrm{T}}, \mathbf{s} = \left[ s_1 \cdots s_{N_s} \right]^{\mathrm{T}}, j = \overline{1, N_f}, \tag{18.37}$$

$$\bar{f}_j(\mathbf{s}) = \left[ \bar{f}_j(s_1) \cdots \bar{f}_j\left(s_{N_s}\right) \right]^{\mathrm{T}}, \tag{18.38}$$

$$\begin{cases} \mathbf{Dc}^1 + \mathbf{V}_1\mathbf{Db} = \bar{f}_1(\mathbf{s}), \\ \mathbf{Dc}^2 + \mathbf{V}_2\mathbf{Db} = \bar{f}_2(\mathbf{s}), \\ \vdots \\ \mathbf{Dc}^{N_f} + \mathbf{V}_{N_f}\mathbf{Db} = \bar{f}_{N_f}(\mathbf{s}). \end{cases} \tag{18.39}$$

In the pole identification part of VF, we actually only interested in $\mathbf{b}$. Let $\mathbf{I} = \left\{1 \cdots N_f\right\}$ be a full set of indexes of functions $\bar{f}_j(s)$, $j = \overline{1, N_f}$. Define two subsets of $\mathbf{I}$ : $\mathbf{I}_{\text{sel}}$ with cardinality $\text{card}(\mathbf{I}_{\text{sel}}) = N_{\text{sel}} < N_f$, and $\mathbf{I}' = \mathbf{I}/\mathbf{I}_{\text{sel}}$. The idea is to directly include in the system (18.39) only functions with indexes in $\mathbf{I}_{\text{sel}}$ and sum the rest of them in a single equation

$$\begin{cases} \mathbf{Dc}^{j_1} + \mathbf{V}_{j_1}\mathbf{Db} = \bar{f}_{j_1}(\mathbf{s}), \\ \mathbf{Dc}^{j_2} + \mathbf{V}_{j_2}\mathbf{Db} = \bar{f}_{j_2}(\mathbf{s}), \\ \vdots \\ \mathbf{Dc}^{j_{N_{\text{sel}}}} + \mathbf{V}_{j_{N_{\text{sel}}}}\mathbf{Db} = \bar{f}_{N_{\text{sel}}}(\mathbf{s}), \\ \mathbf{D}\left(\sum\limits_{j\in\mathbf{I}'} \mathbf{c}^j\right) + \left(\sum\limits_{j\in\mathbf{I}'} \mathbf{V}_j\right)\mathbf{Db} = \sum\limits_{j\in\mathbf{I}'} \bar{f}_j(\mathbf{s}), \end{cases} \quad j_n \in \mathbf{I}_{\text{sel}}, n = \overline{1, N_{\text{sel}}}. \tag{18.40}$$

When system (18.40) is casted into real form like (18.18) its size is $\left(2N_s(N_{\text{sel}}+1) \times N_p(N_{\text{sel}}+2)\right)$. Basically, this procedure reduces the problem of fitting $N_f$ functions to the problem of fitting only some of these functions and *the sum of the rest*. After **b** is obtained, the residue identification part of VF method performed unchanged, and the residues are calculated *for all* $N_f$ *functions*.

The input data of the algorithm for adaptive rational approximation of $N_f$ functions is: $\omega_{\max}$, $t_{\max}$, $\xi$, $N_{\mathrm{sel}}$, the number of starting frequencies $N_s^0$, the maximum number of frequencies $N_s^{\max}$, the number of frequencies for fine sampling $N_s^{\mathrm{fine}}$, the number of additional steps for testing the convergence $N_{\mathrm{steps}}$, and the error tolerance tol.

In the preliminary part, the following steps are performed:

- Calculate $\alpha$ using (18.34)
- Define the starting frequencies $s_l = \alpha + i\omega_{\max}(l-1)/\left(N_s^0-1\right)$, $l=\overline{1, N_s^0}$
- Define the set of fine sampling frequencies $s^{\mathrm{fine}} = \left\{\alpha + i\omega_{\max}(l-1)/\left(N_s^{\mathrm{fine}}-1\right)\right\}$, $l=\overline{1, N_s^{\mathrm{fine}}}$
- Define the set of initial poles with complex conjugate pairs $a_k = \left\{\frac{\omega_{\max}}{2}(0.01+i)\left(1+\frac{k-1}{U_p^0-1}\right), \frac{\omega_{\max}}{2}(0.01-i)\left(1+\frac{k-1}{U_p^0-1}\right)\right\}$, $U_p^0 = \left\lfloor \frac{N_s^0(N_{\mathrm{sel}}+1)}{N_{\mathrm{sel}}+2} \right\rfloor - 1$, where $\lfloor\rfloor$ is the floor function.
- Important note: when functions $\bar{\mathbf{f}}(s)$ are calculated at the frequency sample $s_l$, they subsequently scaled as $\tilde{\bar{\mathbf{f}}}(s) = \left[\tilde{\bar{f}}_j(s)\right] = \left[\bar{f}_j(s)/G_j\right]$, $G_j = \left|\bar{f}_j(\alpha + i\omega_{\max})\right|$, $j=\overline{1, N_f}$, where $\tilde{x}$ denote scaled quantity $x$.
- Define $\mathbf{I}_{\mathrm{sel}} = \left\{1 \cdots N_{\mathrm{sel}}\right\}$
- Obtain rational fitting $\tilde{\mathbf{F}}^0(s_l, a_k, \mathbf{I}_{\mathrm{sel}}) = \left\{p_k, r_k^j\right\}$ using VF with (18.40). Here $p_k$ and $r_k^j$ are the poles and corresponding residues obtained in the VF process.
- Add a pair of complex conjugate poles $p_k = p_k \cup \left\{\frac{\omega_{\max}}{2}(0.01+i), \frac{\omega_{\max}}{2}(0.01-i)\right\}$
- Set $a_k = p_k, v = 1$

Next, the following steps are repeated until convergence is reached or number of frequencies exceeds $N_s^{\max}$ :

1. Obtain rational fitting $\tilde{\mathbf{F}}^v(s_l, a_k, \mathbf{I}_{\mathrm{sel}}) = \left\{p_k, r_k^j\right\}$ using VF with (18.40)
2. Calculate the frequency-wise vectors of absolute errors between $\tilde{\mathbf{F}}^{v-1}(s)$ and $\tilde{\mathbf{F}}^v(s)$

$$\mathbf{AE}_j^v(s_i) = \left[\left|\tilde{F}_j^v(s_i) - \tilde{F}_j^{v-1}(s_i)\right|\right]^{\mathrm{T}}, s_i \in s^{\mathrm{fine}} \tag{18.41}$$

3. Calculate the vector of root mean square errors between $\tilde{\mathbf{F}}^{v-1}(s)$ and $\tilde{\mathbf{F}}^v(s)$ function-wise

$$\mathbf{RMSE}^v = \left[\sqrt{\frac{\sum_{s_i \in s^{\mathrm{fine}}} \mathbf{AE}_j^v(s_i)}{N_s^{\mathrm{fine}}}}\right]^{\mathrm{T}}, j=\overline{1, N_f} \tag{18.42}$$

4. If $\max_{j=\overline{1,N_f}} \mathrm{RMSE}_j^{v-N_{\text{steps}}+q} < \text{tol}$, $q = \overline{1, N_{\text{steps}}}$, then the convergence criterion is satisfied for $N_{\text{steps}}$ successive steps, rational fitting $\mathbf{F}(s) = \left[\tilde{F}_j^v(s) G_j\right]^{\mathrm{T}}$ is obtained with desirable accuracy, and iteration process is terminated
5. If $N_s = \text{card}(s_l)$ is equal to $N_s^{\max}$, then the convergence is not reached, and iteration process is terminated
6. $\mathbf{I}_{\text{sel}}$ for the next iteration is defined as a set of $N_{\text{sel}}$ indexes of functions with the biggest root mean square errors defined in (18.42). This means, that in general, $\mathbf{I}_{\text{sel}}$ can change at each iteration.
7. New sampling frequency is defined as

$$s_{\text{new}} = s_d^{\text{fine}},\ d = \arg\max_{s_i \in s^{\text{fine}}} \mathbf{AE}_m^v(s_i),\ m = \arg\max_{j=\overline{1,N_f}} \mathrm{RMSE}_j^v \tag{18.43}$$

8. Add new sampling frequency $s_l = s_l \bigcup \{s_{\text{new}}\}$, set $N_s = \text{card}(s_l)$
9. Set $N_p = \text{card}(p_k)$, define $N_p^{\text{add}} = \left\lfloor \frac{2N_s(N_{\text{sel}}+1) - N_p(N_{\text{sel}}+2)}{2(N_{\text{sel}}+2)} \right\rfloor$ and if $N_p^{\text{add}} > 0$ then add $N_p^{\text{add}}$ pairs of complex conjugate poles $p_k = p_k \cup \left\{\frac{\omega_{\max}}{2}(0.01 + i)\left(1 + \frac{q-1}{N_p^{\text{add}}-1}\right), \frac{\omega_{\max}}{2}(0.01 - i)\left(1 + \frac{q-1}{N_p^{\text{add}}-1}\right)\right\}, q = \overline{1, N_p^{\text{add}}}$
10. Set $a_k = p_k$, $v = v + 1$, go to step 1

Since the poles are relocated at each iteration, we do not perform iterations of pole identification part in the VF process. Furthermore, results of numerical investigations suggest that iteration of pole identification can actually slow down the convergence of the presented algorithm.

## 18.4 Numerical Examples

The presented algorithm is tested using the Laplace domain solutions of model poroelastodynamic problem. Consider a one-dimensional finite poroelastic column of length $l$ subjected to a Heaviside-type load as depicted in Fig. 18.1. To model the dynamic behavior of poroelastic material, Biot's theory of fully saturated poroelasticity (Biot, 1956a, b) is employed.

The boundary conditions are prescribed as follows

$$u(t, y = 0) = 0 \text{ m},$$

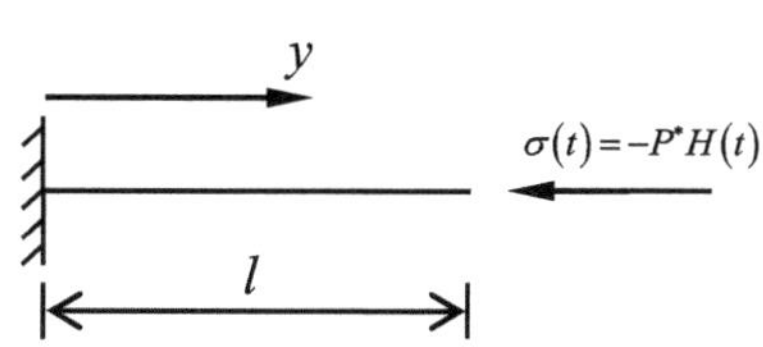


**Fig. 18.1** One-dimensional poroelastic column under Heaviside-type loading

$$\sigma(t, y = l) = -P^* H(t) \text{ Pa},$$
$$p(t, y = l) = 0 \text{ Pa},$$
$$q(t, y = 0) = 0 \text{ m/s},$$

where $u(t, y)$, $\sigma(t, y)$, $p(t, y)$, and $q(t, y)$ are displacement, stress, pore pressure, and flux, respectively. $H(\cdot)$ denotes the Heaviside step function.

For the specified boundary conditions, analytical solutions in the Laplace domain are available (Schanz and Cheng 2000)

$$\bar{u}_{\text{an}}(s, y) = \frac{F}{Es^2(d_1\lambda_2 - d_2\lambda_1)} \left[ \frac{d_2\left(e^{-\lambda_1 s(l-y)} - e^{-\lambda_1 s(l+y)}\right)}{1 + e^{-2\lambda_1 sl}} - \frac{d_1\left(e^{-\lambda_2 s(l-y)} - e^{-\lambda_2 s(l+y)}\right)}{1 + e^{-2\lambda_2 sl}} \right],$$

$$\bar{p}_{\text{an}}(s, y) = \frac{F d_1 d_2}{Es(d_1\lambda_2 - d_2\lambda_1)} \left[ \frac{e^{-\lambda_1 s(l-y)} + e^{-\lambda_1 s(l+y)}}{1 + e^{-2\lambda_1 sl}} - \frac{e^{-\lambda_2 s(l-y)} + e^{-\lambda_2 s(l+y)}}{1 + e^{-2\lambda_2 sl}} \right],$$

$$\bar{\sigma}_{\text{an}}(s, y) = E\frac{\partial \bar{u}_{\text{an}}(s, y)}{\partial y} - \alpha \bar{p}_{\text{an}}(s, y),$$

$$\bar{q}_{\text{an}}(s, y) = -\frac{\beta}{s^2\rho_f}\left( \frac{\partial \bar{p}_{\text{an}}(s, y)}{\partial y} + s^2\rho_f \bar{u}_{\text{an}}(s, y) \right),$$

where

$$\frac{\partial \bar{u}_{\text{an}}(s, y)}{\partial y} = \frac{F}{Es^2(d_1\lambda_2 - d_2\lambda_1)} \left[ \frac{d_2\lambda_1 s\left(e^{-\lambda_1 s(l-y)} + e^{-\lambda_1 s(l+y)}\right)}{1 + e^{-2\lambda_1 sl}} - \frac{d_1\lambda_2 s\left(e^{-\lambda_2 s(l-y)} + e^{-\lambda_2 s(l+y)}\right)}{1 + e^{-2\lambda_2 sl}} \right],$$

$$\frac{\partial \bar{p}_{\text{an}}(s, y)}{\partial y} = \frac{F d_1 d_2}{Es(d_1\lambda_2 - d_2\lambda_1)} \left[ \frac{\lambda_1 s\left(e^{-\lambda_1 s(l-y)} - e^{-\lambda_1 s(l+y)}\right)}{1 + e^{-2\lambda_1 sl}} - \frac{\lambda_2 s\left(e^{-\lambda_2 s(l-y)} - e^{-\lambda_2 s(l+y)}\right)}{1 + e^{-2\lambda_2 sl}} \right],$$

$$\lambda_{1,2}(s) = \sqrt{\frac{B \pm \sqrt{B^2 - 4AC}}{2A}},$$

$$A(s) = \frac{E\beta}{\rho_f},\ B(s) = E\frac{\phi^2}{R} + \left(\rho - \beta\rho_f\right)\frac{\beta}{\rho_f} + (\alpha - \beta)^2,\ C(s) = \frac{\phi^2\left(\rho - \beta\rho_f\right)}{R},$$

$$d_i(s) = \frac{E\lambda_i^2 - \left(\rho - \beta\rho_f\right)}{(\alpha - \beta)\lambda_i},\ i = \overline{1, 2},\ \beta(s) = \frac{\kappa\rho_f\phi s}{\phi + 1.66 s\kappa\rho_f},\ E = K + \frac{4}{3}G,$$

$$R = \frac{\phi^2 K_f K_s^2}{K_f(K_s - K) + \phi K_s(K_s - K_f)}, \alpha = 1 - \frac{K}{K_s},$$

where $\phi$ denotes porosity, $\rho$ and $\rho_f$ are the bulk and fluid densities, respectively, $K$ and $G$ are the compression and shear moduli of the isotropic elastic skeleton, $K_s$ and $K_f$ are the compression moduli of the solid grains and fluid, respectively, $\kappa$ is the permeability.

When the friction between solid and interstitial fluid is negligible, i.e., $\kappa \to \infty$, analytical solutions in the time domain can be found

$$u_{\mathrm{an}}(t, y) = \frac{F}{E(d_1\lambda_2 - d_2\lambda_1)} \sum_{n=0}^{\infty} (-1)^{-n} \{d_2[\gamma_1 H(\gamma_1) - \chi_1 H(\chi_1)] - d_1[\gamma_2 H(\gamma_2) - \chi_2 H(\chi_2)]\},$$

$$p_{\mathrm{an}}(t, y) = \frac{F d_1 d_2}{E(d_1\lambda_2 - d_2\lambda_1)} \sum_{n=0}^{\infty} (-1)^{-n} [H(\gamma_1) - H(\gamma_2) + H(\chi_1) - H(\chi_2)],$$

$$\sigma_{\mathrm{an}}(t, y) = E\frac{\partial u_{\mathrm{an}}(t, y)}{\partial y} - \alpha p_{\mathrm{an}}(t, y),$$

$$q_{\mathrm{an}}(t, y) = -\frac{\beta F}{E(d_1\lambda_2 - d_2\lambda_1)} \left\{ \frac{d_1 d_2}{\rho_f} \sum_{n=0}^{\infty} (-1)^{-n} [\lambda_1(H(\gamma_1) - H(\chi_1)) - \lambda_2(H(\gamma_2) - H(\chi_2))] + \sum_{n=0}^{\infty} (-1)^{-n} [d_2(H(\gamma_1) - H(\chi_1)) - d_1(H(\gamma_2) - H(\chi_2))] \right\},$$

$$\frac{\partial u_{\mathrm{an}}(t, y)}{\partial y} = \frac{F}{E(d_1\lambda_2 - d_2\lambda_1)} \sum_{n=0}^{\infty} (-1)^{-n} \{d_2\lambda_1[H(\gamma_1) + H(\chi_1)] - d_1\lambda_2[H(\gamma_2) + H(\chi_2)]\},$$

$$\gamma_i(n, t, y) = t - \lambda_i(l(2n+1) - y), \chi_i(n, t, y) = t - \lambda_i(l(2n+1) + y), i = \overline{1, 2}.$$

We consider a poroelastic material with the following parameters: $\phi = 0.76$, $\rho = 1396\ \mathrm{kg/m^3}$, $\rho_f = 1000\ \mathrm{kg/m^3}$, $K = 3.7 \cdot 10^7$ Pa, $G = 2.2 \cdot 10^7$ Pa, $K_s = 3.6 \cdot 10^{10}$ Pa, $K_f = 2.3 \cdot 10^9$ Pa, $\kappa = \kappa_{\mathrm{fin}} = 1.0 \cdot 10^{-8}\ \mathrm{m^4/N/s}$. Length of the column is $l = 10$ m.

For convenience, we work with dimensionless quantities and variables:

$$\rho = \frac{\rho p_2^3}{p_3}, \rho_f = \frac{\rho_f p_2^3}{p_3}, K = \frac{K p_1^2 p_2}{p_3}, G = \frac{G p_1^2 p_2}{p_3}, K_s = \frac{K_s p_1^2 p_2}{p_3},$$

$$K_f = \frac{K_f p_1^2 p_2}{p_3}, \kappa = \frac{\kappa p_3}{p_1 p_2^3}, l = \frac{l}{p_2},$$
$$p_1 = 0.15 \text{ s}, p_2 = 10 \text{ m}, p_3 = 8.1 \cdot 10^9 \text{ kg}.$$

In all further calculations, maximum observation time is $t_{\max} = 0.15\,\mathrm{s}/p_1 = 1.0$, maximum imaginary part of the complex frequencies is $\omega_{\max} = p_1 10000\,\mathrm{rad}/\mathrm{s} = 1500$, and amplitude of loading is $P^* = (1\,\mathrm{Pa})\,p_1^2 p_2/p_3$.

To measure accuracy of fitting on each iteration, we use maximum value of vector (18.42) of root mean square errors between two consecutive approximations

$$E_v = \max_{j=\overline{1,N_f}} \mathrm{RMSE}_j^v.$$

To investigate how error $E_v$ between two fittings correlate to the deviation from true solutions, we use $E_v^{\mathrm{an}}$ which is the maximum value of vector of root mean square errors between approximation on current iteration and analytical solution. $E_v^{\mathrm{an}}$ is essentially defined by (18.41)–(18.42), but absolute error is measured not between two fittings, but between a fitting on the current iteration and the analytical solution.

To obtain values of vector function at given frequency sample, we use the following generating procedure:

$$\begin{aligned}
&\text{function } \bar{\mathbf{f}}(s, N_i): \\
N_f &= 4N_i, \\
\text{points} &= \mathrm{linspace}\big(l, 0.0, N_f\big) \\
\text{for } i &= \overline{1, N_i}: \\
\bar{f}_{4i-3}(s) &= \bar{u}_{\mathrm{an}}\big(s, \text{points}_{4i-3}\big) \\
\bar{f}_{4i-2}(s) &= \bar{p}_{\mathrm{an}}\big(s, \text{points}_{4i-2}\big) \\
\bar{f}_{4i-1}(s) &= \bar{\sigma}_{\mathrm{an}}\big(s, \text{points}_{4i-1}\big) \\
\bar{f}_{4i}(s) &= \bar{q}_{\mathrm{an}}\big(s, \text{points}_{4i}\big)
\end{aligned}$$

where *linspace* returns linearly spaced vector including endpoints.

Parameters of the algorithm for adaptive rational approximation: $\xi = 2.5$, $N_{\mathrm{sel}} = 1$, $N_s^0 = 8$, $N_s^{\mathrm{fine}} = \lfloor 5\omega_{\max}/(2\pi p_1) \rfloor$.

For $\kappa = \kappa_{\mathrm{fin}}$, $E_v$, and $E_v^{\mathrm{an}}$ are compared in Fig. 18.2 for $\bar{\mathbf{f}}(s, 1)$ and in Fig. 18.3 for $\bar{\mathbf{f}}(s, 2)$. For $\kappa \to \infty$, corresponding results are shown in Figs. 18.4 and 18.5. As it can be seen in all cases, $E_v$ is an excellent indicator of the fitting accuracy.

In Figs. 18.6 and 18.7, $E_v$ for $\bar{\mathbf{f}}(s, 1)$ $\big(N_f = 4\big)$, $\bar{\mathbf{f}}(s, 2)$ $\big(N_f = 8\big)$, $\bar{\mathbf{f}}(s, 5)$ $\big(N_f = 20\big)$, $\bar{\mathbf{f}}(s, 50)$ $\big(N_f = 200\big)$, $\bar{\mathbf{f}}(s, 500)$ $\big(N_f = 2000\big)$, are shown for $\kappa = \kappa_{\mathrm{fin}}$ and $\kappa \to \infty$, respectively. Though expected inevitable loss of accuracy is observed with increasing $N_f$, overall performance is still impressive, considering that we used

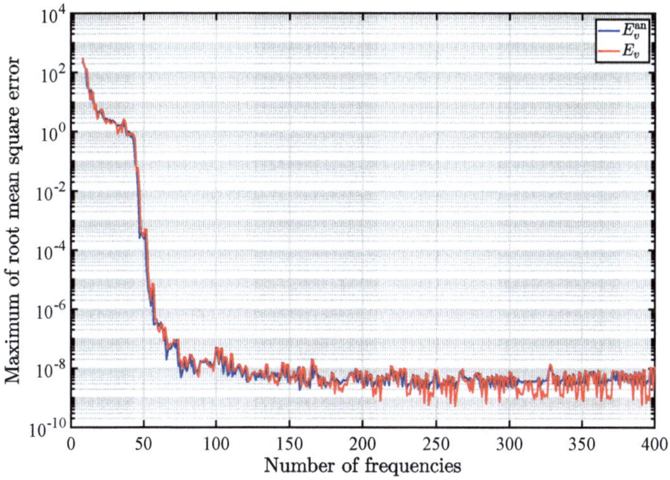


**Fig. 18.2** Comparison of $E_v$ and $E_v^{\rm an}$ for $\kappa = \kappa_{\rm fin}$, $\bar{\mathbf{f}}(s, 1)$

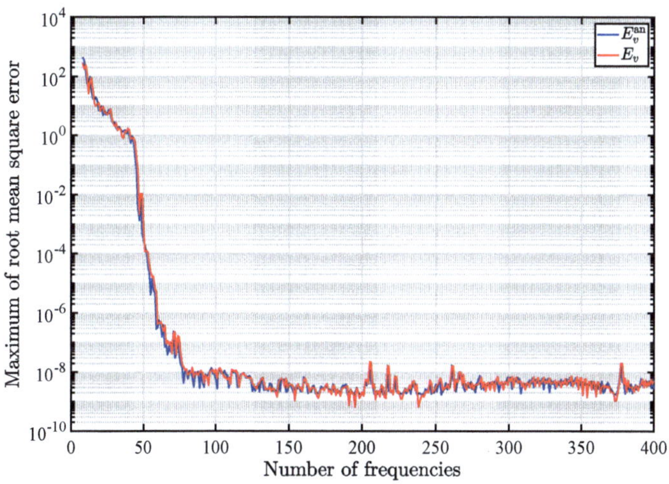


**Fig. 18.3** Comparison of $E_v$ and $E_v^{\rm an}$ for $\kappa = \kappa_{\rm fin}$, $\bar{\mathbf{f}}(s, 2)$

$N_{\rm sel} = 1$, meaning only one function on each iteration was directly included in the linear system (18.40).

Consider vector function $\bar{\mathbf{f}}(s) = \left[\bar{u}_{\rm an}(s, l/4), \bar{p}_{\rm an}(s, l/4), \bar{\sigma}_{\rm an}(s, l/4), \bar{q}_{\rm an}(s, l/4)\right]$, for $\kappa \to \infty$ and $N_s^{\rm max} = 500$, $N_{\rm steps} = 5$, tol $= 10^{-4}$. The rest of parameters remains the same. Algorithm reached convergence with the total number of complex frequencies $N_s = 342$. Obtained partial fraction approximations were analytically inverted to the time domain using (18.33). Figures 18.8, 18.9, 18.10, and 18.11 present obtained time domain results compared with the analytical solu-

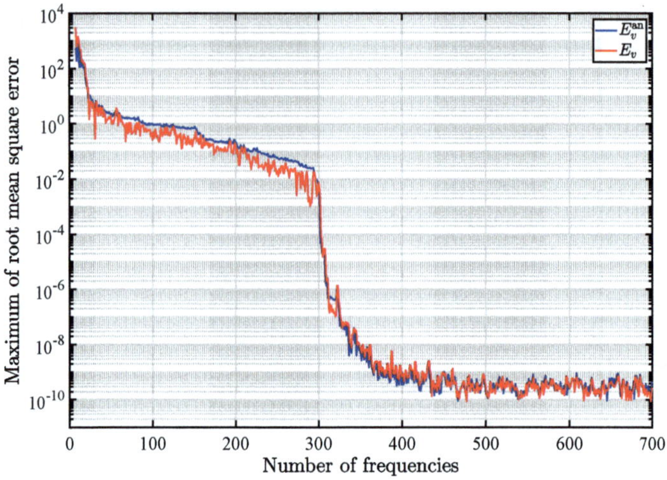


**Fig. 18.4** Comparison of $E_v$ and $E_v^{an}$ for $\kappa \to \infty$, $\bar{\mathbf{f}}(s, 1)$

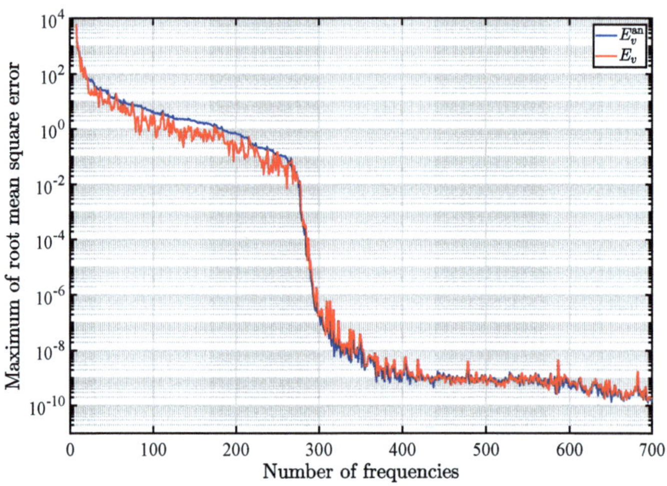


**Fig. 18.5** Comparison of $E_v$ and $E_v^{an}$ for $\kappa \to \infty$, $\bar{\mathbf{f}}(s, 2)$

tions for dimensionless displacement, stress, pore pressure, and flux. High-frequency oscillations (Gibbs phenomenon) occur due to truncation of the frequency range.

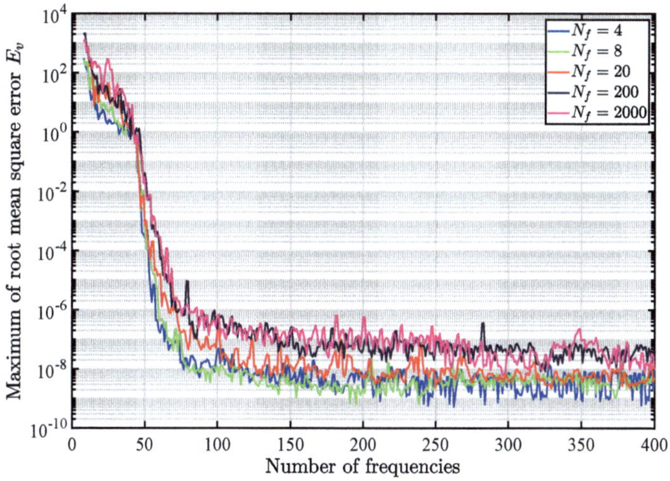


**Fig. 18.6** $E_v$ for $\kappa = \kappa_{\text{fin}}$, $\bar{\mathbf{f}}(s, 1)$, $\bar{\mathbf{f}}(s, 2)$, $\bar{\mathbf{f}}(s, 5)$, $\bar{\mathbf{f}}(s, 50)$, $\bar{\mathbf{f}}(s, 500)$

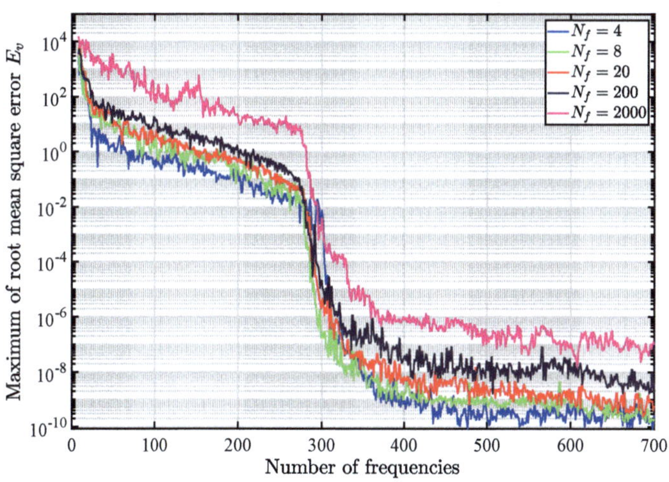


**Fig. 18.7** $E_v$ for $\kappa \to \infty$, $\bar{\mathbf{f}}(s, 1)$, $\bar{\mathbf{f}}(s, 2)$, $\bar{\mathbf{f}}(s, 5)$, $\bar{\mathbf{f}}(s, 50)$, $\bar{\mathbf{f}}(s, 500)$

## 18.5 Conclusions

Obtained error measurements confirm suitability of the proposed algorithm with modification to pole identification part of vector fitting methodology for accurate fitting of vector functions with large number of elements. This would be completely impossible with the original vector fitting process, because least squares matrix would quickly become prohibitively large. With fast implementation (Deschrijver et al. 2008) of VF, it would be feasible but still computationally challenging and less

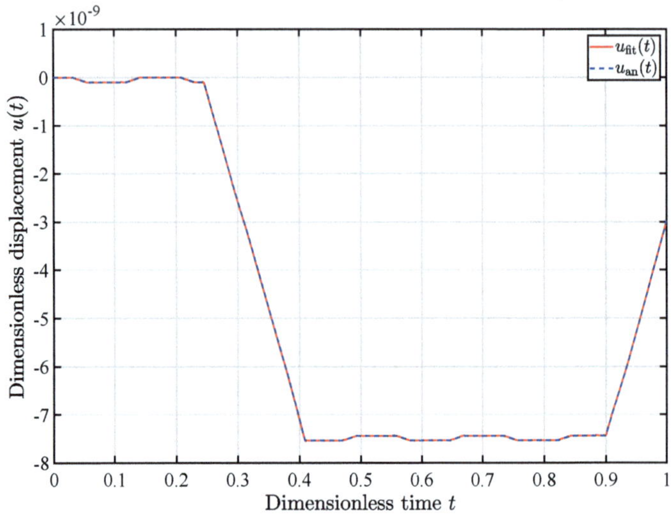


**Fig. 18.8** Comparison of dimensionless displacements $u_{\text{fit}}(t)$ and $u_{\text{an}}(t)$

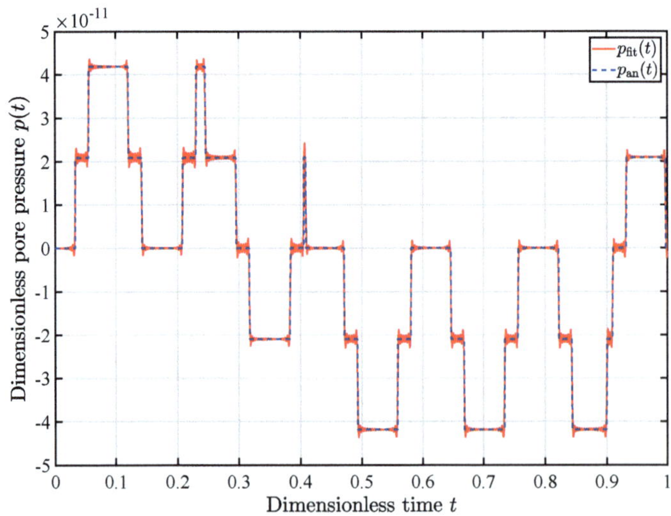


**Fig. 18.9** Comparison of dimensionless pore pressures $p_{\text{fit}}(t)$ and $p_{\text{an}}(t)$

efficient in terms of order of pole-residue model, since fast implementation can accommodate a smaller number of poles for the same number of complex frequencies.

Rational approximations obtained with the presented technique allow for analytic inversion to the original domain. With sufficiently accurate partial fraction approximations in the Laplace domain, accuracy of time domain solutions depends more on things like truncation of high-frequency range and value of real part of complex

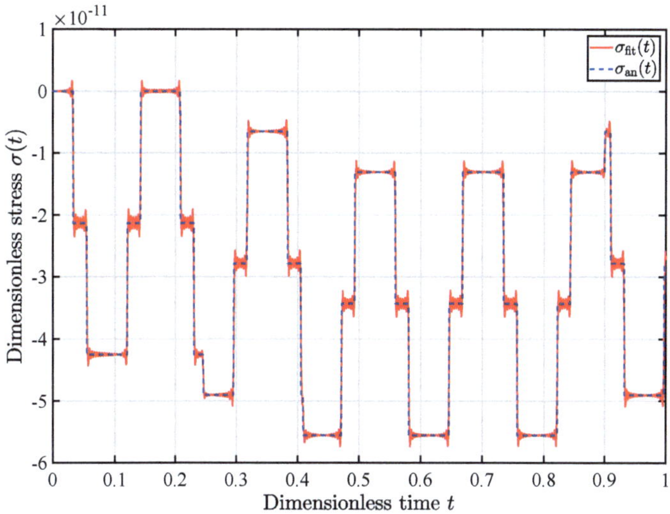


**Fig. 18.10** Comparison of dimensionless stresses $\sigma_{\mathrm{fit}}(t)$ and $\sigma_{\mathrm{an}}(t)$

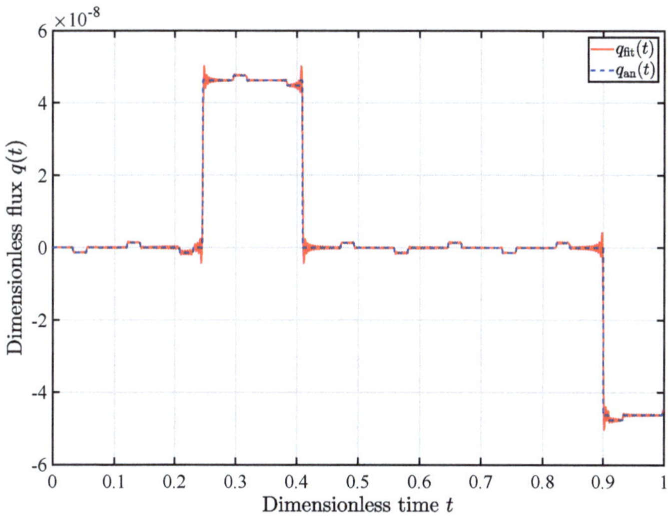


**Fig. 18.11** Comparison of dimensionless fluxes $q_{\mathrm{fit}}(t)$ and $q_{\mathrm{an}}(t)$

frequencies which often influenced by considerations outside of scope of the fitting algorithm.

**Acknowledgements** The reported study was funded by RFBR, project number 20-08-00451.

## References

Biot MA (1956a) Theory of propagation of elastic waves in a fluid-saturated porous solid. I. Low-frequency range. J Acoust Soc America 28(2):168–178. https://doi.org/10.1121/1.1908239
Biot MA (1956b) Theory of propagation of elastic waves in a fluid-saturated porous solid. II. Higher frequency range. J Acoust Soc America 28(2):179–191. https://doi.org/10.1121/1.1908241
Crump KS (1976) Numerical inversion of Laplace transforms using a fourier series approximation. J ACM 23(1):89–96. https://doi.org/10.1145/321921.321931
Deschrijver D, Mrozowski M, Dhaene T, De Zutter D (2008) Macromodeling of multiport systems using a fast implementation of the vector fitting method. IEEE Microw Wireless Compon Lett 18(6):383–385. https://doi.org/10.1109/LMWC.2008.922585
Dubner H, Abate J (1968) Numerical inversion of laplace transforms by relating them to the finite fourier cosine transform. J ACM 15(1):115–123. https://doi.org/10.1145/321439.321446
Durbin F (1974) Numerical inversion of Laplace transforms: an efficient improvement to Dubner and Abate's method. Comput J 17(4):371–376. https://doi.org/10.1093/comjnl/17.4.371
Gustavsen B (2006) Improving the pole relocating properties of vector fitting. IEEE Trans Power Delivery 21(3):1587–1592. https://doi.org/10.1109/TPWRD.2005.860281
Gustavsen B, Semlyen A (1999) Rational approximation of frequency domain responses by vector fitting. IEEE Trans Power Delivery 14(3):1052–1061. https://doi.org/10.1109/61.772353
Gustavsen B, Semlyen A (1998) Simulation of transmission line transients using vector fitting and modal decomposition. IEEE Trans Power Delivery 13(2):605–614. https://doi.org/10.1109/61.660941
Schanz M (2010) On a reformulated convolution quadrature based boundary element method. CMES Comput Model Eng Sci 58(2):109–128. https://doi.org/10.3970/cmes.2010.058.109
Schanz M, Cheng AH-D (2000) Transient wave propagation in a one-dimensional poroelastic column. Acta Mech 145(1–4):1–18. https://doi.org/10.1007/BF01453641
Schanz M, Ye W, Xiao J (2016) Comparison of the convolution quadrature method and enhanced inverse FFT with application in elastodynamic boundary element method. Comput Mech 57(4):523–536. https://doi.org/10.1007/s00466-015-1237-z
Schapery R (1962) Approximate methods of transform inversion for viscoelastic stress analysis. In: Proceedings of the fourth U.S. National congress of applied mechanics, pp 1075–1084
Stehfest H (1970) Algorithm 368: numerical inversion of Laplace transforms [D5]. Commun ACM 13(1):47–49. https://doi.org/10.1145/361953.361969
Talbot A (1979) The accurate numerical inversion of Laplace transforms. IMA J Appl Math 23(1):97–120. https://doi.org/10.1093/imamat/23.1.97
Weeks WT (1966) Numerical inversion of Laplace transforms using laguerre functions. J ACM 13(3):419–429. https://doi.org/10.1145/321341.321351
Xiao J, Ye W, Wen L (2013) Efficiency improvement of the frequency-domain BEM for rapid transient elastodynamic analysis. Comput Mech 52(4):903–912. https://doi.org/10.1007/s00466-013-0852-9
Zhao X (2004) An efficient approach for the numerical inversion of Laplace transform and its application in dynamic fracture analysis of a piezoelectric laminate. Int J Solids Struct 41(13):3653–3674. https://doi.org/10.1016/j.ijsolstr.2004.01.006

# Chapter 19
# Advances in Modeling and Identification of Prestresses in Modern Materials

**Rostislav D. Nedin and Alexander O. Vatulyan**

**Abstract** Within the framework of the linearized model of a prestressed elastic body, we study some statements of two-dimensional inverse problems on the prestressed state restoration. We consider inverse problems of 2 types, in the presence of additional data on the measured displacement field: (1) in a set of points of the region at a fixed vibration frequency, and (2) on a boundary part in some frequency range. We investigate the questions of solution uniqueness for the 1st type inverse problem. We propose and discuss some techniques for solving the stated inverse problems.

**Keywords** Prestresses · Linearized model · Oscillation · Inverse problem · Uniqueness · Iterative method · Regularization · Plane region

## 19.1 Introduction

At the stage of manufacturing modern materials of complex structure, in particular, functionally graded or layered composites, unaccounted prestress fields are often formed, which continue to exist in structures in the absence of external (or visible) influences. The analysis of prestress state in modern materials is usually carried out using various experimental methods—destructive, semi-destructive and non-destructive ones; the most promising of these are non-destructive methods. At the same time, the use of non-destructive technologies requires the construction of adequate, experimentally confirmed, theoretical models of inhomogeneous prestress fields for specific materials. It is also necessary to create a mathematical apparatus

R. D. Nedin (✉) · A. O. Vatulyan
Institute of Mathematics, Mechanics and Computer Sciences named after I.I. Vorovich, Southern Federal University, 8a Mil'chakova Str., 344090 Rostov-on-Don, Russia
e-mail: rdn90@bk.ru

A. O. Vatulyan
e-mail: aovatulyan@sfedu.ru

Vladikavkaz Scientific Center of Russian Academy of Sciences, Southern Mathematical Institute, 22 Markusa Str., 362027 Vladikavkaz, Russia



H. Altenbach et al. (eds.), *Advanced Materials Modelling for Mechanical, Medical and Biological Applications*, Advanced Structured Materials 155,
https://doi.org/10.1007/978-3-030-81705-3_19

that allows one to reliably identify the internal non-uniform prestress distributions in bodies within the framework of various concepts of non-destructive testing.

Prestress fields may also be formed in parts during surface treatment, such as shot peening, laser impact treatment, grinding, etc.; (Salvati and Korsunsky 2018) present a technique based on the eigenstrain method and the finite element method, which allows predicting residual stress fields, arising as a result of surface treatment of an arbitrary three-dimensional body. In Ling et al. (2020), the authors propose some approaches to determine the field of residual stresses formed as a result of grinding samples made of homogeneous materials. The approach to measuring residual stresses on the surface has been tested using the X-ray diffraction method.

Another currently widely used method for determining prestress state parameters is indentation, which is very popular due to its ease of use and applicability at the macro-, micro- and nanoscale. In Suresh and Giannakopoulos (1998), the relationship between indentation parameters and residual stress is formulated; in this case, the residual stress field was considered uniaxial. Pham and Kim (2017) carried out an extensive analysis by the finite element method in order to study the influence of the axisymmetric residual stresses field and plastic parameters on the reaction to indentation into structural steel. In Moharrami and Sanayei (2020), the influence of the yield point and the biaxial prestress components ratio on the accuracy of the procedure for determining residual stresses in metals using indentation was analyzed.

Finite element techniques based on the displacement method has recently been actively used to define prestress fields in structural elements. This method is most suitable for reconstructing residual stress fields by simulating real production processes such as casting, machining and welding. For example, Deng and Kiyoshima (2012) analyzed the distribution of residual stresses arisen during welding by creating a full-size accurate finite element model; in Yonezu et al. (2015), a simulation of full-scale microindentation experiments was carried out to analyze residual stresses in a steel sample. In Wang et al. (2018), a method for predicting the residual stress state from measurements on the surface using the finite element method is described and applied to a beam specimen.

In the present work, we consider a model of oscillations of an inhomogeneous body in the presence of a prestress field. Several statements of inverse problems on the identification of prestress state in bodies are investigated based on the acoustic sounding method. The questions of uniqueness of the inverse problem solutions are discussed in case of additional data on the measured displacement field in the body region for a single-measured vibration frequency. Another formulation is also considered, based on information about measurements of displacements or frequency response on some part of body's boundary in a frequency range. For the second class of problems, research methods are mainly based on iterative-regularization schemes for solving the corresponding ill-posed coefficient inverse problems. The advantages of such approaches are associated with the use of additional data that can be obtained within the framework of non-destructive sounding methods, as well as with the ability to restore variable material properties and changes in prestress components depending on coordinates. Methods and algorithms for solving the formulated inverse problems are proposed and discussed, and the results of some computational experiments are presented.

## 19.2 Direct Problem on Vibrations of a Prestressed Plane Region

One of the most commonly used in the literature linearized statements of the direct problem on vibrations of a prestressed elastic plane region has the form (Guz et al. 1977; Hoger 1986; Robertson 1998; Nedin et al. 2017)

$$\begin{aligned} &\left(\sigma_{ij} + u_{i,k}\sigma_{kj}^0\right)_{,j} + \rho\omega^2 u_i = 0, \\ &\sigma_{ij} = \lambda\delta_{ij}u_{k,k} + \mu\left(u_{i,j} + u_{j,i}\right), \\ &\left(\sigma_{ij} + u_{i,k}\sigma_{kj}^0\right) n_j\Big|_{S_\sigma} = P_i, \quad u_i|_{S_u} = 0, \quad i, j, k = 1, 2, \end{aligned} \tag{19.1}$$

where $\sigma_{ij}$ and $\sigma_{ij}^0$ are the components of the incremental and initial stress tensors, $u_i$ are the components of the small incremental displacements vector, $\rho$ is the body density, $\omega$ is the steady oscillation frequency, $\lambda$, $\mu$ are the Lame coefficients, $n_i$ are the components of the unit outward normal vector of the region boundary $\partial S = l_u \bigcup l_\sigma$, $P_i$ are the vector components of the external mechanical load acting on the boundary part $l_\sigma$, and the area is clamped on the part $l_u$.

Note that there are many approaches to modeling prestress in solids. A number of models based on the linearization of nonlinear relations in continuum mechanics were proposed and described, for instance, in Truesdell (1977), Hoger (1986), Robertson (1998), Guz et al. (1977); see also the review of different prestress linearized models in Nedin et al. (2017). On the basis of such an approach, in Vatulyan et al. (2015), Nedin and Vatulyan (2013), Nedin et al. (2014, 2017, 2018), Nedin and Vatulyan (2014), Vatulyan and Nedin (2020), Nedin (2020), Yurov et al. (2020), we have proposed theoretical foundations for determining inhomogeneous prestress fields in elastic bodies, including those made of materials with a complex structure (e.g., functional-gradient composites), using the acoustic sounding method. The proposed methods are mainly based on iterative-regularization schemes for solving the corresponding ill-posed coefficient inverse problems (IP). The advantages of such approaches are associated with the use of additional data on the measured displacement fields on a part of the body surface, which can be obtained using non-destructive sensing methods, as well as the ability to restore the variable mechanical properties of modern materials and changes in the prestress components by coordinates.

## 19.3 First-Type Inverse Problems on Prestress Reconstruction

Let us study the possibilities of restoring components of prestress field using some additional information. In this case, the most adequate statement to the procedure of prestress reconstruction using acoustic sounding is that in which the components of the displacements arising from the action of a given load are specified (measured) on the body surface in a certain frequency range (Vatulyan et al. 2015).

Next, we shall consider the formulation of 2D problem (19.1) on steady-state oscillations of some region and set IP on determining self-balanced prestress field $\sigma_{ij}^0(x_1, x_2)$ (i.e. satisfying the equilibrium equations $\sigma_{ij,j}^0 = 0$) by the additional data on the specified displacement field $u_i\,(x_1, x_2)$ in the whole region $S$:

$$u_i|_S = f_i. \tag{19.2}$$

For such a statement, the IP is linear; as in other coefficient IPs of this type, it is possible to use the established schemes for their solution, however, there are some specific features associated with the ill-posedness of the problems under study (Vatulyan and Nedin 2020). One of the most important is the question of the uniqueness of the IP solution.

### *19.3.1 On the Uniqueness of the Prestress Field Reconstruction for the First Inverse Problem Statement*

Let us consider the following.

**Lemma 1** *Assume that the data of the problem* (19.1), (19.2) *correspond to 2 solutions* $\sigma_{ij}^{0(1)}$ *and* $\sigma_{ij}^{0(2)}$. *Then for*

$$d_{ij}^0 = \sigma_{ij}^{0(1)} - \sigma_{ij}^{0(2)}$$

*the following equalities hold*

$$u_{i,kj} d_{kj}^0 = 0, \quad u_{i,k} d_{kj}^0 n_j\big|_{l_\sigma} = 0. \tag{19.3}$$

***Proof*** Let $\sigma_{ij}^{0(1)}$ and $\sigma_{ij}^{0(2)}$ be the solutions of the 1st type IP for the load $P_i$, and $u_i$ is the corresponding vector field. Consider the motion equations

$$\left(\sigma_{ij} + u_{i,k}\sigma_{kj}^{0(1)}\right),_j + \rho\omega^2 u_i = 0, \quad \left(\sigma_{ij} + u_{i,k}\sigma_{kj}^{0(2)}\right),_j + \rho\omega^2 u_i = 0$$

and find their difference:

$$\left(\sigma_{ij} + u_{i,k}\sigma_{kj}^{0(1)}\right),_j - \left(\sigma_{ij} + u_{i,k}\sigma_{kj}^{0(2)}\right),_j = 0,$$
$$\left(u_{i,k}d_{kj}^0\right),_j = 0,$$

or

$$u_{i,kj}d_{kj}^0 + u_{i,k}d_{kj,j}^0 = 0.$$

But, due to the self-balance of the prestress field, we have

$$d_{kj,j}^0 = \sigma_{kj,j}^{0(1)} - \sigma_{kj,j}^{0(2)} = 0,$$

hence

$$u_{i,kj}d_{kj}^0 = 0.$$

The boundary conditions are:

$$\left(\sigma_{ij} + u_{i,k}\sigma_{kj}^{0(1)}\right)n_j\Big|_{l_\sigma} = P_i, \quad \left(\sigma_{ij} + u_{i,k}\sigma_{kj}^{0(2)}\right)n_j\Big|_{l_\sigma} = P_i.$$

Subtracting one from the other, we get:

$$u_{i,k}d_{kj}^0 n_j\Big|_{l_\sigma} = 0.$$

Thus, the relations and boundary conditions from (19.3) are satisfied, which was required to establish.

Now we use this lemma to study the question of the uniqueness of finding the prestress state in the framework of the 2D statement. The first relations in (19.3) are a system of two linear equations for three components $d_{ij}^0$:

$$\begin{aligned} u_{1,11}d_{11}^0 + 2u_{1,12}d_{12}^0 + u_{1,22}d_{22}^0 &= 0, \\ u_{2,11}d_{11}^0 + 2u_{2,12}d_{12}^0 + u_{2,22}d_{22}^0 &= 0. \end{aligned} \tag{19.4}$$

Below we consider 2 cases:

1. Assume that one of the three prestress components $\sigma_{ij}^0$ equals 0 (and hence the corresponding difference $d_{ij}^0$ is also zero). For example, let $\sigma_{12}^0 = 0$, and $d_{12}^0 = 0$. Then from (19.4) we have

$$\begin{aligned} u_{1,11}d_{11}^0 + u_{1,22}d_{22}^0 &= 0, \\ u_{2,11}d_{11}^0 + u_{2,22}d_{22}^0 &= 0. \end{aligned} \tag{19.5}$$

The condition for the existence of only *zero* solution of this homogeneous SLAE is

$$\Delta = \begin{vmatrix} u_{1,11} & u_{1,22} \\ u_{2,11} & u_{2,22} \end{vmatrix} = u_{1,11}u_{2,22} - u_{1,22}u_{2,11} \neq 0.$$

Under this condition, we have $d_{11}^0 = d_{22}^0 = 0$, and the uniqueness of the IP in the considered statement is established. Similarly, we can obtain uniqueness conditions in the presence of a priori information $\sigma_{11}^0 = 0$ or $\sigma_{22}^0 = 0$.

2. Let the prestress field be of general form (there is no information about zero prestress components), then the relations from (19.3) are insufficient to establish uniqueness. In this case, from the system (19.4) we may express

$$d_{11}^0 = \frac{\Delta_1}{\Delta} d_{12}^0, \quad d_{22}^0 = \frac{\Delta_2}{\Delta} d_{12}^0,$$

where

$$\Delta_1 = \begin{vmatrix} -2u_{1,12} & u_{1,22} \\ -2u_{2,12} & u_{2,22} \end{vmatrix} = 2\left(u_{2,12}u_{1,22} - u_{1,12}u_{2,22}\right),$$

$$\Delta_2 = \begin{vmatrix} u_{1,11} & -2u_{1,12} \\ u_{2,11} & -2u_{2,12} \end{vmatrix} = 2\left(u_{1,12}u_{2,11} - u_{2,12}u_{1,11}\right).$$

In this case, nothing can be said about the uniqueness of the zero solution; additional information is required. This can be an information on the displacement field $u_i^{(2)}$ when subjected to another load $P_i^{(2)}$. Then, under one of the two conditions

$$\frac{\Delta_1\left(u^{(1)}\right)}{\Delta\left(u^{(1)}\right)} \neq \frac{\Delta_1\left(u^{(2)}\right)}{\Delta\left(u^{(2)}\right)} \quad \text{or} \quad \frac{\Delta_2\left(u^{(1)}\right)}{\Delta\left(u^{(1)}\right)} \neq \frac{\Delta_2\left(u^{(2)}\right)}{\Delta\left(u^{(2)}\right)}$$

the solution $d_{11}^0 = d_{22}^0 = d_{12}^0 = 0$, and the uniqueness is proven.

**Remark 1** Checking the conditions of the lemma requires finding the second derivatives of the displacement components, which is an ill-posed problem that is usually treated using spline approximations.

### *19.3.2 Approach to Reconstruction of an Inhomogeneous Prestress State in a Plane Region Using the Cauchy Problems*

Considering two motion equations from (19.1) and taking into account the equilibrium equations for the initial state, we obtain an overdetermined system of 4 equations for determining only 3 prestress components $\sigma_{ij}^0$ together with two corresponding natural boundary conditions written in terms of the components $\sigma_{ij}^0$. This problem is initially overdetermined; therefore, to construct an approximate solution, we may use the following approach (Vatulyan and Nedin 2020):

1. We exclude the component $\sigma_{22}^0$ from the motion equations using standard transformations and express

$$\sigma_{12}^0 = \tilde{D}_1 + \sigma_{11}^0 \tilde{A}_1, \tag{19.6}$$

where $\tilde{D}_1 = -D_1/C_1$, $\tilde{A}_1 = -A_1/C_1$, $D_1 = u_{2,22}\left(\sigma_{11,1} + \sigma_{12,2}\right) - u_{1,22}$ $\left(\sigma_{12,1} + \sigma_{22,2}\right) + \rho\omega^2\left(u_1 u_{2,22} - u_2 u_{1,22}\right)$,

$$A_1 = u_{1,11}u_{2,22} - u_{2,11}u_{1,22},\, C_1 = 2\Delta_1,\, \Delta_1 = u_{1,12}u_{2,22} - u_{2,12}u_{1,22}.$$

Substituting (19.6) into the first equilibrium equation for the initial medium, we obtain:

$$\sigma_{11,1}^0 + \tilde{A}_1\sigma_{11,2}^0 + \tilde{A}_{1,2}\,\sigma_{11}^0 + \tilde{D}_{1,2} = 0, \tag{19.7}$$

– linear inhomogeneous first order partial differential equation with respect to the function $\sigma_{11}^0(x_1, x_2)$. At that we assume $\Delta_1 \neq 0$. Substituting the component $\sigma_{12}^0$ expression through $\sigma_{11}^0$ (19.6) into the natural boundary conditions (19.1), the boundary condition can be expressed

$$\sigma_{11}^0|_{l_\sigma} = F_1(P_i, u_i). \tag{19.8}$$

Therefore, (19.7)–(19.8) forms the Cauchy problem of finding the function $\sigma_{11}^0$, which can be solved either using difference approximations of the operator in (19.7), or using the scheme described by Vatulyan and Gukasyan (2019) when solving the Cauchy problem.

2. Eliminating in a similar way from Eq. (19.5) the component $\sigma_{11}^0$, we get

$$\sigma_{12}^0 = \tilde{D}_2 + \sigma_{22}^0 \tilde{A}_2, \tag{19.9}$$

where

$$\begin{aligned}
&\tilde{D}_2 = -D_2/C_2, \quad \tilde{A}_2 = -A_2/C_2,\\
&D_2 = u_{2,11}\left(\sigma_{11,1} + \sigma_{12,2}\right) - u_{1,11}\left(\sigma_{12,1} + \sigma_{22,2}\right) + \rho\omega^2\left(u_1 u_{2,11} - u_2 u_{1,11}\right),\\
&A_2 = u_{2,11}u_{1,22} - u_{1,11}u_{2,22}, \quad C_2 = 2\Delta_2,\, \Delta_2 = u_{1,12}u_{2,11} - u_{2,12}u_{1,11} \neq 0.
\end{aligned}$$

Satisfying the second equilibrium equation, we finally find

$$\sigma_{22,2}^0 + \tilde{A}_2\sigma_{22,1}^0 + \tilde{A}_{2,1}\sigma_{22}^0 + \tilde{D}_{2,1} = 0. \tag{19.10}$$

Similarly, we formulate the boundary condition

$$\sigma_{22}^0|_{l_\sigma} = F_2(P_i, u_i). \tag{19.11}$$

Solving the two Cauchy problems (19.7)–(19.8) and (19.10)–(19.11), one may uniquely find two prestress components $\sigma_{11}^0$ and $\sigma_{22}^0$ independently. The remaining 3rd component of the tangential prestress $\sigma_{12}^0$, which is expressed by 2 different relations $\sigma_{12}^{0(1)}$ and $\sigma_{12}^{0(2)}$ according to (19.6) and (19.9), may be determined by minimizing the residual functional

$$G = \int\limits_{S} \left(d_1^2 + d_2^2\right) \mathrm{d}S,$$

where

$$d_1 = \sigma_{12}^0 - \sigma_{12}^{0(1)}, \quad d_2 = \sigma_{12}^0 - \sigma_{12}^{0(2)}.$$

### *19.3.3 Level Restoration of the Inhomogeneous Prestress State*

One of the first type IP statements consists in determining the level of an inhomogeneous prestress state based on the measurement data on the displacement fields in the region *S*, given the information that the prestress state was formed as a result of applying an initial mechanical load of a known type; in this case, the prestress field is linearly dependent on it, and we may represent $\sigma_{ij}^0 = p\tilde{\sigma}_{ij}^0$, where $\tilde{\sigma}_{ij}^0$ are the components of the prestress field, which can be obtained by solving the corresponding statics direct problem for a given load of a *unit value*; *p* in this case characterizes the desired prestress level and, at the same time, the value of the initial load. In this case, it is possible to determine the parameter *p* from the motion equations using the technique of minimizing the residual functional in the region *S* based on the measurements of the in-plane displacements in a set of points of the plate region. In Nedin (2019, 2020), some results of the reconstruction of the level of inhomogeneous prestress state in functionally graded plates, including those with holes and inclusions, are presented, based on the measurement data on the displacement fields in the region.

## 19.4 Second-Type Inverse Problems on Prestress Reconstruction

Let us consider the second IP statement as follows: it is required to identify the prestress state based on additional acoustic sounding data on the measured displacement amplitudes on a part of the region boundary

$$u_i|_{S_\sigma} = f_i \tag{19.12}$$

in a certain frequency range $\omega \in \left[\omega_-, \omega_+\right]$.

Note that when considering IP on the reconstruction of material parameters, such as the variable Young modulus, there is a priori information about its positivity, which simplifies the reconstruction process. In the case of identifying the prestress state components, there is no such a priori information; however, an additional condition on the prestress field self-balance can be set, leading to an overdetermined problem (Vatulyan and Nedin 2020).

**Remark 2** Mainly, the techniques for studying the problems of the prestress state reconstruction in the second statement are based on the use of iterative-regularization schemes for solving the corresponding ill-posed coefficient IP. Such iterative processes are considered in detail in Vatulyan et al. (2015), Nedin and Vatulyan (2013), Nedin et al. (2017, 2018); in these works, no restrictions were imposed on the smallness of the prestress components. At the same time, special attention should be paid to the technique of investigating the formulated IP *without carrying out the iterative process*, when the assumption of *prestress smallness* is assumed, i.e., when the maximum values of the components of the prestress fields are several orders of magnitude less than the characteristic values of the elastic modulus; the use of this fact greatly simplifies the IP solution.

## *19.4.1 Restoration of Small Prestress Components*

By representing $\sigma_{ij}^0 = \tau s_{ij}$, where $\tau$ is small parameter, one can obtain an integral equation for the stress components $s_{ij}$ in the form (Vatulyan et al. 2015)

$$\int\limits_S u_{i,k}^0 u_{i,j}^0 s_{kj} \mathrm{d}S = \int\limits_{l_\sigma} P_i \left(f_i - u_i^0\right) \mathrm{d}l = F(\omega), \quad \omega \in \left[\omega_-, \omega_+\right], \tag{19.13}$$

where $u_i^0$ represent displacements for the problem without a prestress. Such an approach implies a simplification of the iterative process; only one iteration is considered, and a zero initial approximation for the prestress is chosen.

To formulate a system of operator equations, we may use 3 sets of boundary conditions, describing 3 experiments (tests) on acoustic sounding of the region using loads applied to different parts of the boundary:

$$T_{ij} n_j \big|_{l_k} = P_i^{(k)}, \quad u_i|_{l_k} = f_i^{(k)}, \quad k = \overline{1, 3}.$$

In this case, the border parts $l_1$ , $l_2$ , $l_3$ may intersect. The corresponding right-hand sides of the integral Eq. (19.13) take the form

$$F_k(\omega) = \int\limits_{l_k} P_i^{(k)} \left(f_i^{(k)} - u_i^0\right) \mathrm{d}l_k, \quad k = \overline{1, 3}.$$

Here, the frequency of steady-state oscillations $\omega$ is indicated as a formal argument. The system of Fredholm's integral equations of the first kind with a smooth kernel can be solved on the basis of the A. N. Tikhonov regularization method (Tikhonov and Arsenin 1977).

**Remark 3** This approach allows quite simply, in one iteration, to identify the prestress state components, taking into account their smallness. Note that in this case, the accuracy of the found approximations in a number of situations may not be high enough; however, they can be used as estimates and initial approximations. In order to further refine the specified reconstruction, one may use IP research techniques based on the application of iterative processes by using regularization procedures.

### *19.4.2 Iterative-Regularization Scheme*

In Nedin and Vatulyan (2013), Vatulyan et al. (2015), a linearized reciprocity relation is constructed for studying the considered 2D IP on identifying prestress state in the second statement. This relationship makes it possible to implement an iterative process for determining the corrections of the prestress tensor components with respect to some initial state selected:

$$\int_S \delta\sigma_{kj}^{0(n)} u_{i,j}^{(n-1)} u_{i,k}^{(n-1)} \mathrm{d}S + \int_{l_\sigma} P_i \left( f_i - u_i^{(n-1)} \right) \mathrm{d}l = 0, \quad \omega \in [\omega_-, \omega_+]. \quad (19.14)$$

Here, the indices $(n)$, $(n-1)$ are the iteration numbers; $\delta\sigma_{kj}^{0(n)}$ are the corrections to the corresponding prestress components at the current iteration; $f_i|_{l_\sigma}$ are the displacements measured under probing loads forming the additional data in the IP statement. Thus, it is possible to construct an iterative process, at each step of which the boundary value problem is solved for the current approximation for the prestress $\sigma_{ij}^{0(n-1)}$, and the corresponding displacement field $u_i^{(n-1)}$ is calculated; after that from the Fredholm integral equation of the first kind (19.14), the corrections $\sigma_{ij}^{0(n)}$ are found, and the current approximation is refined:

$$\sigma_{ij}^{0(n)} = \sigma_{ij}^{0(n-1)} + \delta\sigma_{ij}^{0(n)}.$$

This approach requires knowledge of the initial approximation $\sigma_{ij}^{0(0)}$, which can be searched for in some narrow class of functions (for example, linear or constant ones) by minimizing the residual functional in a space of small dimension (Vatulyan et al. 2015).

Next, we will consider various two-dimensional formulations of the considered IP and ways of studying them. Let us represent Eq. (19.14) in the form

$$\int\limits_S \left(\delta\sigma_{11}^{0(n)} K_{11}^{(n-1)} + \delta\sigma_{12}^{0(n)} K_{12}^{(n-1)} + \delta\sigma_{22}^{0(n)} K_{22}^{(n-1)}\right) \mathrm{d}S = F^{(n-1)}(\omega),\ \omega \in \left[\omega_-, \omega_+\right], \tag{19.15}$$

where

$$F^{(n-1)} = \int\limits_{l_\sigma} P_i \left(u_i^{(n-1)} - f_i\right) \mathrm{d}S.$$

Below we list the expressions for the kernels $K_{ij}^{(n-1)}$ and the right-hand side $F^{(n-1)}$ for various vibration modes of thin plates:

(A) In-plane vibrations of a plate (Nedin and Vatulyan 2013, 2014)

$$\begin{aligned}
K_{11}^{(n-1)} &= \left[u_{1,1}^{(n-1)}\right]^2 + \left[u_{2,1}^{(n-1)}\right]^2, \quad K_{22}^{(n-1)} = \left[u_{1,2}^{(n-1)}\right]^2 + \left[u_{2,2}^{(n-1)}\right]^2, \\
K_{12}^{(n-1)} &= 2\left[u_{1,1}^{(n-1)} u_{1,2}^{(n-1)} + u_{2,1}^{(n-1)} u_{2,2}^{(n-1)}\right], \\
F^{(n-1)}(\omega) &= \int\limits_{l_\sigma} \left[P_1(u_1^{(n-1)} - f_1) + P_2(u_2^{(n-1)} - f_2)\right] \mathrm{d}l_\sigma,
\end{aligned}$$

where $P_1$, $P_2$ are the in-plane load components acting on the boundary part $l_\sigma$.

(B) Bending vibrations of a plate, Kirchhoff's model (Nedin and Vatulyan 2013)

$$\begin{aligned}
K_{11}^{(n-1)} &= \frac{h^3}{12}\left[\left(w,_{11}^{(n-1)}\right)^2 + \left(w,_{12}^{(n-1)}\right)^2\right] + h\left(w,_1^{(n-1)}\right)^2, \\
K_{22}^{(n-1)} &= \frac{h^3}{12}\left[\left(w,_{12}^{(n-1)}\right)^2 + \left(w,_{22}^{(n-1)}\right)^2\right] + h\left(w,_2^{(n-1)}\right)^2, \\
K_{12}^{(n-1)} &= \frac{h^3}{6} w,_{12}^{(n-1)} \Delta w + 2hw,_1^{(n-1)} w,_2^{(n-1)}, \\
F^{(n-1)}(\omega) &= \int\limits_{l_\sigma} p(w^{(n-1)} - f)\mathrm{d}l_\sigma,
\end{aligned}$$

where $w^{(n-1)}$ is the plate deflection function, $p$ is the bending load component on $l_\sigma$, $h$ is the plate thickness, $w|_{l_\sigma} = f$ is the additional data of the IP.

(C) Bending vibrations of a plate, Timoshenko's model (Nedin et al. 2018)

$$\begin{aligned}
K_{11}^{(n-1)} &= \frac{h^3}{12}\left[\left(\theta_{1,1}^{(n-1)}\right)^2 + \left(\theta_{2,1}^{(n-1)}\right)^2\right] + h\left(w,_1^{(n-1)}\right)^2, \\
K_{22}^{(n-1)} &= \frac{h^3}{12}\left[\left(\theta_{1,2}^{(n-1)}\right)^2 + \left(\theta_{2,2}^{(n-1)}\right)^2\right] + h\left(w,_2^{(n-1)}\right)^2, \\
K_{12}^{(n-1)} &= \frac{h^3}{6}\left(\theta_{1,1}^{(n-1)}\theta_{1,2}^{(n-1)} + \theta_{2,1}^{(n-1)}\theta_{2,2}^{(n-1)}\right) + 2hw,_1^{(n-1)} w,_2^{(n-1)},
\end{aligned}$$

where $\theta_1^{(n-1)}, \theta_1^{(n-1)}$ are the functions of the rotation angles of the normal relative to the axes $x_1, x_2$; the right-hand side $F^{(n-1)}(\omega)$ has the same form as in the case (B).

Assuming the prestress field to be self-balanced, we now introduce the stress function and express the corrections to the prestress at the $n$-th iteration through it:

$$\delta\sigma_{11}^{0(n)} = \delta\Phi,_{22}^{(n)}, \quad \delta\sigma_{22}^{0(n)} = \delta\Phi,_{11}^{(n)}, \quad \delta\sigma_{12}^{0(n)} = -\delta\Phi,_{12}^{(n)}.$$

Then, Eq. (19.15) takes the form

$$\int_S \left(\delta\Phi,_{22}^{(n)} K_{11}^{(n-1)} - \delta\Phi,_{12}^{(n)} K_{12}^{(n-1)} + \delta\Phi,_{11}^{(n)} K_{22}^{(n-1)}\right) \mathrm{d}S = F^{(n-1)}(\omega),\, \omega \in [\omega_-, \omega_+]. \tag{19.16}$$

Equation (19.16) is an integral-differential equation for the stress function $\delta\Phi^{(n)}$. The study of such an operator equation is a much more complicated problem than the study of the classical Fredholm equation of the 1st kind with a continuous kernel arising when investigating coefficient IP of finding variable elasticity modulus or variable Lamé parameters (Vatulyan 2019). Below we outline some approaches to the study of this problem.

#### 19.4.2.1 Using Finite Element Approximations

Let us split the flat region $S$ into $N$ subdomains $S = \bigcup_{k=1}^{N} S_k$, forming a set of superelements, on each of which we approximate the required function by a quadratic polynomial:

$$\delta\Phi_k^{(n)} = a_k^{(n)} x_1^2 + b_k^{(n)} x_1 x_2 + c_k^{(n)} x_2^2,$$

which corresponds to constant prestresses on each of the considered superelements:

$$\delta\sigma_{11}^{0(n)}\Big|_{S_k} = 2c_k^{(n)}, \quad \delta\sigma_{22}^{0(n)}\Big|_{S_k} = 2a_k^{(n)}, \quad \delta\sigma_{12}^{0(n)}\Big|_{S_k} = -b_k^{(n)}.$$

Then, Eq. (19.15) takes the form

$$\sum_{k=1}^{N}\left[2c_k^{(n)} \int_{S_k} K_{11}^{(n-1)} \mathrm{d}S_k + 2a_k^{(n)} \int_{S_k} K_{22}^{(n-1)} \mathrm{d}S_k - b_k^{(n)} \int_{S_k} K_{12}^{(n-1)} \mathrm{d}S_k\right] = F^{(n-1)}. \tag{19.17}$$

Based on the presented linear operator Eq. (19.17), it is possible to obtain systems of linear algebraic equations (SLAE) for the set of sought-for coefficients $a_k^{(n)}, b_k^{(n)}, c_k^{(n)}$ using various additional IP information. For example, we may consider a series of experiments to measure displacements or frequency response on a boundary part; as a result, we obtain a SLAE from (19.16) with respect to $3N$ unknowns $a_k^{(n)}, b_k^{(n)}, c_k^{(n)}$.

The initial approximation can be defined in the class of constant or linear functions. Note that, generally speaking, the number of SLAE equations may not coincide with the number of unknown coefficients. The resulting ill-conditioned SLAE is solved using the A. N. Tikhonov regularization method (Tikhonov and Arsenin 1977). In Nedin et al. (2017, 2018), some reconstruction results were published within the framework of the described approach when considering just one loading mode in a frequency range in the vicinity of the second resonance.

#### 19.4.2.2 Using the Projection Method

Another efficient way to study Eq. (19.15) is the projection method. Its advantage over the difference approximations methods is that there is no additional error in calculating the derivatives of numerically specified functions. The projection method for studying Eq. (19.15) is based on finding the function $\Phi$ in a form of a combination of linearly independent functions. Let us represent the required stress function as a linear combination of polynomials of degrees higher than 2:

$$\delta\Phi^{(n)} = \sum_{k,l=0}^{K} a_{kl}^{(n)} x_1^k x_2^l,$$

where $a_{kl}^{(n)}$ are the required expansion coefficients for the n-th iteration, the indices $k, l$ correspond to the powers of $x_1^k, x_2^l$. Writing out the left-hand side of Eq. (19.15), we obtain

$$\int\limits_S \left(\delta\Phi,_{22}^{(n)} K_{11}^{(n-1)} - \delta\Phi,_{12}^{(n)} K_{12}^{(n-1)} + \delta\Phi,_{11}^{(n)} K_{22}^{(n-1)}\right) \mathrm{d}S$$
$$= \int\limits_S \left( \left[ \sum_{k,l=0}^{K} a_{kl}^{(n)} l(l-1) x_1^k x_2^{l-2} \right] K_{11}^{(n-1)} - \left[ \sum_{k,l=0}^{K} a_{kl}^{(n)} kl x_1^{k-1} x_2^{l-1} \right] K_{12}^{(n-1)} \right.$$
$$\left. + \left[ \sum_{k,l=0}^{K} a_{kl}^{(n)} k(k-1) x_1^{k-2} x_2^{l} \right] K_{22}^{(n-1)} \right) \mathrm{d}S = \sum_{k,l=0}^{K} a_{kl}^{(n)} A_{kl}^{(n-1)},$$

where

$$A_{kl}^{(n-1)} = \int_S \left( l(l-1) x_1^k x_2^{l-2} K_{11}^{(n-1)} - kl x_1^{k-1} x_2^{l-1} K_{12}^{(n-1)} + k(k-1) x_1^{k-2} x_2^l K_{22}^{(n-1)} \right) \mathrm{d}S$$

are coefficients calculated at the n-th iteration; kernels $K_{ij}^{(n-1)}$ are calculated by one of the ways described previously, depending on the type of sounding of the plate. Thus, the operator Eq. (19.15) takes the form

$$\sum_{k,l=0}^{K} A_{kl}^{(n-1)} a_{kl}^{(n)} = F^{(n-1)}. \tag{19.18}$$

Based on (19.18), one can obtain a SLAE with respect to the set of unknown coefficients $a_{kl}^{(n)}$ using various additional IP data.

One of the ways to reduce the IP to SLAE based on (19.18) is to consider several vibration frequencies $\omega_k \in [\omega_-, \omega_+]$ for the selected dynamic loading mode. In addition, various loading regimes can be considered, differing in the type and location of the loads applied. Thus, in the general case, we can consider a series of $M$ different tests on static or dynamic loading in different frequency ranges, and the IP is reduced to the study of the following SLAE at each iteration:

$$\sum_{k,l=0}^{K} A_{klm}^{(n-1)} a_{kl}^{(n)} = F_m^{(n-1)}, \quad m = \overline{1, M}, \tag{19.19}$$

where the subscript $m$ corresponds to the test number for static or frequency sensing at a given frequency. In this case, the number of equations in SLAE (19.19) may not be equal to the number of the unknowns $a_{kl}^{(n)}$; this system is solved using the regularization method of A. N. Tikhonov (Tikhonov and Arsenin 1977). The initial approximation can also be selected in the class of constant functions by minimizing the residual functional on the grid, using information about the boundedness of prestress functions. In particular, one can use the zero initial approximation.

In Nedin and Vatulyan (2014), Nedin et al. (2018), Vatulyan and Nedin (2020), the results of using this technique to reconstruct various laws of prestress inhomogeneity *for a single loading mode* in a set of frequencies are presented. Note that considering only one loading mode allows us to reconstruct the prestress fields at best in the class of linear functions. At the same time, consideration of *several loading modes* can significantly improve the reconstruction accuracy.

#### 19.4.2.3 Computational Experiments

Below we give 2 examples of reconstruction when considering several static experiments on in-plane loading of the rectangular plate occupying the flat region $\{x_1 \in [0, l], x_2 \in [-h/2, h/2]\}$ with the parameters $l = 1.3$ m, $h = 0.5$ m, $E = 1.98 \times 10^{11}$ Pa, $\nu = 0.28$, $\rho = 7884$ kg/m$^3$, clamped by one face $x_1 = 0$. The true prestress fields were set within the range

$$\frac{\tilde{\sigma}_{ij}^0}{E} = 10^{-4} \div 10^{-3}.$$

1. Prestress reconstruction in the class of constant approximations.
   Consider the true stress function $\Phi$ in the form of a linear combination of polynomials of not higher than the fourth degree

$$\begin{aligned}\Phi &= a_{20}x_1^2 + a_{11}x_1x_2 + a_{02}x_2^2\\ &+ a_{30}x_1^3 + a_{21}x_1^2x_2 + a_{12}x_1x_2^2 + a_{03}x_2^3\\ &+ a_{40}x_1^4 + a_{31}x_1^3x_2 + a_{22}x_1^2x_2^2 + a_{13}x_1x_2^3 + a_{04}x_2^4,\end{aligned}$$

which corresponds to the quadratic laws of prestress:

$$\begin{aligned}\sigma_{11}^0 &= 2a_{02} + 2a_{12}x_1 + 6a_{03}x_2 + 2a_{22}x_1^2 + 6a_{13}x_1x_2 + 12a_{04}x_2^2,\\ \sigma_{22}^0 &= 2a_{20} + 6a_{30}x_1 + 2a_{21}x_2 + 12a_{40}x_1^2 + 6a_{31}x_1x_2 + 2a_{22}x_2^2,\\ \sigma_{12}^0 &= -\left(a_{11} + 2a_{21}x_1 + 2a_{12}x_2 + 3a_{31}x_1^2 + 4a_{22}x_1x_2 + 3a_{13}x_2^2\right).\end{aligned} \tag{19.20}$$

The values of the coefficients are taken equal to

$$\begin{aligned}a_{20} &= \;\;0.00034\;E,\\ a_{11} &= \;\;0.00051\;E,\\ a_{02} &= -0.00037\;E,\\ a_{30} &= \;\;0.4\;a_{20}/6l,\\ a_{21} &= \;\;0.6\;a_{20}/2h,\\ a_{12} &= \;\;0.7\;a_{02}/2l,\\ a_{03} &= \;\;0.5\;a_{02}/6h\end{aligned}$$

(the corresponding prestress values are measured in Pa).
We shall search for $\Phi$ in the class of quadratic polynomials, which corresponds to the constants prestresses:

$$\delta\sigma_{11}^{0(n)} = 2a_{02}^{(n)},\quad \delta\sigma_{22}^{0(n)} = 2a_{20}^{(n)},\quad \delta\sigma_{12}^{0(n)} = -a_{11}^{(n)}. \tag{19.21}$$

Figure 19.1 depicts the reconstruction of the prestress of the form (19.21). Here, 3 static loading experiments were considered with measuring the displacements on the boundary parts that differ in the type and location of the loads applied. The used loading modes correspond to the tension and bending of the plate, with the components of tangential and normal loads on different faces specified.

2. Prestress reconstruction in the class of linear approximations.
Figure 19.2 shows the result of quadratic prestress reconstruction in the class of linear approximations at 3 iterations. The relative error was calculated by the formula

$$\delta_{ij} = \frac{\left|\tilde{\sigma}_{ij}^0 - \sigma_{ij}^0\right|}{\max\limits_{x,y}\left|\tilde{\sigma}_{ij}^0\right|}100\%,$$

where $\tilde{\sigma}_{ij}^0$ are the true prestresses, and $\sigma_{ij}^0$ are the recovered ones. The average error values were searched for using the formula

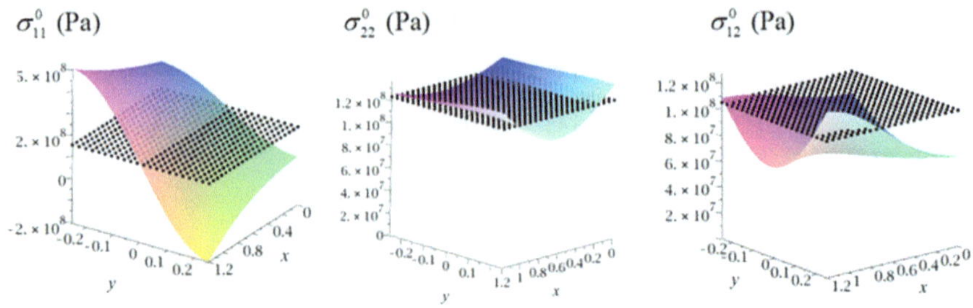


**Fig. 19.1** Reconstruction of the prestress components in the class of constants. Surfaces—the true prestress functions, points—the reconstruction result ($\sigma_{11}^0 = -1.79 \times 10^8$ Pa, $\sigma_{22}^0 = 1.55 \times 10^8$ Pa, $\sigma_{12}^0 = -1.65 \times 10^8$ Pa)

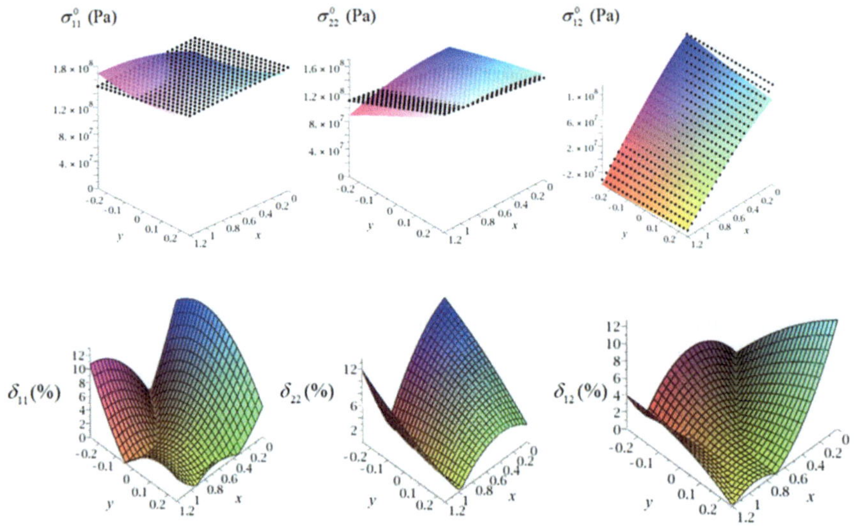


**Fig. 19.2** Above: reconstruction of the prestress components in the linear function class. Surfaces—the true prestress functions, points—the reconstruction result. Bottom: plots of the corresponding relative error

$$\bar{\delta}_{ij} = \frac{1}{S} \int\limits_S \delta_{ij}(x_1, x_2) \mathrm{d}x_1 \mathrm{d}x_2$$

and equal $\bar{\delta}_{11} = 3.66\%$, $\bar{\delta}_{22} = 5.59\%$ and $\bar{\delta}_{12} = 2.05\%$, correspondingly. To obtain this reconstruction, 6 simple plate loads were considered using normal and tangential loads applied to different faces.

## 19.5 Conclusions

A linearized formulation of the problem on vibrations of a prestressed elastic body is considered. On its basis, the inverse problems of prestress recovery are studied; IP of 2 types are considered, in the presence of additional information on the measured displacement field: (1) in a set of points of the region at a fixed vibration frequency; (2) on a part of the boundary of the region in some frequency range. The questions of uniqueness of the inverse problem solution of the first type are investigated. Methods and algorithms for studying two-dimensional inverse problems are proposed and discussed, and the results of some computational experiments are presented.

**Acknowledgements** This work was financially supported by a grant from the Government of the Russian Federation (Agreement No. 075-15-2019-1928).

## References

Deng D, Kiyoshima S (2012) Numerical simulation of welding temperature field, residual stress and deformation induced by electro slag welding. Comput Mater Sci 62:23–34

Guz A, Makhort F, Gushcha O (1977) Introduction to acoustoelasticity. Naukova Dumka, Kiev (in Russ.)

Hoger A (1986) On the determination of residual stress in an elastic body. J Elast 16:303–324

Ling H, Yang C, Feng S, Lu H (2020) Predictive model of grinding residual stress for linear guideway considering straightening history. Int J Mech Sci 176:105536

Moharrami R, Sanayei M (2020) Numerical study of the effect of yield strain and stress ratio on the measurement accuracy of biaxial residual stress in steels using indentation. J Mater Res Technol 9(3):3950–3957

Nedin R (2019) Modeling and frequency analysis of prestressed functionally graded plates with holes. Comput Continuum Mech (Vychislitel'naya mekhanika sploshnykh sred) 12(2):192–201 (in Russ.)

Nedin R (2020) Analysis of non-uniform residual stress state in plates with in-plane and transverse heterogeneities. IOP Conf Ser J Phys Conf Series 1479:012119

Nedin R, Vatulyan A (2013) Inverse problem of non-homogeneous residual stress identification in thin plates. Int J Solids Struct 50:2107–2114

Nedin R, Vatulyan A (2014) Concerning one approach to the reconstruction of heterogeneous residual stress in plate. ZAMM Z Angew Math Mech 94(1–2):142–149

Nedin R, Dudarev V, Vatulyan A (2014) Nondestructive identification of inhomogeneous residual stress state in deformable bodies on the basis of the acoustic sounding method. Adv Mater Res 996:409–414

Nedin R, Dudarev V, Vatulyan A (2017) Some aspects of modeling and identification of inhomogeneous residual stress. Eng Struct 151:391–405

Nedin R, Vatulyan A, Bogachev I (2018) Direct and inverse problems for prestressed functionally graded plates in the framework of the Timoshenko model. Math Methods Appl Sci 41(4):1600–1618

Pham TH, Kim SE (2017) Determination of equi-biaxial residual stress and plastic properties in structural steel using instrumented indentation. Mater Sci Eng A 688:352–363

Robertson R (1998) Determining residual stress from boundary. Measurements: a linearized approach. J Elast 52:63–73

Salvati E, Korsunsky A (2018) A simplified FEM eigenstrain residual stress reconstruction for surface treatments in arbitrary 3D geometries. Int J Mech Sci 138–139:457–466

Suresh S, Giannakopoulos A (1998) A new method for estimating residual stresses by instrumented sharp indentation. Acta Mater 46(16):5755–5767

Tikhonov A, Arsenin V (1977) Solution of ill-posed Problems. Winston & Sons, Washington, D.C

Truesdell C (1977) A first course in rational continuum mechanics, vol 1. Academic Press, New York, San Francisco, London

Vatulyan A (2019) Coefficient inverse problems of mechanics. Fizmatlit, Moscow (in Russ.)

Vatulyan A, Gukasyan L (2019) On the Cauchy problem for a partial differential equation of the first order and its applications in the theory of inverse problems. DSTU Bull (Vestnik DGTU) 12(7):11–20 (in Russ.)

Vatulyan A, Nedin R (2020) On the reconstruction of the characteristics of the plane initial stress state. Mech Solids 55(5):624–632

Vatulyan A, Dudarev V, Nedin R (2015) Initial stresses: modeling and identification. SFedU Publishing House, Rostov-on-Don (in Russ.)

Wang F, Mao K, Li B (2018) Prediction of residual stress fields from surface stress measurements. Int J Mech Sci 140:68–82

Yonezu A, Kusano R, Hiyoshi T, Chen X (2015) A method to estimate residual stress in austenitic stainless steel using a microindentation test. J Mater Eng Perform 24:362–372

Yurov V, Nedin R, Vatulyan A (2020) Oscillations of a non-uniform finite hollow cylinder under conditions of complex prestressed state. Eng Struct 221:111019

# Chapter 20
# Three-Dimensional Dynamic Response of a Cylindrical Cavity in a Half-Space of Partially Saturated Soil Under Internal Step-Loading

**Andrey N. Petrov, Mikhail V. Grigoryev, and Leonid A. Igumnov**

**Abstract** In this paper, three-dimensional dynamic response of a cylindrical cavity in a half-space of soil under uniform normal traction applied at the cavity surface is studied numerically. The surrounding soil is treated as a partially saturated porous medium, and the governing equations for the dynamic problem are presented. Analysis of the problem is preformed in the Laplace domain using the direct boundary element technique. The technique is employed together with the Laplace inverse transform by the stepped method to obtain the time-domain solutions. A parametric study is presented to demonstrate the effect of poroelastic material parameters and geometrical parameters on the response of the cavity.

**Keywords** Poroelastic half-space · Embedded cylindrical cavity · Gradually applied step load · Boundary element method · Laplace transform

## 20.1 Introduction

Because of the great practical usefulness in engineering, evaluation of the response of a cavity due to applied internal transient loads is the subject of numerous studies. In practical applications, the case of structures like the underground caverns or the mines subjected to explosions and sudden excavations can be idealized by the model of the semi-infinite medium involving a cylindrical cavity subjected to dynamic load acting on the cavity surface.

A. N. Petrov (✉) · M. V. Grigoryev · L. A. Igumnov
Research Institute for Mechanics, National Research Lobachevsky State University of Nizhny Novgorod, 23, Gagarin Av., Bld. 6, 603950 Nizhny Novgorod, Russia
e-mail: andrey.petrov@mech.unn.ru

M. V. Grigoryev
e-mail: grimok@live.ru

L. A. Igumnov
e-mail: igumnov@mech.unn.ru



H. Altenbach et al. (eds.), *Advanced Materials Modelling for Mechanical, Medical and Biological Applications*, Advanced Structured Materials 155,
https://doi.org/10.1007/978-3-030-81705-3_20

In this type of problems discrete formulations such as Finite Element Method (FEM) and Boundary Element Method (BEM) are mostly used (Akhlaghi and Nikkar 2014; Gao et al. 2011; Amorosi and Boldini 2009; Rajapakse and Senjuntichai 1995; Kattis et al. 2003). Especially, use of the boundary element method in analyzing the class of problems in question is justified by its key advantages: its numerical-analytical nature, the relatively random geometry of boundary surfaces, automatic provision of the conditions of behavior over infinity in semi-infinite bodies. However, the numerical results published in previous studies pertain only to either two-dimensional problems (He et al. 2017) or to wave scattering problems (Yuan et al. 2017), or were obtained by treating the soil surrounding the cavity as a fully saturated poroelastic media (Senjuntichai and Rajapakse 1993). In reality soil is a three-phased material, comprising solid particles, water and air and it should be more properly modeled as a porous partially saturated media.

Fundamental solutions of the three-dimensional dynamic theory of partially saturated poroelastic media, required for implementing the boundary-element methodology, are presented in Ashayeri et al. (2011), but without the results of the analysis of the problem. Based on somewhat different dynamic equations of poroelastic media, Li and Schanz (2013) also derives fundamental solutions and analyzes problems of wave propagation as a result of a load acting on a half-space surface. Results of boundary-element modeling of wave processes in partially saturated poroelastic bodies are also given in Igumnov et al. (2019a, 2019b).

The present paper uses the boundary element method to analyze the problem of dynamic loading acting inside a cylindrical cavity located in a partially saturated poroelastic half-space. The effect of the permeability and of the cavity depth on the dynamic responses of the pore pressure and the displacement is studied.

## 20.2 Poroelastic Model

The mathematical model of the problem will be written following the approach of Li and Schanz (2013). A poroelastic medium is represented using the following mathematical model of a heterogeneous material: an elastic skeleton phase ($s$) and two fluid phases—pore water ($w$) and pore air ($a$) filling the pore system. The sealed pores are considered as part of the solid. All the three phases are assumed to be compressible. Temperature variations are neglected. Such a poroelastic material can be considered as partially saturated, and its model is called a three-phase model. The relative proportions of constituent volumes are characterized by its porosity $\phi$ and saturation degree $S_f$ of each phase $f \in \{w, a\}$ as:

$$\phi = \frac{V_v}{V}, S_f = \frac{V_f}{V_v}, \tag{20.1}$$

where $V_v$ is the volume of interconnected pores in the specimen, $V$ is the total volume of the material, $V_f$ is the volume of the $f$-phase. This gives the average density of the mixture $\rho$ as:

$$\rho = (1-\phi)\rho_s + \phi S_w \rho_w + \phi S_a \rho_a, \tag{20.2}$$

where $\rho_s$, $\rho_w$ and $\rho_a$ are partial densities of each phase. Consider a case where the pores are completely filled:

$$S_a + S_w = 1. \tag{20.3}$$

The following equation suggested by Bishop (1960) for the effective stress in partially saturated porous media is used assuming the Bishop's parameter is identical to the saturation degree $S_w$:

$$\sigma_{ij} = \sigma_{ij}^{\text{eff}} - \alpha \left( S_w p^w + S_a p^a \right) \delta_{ij}, \tag{20.4}$$

where $\sigma_{ij}^{\text{eff}}$ is the effective stress, $\sigma_{ij}$ represents the total stress, $p^w$ and $p^a$ are the pore water and the pore air pressures, and $\delta_{ij}$ is Kronecker's delta. The Biot's coefficient $\alpha$ is described by the relationship:

$$\alpha = 1 - \frac{K}{K_s}, \tag{20.5}$$

where $K$ is the bulk modulus of the porous media and $K_s$ is the bulk modulus of the solid grains. The relation between stress and strain in the solid skeleton under assumption of its isotropic linear elastic behaviour is given by:

$$\sigma_{ij}^{\text{eff}} = 2G\varepsilon_{ij}^s + \left( K - \frac{2}{3}G \right) \varepsilon_{kk}^s \delta_{ij}, \tag{20.6}$$

where $G$ denotes the shear modulus. The linear strain tensor $\varepsilon_{ij}^s$ is defined as

$$\varepsilon_{ij}^s = \frac{1}{2} \left( u_{i,j}^s + u_{j,i}^s \right), \tag{20.7}$$

where $u_i^s$ is the displacement vector in the solid phase of the porous medium.

To construct equations of motion, constitutive equations must be combined with the related balance equations of momentum and balance equations of mass of each of the phases (for details see Li and Schanz 2013). The linear momentum balance equation for each fluid phase results in the generalised Darcy equation for multiphase flow as:

$$\phi S_w q_i^f = -\kappa_f \left( p_{,i}^f + \rho_f \ddot{u}_i^s + \rho_f \dot{q}_i^f \right), \tag{20.8}$$

where $q_i^f$ represents the Darcy velocity and $\kappa_f$ is the phase permeability. The phase permeability $\kappa_f$ is defined as:

$$\kappa_f = K_{\text{rf}} \frac{k}{\eta_f}, \tag{20.9}$$

where $k$ is the intrinsic fluid permeability of the porous medium, $K_{\mathrm{rf}}$ is the relative fluid phase permeability, $\eta_f$ is the viscosity of the fluid. $K_{\mathrm{rf}}$ has the following form for water and air, respectively:

$$K_{\mathrm{rw}} = S_e^{(2+3\theta)/\theta},\ K_{\mathrm{ra}} = (1 - S_e)^2 \left(1 - S_e^{(2+\theta)/\theta}\right), \tag{20.10}$$

where $\theta$ is the pore size distribution index and $S_e$ is the effective saturation degree given by

$$S_e = \frac{S_w - S_{rw}}{S_{\mathrm{ra}} - S_{\mathrm{rw}}}, \tag{20.11}$$

where $S_{\mathrm{rw}}$ is the residual water saturation degree and $S_{\mathrm{ra}}$ is the residual air saturation degree.

In a partially saturated porous material containing two non-mixing fillers the interface surface between then is curved as a result of intermolecular interaction forces. The accompanying pressure difference in both of the interfacing phases is called capillary pressure $p^c$. Capillary pressure can be represented as a function of the saturation degree as follows:

$$p^c = p^a - p^w = p^d S_e^{-1/\theta}, \tag{20.12}$$

where $p^d$ is the air pressure required for driving water out of the pores.

Using a Laplace transform assuming vanishing initial conditions makes it possible to write down a precise form of the governing equations in Laplace transform domain in terms of spatial derivatives of displacements $u_i^s$ as well as pore pressures $p^w$ and $p^a$. The final differential equations in the Laplace domain are arranged into matrix product form as follows:

$$\begin{bmatrix} B_1\delta_{ij} + B_2\partial_i\partial_j & B_3\partial_i & B_4\partial_i \\ B_5\partial_j & B_6 & B_7 \\ B_8\partial_j & B_9 & B_{10} \end{bmatrix} \begin{bmatrix} \hat{u}_i \\ \hat{p}^w \\ \hat{p}^a \end{bmatrix} = \begin{bmatrix} -\hat{F}_i \\ -\hat{I}^w \\ -\hat{I}^a \end{bmatrix}, \mathbf{x} \in \Omega, \Omega \subset \mathbf{R}^3, \tag{20.13}$$

where

$$B_1 = G\nabla^2 - (\rho - \beta S_w \rho_w - \gamma S_a \rho_a)\, s^2,\ B_2 = K + \frac{G}{3},\ B_3 = -(\alpha - \beta)\, S_w, \tag{20.14}$$

$$B_4 = -(\alpha - \gamma)\, S_w,\ B_5 = -(\alpha - \beta)\, S_w s,\ B_6 = -\left(\zeta S_{ww} S_w + \frac{\phi}{K_w} S_w - S_u\right), \tag{20.15}$$

$$B_7 = -(\zeta S_{aa} S_w + S_u)\, s,\ B_8 = -(\alpha - \gamma)\, S_a\, s,\ B_9 = -(\zeta S_{ww} S_a + S_u)\, s, \tag{20.16}$$

$$B_{10} = -\left(\zeta S_{aa} S_a + \frac{\phi}{K_a} S_a - S_u\right) s + \frac{\gamma S_a}{\rho_a s}\nabla^2,\ S_u = -\frac{\theta\,(S_{ra} - S_{rw})}{p^d} S_e^{(\theta+1)/\theta}, \tag{20.17}$$

$$\beta = \frac{\kappa_w \phi \rho_w s}{\phi S_w + \kappa_w \rho_w s}, \gamma = \frac{\kappa_a \phi \rho_a s}{\phi S_a + \kappa_a \rho_a s}, \tag{20.18}$$

$$S_{ww} = S_w - \theta\left(S_w - S_{rw}\right), S_{aa} = S_a + \theta\left(S_w - S_{rw}\right). \tag{20.19}$$

In (20.13) $\hat{F}_i$ denotes the bulk body force per unit volume, $\hat{I}^w$ and $\hat{I}^a$ denote fluid source terms, and symbol $\wedge$ denotes Laplace transform with complex variable $s$.

A generalized displacement vector and a generalized force vector are additionally introduced as

$$\mathbf{u}(\mathbf{x}, s) = \left(\hat{u}_1^s, \hat{u}_2^s, \hat{u}_3^s, \hat{p}^w, \hat{p}^a\right), \tag{20.20}$$

$$\mathbf{t}(\mathbf{x}, s) = \left(\hat{t}_1^s, \hat{t}_2^s, \hat{t}_3^s, \hat{q}^w, \hat{q}^a\right), \tag{20.21}$$

where $\hat{t}_i^s = \hat{\sigma}_{ij} n_j$ and $\hat{q}^f = \hat{q}_i^f n_i$, $n_i$ is the outward unit normal vector to the boundary $\partial\Omega$.

Equation (20.13), supplemented with boundary conditions:

$$\mathbf{u}(\mathbf{x}, s) = \tilde{\mathbf{u}}, \mathbf{x} \in \Gamma^u, \tag{20.22}$$

$$\mathbf{t}(\mathbf{x}, s) = \tilde{\mathbf{t}}, \mathbf{x} \in \Gamma^\sigma, \tag{20.23}$$

where $\Gamma^u$ is the Dirichlet's boundary and $\Gamma^\sigma$ is the Neumann's boundary, fully describe the boundary problem in the representations of the 3D isotropic dynamic theory of poroelasticity.

## 20.3 Boundary Element Method Framework

Boundary-value problem (20.13), (20.22), (20.23) is solved using the direct boundary element method, based on a combined use of integral Laplace transform and boundary integral equation (BIE) of the 3D isotropic theory of poroelasticity:

$$\mathbf{C}(\mathbf{y})\mathbf{u}(\mathbf{y}, s) + \int_\Gamma \mathbf{T}(\mathbf{x}, \mathbf{y}, s)\mathbf{u}(\mathbf{x}, s)d\Gamma = \int_\Gamma \mathbf{U}(\mathbf{x}, \mathbf{y}, s)\mathbf{t}(\mathbf{x}, s)d\Gamma, \mathbf{x}, \mathbf{y} \in \Gamma, \tag{20.24}$$

where **U**(**x**, **y**,s) and **T**(**x**, **y**,s) are matrices of fundamental and singular solutions, respectively, **x** is an integration point, **y** is an observation point. The values of the coefficients of matrix **C** are defined by the geometry of boundary $\Gamma$ . A procedure for obtaining BIE's, based on the weighted residual method can be found in Schanz (2001). Some of the problems of the arising kernels of BIE's are discussed in Amenitsky et al. (2009).

Equation (20.24) comprise singular integrals in the sense of Cauchy, which are quite difficult to compute. Use of the boundary properties of retarded potentials makes it possible, based on Igumnov et al. (2016), to write down a regular representation of Eq. (20.18):

$$\int_{\Gamma} \left(\mathbf{T}(\mathbf{x}, \mathbf{y}, s)\mathbf{u}(\mathbf{x}, s) - \mathbf{T}^0(\mathbf{x}, \mathbf{y}, s)\mathbf{u}(\mathbf{y}, s) - \mathbf{U}(\mathbf{x}, \mathbf{y}, s)\mathbf{t}(\mathbf{x}, s)\right) \mathrm{d}\Gamma = 0, \mathbf{x}, \mathbf{y} \in \Gamma, \tag{20.25}$$

where $\mathbf{T}^0(\mathbf{x}, \mathbf{y}, s)$ is singularity matrix. Using Eq. (20.25), it is possible to construct a boundary-element solution of the BIE.

In the result of spatial discretization, boundary $\Gamma$ is represented with a set of $K_E$ quadrangular eight-node boundary elements. The geometry of each element $E_K$ is defined by biquadratic functions of form $N_m$ and the global coordinates of nodes $\mathbf{x}_m^k$, related as Bazhenov and Igumnov (2008)

$$\mathbf{x}(\zeta) = \sum_{m=1}^{8} N_m(\xi)\mathbf{x}_m^k, \quad k = 1, \ldots, K \tag{20.26}$$

where $\xi = (\xi_1, \xi_2) \in (-1, 1)^2$ are local coordinates. According to correlated interpolation model (Goldshteyn 1978), displacements are described using bilinear elements with the related bilinear functions of form $R_l(\xi)$, and surface generalized forces are described with constant boundary elements:

$$\mathbf{u}(\xi) = \sum_{l=1}^{4} R_l(\xi)\mathbf{u}_l^k, \tag{20.27}$$

$$\mathbf{t}(\xi) = \mathbf{t}^k, \tag{20.28}$$

where $\mathbf{u}_m^k$ and $\mathbf{t}^k$ are nodal values of displacements and tractions, respectively, over element $E_k$ .

A discrete representation of BIE's written at the nodes of the approximation of boundary functions $\mathbf{y}^i$ , using the collocation method and accounting for (20.26)–(20.28), is of the following form:

$$\sum_{k=1}^{K_E} \sum_{m=1}^{4} \Delta\mathbf{T}_{mi}^k \mathbf{u}_m^k = \sum_{k=1}^{K} \Delta\mathbf{U}_{mi}^k \mathbf{t}^k, \tag{20.29}$$

$$\Delta\mathbf{U}_{mi}^k = \int_{-1}^{1} \mathbf{U}(\mathbf{x}^k(\xi), \mathbf{y}^i, s) J^k(\xi)\mathrm{d}\xi, \tag{20.30}$$

$$\Delta\mathbf{T}_{mi}^k = \int_{-1}^{1} \left(R_m(\xi)\mathbf{T}(\mathbf{x}^k(\xi), \mathbf{y}^i, s) - \mathbf{I} \cdot \mathbf{T}^0(\mathbf{x}^k(\xi), \mathbf{y}^i)\right) J^k(\xi)\mathrm{d}\xi, \tag{20.31}$$

where $\mathbf{I}$ is unit matrix, $J^k$ is Jacobian of the local coordinates into global ones.

The elements of matrices $\Delta\mathbf{U}_{mi}^{k}$, $\Delta\mathbf{T}_{mi}^{k}$ are computed using numerical integration schemes depending on the kind of integral (nonsingular or singular). Nonsingular integrals arise, when the collocation point does not belong to the element. Here, standard Gaussian-type formula is used in combination with a hierarchical subdivision of the elements (Lachat and Watson 1976). Singular integrals arise, when the collocation point is situated on the element being integrated over. In this case, new local coordinates are introduced, making it possible to avoid a singularity in the integrand and to use Gaussian integration.

## 20.4 Laplace Transform Inversion

Consider a method based on the fundamental integration theorem—the stepped method of numerical inversion of Laplace transform. Consider the following integral:

$$y(t) = \int_{0}^{1} f(\tau)\mathrm{d}\tau \tag{20.32}$$

Integral (20.32) gives rise to Cauchy problem for an ordinary differential equation:

$$\frac{\mathrm{d}}{\mathrm{d}t}x(t) = sx(t) + C, x(0) = 0. \tag{20.33}$$

Integral (20.32) is substituted for by a quadrature sum, weighting factors of which are determined using Laplace representation and the linear multistep method (Igumnov and Petrov 2016). Further derivation is based on the results of this work. The traditional stepped method of integrating the original is that integral (20.32) is calculated using the following relation:

$$y(0) = 0, y(n\Delta t) = \sum_{k=1}^{n} \omega_k(\Delta t), n = 1..N, \tag{20.34}$$

$$\omega_n(\Delta t) = \frac{R^{-n}}{L} \sum_{l=0}^{L-1} \hat{f}\left(\frac{\gamma(R\mathrm{e}^{\mathrm{i}l\frac{2\pi}{L}})}{\Delta t}\right) \mathrm{e}^{-\mathrm{i}nl\frac{2\pi}{L}}, \tag{20.35}$$

where $\Delta t$ is time step, $\gamma(z) = 3/2 - 2z + z^2/2$, $N$ is number of time steps, $R$ is the parameter of the method.

## 20.5 Numerical Results and Analysis

Using the above methodology and the related software realized on its basis, a problem of a load acting inside a cylindrical cavity of radius $r = 3$ m embedded in a partially saturated poroelastic half-space is analyzed (Fig. 20.1). The length of the cavity is $l = 6 \times r$, and the depth of the cavity is $h$. A load applied to the cavity surface linearly increases up to time 0.1 s and then acquires a constant value of $10^6$ N/m$^2$. Zero boundary conditions for filler flows are also assigned over the cavity surface. The half-space surface is permeable and load-free. The material parameters used in calculation are summarized in Table 20.1. The computation values of all parameters in (20.34)–(20.35) are taken $R = 0.997$, $L = N = 500$, $\Delta t = 0.002$ s.

The effect of the depth $h$ on displacements and pore pressures at the cavity surface is examined firstly and six values of $h$, i.e. $h = 1.25 \times r$, $h = 1.5 \times r$, $h = 1.75 \times r$, $h = 2 \times r$, $h = 2.5 \times r$ and $h = 3 \times r$ are discussed. The results of the corresponding numerical analyses are presented in Figs. 20.2, 20.3, 20.4 and 20.5.

It can be seen in Figs. 20.2 and 20.3 that the magnitude of normal displacement at points $A$, $B$, $C$, $D$ increases gradually with time reaching the maximum value near $t = 0.1$s and thereafter remains constant for every fixed $h$. Maximum displacement occurs at point $C$ for the case $h = 3 \times r$, i.e., when the cavity is farthest from the free surface. It is also observed that the difference between the magnitudes of normal displacement at points $A$, $B$, $C$ becomes more and more negligible with increasing $h$. At the same time, the maximum displacement at point $D$ for $h = 2.5 \times r$ is almost 40% higher than the maximum displacement at the same point for $h = 3 \times r$.

Figures 20.4 and 20.5 show the effect of varying the cavity depth $h$ on the pore water pressure and the pore air pressure at points $A$, $B$, $C$, $D$. The solid and dashed lines correspond to the water and air pore pressures, respectively. It can be seen that the pore pressures increase rapidly with time in the interval $0\,\text{s} < t < 1.0\,\text{s}$ at all four points and reaches its peak value near $t = 0.1$ s for every fixed $h$. Thereafter, the pore pressures decrease with time and tend to an asymptotic value when $t > 0.2$ s.

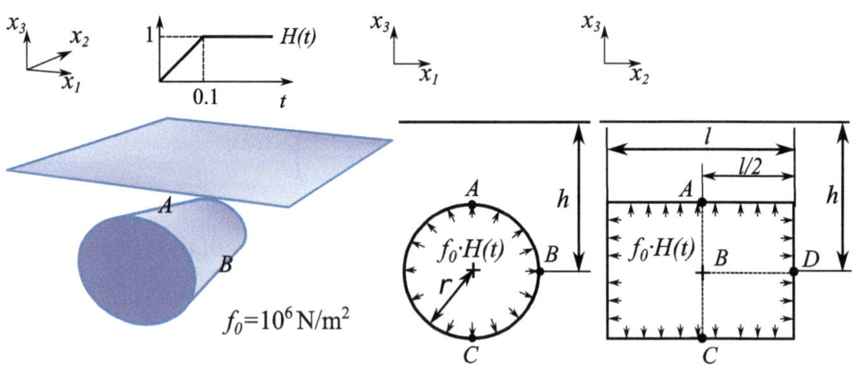


**Fig. 20.1** Problem description

**Table 20.1** General material parameters

| Parameter type | Symbol | Value | Unit |
|---|---|---|---|
| Porosity | $\phi$ | 0.23 | – |
| Density of the solid skeleton | $\rho_s$ | 2650 | $kg/m^3$ |
| Density of the water | $\rho_w$ | 997.0 | $kg/m^3$ |
| Density of the air | $\rho_a$ | 1.10 | $kg/m^3$ |
| Drained bulk modulus of the mixture | $K$ | $1.02 \times 10^9$ | $N/m^2$ |
| Shear modulus of the mixture | $G$ | $1.44 \times 10^9$ | $N/m^2$ |
| Bulk modulus of the solid grains | $K_s$ | $3.5 \times 10^{10}$ | $N/m^2$ |
| Bulk modulus of the water | $K_w$ | $2.25 \times 10^9$ | $N/m^2$ |
| Bulk modulus of the air | $K_a$ | $1.10 \times 10^5$ | $N/m^2$ |
| Intrinsic permeability | $k$ | $2.5 \times 10^{-12}$ | $m^2$ |
| Viscosity of the water | $\eta_w$ | $1.0 \times 10^{-3}$ | $N\,s/m^2$ |
| Viscosity of the air | $\eta_a$ | $1.8 \times 10^{-5}$ | $N\,s/m^2$ |
| Gas entry pressure | $p^d$ | $5 \times 10^4$ | $N/m^2$ |
| Water saturation | $S_w$ | 0.9 | – |
| Residual water saturation | $S_{rw}$ | 0 | – |
| Residual air saturation | $S_{ra}$ | 1 | – |
| Pore size distribution index | $\theta$ | 1.5 | – |

The maximum pressure occurs at point $A$ for $h = 1.25 \times r$. The magnitudes of pore pressures at each of points $A$, $B$, $C$ become closer one another with increasing $h$. In the same time, the peak values of pore pressure as well as the its asymptotic values increasing with increasing $h$ at point $D$ and exceed the corresponding values at three other points for $h = 3 \times r$. Thus, it is found that depth of the cavity has a significant effect on the response.

To investigate the influence of the permeability on the response of the cavity, the case where $h = 2 \times r$ is considered. Figures 20.6 and 20.7 show the response of pore pressures for different permeabilities $k = 2.5 \times 10^{-12}\,m^2$, $k = 5 \times 10^{-12}\,m^2$, $k = 12.5 \times 10^{-12}\,m^2, k = 2.5 \times 10^{-11}\,m^2, k = 2.5 \times 10^{-10}\,m^2$. The displacements are not presented here since the effect of permeability on them is negligible in the case of the impermeable cavity surface.

It can be seen in Figs. 20.6 and 20.7, that for all the considered values of $k$, the pressures at points $A, B, C, D$ behave in the same way as was described earlier. Also it is observed that the pore pressures decrease as $k$ increases and that the maximum pore pressure occurs at point $D$.

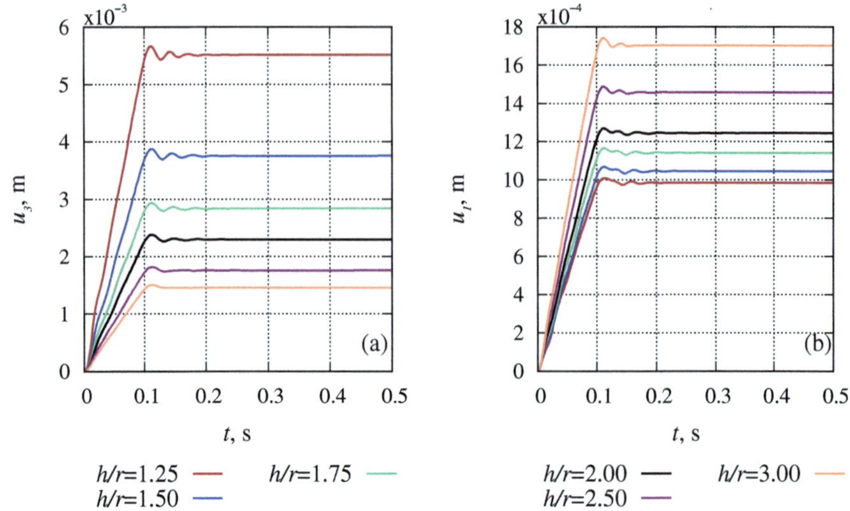


**Fig. 20.2** Effect of $h$ on the normal displacements at points $A$ (**a**) and $B$ (**b**)

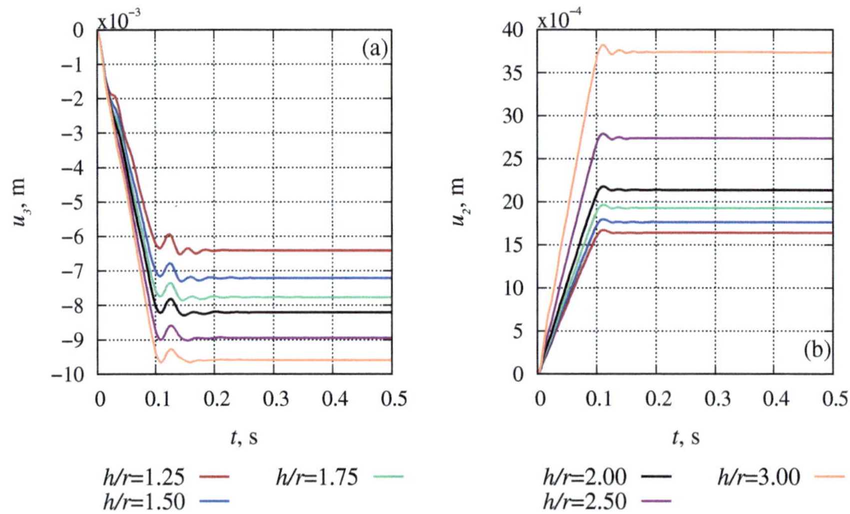


**Fig. 20.3** Effect of $h$ on the normal displacements at points $C$ (**a**) and $D$ (**b**)

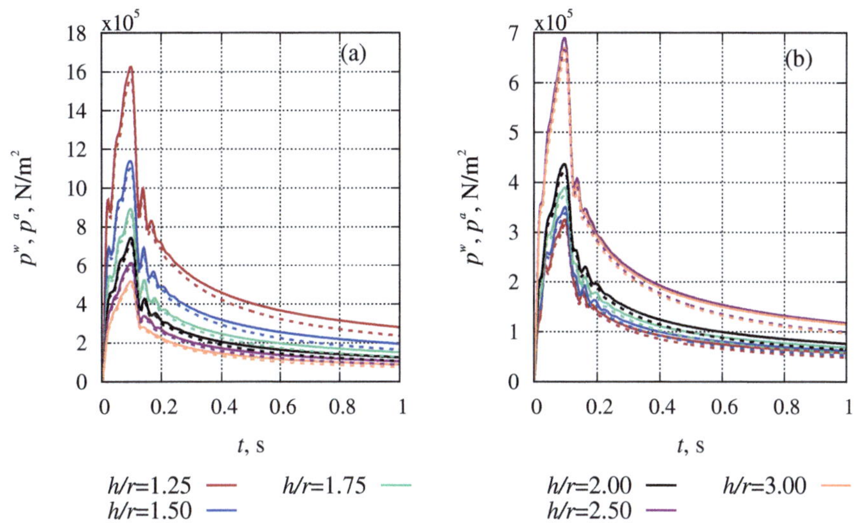


**Fig. 20.4** Effect of $h$ on the pore pressures at points $A$ (**a**) and $B$ (**b**)

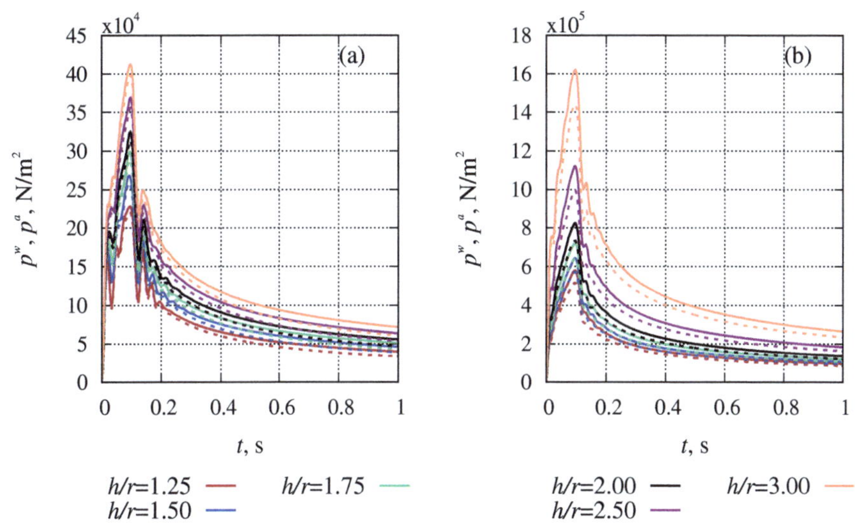


**Fig. 20.5** Effect of $h$ on the pore pressures at points $C$ (**a**) and $D$ (**b**)

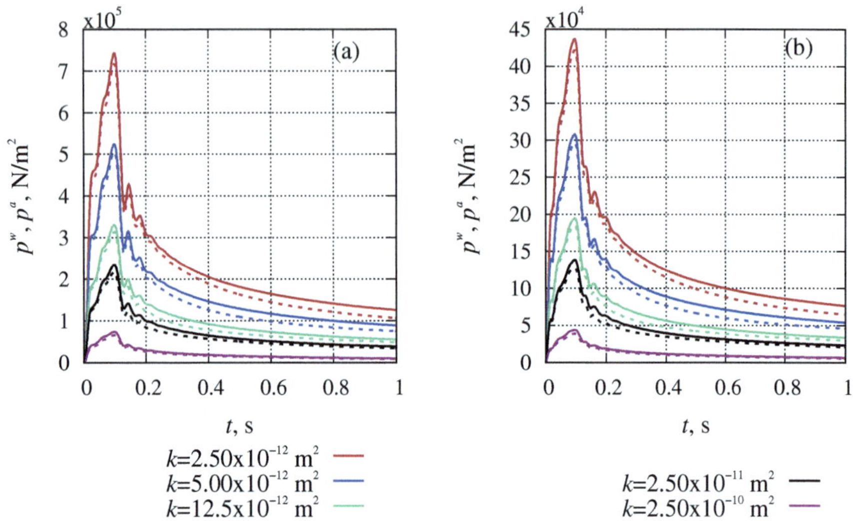


**Fig. 20.6** Effect of $k$ on the pore pressures at points $A$ (**a**) and $B$ (**b**)

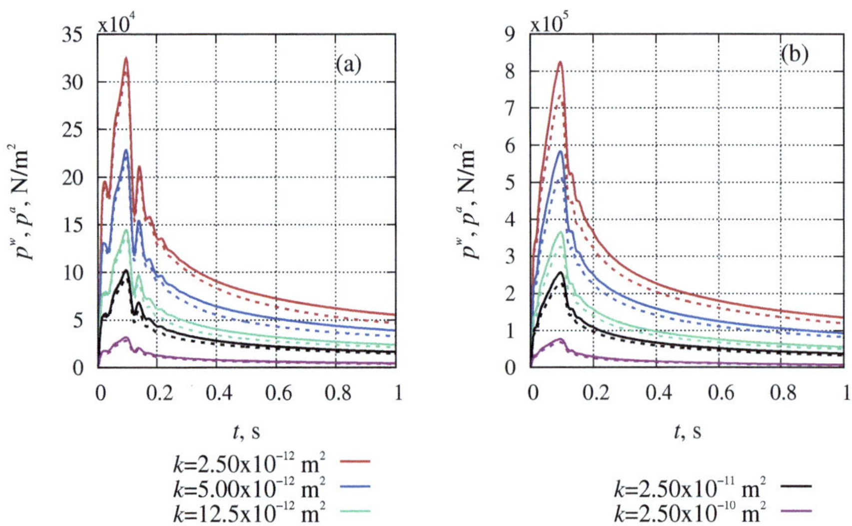


**Fig. 20.7** Effect of $h$ on the pore pressures at points $C$ (**a**) and $D$ (**b**)

## 20.6 Concluding Remarks

The boundary-element approach and the apparatus of Laplace transform have been used in the numerical parametric study of the analysis of the problem of a transient gradually applied step load acting inside a cylindrical cavity situated in a partially saturated poroelastic half-space. The effect of the values of permeability and the depth of the position of the cavity on dynamic responses of displacements and pore pressures at points on the cavity surfaces has been analyzed. In general, it is found that maximum normal displacements at the cavity surface occurred at the top and bottom points, i.e., at the nearest and farthest points from the free surface of the half-space. With the increase cavity depth, these values decrease at the top of the cavity, whereas they increase bottom of the cavity. It is also found that the pore pressures decrease with the increasing the cavity depth at the nearest point to the free surface of the half-space, whereas they increase at the others considered points. For a fixed cavity depth the normal displacements on the cavity surface are found to change insignificantly as the permeability increases. In the same case, it is found that the pore pressures on the cavity surface decrease as the permeability increases and that the maximum pore pressure occurs at the farthest point from the free surface of the half-space. It is concluded that both the displacements and pore pressures at the cavity surface are significantly influenced by the permeability and the proximity of the cavity to the free surface.

**Acknowledgements** The work was employed under financial support the Ministry of Science and Higher Education of the Russian Federation (task 0729–2020-0054) at the part of boundary-value problem statement and boundary-element formulation. The work on numerical experiments in the framework of the poroelastic model of the material was done with the financial support of RFBR (project number 19-38-90224).

## References

Akhlaghi T, Nikkar A (2014) Effect of vertically propagating shear waves on seismic behavior of circular tunnels. Sci World J 2014:806092

Amenitsky AV, Belov AA, Igumnov LA, Karelin IS (2009) Granichnye integral'nye uravneniya dlya resheniya dinamicheskikh zadach trekhmernoi teorii porouprugosti (Boundary integral equations for analyzing dynamic problems of 3-D porouselasticity). Probl Strength Plast 71:164–171 (in Russian)

Amorosi A, Boldini D (2009) Numerical modelling of the transverse dynamic behaviour of circular tunnels in clayey soils. Soil Dyn Earthq Eng 29(6):1059–1072

Ashayeri I, Kamalian M, Jafari MK, Gatmiri B (2011) Analytical 3D transient elastodynamic fundamental solution of unsaturated soils. Int J Numer Anal Methods Geomech 35(17):1801–1829

Bazhenov VG, Igumnov LA (2008) Metody granichnyh integral'nyh uravnenij i granichnyh elementov v reshenii zadach trekhmernoj dinamicheskoj teorii uprugosti s sopryazhennymi polyami (Boundary integral equations and boundary element methods in treating the problems of 3D elastodynamics with coupled fields). PhysMathLit (in Russian)

Bishop A (1960) The principles of effective stress. Norges Geotekniske Institute

Gao M, Gao GY, Li DY (2011) Transient response of lining structure subjected to sudden internal uniform loading considering effects of coupling mass. Yantu Gongcheng Xuebao 33(6):862–868

Goldshteyn RV (1978) Metod granichnyh integral'nyh uravnenij—sovremennyj vychislitel'nyj metod prikladnoj mekhaniki (Boundary integral equations method: numerical aspects and application in mechanics. MIR (in Russian)

He C, Zhou S, Di H, Xiao J (2017) A 2.5-D coupled FE-BE model for the dynamic interaction between tunnel and saturated soil. Lixue Xuebao 49(1):126–136

Igumnov LA, Petrov AN (2016) Modelirovanie dinamiki chastichno nasyshchennyh porouprugih tel na osnove metoda granichno-vremennyh elementov (Dynamics of partially saturated poroelastic solids by boundary-element method). PNRPU Mech Bull 3:47–61 (in Russian)

Igumnov LA, Litvinchuk SY, Petrov AN, Ipatov AA (2016) Numerically analytical modeling the dynamics of a prismatic body of two- and three-component materials. In: Parinov IA, Chang SH, Topolov VY (eds) Advanced materials, vol 175. Springer International Publishing, Cham, pp 505–516

Igumnov LA, Petrov AN, Belov AA, Mironov AA, Lyubimov AK, Dianov DY (2019a) Numerically-analytically studying fundamental solutions of 3-D dynamics of partially saturated poroelastic bodies. Mater Phys Mech 42(5):596–601

Igumnov LA, Petrov AN, Vorobtsov IV (2019b) The time-step boundary-element scheme on the nodes of the Lobatto method in problems of 3-D dynamic poroelasticity. Mater Phys Mech 42(1):103–111

Kattis SE, Beskos DE, Cheng AHD (2003) 2D dynamic response of unlined and lined tunnels in poroelastic soil to harmonic body waves. Earthq Eng Struct Dyn 32(1):97–110

Lachat JC, Watson JO (1976) Effective numerical treatment of boundary integral equations: a formulation for three-dimensional elastostatics. Int J Numer Meth Eng 10(5):991–1005

Li P, Schanz M (2013) Time domain boundary element formulation for partially saturated poroelasticity. Eng Anal Boundary Elem 37(11):1483–1498

Rajapakse RKND, Senjuntichai T (1995) An indirect boundary integral equation method for poroelasticity. Int J Numer Anal Methods Geomech 19(9):587–614

Schanz M (2001) Wave propagation in viscoelastic and poroelastic continua: a boundary element approach. Springer, Berlin

Senjuntichai T, Rajapakse R (1993) Int J Numer Anal Methods Geomech 17(6):357–383

Yuan Z, Boström A, Cai Y, Cao Z (2017) Closed-form analytical solution for vibrations from a tunnel embedded in a saturated poroelastic half space. J Eng Mech 143(9):04017079

# Chapter 21
# Correlating the Mechanical Properties to the Mineral Density of Brown Spot Lesion in Dentine Using Nanoindentation and X-ray Micro-tomography

**Evgeniy V. Sadyrin**

**Abstract** Caries in the stage of brown spot lesion is an advanced phase of the disease thus causing serious problems for patients. Having passed the dentine-enamel junction, carieogenic bacteria begin demineralization of dentine in its vicinity. If not treated promptly, the affected area may reach the pulp of the tooth, creating a gradient of strength characteristics in dentine. In the present work, the mechanical properties and mineral density in the direction from the dentine-enamel junction to the dental pulp were obtained using nanoindentation and X-ray computed micro-tomography and correlated. The results show that the mechanical properties of dentine are dependent on its mineral content. The approach can be used as control technique for the further research of remineralization procedures in clinical practice.

## 21.1 Introduction

A by-product of the metabolism of plaque cariogenic bacteria (primarily of the *Streptococcus mutans* and *Lactobacillus* groups) in the presence of carbohydrates in the oral cavity is organic acid (Leverett et al. 1993a, b; Jia et al. 2021), mainly the lactic one. It locally lowers Ph in the area of plaque and penetrates into the tooth enamel, dissolving its main structural components—hydroxyapatite crystallites (Hannig and Hannig 2010; Harper et al. 2021; Takahashi and Nyvad 2016; Yanagisawa and Miake 2003; Sadyrin et al. 2020a). This phenomenon is called demineralization, thus, caries begins. Bacteria are able to penetrate the enamel, reaching the dentine-enamel junction and then the dentinal tubules, propelling themselves down them. This then causes the breakup of the main organic component of dentine—the collagen. In this case, a density gradient arises in dentine (from the pulp the density decreases towards the dentine-enamel junction). The lower the density, the more severely is damaged dentine. Usually at this phase of caries, the dental clinician may observe the brown area; thus, this phenomenon is called the brown spot lesion (BSL).

E. V. Sadyrin (✉)
Laboratory for Mechanics of Biomaterials, Research and Education Center Materials, Don State Technical University, 344003 Rostov-on-Don, Russia
e-mail: e.sadyrin@sci.donstu.ru

 
H. Altenbach et al. (eds.), *Advanced Materials Modelling for Mechanical, Medical and Biological Applications*, Advanced Structured Materials 155,
https://doi.org/10.1007/978-3-030-81705-3_21

The hard tissues of teeth were reported to demonstrate the decrease values of both mineral density (Featherstone et al. 1983; Banerjee and Boyde 1998; Kinney et al. 1994) and mechanical properties (Sadyrin et al. 2021, 2020b; Huang et al. 2010; Hosoya et al. 2000; Kislyakov et al. 2020) caused by natural and artificial demineralization. The study of caries from a materials and microstructural point of view provides a better understanding of the changes in the carious enamel and affected dentine. Various approaches can be applied in this regard, each with its benefits and drawbacks. Hardness measurements, as a tool for detecting such changes, have come a long way from the pioneering studies more than 50 years ago (Avery 1962) to nanotechnological research involving instrumented estimation of both indentation hardness and reduced Young's modulus at the sub-micron level (Angker et al. 2005; Biswas et al. 2018). Computed X-ray micro-tomography (micro-CT) is another powerful tool to study the biological tissues (Mashiatulla et al. 2017; Kogan et al. 2020; Domenyuk et al. 2018), in particular the processes of demineralization and remineralization of teeth (Cochrane et al. 2010; Pires et al. 2018), and it can also be applied for testing efficiency of caries treatments (Songsiripradubboon et al. 2014). This work aims to study mineral density and reduced Young's modulus by the depth of carious dentine in case of the liquid saturated BSL. The results of the study may serve as reference data for researches of efficacy of minimally invasive methods for clinical treatment of caries.

## 21.2 Materials and Methods

A permanent human molar was extracted for orthodontic reasons from an individual in the dental department of Rostov State Medical University clinic (Rostov-on-Don, Russia). The patient prior to the procedure provided informed consent. The local independent ethics committee of Rostov State Medical University approved the study (statement № 13/20 from 10.09.2020). The fissure BSL was assessed by a clinician. Following extraction, the sample was kept in 1 wt. % NaClO solution for 10 min. Then the sample was stored in Hanks Balanced Salt Solution (HBSS) at 4 °C with thymol granules (Unifarm, Russia), added to prevent fungal growth and for disinfection reasons. The ratio of thymol to HBSS was 1:1000.

For a non-destructive estimation of mineral density in the BSL area, the micro-CT device (Xradia Versa 520, Carl Zeiss X-ray Microscopy, USA) was used. The tooth was placed inside the chamber of the device in the container filled with distilled water with the calibration sample (phantom), placed above the tooth. A calibration sample was made from four parallelepipeds of 15 mm high, base 50 × 50 mm with measured densities. The densities were as follows for each parallelepiped: PET ($\rho = 1.35$ g/cm$^3$), magnesium alloy Ma2-1 M ($\rho = 1.78$ g/cm$^3$), aluminium alloy SAS1-400 ($\rho = 2.69$ g/cm$^3$) and fluorite ($\rho = 3.09$ g/cm$^3$). The phantom was used for building a linear relationship between grey value and mineral density that allowed conversion of the measured grey value of the tooth slice to an estimated mineral density (Zou et al. 2011; Alyahya et al. 2019; Tkachev et al. 2021). Thus, in VGStudio

MAX 3.4 software (Volume Graphics GmbH, Heidelberg, Germany), the virtual region of interest as maximum cylinder in width, length and height was taken, which could be inserted into each parallelepiped (the same for all parallelepipeds). Inside it the grey value was averaged.

The following acquisition parameters were selected: magnification objective 0.4× , voltage 60 kV, power 4.5 W, source filter HE6 (according recommendations of the manufacturer), exposure time 16 s (to maintain intensity values greater than 5000 with the selected source parameters and filter) and voxel size 19.1 μm. Totally, 1601 X-ray projections were acquired, which were afterwards reconstructed using XRMReconstructor 12.0.8086.19558 software (Carl Zeiss AG, Germany) with automatically adjusted centre shift values, $\sigma = 0.5$ Gauss blurring filter, and beam hardening constant 0.19. For each scan, the sample was placed at the closest possible distance to the X-ray source in such a way that the whole sample with the calibration phantom attached could be acquired. The 2048 × 2048 pixels CCD camera was kept at − 59 °C. The acquisition was conducted with camera binning factor = 2, which resulted in 1024 × 1024 pixels sized projection images. The dynamic ring removal option, which enables small random motions of the sample during acquisition, was enabled for each projection. In total, 10 reference X-ray images were acquired with equal time intervals between them (air was used as reference). The average of these references was applied to each projection. Prior acquisition, up to one-hour warm-up scan was performed with the same source parameters. After tomography acquisition, the images were exported in DICOM format for volume rendering in VGStudio MAX 3.4 software.

Following micro-CT study, the sample was prepared for the nanoindentation test. Thus, a longitudinal cross-section through the region containing a BSL was performed using a precision saw (Isomet 4000, Buehler, USA) with an abrasive SiC disc (MetAbrase, Buehler, USA). The cross-section was carefully ground using the SiC-based abrasive papers with the grit sizes: P800, P1200, P2000, P2500 with distilled water used as lubricant. After that, polishing was conducted with diamond oil-based suspensions with the particles of 15, 6, 3, and 1 μm diameters. Next, the final polishing was made using de-agglomerated alumina 0.05 μm diameter suspension. The lubricant hexilene glycole GreenLube (Allied, USA) was applied during polishing steps. All suspensions were applied on no nap cloth according to manufacturer's recommendations. Each grinding and polishing step was followed by ultrasonic cleaning of the sample in distilled water in the bath (Sonorex RK 31, Bandelin, Germany) for 10 min.

An optical stereomicroscope (Stemi 305, Zeiss, PRC) operating by the Greenough scheme with a colour video camera (Axiocam 105, Zeiss, Germany) was used for visual observation of BSL area of prepared tooth in reflected light.

The distribution of the Young's modulus by depth of the BSL was studied using the nanoindentation device (NanoTest 600 Platform 3, Micro Materials, UK). The experiments were carried out at a constant temperature of 27.0 ± 0.1 °C in a closed chamber, and a calibrated diamond pyramidal Berkovich indenter was used. The maximum load was 5 mN. The loading and unloading rates were 1 mN/s, and dwell period at maximum load was 30 s. The thermal drift was recorded and corrected using the software of the device. To prevent shrinkage of dentine tissue, the sample was maintained wet with saline droplets using a syringe pump (Terufusion TE-332, Terumo, Belgium). Droplets were applied on the sample in the time period between indentations in order to prevent the influence of the droplet impact on the indenter tip. The values of the reduced Young's modulus $E_r$ were calculated using the Oliver–Pharr method (Oliver and Pharr 1992, 2004).

## 21.3 Results and Discussion

Four segments (lines) were marked on the image obtained by optical microscopy (Fig. 21.1). The same plane (in a form of the virtual cross-section) was found on the non-destructive map of mineral density obtained using micro-CT.

On this virtual cross-section, the lines in the same coordinates were virtually marked (Fig. 21.2). Across these segments, the values of the reduced Young's modulus, indentation hardness, and mineral density were measured in this research, the correlated diagrams of which are presented on Fig. 21.3. Note that for the mineral density measurements, the segments were not just lines, but parallelepipeds with actual width and depth, 10 μm each.

Analysing behaviour of the reduced Young's modulus and mineral density, one should bear in mind that the BSL in the sample presented in this research had a serious amount of time to progress prior tooth extraction. It is safe to assume that dentine managed to form new layers of materials inside the pulp chamber (the formation of the denticle clearly visible on both optical image and the map of mineral density). The odontoblast inside the dentinal tube, sensing the appearance of a bacterium, gives a signal about the invasion of an alien microorganism into the pulp, which in turn releases growth factors that make the odontoblast to pump more minerals from the tooth pulp [building of secondary dentine (Arana-Chavez and Massa 2004; Ricucci et al. 2020, 2014; Goldberg et al. 2008)].

Figure 21.3a shows segment 1 under research corresponding to sound dentine. These control results are typical for dentine with a small scattering: $E_r = 22.34 \pm 4.02$ GPa if not counted the last three points on the graph, most close to the dentine-enamel junction where a decrease in the values was detected, whereas mineral density is $2.24 \pm 0.02$ g/cm$^3$ for the whole investigated range. The part of the segment in the vicinity of dentine-enamel junction demonstrates slight decrease not only for the Young's modulus, but also for mineral density.

Segment 2 demonstrates a gradual decrease in the values of both mechanical properties and density from the pulp to the dentine-enamel junction. In general, this is predictable since in the vicinity of the pulp dentine is least affected by demineralization.

Segment 3 as well demonstrates a decrease in the values of both the reduced elastic modulus and the mineral density from the pulp to the dentine-enamel junction, however, the character of the diagrams is different. The drop of the Young's modulus is sharp from the very beginning of the investigated range, and on the distance about

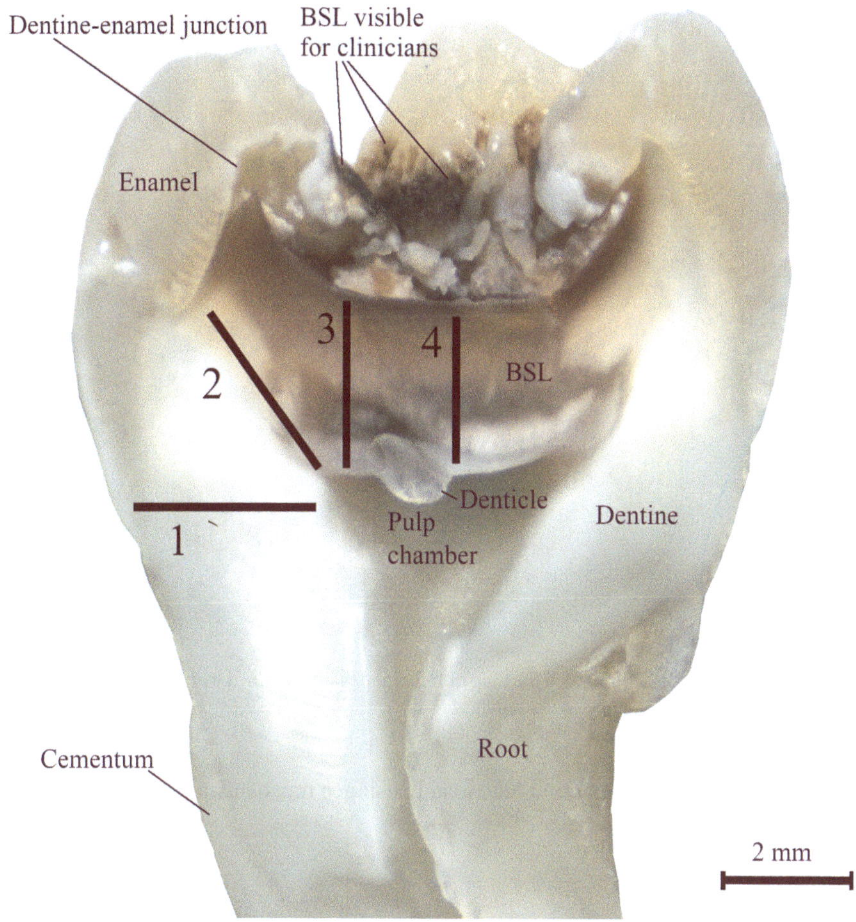


**Fig. 21.1** Tooth cross-section image obtained by optical microscopy. Note the segments (lines) across which the mechanical properties and mineral density are measured

400 μm, the values tend to stabilize. From the distance of 400–1000 μm, a step can be observed on the diagram following by the further decrease. The mineral density decreases more gradually, resembling the behaviour of the values from the segment 2.

Segment 4 shows the sudden decrease in the values of the Young's modulus very similar to the one on the segment 3. It is not surprise as both segments are situated in the area affected by BSL in more extent than the segment 2. Moreover, segment 4 crosses the area under the fissure, where the caries presumably originated. This may be the reason of absence of the step part of the diagram of the Young's modulus on Fig. 21.3c: the drop in it is most severe from all the cases studied in the paper. At the same time, the drop in density is also strong, but not so sharp. Judging by the density of dentine in the first half of the segment, it is only slightly affected by

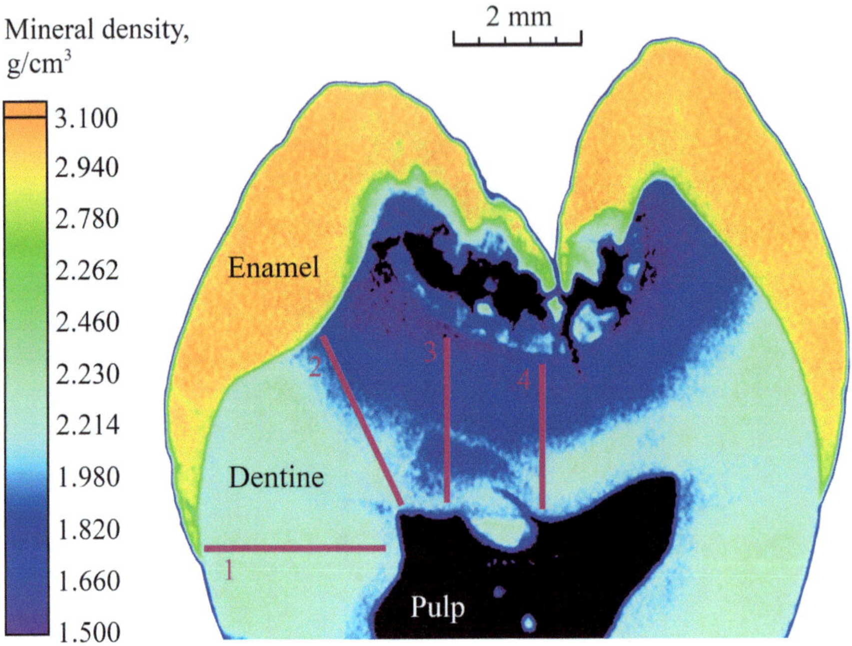


**Fig. 21.2** Virtual cross-section with the map of mineral density obtained using micro-CT. Note the same segments (lines) as on the optical image

carious process; at the same time, the load resistance mechanism is already severely impaired.

## 21.4 Conclusion

The reduced Young's modulus obtained using nanoindentation and mineral density estimated using micro-CT in the direction from the tooth pulp to the dentine-enamel junction was studied across three segments of the BSL on the human molar and one control sound segment. The research was conducted in the presence of liquid. The features connected to the abnormality of the indentation response were recorded. In all three segments situated inside the BSL region, the decrease of both mineral density and Young's modulus was detected and described. It is planned to include prediction of behaviour of carious tissues using either analytical (Vasiliev et al. 2020, 2019; Argatov and Sabina 2016; Krenev et al. 2018; Sadyrin et al. 2017; Aizikovich et al. 2013; Zelentsov et al. 2018) or finite-elemental (Marteau et al. 2015; Wang et al. 1992; Ecker et al. 2020; Xu and Rowcliffe 2004) mathematical model describing indentation into inhomogeneous media. The results can serve as reference data for researches of efficacy of minimally invasive methods for clinical treatment.

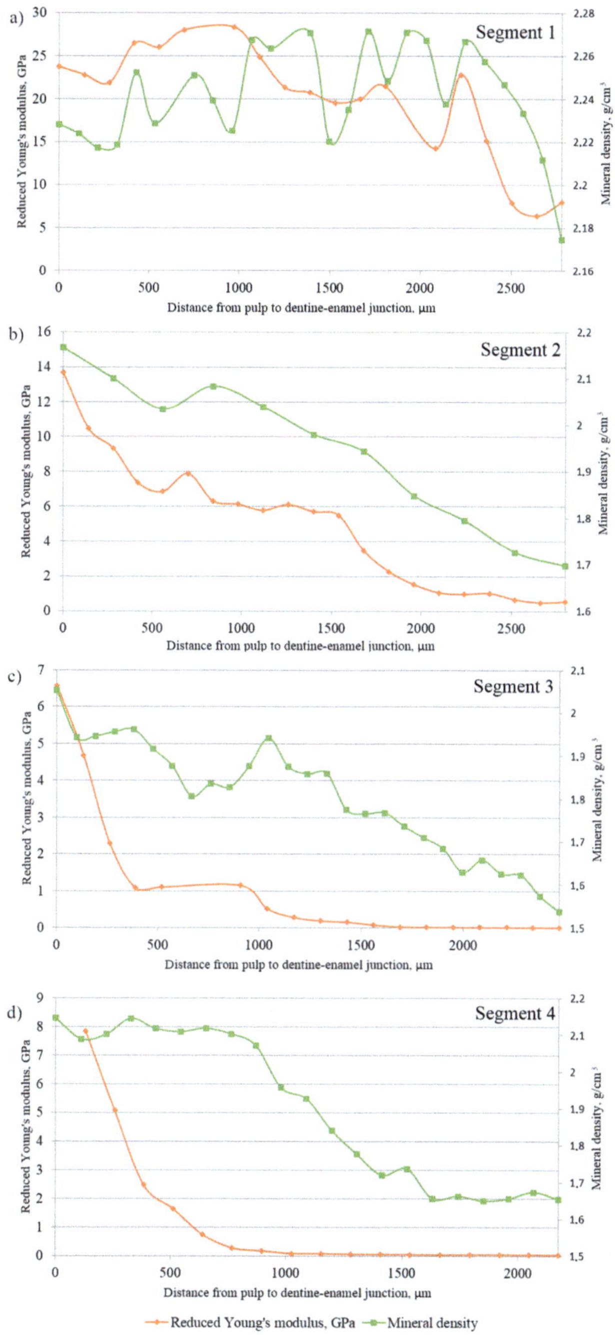


**Fig. 21.3** Results of measurements of the Young's modulus and the mineral density along the segments: 1 (**a**); 2 (**b**); 3 (**c**); 4 (**d**)

**Acknowledgements** This work was supported by the Russian Science Foundation (RSF) through grant No. 19-19-00444. The author also thanks Professor M.V. Swain (University of Sydney, Sydney, Australia) and B.I. Mitrin (Don State Technical University, Rostov-on-Don, Russia) for help in processing of the experimental data, Professor S.Yu. Maksyukov and D.V. Yogina (Rostov State Medical University, Rostov-on-Don, Russia) for providing the sample. The research was conducted in the Research and Education Center of Don State Technical University (https://nano.donstu.ru/).

## References

Aizikovich SM, Vasiliev AS, Volkov SS, Mitrin BI, Sadyrin EV (2013) An analytical solution to the problem of interaction of a circular plate with an inhomogeneous soft layer. In: Pietraszkiewicz W, Gorski J (eds) Shell structures: theory and applications—proceedings of the 10th SSTA 2013 conference, vol 3. CRC Press, Croydon, pp 45–48

Alyahya A, Alqareer A, Swain M (2019) Microcomputed tomography calibration using polymers and minerals for enamel mineral content quantitation. Med Princ Pract 28(3):247–255

Angker L, Swain MV, Kilpatrick N (2005) Characterising the micro-mechanical behaviour of the carious dentine of primary teeth using nano-indentation. J Biomech 38(7):1535–1542

Arana-Chavez VE, Massa LF (2004) Odontoblasts: the cells forming and maintaining dentine. Int J Biochem Cell Biol 36(8):1367–1373

Argatov II, Sabina FJ (2016) Small-scale indentation of an elastic coated half-space: the effect of compliant substrate. Int J Eng Sci 104:87–96

Avery JK (1962) Microradiographic and microhardness studies of developing enamel. Arch Oral Biol 7:245–256

Banerjee A, Boyde A (1998) Autofluorescence and mineral content of carious dentine: scanning optical and backscattered electron microscopic studies. Caries Res 32(3):219–226

Biswas N, Mukhopadhyay AK, Dey A (2018) Nanoindentation of teeth: a hard but tough hybrid functionally graded composite. Nanoindent Nat Mater Hierarchical Funct Graded Microstr 13:59

Cochrane NJ, Cai F, Huq NL, Burrow MF, Reynolds EC (2010) New approaches to enhanced remineralization of tooth enamel. J Dent Res 89(11):1187–1197

Domenyuk DA, Zelensky VA, Rzhepakovsky IV, Anfinogenova OI, Pushkin SV (2018) Application of laboratory and x-ray gentral studies un early diagnostics of metabolic disturbances of bone tissue in children with autoimmune diabetes mellitus. Entomol Appl Sci Lett 5(4):1–12

Ecker W, Keckes J, Krobath M, Zalesak J, Daniel R, Rosenthal M, Todt J (2020) Nanoscale evolution of stress concentrations and crack morphology in multilayered CrN coating during indentation: experiment and simulation. Mater Des 188:108478

Featherstone JDB, Ten Cate JM, Shariati M, Arends J (1983) Comparison of artificial caries-like lesions by quantitative microradiography and microhardness profiles. Caries Res 17(5):385–391

Goldberg M, Lacerda-Pinheiro S, Priam F, Jegat N, Six N, Bonnefoix M, Septier D, Chaussain-Miller C, Veis A, Denbesten P, Poliard A (2008) Matricellular molecules and odontoblast progenitors as tools for dentin repair and regeneration. Clin Oral Invest 12(2):109–112

Hannig M, Hannig C (2010) Nanomaterials in preventive dentistry. Nat Nanotechnol 5(8):565–569

Harper RA, Shelton RM, James JD, Salvati E, Besnard C, Korsunsky AM, Landini G (2021) Acid-induced demineralisation of human enamel as a function of time and pH observed using X-ray and polarised light imaging. Acta Biomater 120:240–248

Hosoya Y, Marshall SJ, Watanabe LG, Marshall GW (2000) Microhardness of carious deciduous dentin. Oper Dent 25(2):81–89

Huang TTY, He LH, Darendeliler MA, Swain MV (2010) Nano-indentation characterisation of natural carious white spot lesions. Caries Res 44(2):101–107

Jia L, Tao S, Yang J, Liang K, Yu Z, Gao Y, Fan M, Zhang M, He L, Li J (2021) Adhesion of Streptococcus mutans on remineralized enamel surface induced by poly (amido amine) dendrimers. Colloids Surf B Biointerf 197:111409

Kinney JH, Marshall GW Jr, Marshall SJ (1994) Three-dimensional mapping of mineral densities in carious dentin: theory and method. Scanning Microsc 8(2):6

Kislyakov EA, Karotkiyan RV, Sadyrin EV, Mitrin BI, Yogina DV, Kheygetyan AV, Maksyukov SY (2020) Nanoindentation derived mechanical properties of human enamel and dentine subjected to etching with different concentrations of citric acid. In: Aizikovich SM, Altenbach H, Eremeyev VA, Swain MV, Galybin A (eds) Modeling, synthesis and fracture of advanced materials for industrial and medical applications. Springer, Cham, pp 75–83

Kogan M, Popov I, Mitrin B, Popov I, Sadyrin E, Pasetchnik D, Ermakov A, Ugrekhelidze N, Kulikova N (2020) Visualisation of penile structures of laboratory rabbit: ultrasound, histology, and micro-CT. E3S Web Conf 210:06016

Krenev LI, Volkov SS, Sadyrin EV, Chizhik SA (2018) Mechanical material tests by the nanoindentation method at various indenter and specimen temperatures. J Eng Phys Thermophys 91(3):594–600

Leverett DH, Proskin HM, Featherstone JDB, Adair SM, Eisenberg AD, Mundorff-Shrestha SA, Shields CP, Shaffer CL, Billings RJ (1993a) Caries risk assessment in a longitudinal discrimination study. J Dent Res 72(2):538–543

Leverett DH, Featherstone JDB, Proskin HM, Adair SM, Eisenberg AD, Mundorff-Shrestha SA, Shields CP, Shaffer CL, Billings RJ (1993b) Caries risk assessment by a cross-sectional discrimination model. J Dent Res 72(2):529–537

Marteau J, Bouvier S, Bigerelle M (2015) Review on numerical modeling of instrumented indentation tests for elastoplastic material behavior identification. Arch Comput Method Eng 22(4):577–593

Mashiatulla M, Ross RD, Sumner DR (2017) Validation of cortical bone mineral density distribution using micro-computed tomography. Bone 99:53–61

Oliver WC, Pharr GM (1992) An improved technique for determining hardness and elastic modulus using load and displacement sensing indentation experiments. J Mater Res 7(6):1564–1583

Oliver WC, Pharr GM (2004) Measurement of hardness and elastic modulus by instrumented indentation: advances in understanding and refinements to methodology. J Mater Res 19(1):3–20

Pires PM, Dos Santos TP, Fonseca-Gonçalves A, Pithon MM, Lopes RT, de Almeida NA (2018) A dual energy micro-CT methodology for visualization and quantification of biofilm formation and dentin demineralization. Arch Oral Biol 85:10–15

Ricucci D, Loghin S, Lin LM, Spångberg LS, Tay FR (2014) Is hard tissue formation in the dental pulp after the death of the primary odontoblasts a regenerative or a reparative process? J Dent 42(9):1156–1170

Ricucci D, Siqueira Jr, JF, Rôças IN, Lipski M, Shiban A, Tay FR (2020) Pulp and dentine responses to selective caries excavation: a histological and histobacteriological human study. J Dent 100:103430

Sadyrin EV, Mitrin BI, Krenev LI, Nikolaev AL, Aizikovich SM (2017) Evaluation of mechanical properties of the two-layer coating using nanoindentation and mathematical modeling. In: Parinov IA, Chang S-H, Gupta VK (eds) International conference on physics and mechanics of new materials and their applications. Springer, Cham, pp 495–502

Sadyrin EV, Kislyakov EA, Karotkiyan RV, Yogina DV, Drogan EG, Swain MV, Maksyukov SYu, Nikolaev AL, Aizikovich SM (2020a) Influence of citric acid concentration and etching time on enamel surface roughness of prepared human tooth: in vitro study. In: Altenbach H, Brünig M, Kowalewski ZL (eds) Plasticity, damage and fracture in advanced materials. Springer, Cham, pp 135–150

Sadyrin E, Swain M, Mitrin B, Rzhepakovsky I, Nikolaev A, Irkha V, Yogina D, Lyanguzov N, Maksyukov S (2020b) Characterization of enamel and dentine about a white spot lesion: mechanical properties, mineral density. Microstruct Molecul Composit Nanomater 10(9):1889

Sadyrin EV, Mitrin BI, Yogina DV, Swain MV (2021) Preliminary study of distribution of mechanical properties and mineral density by depth of liquid saturated carious dentine. IOP Conf Ser Mater Sci Eng 1029(1):012056
Songsiripradubboon S, Hamba H, Trairatvorakul C, Tagami J (2014) Sodium fluoride mouthrinse used twice daily increased incipient caries lesion remineralization in an in situ model. J Dent 42(3):271–278
Takahashi N, Nyvad B (2016) Ecological hypothesis of dentin and root caries. Caries Res 50(4):422–431
Tkachev SY, Mitrin BI, Karnaukhov NS, Sadyrin EV, Voloshin MV, Maksimov AY, Goncharova AS, Lukbanova EA, Zaikina EV, Volkova AV, Khodakova DV, Mindar MV, Yengibarian MA, Protasova TP, Kit SO, Ermakov AM, Chapek SV, Tkacheva MS (2021) Visualization of different anatomical parts of the enucleated human eye using X-ray micro-CT imaging. Exp Eye Res 203:108394
Vasiliev AS, Sadyrin EV, Volkov SS, Kislyakov EA, Sevostianov I (2019) Construction of the simplified analytical solution of the flat punch indentation contact problem. In: Mladenovic V, Tung PD, Sevostyanov IB, Najafabadi TA, Petkovic M (eds) AIP conference proceedings, vol 2188, issue 1, p 040017
Vasiliev AS, Volkov SS, Sadyrin EV, Aizikovich SM (2020) Simplified analytical solution of the contact problem on indentation of a coated half-space by a conical punch. Mathematics 8(6):983
Wang HF, Yang X, Bangert H, Torzicky P, Wen L (1992) Two-dimensional finite element method simulation of Vickers indentation of hardness measurements on TiN-coated steel. Thin Solid Films 214(1):68–73
Xu ZH, Rowcliffe D (2004) Finite element analysis of substrate effects on indentation behaviour of thin films. Thin Solid Films 447:399–405
Yanagisawa T, Miake Y (2003) High-resolution electron microscopy of enamel-crystal demineralization and remineralization in carious lesions. Microscopy 52(6):605–613
Zelentsov VB, Sadyrin EV, Sukiyazov AG, Shubchinskaya NY (2018) On a method for determination of Poisson's ratio and Young modulus of a material. In: Gross H-G, Khaled H, Najafabadi TA, Nguyen DA, Kudish I, Sevostyanov IB, Cen YG, Egiazarian K, Mladenovic V, Karapetyan G, Minkin M (eds) MATEC web of conferences 2018, vol 226. EDP Sciences, Les Ulis, p 03027
Zou W, Hunter N, Swain MV (2011) Application of polychromatic μCT for mineral density determination. J Dent Res 90(1):18–30

# Chapter 22
# Characterization of Physical Properties of a Porous Material in Terms of Tortuosity of the Porous Space: A Review

**Igor Sevostianov**

**Abstract** The overall properties—thermal, electric, elastic, etc.—of any porous material are strongly dependent on their three-dimensional (3D) microstructures, which include the porosity, pore sizes, and shapes and connectivity of the porous space. These microstructural parameters can be collectively described as the "tortuosity of the porous space" (see, e.g., Chen et al., J Power Sources 273:486–494 2013). We propose to use the tortuosity parameter to characterize the overall material properties of a porous material with interconnected porous space. In the text to follow, we discuss the concept of this parameter and its application to characterize, elastic, electric, and mass transport properties as well as the cross-property connections.

**Keywords** Tortuosity · Effective properties · Porous material · Conductivity · Elasticity

## 22.1 Introduction: The Concept of Tortuosity

The term tortuosity, to the best of our knowledge, has been first introduced by Thomson and Tait (1879) in the context of curvature of a non-plane curve (see Sects. 7–9 of their book). Noting $\delta\varphi$ the angle between the osculating planes at two points at a distance $\delta s$ from one another along the curve, they defined tortuosity $\tau$ of a curve as a derivative

$$\tau_{\text{T-T}} = \frac{\mathrm{d}\varphi}{\mathrm{d}s} \tag{22.1.1}$$

(we use subscript "T-T" to identify the definition of Thomson and Tait). At the beginning of the twentieth-century term tortuosity was adopted in medicine to describe (qualitatively) spatial curvature of blood vessels (see, e.g., Edington 1901, or Cairney

I. Sevostianov (✉)
Department of Mechanical and Aerospace Engineering, New Mexico State University, Las Cruces, NM 88003, USA
e-mail: igor@nmsu.edu



H. Altenbach et al. (eds.), *Advanced Materials Modelling for Mechanical, Medical and Biological Applications*, Advanced Structured Materials 155,
https://doi.org/10.1007/978-3-030-81705-3_22

1924). Later, Carman (1939) suggested to use this term to describe curvilinear character of porous space in the context of hydraulic conductivity. Carman defined it as the ratio of the effective length ($L_e$) of the fluid flow path to the apparent length ($L_a$) of a specimen:

$$\tau_C = L_e / L_a \tag{22.1.2}$$

(subscript C stays for definition of Carman).

Due to the obvious uncertainty of this definition and difficulties associated with its practical implications, Wyllie and Rose (1950), Winsauer et al. (1951), and Cornell and Katz (1953) suggested to use different kinds of tortuosity $\tau$ related to the resistivity factor $F$ (sometimes this factor is also called electrical formation factor or electrical retardation factor):

$$F \equiv k^0 / k^{\text{eff}} = \tau / \psi, \tag{22.1.3}$$

where $k^{\text{eff}}$ is the bulk (effective) electrical conductivity of the porous material completely saturated by electrolyte, $k^0$ is the conductivity of the electrolyte, and $\psi$ is the ratio of the apparent cross-sectional area of the conducting electrolyte to the total cross-sectional area of the specimen. This ratio, however, varies from one cross section to another, and it is unclear, which value is appropriate—maximum, minimum, average, or anything else (see Sevostianova et al. 2010).

The tortuosity $\tau$ in (22.1.3), unlike parameter $T$ introduced by Carman (1939), depends on the degree of constriction of the capillaries. It was demonstrated by Dullien (1975) using the example of a porous medium consisting of identical parallel capillaries with periodic step changes in their diameter. Dullien suggested an expression for $\tau$ that is analogous to one obtained by Michaels (1959) in the context of diffusion problem:

$$\tau = \frac{\left(1 + x^2 y\right)\left(1 + y / x^2\right)}{(1 + y)^2}, \tag{22.1.4}$$

where $x = D_l / D_s$ is the ratio of the diameters of large and small capillary segments and $y = l_l / l_s$ is the ratio of corresponding lengths. However, since "large" and "small" capillary segments are not clearly defined; (22.1.4) does not allow one to estimate tortuosity directly from the image analysis. Dullien also suggested representing tortuosity $\tau$ as a product of classical tortuosity (introduced by Carman) that accounts for increased path length and shape factor $S$, which he called *constriction factor*:

$$\tau = T\,S \tag{22.1.5}$$

Perkins et al. (1956) suggested an experimental procedure to estimate tortuosity $\tau$ and showed that, for completely saturated sandstone, the resistivity factor, tortuosity, and porosity $p$ are interrelated by

$$F = \tau^2 / p = k^0 / k^{\text{eff}}, \tag{22.1.6}$$

It means, in particular, that parameter $\psi$ in (22.1.3), porosity, and tortuosity are interrelated by

$$\psi = p / \tau \tag{22.1.7}$$

If only some of the pores are saturated with electrolyte, this expression has to be modified as follows:

$$F = \tau^2 / \theta\, p, \tag{22.1.8}$$

where $\theta$ is electrolyte content. The experimental procedure was, however, criticized by Wyllie and de Witte (1956). According to their comments, Perkins et al. (1956) actually measured porosity of the sample rather than its tortuosity and they introduced restrictive assumptions into their model.

Amyx et al. (1960) suggested evaluating tortuosity from

$$(Fp)^q = \tau, \tag{22.1.9}$$

where $q$ is the fitting parameter, varying from 0.5 to 1.5. The minimum value of tortuosity according to Amyx et al. (1960) is approached when the exponent in (22.1.9) is equal to 0.5, i.e., when (22.1.9) coincides with (22.1.6). A number of other models containing fitting parameters have been proposed later. Based on a two-dimensional computer simulation of the flow of a Newtonian uncompressible fluid constructed by randomly placed rectangles of equal size, Koponen et al. (1996) suggested the following fitting equation that links tortuosity and porosity:

$$\tau = 1 + 0.8(1 - p). \tag{22.1.10}$$

This equation, however, does not distinguish between tortuosities of rock and sandy soil, which differ substantially at the same porosity level. Revil et al. (1998) tried to clarify the situation by developing an equation for electrical conductivity that accounts for the different behavior of ions in the pore space. They show that tortuosity of the transport of anions is independent of the salinity and corresponds to the bulk tortuosity of the pore space which is given by the product of the resistivity factor F and the porosity. For cations, the situation is different. At high salinities, the dominant paths for the electromigration of cations are located in the interconnected pore space, and the tortuosity for the transport of cations is therefore the bulk tortuosity. Weerts et al. (2001) compared models for gas diffusion coefficient and bulk electrical conductivity of sandy soil. They show that parameters for measured gas and water phase tortuosity coincide. They noted, however, that the fitted parameters have different physical meaning: The tortuosity parameter of the water phase is mainly linked with the tortuosity of the pathway, while the tortuosity parameter of the gas phase is linked with the connectivity of the pathway. These two parameters coincide if the sample is completely dry and gas diffusion is not hindered by water presence. They also provided a lattice Boltzmann computer simulation to compute the tortuosity parameters. Only qualitative agreement with the experimental data is reported. Using Berea and synthetic sandstones cores, Attia (2005) studied the dependence of tortuosity on various petrophysical rock properties, i.e., amount of fine grains, porosity, cementation factor, formation resistivity factor, electrolyte concentrations, and degree of brine saturation. He formulated several fitting expressions for these dependences.

***Remark*** An important consequence of the tortuosity definition according to (22.1.6) is its tensor character. Indeed, since electrical conductivity is a symmetric second-rank tensor, it immediately follows from (22.1.10) that tortuosity is also a symmetric second-rank tensor such that

$$\tau_{ij}\tau_{jk} = k_{ij}^{0}\left(k_{jk}^{\text{eff}}\right)^{-1} \Big/ p. \tag{22.1.11}$$

For isotropic materials, $\tau_{ij} = \tau\delta_{ij}$, and the tortuosity can be treated as a scalar.

Definition (22.1.3) was adopted (with slight modification) by Helmer et al. (1995) to describe mass transfer in tumor:

$$\tau_H \equiv D^0 \big/ D^{\text{eff}}, \tag{22.1.12}$$

where $D^{\text{eff}}$ is the effective mass transfer coefficient of the material and $D^0$ is the mass transfer coefficient of the conducting phase (the second phase is assumed to be impenetrable). Namely, definition (22.1.12) is presently used in the biomedical applications for analysis of the properties of tumor. This definition is slightly inconsistent with (22.1.3)—$\tau_H$ in (22.1.12) is actually a retardation factor rather than tortuosity, as discussed by Clennell (1997).

Pride (1994) has shown that electrical tortuosity (22.1.3) is one of four fundamental properties of a porous medium that are measurable, and rigorously interrelated (the others are porosity, steady hydrodynamic permeability and the electrical length-scale lambda). This rigorous definition does not involve the concept of an effective path length (22.1.2). Steady-state tortuosity represents an average of transport through all available flow pathways. The details of pore structure are only resolved if we consider unsteady transport processes.

Application of tortuosity concept to different phenomena has been discussed in detail by Clennell (1997). He pointed out that tortuosity means different things to different people. He distinguished between geometrical tortuosity (objective characteristic of the pore structure). The second class of tortuosity measures is "retardation factors" extracted from the transport properties of the porous medium. Electrical tortuosity (22.1.3) and the mass transport tortuosity (22.1.12) are examples. Third, he considers tortuosity parameters that enter into some simplifications of a real pore space, such as a network model. Finally, he mentioned tortuosity as an adjustable parameter in empirical models. He pointed out that, different tortuosities can be compared if one converts the transport to an overall flux and compare the efficiency of the transport in the specimen with an idealized case that has maximum efficiency. Due to that, it is possible to prove that the electrical tortuosity (22.1.3) and diffusional tortuosity (22.1.12) are identical in the steady state, i.e.,

$$F = \tau_H \tag{22.1.13}$$

## 22.2 Micromechanical Meaning of Tortuosity

In this section, we use the expression (22.1.3) that connects tortuosity with resistivity factor. The latter can be calculated using various micromechanical models (see review of Markov 2000). Let us consider an electrically conducting material (matrix) containing multiple insulating inhomogeneities (inclusions). We have to evaluate the contribution of insulating inhomogeneities into overall resistivity (Sevostianov et al. 2006).

### *22.2.1 Electrical Resistivity Contribution Tensors*

We first consider a reference volume $V$ of an infinite media with the isotropic electrical conductivity $k_0$ which contains an insulating inhomogeneity. Assuming a linear conduction law (linear relation between the gradient of electric potential $\mathbf{U}$ and the electric current vector $\mathbf{Q}$ per volume $V$), the resistivity contribution tensor of the inhomogeneity $\mathbf{H}^R$ is defined by the following relation (Sevostianov and Kachanov 2002):

$$\mathbf{U} = \frac{1}{k_0}\mathbf{Q} + \Delta\mathbf{U} = \frac{1}{k_0}\mathbf{Q} + \frac{1}{V}\mathbf{H}^R \cdot \mathbf{Q}. \tag{22.2.1}$$

If the inhomogeneity has a spheroidal shape, or ellipsoidal shape, tensors $\mathbf{H}^R$ can be expressed in terms of Eshelby's tensor for conductivity problem $\mathbf{s}^C$ as follows:

$$\mathbf{H}^R = \frac{V^*}{V}\frac{1}{k_0}((1 - f_0(\gamma))\mathbf{I} + (3f_0(\gamma) - 1)\mathbf{nn})^{-1} = \frac{V^*}{V_0}\frac{1}{k_0}(A_1\boldsymbol{I} + A_2\boldsymbol{nn}), \tag{22.2.2}$$

where $\mathbf{I}$ is the second-rank unit tensor,

$$A_1 = \frac{1}{1 - f_0(\gamma)},\ A_2 = \frac{1 - 3f_0(\gamma)}{2f_0(\gamma)\left[1 - f_0(\gamma)\right]},\ f_0 = \frac{\gamma^2(1 - g)}{2(\gamma^2 - 1)}, \tag{22.2.3}$$

and the shape factor g is expressed in terms of the aspect ratio $\gamma$ as follows

$$g(\gamma) = \begin{cases} \frac{1}{\gamma\sqrt{1-\gamma^2}} \arctan\frac{\sqrt{1-\gamma^2}}{\gamma}, & \text{oblate shape } (\gamma < 1) \\ \frac{1}{2\gamma\sqrt{\gamma^2-1}} \ln\frac{\gamma+\sqrt{\gamma^2-1}}{\gamma-\sqrt{\gamma^2-1}}, & \text{prolate shape } (\gamma > 1) \end{cases} \tag{22.2.4}$$

Representation (22.2.2) serves as a basic block for most of the approximate schemes used to calculate effective conductive properties. Below we consider a heterogeneous material containing insulating inhomogeneities of volume fraction of $(1 - p)$ (so that volume fraction of the conducting phase is $p$).

### 22.2.2 Non-interaction Approximation

This approximation neglects interaction between inhomogeneities so that it can be assumed that each inhomogeneity is subjected to the same remotely applied electric current field. The contribution of inhomogeneity into the change in voltage can be treated separately, and the total gradient of the electric potential is

$$\mathbf{U} = \frac{1}{k_0}\mathbf{Q} + \frac{1}{V}\left(\sum_i \mathbf{H}^R\right) \cdot \mathbf{Q}. \tag{22.2.5}$$

In particular, in the case of randomly oriented spheroidal inhomogeneities of the aspect ratio $\gamma$ with volume concentration $(1 - p)$

$$\mathbf{U} = \frac{1}{k_0}\left[1 + (1-p)\left(A_1(\gamma) + A_2(\gamma)/3\right)\right]\mathbf{Q}. \tag{22.2.6}$$

It leads to the following relation for the effective electrical conductivity of a material containing multiple insulating inhomogeneities.

$$k_{\text{eff}}^{\text{NI}}/k_0 = \frac{1}{1 + (1-p)\eta_{\text{NI}}(\gamma)}, \tag{22.2.7}$$

where $\eta_{\text{NI}}(\gamma) = p\left(A_1(\gamma) + A_2(\gamma)/3\right)$ is shape-dependent parameter. Figure 22.1 shows its dependence on the aspect ratio $\gamma$. Since $k_0/k_{\text{eff}} = \sigma_{\text{eff}}/\sigma_0$, combining (22.2.7) and (22.1.5) yields the following expression for tortuosity in the framework of non-interaction approximation:

$$\tau_{\text{NI}}^2 = p\left[1 + (1-p)\eta_{\text{NI}}(\gamma)\right]. \tag{22.2.8}$$

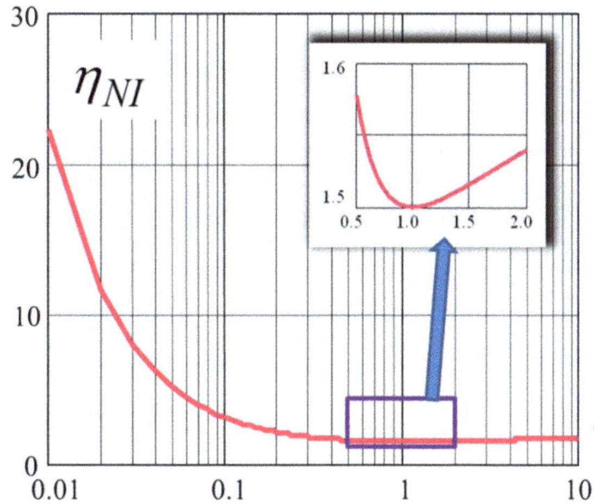


**Fig. 22.1** Dependence of the parameter $\eta_{\mathrm{NI}}$ entering expression for the effective electrical conductivity on the aspect ratio $\gamma$

## 22.2.3 Differential Scheme

The differential scheme proposed by Bruggeman (1935, 1937) assumes that the inhomogeneities are incrementally added to the material until the final volume fraction is reached. On each increment, a set of non-interacting inhomogeneities is added to the homogenized material with the properties determined by the previously embedded inhomogeneities. As shown by Weber et al. (2003) and Sevostianov et al. (2006), this scheme gives the most accurate results on the electrical conductivity of composites containing insulating inhomogeneities. For isotropic distribution of spheroidal inhomogeneities, the differential scheme gives the effective electrical conductivity as:

$$k_{\mathrm{eff}}^{\mathrm{DS}} / k_0 = p^{\eta_{\mathrm{NI}}(\gamma)}. \tag{22.2.9}$$

So that tortuosity in the framework of the differential scheme is

$$\tau_{\mathrm{DS}}^2 = p^{1-\eta_{\mathrm{NI}}(\gamma)}. \tag{22.2.10}$$

## 22.2.4 Effective Field Method

In this scheme, each particle is treated as isolated but embedded into the matrix material; the effect of interaction is accounted for through the assumption that each particle lies within a certain effective field that differs from the applied macroscopic one. The basic idea of the method has roots in the works of Mossotti (see Feynman et al. 1964, Chap. 11). Below, we consider two variants of the effective field method.

#### 22.2.4.1 Mori–Tanaka Scheme (MTS)

This approach (Mori and Tanaka 1973) is based on the assumption that the effective field acting on each inhomogeneity is equal to the average over the matrix. The macroscopic properties may then be calculated from the non-interaction approximation with appropriate change of the remotely applied field. For isotropic distribution of spheroidal insulating particles, the overall electrical conductivity according to the MTS is

$$k_{\text{eff}}^{\text{MT}} / k_0 = \left[ 1 + \frac{1-p}{p} \eta_{\text{NI}}(\gamma) \right]^{-1}. \qquad (22.2.11)$$

So that tortuosity calculated by MTS is given by

$$\tau_{\text{MT}}^2 = p \left[ 1 + \frac{1-p}{p} \eta_{\text{NI}}(\gamma) \right]. \qquad (22.2.12)$$

#### 22.2.4.2 Kanaun–Levin Scheme (KLS)

Kanaun–Levin scheme (KLS) represents a more refined variant of the effective field method. The main assumption is that the effective field acting on each inhomogeneity is equal to the average over the entire material (i.e., over matrix and inhomogeneities). The method was originally developed for elastic properties of composites in the 1970s (see references in the book of Kanaun and Levin 2008). The conductivity problem was addressed much later (Kanaun 2003).

For the case under consideration (isotropic distribution of spheroidal insulating particles), KLS gives the following relation for the effective electrical conductivity:

$$k_{\text{eff}}^{\text{KL}} / k_0 = \left[ 1 + \frac{(1-p)\eta_{\text{NI}}(\gamma)\eta_{\text{NI}}(1)}{\eta_{\text{NI}}(1) - (1-p)\eta_{\text{NI}}(\gamma)} \right]^{-1}. \qquad (22.2.13)$$

Note that for *spherical* pores ($\gamma = 1$) $\eta_{\text{NI}}(1) = 1.5$ and Mori–Tanaka scheme coincides with the one of Kanaun–Levin (and with Hashin–Shtrikman bound). Equation (22.2.13) yields the following relation for tortuosity

$$\tau_{\text{KL}}^2 = p \left[ 1 + \frac{(1-p)\eta_{\text{NI}}(\gamma)\eta_{\text{NI}}(1)}{\eta_{\text{NI}}(1) - (1-p)\eta_{\text{NI}}(\gamma)} \right]. \qquad (22.2.14)$$

Figure 22.2 illustrates dependence of $\tau^2$ calculated by the methods discussed above on porosity and average aspect ratio of the particles.

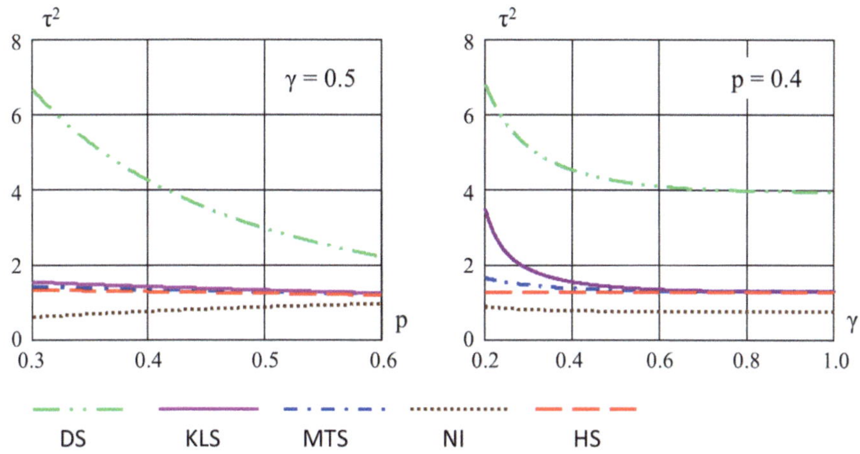


**Fig. 22.2** Tortuosity calculated according to various methods—differential scheme, Kanaun–Levin scheme, Mori–Tanaka scheme, non-interaction approximation—in comparison with Hashin–Shtrikman bound: (a) all the solid particles are considered as spheroids of aspect ratio 0.5 and porosity varies; (b) porosity is fixed as 0.4 and tortuosity is calculated as a function of the average aspect ratio of insulating soil particles

## 22.3 Variational Bounds in Terms of Tortuosity

Berryman and Milton (1988) constructed elasticity–conductivity bounds for a two-phase isotropic composite. They utilize bounds derived separately for the effective conductivity (Beran 1965) and for the effective elastic constants (Beran and Molyneux 1966; McCoy 1970). These two sets of bounds take into account not only volume fractions of phases but the information on microgeometry provided by three-point correlation functions. An important observation is that this information is utilized by the two bounds in a similar way, reflecting similarities between the effects of microgeometry on the two properties. This leads to cross-property bounds that are independent of microgeometry.

The three-point correlation function provides an approximate characterization of microgeometry: It gives the probability that material belonging to a given phase of a composite is located at three given points. For phase 1, this function is defined by

$$s_3(\mathbf{r}_1, \mathbf{r}_2, \mathbf{r}_3) = \left\langle V^{(1)}(\mathbf{x}+\mathbf{r}_1) V^{(1)}(\mathbf{x}+\mathbf{r}_2)\, V^{(1)}(\mathbf{x}+\mathbf{r}_3) \right\rangle, \tag{22.3.1}$$

where $V^{(1)}(\mathbf{x})$ is the characteristic function of the first phase. Based on the correlation function, the following two dimensionless microstructural parameters are introduced:

$$\zeta = \frac{9}{2c_0 c_1} \lim_{\Delta \to 0} \lim_{\Delta' \to \infty} \int_{\Delta}^{\Delta'} \mathrm{d}r \int_{\Delta}^{\Delta'} \mathrm{d}s \int_{-1}^{+1} \frac{S_3(r, s, \mu)}{rs} P_2(\mu) \mathrm{d}\mu,$$

$$\eta = \frac{5\zeta_1}{21} + \frac{150}{7c_0c_1} \lim_{\Delta\to 0} \lim_{\Delta'\to\infty} \int_{\Delta}^{\Delta'} dr \int_{\Delta}^{\Delta'} ds \int_{-1}^{+1} \frac{S_3(r,s,\mu)}{rs} P_4(\mu) d\mu, \tag{22.3.2}$$

where $c_0$ and $c_1$ are volume concentrations of two phases, $P_2(\mu) = \left(3\mu^2 - 1\right)/2$ and $P_4(\mu) = \left(35\mu^4 - 30\mu^2 + 3\right)/8$ are Legendre polynomials of orders 2 and 4 and

$$S_3\left(r_{12}, r_{13}, \mu_{12,13}\right) = s_3(\mathbf{r}_1, \mathbf{r}_2, \mathbf{r}_3), \tag{22.3.3}$$

with

$$r_{ij} = \left|\mathbf{r}_j - \mathbf{r}_i\right|, \mu_{12,13} = \frac{(\mathbf{r}_2 - \mathbf{r}_1)\cdot(\mathbf{r}_3 - \mathbf{r}_1)}{r_{12}r_{13}}. \tag{22.3.4}$$

This form of the correlation function emphasizes the statistical homogeneity of the microgeometry (invariance with respect to translations in any direction). Parameters $\zeta$ and $\eta$ obey the following inequalities (Milton and Phan-Thien 1982):

$$5\zeta/21 \le \eta \le (16 + 5\zeta)/21. \tag{22.3.5}$$

The key observation of Berryman and Milton (1988) is that bounds for different physical properties—magnetic permeability, electrical or thermal conductivity, and elastic constants—can be expressed in terms of the same two parameters, $\zeta$ and $\eta$. Focusing attention on the elasticity–conductivity pair and excluding the two parameters leads to bounds on the effective bulk and shear moduli in terms of the effective conductivity. The basic arguments are as follows.

In addition to the usual volume average $\langle\Pi\rangle = \Pi_0 + (\Pi_1 - \Pi_0)c_1$ of a certain physical property $\Pi$, Berryman and Milton (1988) introduced averages that involve microstructural parameters (22.3.2):

$$\langle\Pi\rangle_\zeta = \Pi_0 + (\Pi_1 - \Pi_0)\zeta, \langle\Pi\rangle_\eta = \Pi_0 + (\Pi_1 - \Pi_0)\eta. \tag{22.3.6}$$

In terms of these averages, bounds of Beran (1965) for the effective conductivity and bounds of Beran and Molyneux (1966) and McCoy (1970) for the elastic moduli take the form

$$\mathrm{A}_1\left(\langle 1/k\rangle_\zeta^{-1}\right) \le k \le \mathrm{A}_1\left(\langle k\rangle_\zeta\right), \tag{22.3.7a}$$

$$\mathrm{A}_2\left(\langle 1/G\rangle_\zeta^{-1}\right) \le K \le \mathrm{A}_2\left(\langle G\rangle_\zeta\right), \tag{22.3.7b}$$

$$\mathrm{A}_3(\Theta/6) \le G \le \mathrm{A}_3(1/6\Xi), \tag{22.3.7c}$$

where functions $A_{1-3}(x)$ are defined as follows

$$A_1(x) = \left\langle \frac{1}{k+2x} \right\rangle^{-1} - 2x,\ A_2(x) = \left\langle \frac{1}{K+4x/3} \right\rangle^{-1} - 4x/3,$$
$$A_2(x) = \left\langle \frac{1}{K+4x/3} \right\rangle^{-1} - 4x/3, \tag{22.3.8}$$

and the following notations are used

$$\Theta = \frac{10\langle G\rangle^2\langle K\rangle_\zeta + 5\langle G\rangle\langle 2K+3G\rangle\langle G\rangle_\zeta + \langle 3K+G\rangle^2\langle G\rangle_\eta}{\langle K+2G\rangle^2},$$
$$\Xi = \frac{10\langle K\rangle^2\langle 1/K\rangle_\zeta + 5\langle G\rangle\langle 2K+3G\rangle\langle 1/G\rangle_\zeta + \langle 3K+G\rangle^2\langle 1/G\rangle_\eta}{\langle 9K+8G\rangle^2}. \tag{22.3.9}$$

Note that bounds for the conductivity and for the bulk modulus contain parameter $\zeta$ only, whereas bounds for the shear modulus contain both microstructural parameters $\zeta$ and $\eta$.

Substituting $\zeta$ from inequality (22.3.7a) for conductivity, as well as (22.3.2), into inequalities (22.3.7b, c) for the bulk and shear moduli would lead to the elasticity–conductivity cross-property bounds. For a general composite material, this procedure is quite cumbersome and, to our knowledge, has not been implemented in explicit form.

For a non-conducting porous material filled with conductive fluid, results can be given in the explicit form. In this case, $c_1 = p$ and Beran's bounds degenerate into one inequality:

$$k_1/k \geq 1 + \left(1 + \frac{1}{2\zeta}\right)\frac{1-p}{p},\ \text{or,}\ \zeta \geq \frac{1}{2}\frac{(1-p)k}{pk_1 - k}, \tag{22.3.10}$$

and Beran–Molyneux bounds are

$$\zeta \leq 1 - \frac{3pK/4G_0}{(1-p) - K/K_0}, \tag{22.3.11}$$

$$\eta \leq 1 - \frac{6R(K_0+2G_0)^2 - 5G_0(1-\zeta)(4K_0+3G_0)}{(3K_0+G_0)^2}, \tag{22.3.12}$$

where $R = (pG/G_0)/(1-p-G/G_0)$. This leads to cross-property bounds interrelating the effective bulk modulus and the effective conductivity:

$$\frac{1}{2}\frac{(1-p)k}{pk_1 - k} \leq \zeta \leq 1 - \frac{3pK/4G_0}{1-p-K/K_0}. \tag{22.3.13}$$

To obtain bounds for the effective shear modulus, we introduce dimensionless combinations of elastic constants of phase "0"

$$A_0 = \frac{6(K_0 + 2G_0)^2}{(3K_0 + G_0)^2}, \; B_0 = \frac{5G_0(4K_0 + 3G_0)}{(3K_0 + G_0)^2}, \tag{22.3.14}$$

Using (22.3.7a–c) and rewriting (22.3.12) in the form

$$\frac{5}{21}\zeta \le \eta \le 1 + B_0 - B_0\zeta - A_0\frac{pG/G_0}{1 - p - G/G_0} \tag{22.3.15}$$

imply the following cross-property bound

$$\frac{1}{2}\frac{(1-p)k}{pk_1 - k} \le \zeta \le 21\left(\frac{1 + B_0}{5 - 21B_0}\right) - \frac{21A_0}{5 - 21B_0}\frac{6pG/G_0}{1 - p - G/G_0}. \tag{22.3.16}$$

***Remark 2***. Berryman and Milton (1988) noted that, since inequalities for parameters $\zeta$ and $\eta$ are independent of the properties of phase "1", they can be specialized, for example, for a porous material and then used for generating bounds for material of the same microgeometry but with pores replaced by a constituent with different properties. Similar ideas were developed by Bergman (1978) and Kantor and Bergman (1984).

Note, that definition of tortuosity via electrical conductivity allows one to easily obtain variational bounds for tortuosity using known results for conductivity. Namely, Hashin–Shtrikman bound for conductivity of a porous material is (Hashin and Shtrikman 1962a, b)

$$\frac{2p}{3-p} \ge \frac{k^{\text{eff}}}{k^0} = \frac{1}{F} = \frac{p}{\tau^2}. \tag{22.3.17}$$

It immediately yields the corresponding bound for tortuosity:

$$\tau^2 \ge 1 + \frac{1-p}{2}. \tag{22.3.18}$$

Similarly, Beran's bound for conductivity has the following form (Beran 1965, 1968)

$$\frac{2\zeta\, p}{1 + 2\zeta - p} \ge \frac{k^{\text{eff}}}{k^0} = \frac{1}{F}. \tag{22.3.19}$$

Inequality (22.3.19) yields the corresponding relation for tortuosity:

$$\tau^2 \geq 1 + \frac{1-p}{2\zeta}. \tag{22.3.20}$$

Then, Berryman–Milton bounds yield the following inequality

$$\tau^2 \geq 1 + \frac{(1/2)(1-p)\left(1-p-K/K_0\right)}{1-p-K\left(1/K_0+3p/4G_0\right)}. \tag{22.3.21}$$

The Hashin–Shtrikman bound for tortuosity (22.3.18) is achieved when effective bulk modulus $K = 0$. Since the right-hand side of (22.3.21) is easily shown to be a monotonically increasing function of $K$, we see that (22.3.21) is an improvement on the Hashin–Shtrikman bound whenever $K > 0$. The upper Hashin–Shtrikman bound on the bulk modulus is (Hashin and Shtrikman 1963):

$$K \leq \frac{(4/3)(1-p)G_0K_0}{pK_0+(4/3)G_0} \equiv K_{HS}^{+}. \tag{22.3.22}$$

It is known that (22.3.22) is an optimum bound (when no geometrical information is available) and that the equality is satisfied for the coated-sphere geometry. When the equality in (22.3.22) is satisfied, the right-hand side of (22.3.21) is infinite which is physically correct since the void phase is unconnected for this geometry and, therefore, the tortuosity is infinite. No other value of $K$ will make the right-hand side of (22.3.21) infinite. Nevertheless, the fact that this choice of $K$ uses (22.3.21) to its limiting value suggests that this bound will sometimes be a significant improvement on the Hashin–Shtrikman bound.

We can also obtain an improved bound on the bulk modulus by making a measurement of the formation factor or, equivalently, the tortuosity, so that

$$K \leq \frac{4}{3}(1-p)G_0K_0\frac{\tau^2-1-(1/2)(1-p)}{\left(\tau^2-1\right)pK_0+\left[\tau^2-1-(1/2)(1-p)\right](4/3)G_0}. \tag{22.3.23}$$

Inequality (22.3.23) is an improvement over (22.3.22) for any value of $\tau > \infty$. To illustrate the behavior of Eq. (22.3.23), consider a porous material with $p = 0.2$ and $G_0/K_0 = 3/4$. Then the Hashin–Shtrikman upper bound on the bulk modulus obtained from (22.3.23) is $K/K_0 = 3$. In Fig. 22.3, the upper bound $K^+/K_0$ is plotted as a function of $\tau^2$. We see that, for values of $\tau^2$ toward the low end of the range $1.4 \leq \tau^2 \leq 3.0$, the new bound can be a substantial improvement on the Hashin–Shtrikman bound. The new bound may be expected to find some use, since values of $\tau^2$ are quite commonly found to lie in this range for real materials.

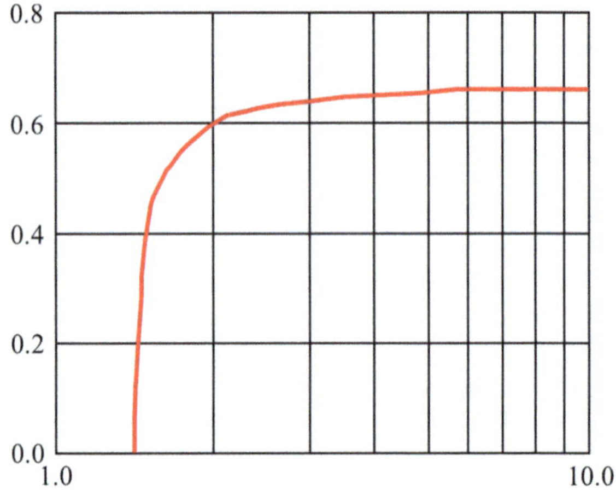


**Fig. 22.3** Plot of the upper bound on the normalized bulk modulus $K^+/K_0$ of a porous material given by (22.3.23) as a function of the squared tortuosity $\tau^2$. The parameters are given by $p = 0.2$ and $G_0/K_0 = 3/4$.

## 22.4 Cross-Property Connections Between Overall Electric Conductivity and Fluid Permeability of a Random Porous Media with Conducting Skeleton

To derive the said cross-property connections, we need to consider a porous material from somewhat unusual point of view. Let us take a look, for example, at a two-phase material in Fig. 22.4.

If the solid phase is electrically conductive with conductivity $k_s$ and its volume concentration is $(1 - p)$, we can formally define the tortuosity $\tau_s$ of this phase according to (22.1.6):

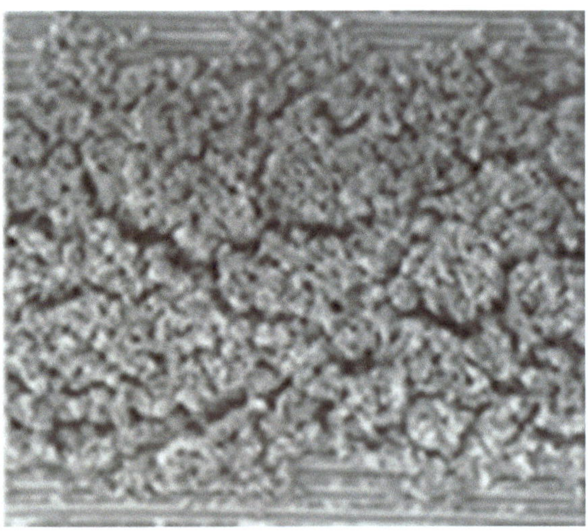

**Fig. 22.4** Typical porous microstructure of sintered copper

$$\tau_s^2 = (1 - p)k_s / k^{\text{eff}}. \tag{22.4.1}$$

Now we can say, that this phase represents the solid skeleton and, thus, we have formal definition of the tortuosity of the solid phase. To define the tortuosity for electrically non-conductive porous space, we can use the cross-property connection derived by Avellaneda and Torquato (1991). It yields the following expression for the tortuosity in terms of the effective fluid permeability $\lambda$ and porosity $p$:

$$\tau_p^2 = \frac{L^2}{8}\frac{p}{\lambda}, \tag{22.4.2}$$

where $L$ is the length parameter which is the weighted sum over the viscous relaxation times $\Theta_n$ associated with the time-dependent Stokes equation (i.e., inversely proportional to the eigenvalues of the Stokes operator).

### 22.4.1 Cross-Property Bounds

For tortuosities (22.4.1) and (22.4.2), Hashin–Shtrikman bounds can be written in the following form:

$$\tau_p^2 \geq 1 + \frac{1-p}{2}; \ \tau_s^2 \geq 1 + \frac{p}{2}. \tag{22.4.3}$$

Eliminating porosity, we get

$$\tau_p^2 \geq \frac{5}{2} - \tau_s^2. \tag{22.4.4}$$

Comparing the latter inequality with the expressions (22.4.3), one can see that (22.4.4) does not lead to any new results as compared to Hashin–Shtrikman bounds. More reasonable inequality can be derived from Beran bounds

$$\tau_p^2 \geq 1 + \frac{1-p}{2\zeta}; \ \tau_s^2 \geq 1 + \frac{p}{2(1-\zeta)}.$$

Eliminating $\zeta$ given by (22.3.2) and keeping porosity as a parameter, we receive the following inequality for the tortuosity of porous space:

$$\tau_p^2 \geq 1 + \frac{1-p}{2}\left(\frac{\tau_s^2 - 1}{\tau_s^2 - 1 - p/2}\right). \tag{22.4.5}$$

As shown in Fig. 22.5, expression (22.4.5) is indeed an improvement of Hashin–Shtrikman bound whenever $0 < p < 1$.

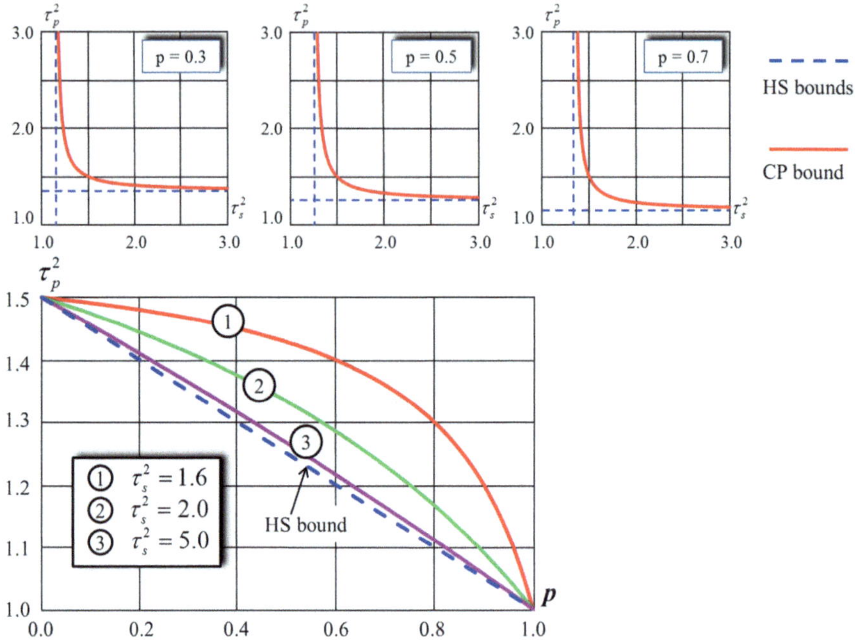


**Fig. 22.5** Comparison of cross-property bound (22.4.5) (C-P bound) with Hashin–Shtrikman bounds for tortuosities (H–S bounds). C-P bound provides substantial improvement for small values of $\tau_p$

***Remark*** Attempts to derive cross-property connections from Hashin–Shtrikman bounds have been also done in relation to elastic and conductive properties (see, e.g., Zhao et al. 2004). As discussed by Sevostianov and Kachanov (2008), this approach always leads to worseness of the initial inequalities.

Expressing now $\tau_s^2$ and $\tau_p^2$ in terms of electrical conductivity and fluid permeability according to (22.4.1) and (22.4.2) yields the following connection between the said physical properties:

$$\frac{1}{\lambda} \geq \frac{8}{L^2 p}\left[1 + \frac{(1-p)}{2}\frac{\left[(1-p) - \left(k/k_s\right)\right]}{(1-p) - \left(k/k_s\right)\left(1 + p/2\right)}\right]. \tag{22.4.6}$$

***Remark*** Cross-property connections (22.4.5) and (22.4.6) are much easier than the connection between bulk modulus of the porous solid and tortuosity of the porous space (22.3.23) derived by Berryman and Milton (1988). Since elastic properties are described by fourth-rank tensor, two microgeometrical parameters are required in the latter case—parameters $\zeta$ and $\eta$, introduced by (22.3.2). Therefore, connection between these two parameters has to be first derived (Milton and Phan-Tien 1982).

An interesting observation is that relations (22.4.5) and (22.4.6) can be obtained from (22.3.23) if we replace bulk moduli by electric conductivities and set Poisson's

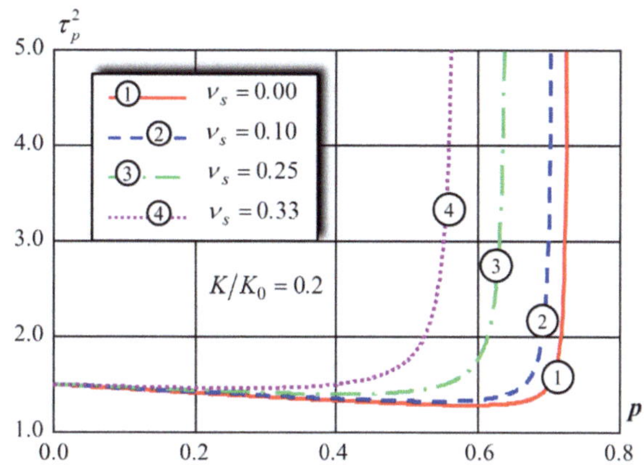


**Fig. 22.6** Berryman–Milton bounds for tortuosity at different levels of Poisson's ratio. Curve 1 coincides with the bound (22.4.6) for tortuosity in terms of electrical conductivity

ratio $\nu_0 = 0$. It echoes with the results of Lutz and Zimmerman (2005) who derived *exact* explicit connections between bulk modulus and electrical conductivity of a material containing spherical inclusions. Figure 22.6 shows cross-property bounds (22.4.6) and (22.3.23).

### 22.4.2 Explicit Cross-Property Connections

The cross-property bounds derived in the previous subsection are accessible and cannot be improved without additional information on material's microstructure. Practical needs of materials science, however, call for the cross-property connections that preferably have an explicit form. For this aim, we have to accept certain assumption about the distribution of the phases. We can use the results of Sevostianova et al. (2010) that clarify the micromechanical meaning of expression (22.1.7). Namely, it was shown that for a porous material of random microstructure filled with electrolyte, $\tau = p/\psi_{\min}$, where $\psi_{\min}$ is the *minimum* ratio of the apparent cross-sectional area of the conducting electrolyte to the total cross-sectional area. Then, the following expressions for tortuosities of porous and solid phases can be written:

$$\tau_p = p/\psi_{\min};\ \tau_s = (1-p)/(1-\psi_{\max}). \tag{22.4.7}$$

Assuming now symmetric spatial distribution of pores around $\langle\psi\rangle = p$ (Gaussian distribution, e.g., Fig. 22.7), one can write

$$\tau_p = p/(p-\alpha);\ \tau_s = (1-p)/(1-p-\alpha) \tag{22.4.8}$$

In particular, it leads to $\tau_p = \tau_s$ for $p = 0.5$; i.e. the phases of the same volume fraction are geometrically indistinguishable. Eliminating $\alpha$ from (22.4.8), one can derive the relation between tortuosities of the two phases:

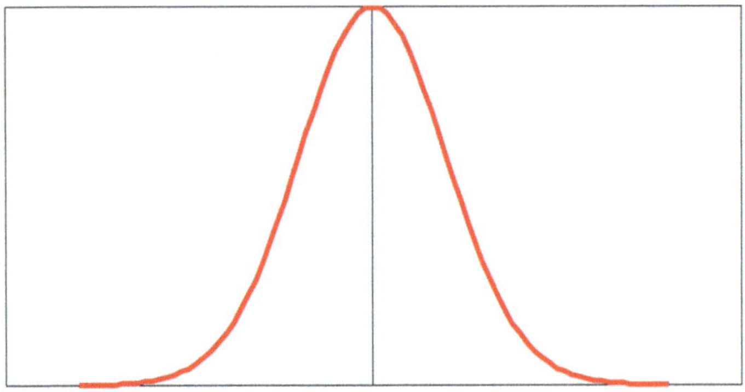


**Fig. 22.7** Sketch of the symmetric spatial distribution of porosity in a porous material—Gaussian distribution

$$\frac{\tau_s - 1}{\tau_s} = \frac{p}{1-p}\frac{\tau_p - 1}{\tau_p}. \tag{22.4.9}$$

For cross-property connection between electric conductivity and fluid permeability, we need expressions for the tortuosities squared:

$$\frac{1}{\tau_p^2} = \left(1 - \frac{\alpha}{p}\right)^2; \frac{1}{\tau_s^2} = \left(1 - \frac{\alpha}{1-p}\right)^2. \tag{22.4.10}$$

Using (22.4.1), the latter relation yields

$$\alpha = (1-p)\left(1 - \sqrt{\frac{1}{1-p}\frac{k}{k_s}8\lambda/L^2}\right). \tag{22.4.11}$$

Substitution of (22.4.11) into the first Formulas (22.4.10) and (22.4.2) allows us to express explicitly fluid permeability $\lambda$ in terms of the ratio $\sigma^{\text{eff}}/\sigma^s$:

$$\lambda = \frac{L^2}{8}p\left[1 - \frac{1-p}{p}\left(1 - \sqrt{\frac{1}{1-p}\frac{k}{k_s}}\right)\right]^2. \tag{22.4.12}$$

Figure 22.8 shows cross-property connection (22.4.12) for several values of porosity.

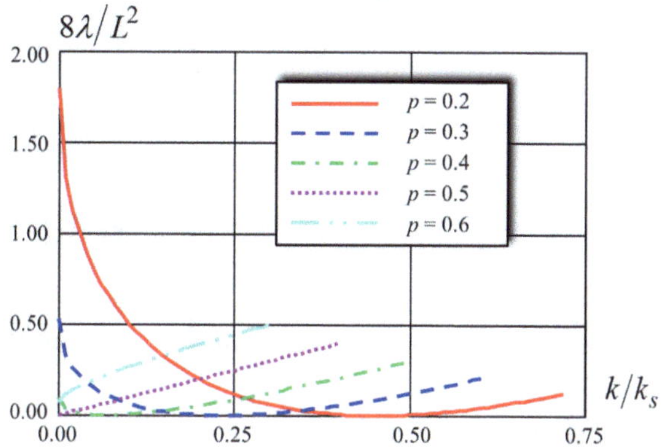


**Fig. 22.8** Cross-property connection between normalized fluid permeability and normalized effective conductivity of a porous material with conducting skeleton at various porosity levels

## 22.5 Connection Between Electrical Conductivity and Mass Transport Coefficient of a Conductive Porous Material Filled with Electrolyte

### *22.5.1 Two-Phase Material with Conductive Constituents: Replacement Relations for Electrical Conductivity*

In this section, we extend the applicability of the tortuosity parameter to the case, when both the phases are electrically conductive and penetrable for diffusion. For this goal, we consider the problem of the change in overall electric properties of a material upon the changing properties of one of its phases (replacement relations) (see Fig. 22.9).

For elastic properties, this problem was first addressed by Gassmann (1951) who derived relation expressing bulk and shear moduli of fully saturated rock in terms of the elastic properties of dry rock. Further development of Gassmann's equation was done in the works of Han and Batzle (2004) and Ciz and Shapiro (2007). Sevostianov and Kachanov (2007) formulated the replacement relations in terms of property contribution tensors. Their results are valid for anisotropic materials (including anisotropic constituents) and can be rewritten for other physical properties as well. For the thermal conductivity problem, replacement relations have been first formulated by Zimmerman (1989) for isotropic air and water-saturated rock. It was generalized to the case of anisotropic two-phase material by Chen et al. (2017a) as:

$$\mathbf{r}_{\text{eff}} = \mathbf{r}_0 + \varphi\left[(\mathbf{r}_1 - \mathbf{r}_0)^{-1} + \varphi\left(\mathbf{r}_{\text{ins}} - \mathbf{r}_0\right)^{-1}\right]^{-1}, \qquad (22.5.1)$$

where $\varphi$ is the volume fraction of phase "1", $\mathbf{r}_0$ and $\mathbf{r}_1$ are electrical resistivities of two phases, $\mathbf{r}_{\text{eff}}$ is overall resistivity of the composite, and $\mathbf{r}_{\text{ins}}$ is the resistivity of the comparison two-phase material that has the same morphology and $\mathbf{r}_0$, but $\mathbf{r}_1 \to$

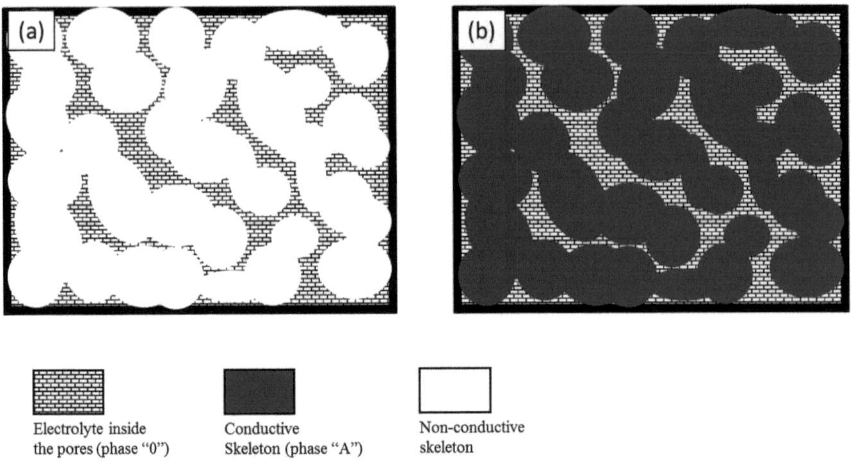


**Fig. 22.9** Sketch of the two materials of the same morphology having insulating (**a**) and conductive (**b**) skeletons and filled with the same electrolyte

$\infty$. The replacement relation (22.5.1) has the same form for any homogenization method (assuming that the properties of materials with insulating and conducting inhomogeneities are calculated with the same method, of course).

In the case of the isotropic mixture of two isotropic phases, expression (22.5.1) can be rewritten in the form:

$$\varphi \frac{k_{\text{eff}}}{k_0 - k_{\text{eff}}} - \varphi \frac{k_{\text{ins}}}{k_0 - k_{\text{ins}}} = \frac{k_1}{k_0 - k_1}, \tag{22.5.2}$$

where $k_{\text{eff}}$ is overall conductivity of the composite with both phases being conductive and having conductivities $k_0$ and $k_1$ (Fig. 22.9a), and $k_{\text{ins}}$ is the overall conductivity of the comparison composite having the same morphology with one phase of conductivity $k_0$ and another phase being a perfect insulator.

Similar replacement relation can be derived for the mass transfer coefficient as well. Indeed, following Knyazeva et al. (2015), let us consider a reference volume $V$ of a material with the isotropic diffusion coefficient $D_0$ containing inhomogeneity with diffusion coefficient $D_1$ occupying domain $V_1 \ll V$. We assume that both inhomogeneity and the surrounding material satisfies the linear Fick's law connecting concentration gradient with the mass transfer rate (molar flux). The homogeneous boundary conditions (Hill 1963) are assumed: the "remotely applied" concentration gradient, or molar flux, is uniform in the absence of the inhomogeneity. Let, for example, the molar flux $\mathbf{J}_0$ be prescribed at the boundary of $V$. Then, the average over $V = V_0 \cup V_1$ concentration gradient $\nabla c$ of the diffusant is related to $\mathbf{J}_0$ by

$$\langle \nabla c \rangle_V = \left[ D_0 \mathbf{I} + \frac{V_1}{V} \mathbf{H}^{DR} \right] \cdot \mathbf{J}_0., \tag{22.5.3}$$

where tensor $\mathbf{H}^{DR}$ can be called the diffusion resistance contribution tensor. For a spheroidal inhomogeneity, in the absence of segregation at the interface,

$$\mathbf{H}^{DR} = \frac{1}{D_0}\left[\frac{(D_0 - D_1)}{D_0 + (D_1 - D_0) f_0}(\mathbf{I} - \mathbf{nn}) + \frac{(D_0 - D_1)}{D_1 - 2(D_1 - D_0) f_0}\mathbf{nn}\right] \quad (22.5.4)$$

and shape factor $f_0$ is expressed in terms of the spheroid aspect ratio $\gamma$ by (22.2.3).

We can rewrite expression (22.5.4) as

$$\left(\mathbf{H}_1^{DR}\right)^{-1} = \frac{D_1 D_0}{D_1 - D_0}\mathbf{I} + D_0[(f_0 - 1)(\mathbf{I} - \mathbf{nn}) - 2 f_0 \mathbf{nn}]. \quad (22.5.5)$$

If phase 1 is impenetrable ($D_1 \to 0$),

$$\left(\mathbf{H}_{imp}^{DR}\right)^{-1} = D_0[(f_0 - 1)(\mathbf{I} - \mathbf{nn}) - 2 f_0 \mathbf{nn}]. \quad (22.5.6)$$

Subtracting (22.5.6) from (22.5.5)

$$\left(\mathbf{H}_1^{\mathrm{DR}}\right)^{-1} - \left(\mathbf{H}_{\mathrm{imp}}^{\mathrm{DR}}\right)^{-1} = \frac{D_A D_0}{D_A - D_0} = \left(\frac{1}{D_A} - \frac{1}{D_0}\right)^{-1}. \quad (22.5.7)$$

In the case of the isotropic mixture of two isotropic phases, expression (22.5.7) yields the following replacement relation for diffusion coefficients [that has the same form as (22.5.2)]:

$$\varphi\frac{D_{\mathrm{eff}}}{D_0 - D_{\mathrm{eff}}} - \varphi\frac{D_{\mathrm{imp}}}{D_0 - D_{\mathrm{imp}}} = \frac{D_1}{D_0 - D_1}, \quad (22.5.8)$$

where $D_{\mathrm{eff}}$ is effective mass transfer coefficient of the composite with both phases being penetrable for diffusant and having coefficients $D_0$ and $D_1$ (Fig. 22.9a), and $D_{\mathrm{imp}}$ is the effective mass transfer coefficient of the composite having the same microstructure and $D_0$, while the second phase is impenetrable for the diffusant (Fig. 22.9b). Expression (22.5.8) completely coincides with (22.5.2). As discussed by Chen et al. (2017b), replacement relation has the same shape in all the homogenization schemes based on the concept of effective field—Mori–Tanaka-Benveniste, Kanaun–Levin, Maxwell, etc.

Note, that the replacement relations (22.5.2) and (22.5.8) are approximate—they turn to be exact only in the case when one of the phases is represented by isolated ellipsoidal inhomogeneities. This approximation, however, shows good accuracy for rather irregular microstructures, as shown by the application of its more complex elastic analogy—Gassmann's equation—in geophysics (Berryman 1999; Avseth et al. 2006).

### 22.5.2 *Connection Between Electric Conductivity and Diffusion Coefficient*

In this section, we derive the connection between electric conductivity and diffusion coefficient of an isotropic two-phase material when both phases are electrically conductive and penetrable for a diffusant (to evaluate the overall diffusion coefficient from the electric conductivity measurements). For this goal, we express both properties using replacement relations (22.5.2) and (22.5.8) and tortuosity parameter that is common for both the properties.

We start with Eq. (22.1.6) that expresses retardation factor for conductivity as $F \equiv k^0 / k^{\text{eff}}$. Substitution of this expression into (22.5.2) yields

$$\frac{k_{\text{eff}}}{k_0 - k_{\text{eff}}} - \frac{1}{F-1} = \frac{1}{\varphi}\frac{k_1}{k_0 - k_1}. \tag{22.5.9}$$

Thus, the resistivity factor (retardation factor for conductivity) $F$ is expressed in terms of effective conductivity of the composite $k_{\text{eff}}$ and conductivities of its two phases, $k_0$ and $k_1$ as

$$F = 1 + \frac{\varphi(k_0 - k_1)(k_0 - k_{\text{eff}})}{\varphi k_{\text{eff}}(k_0 - k_1) - k_1(k_0 - k_{\text{eff}})}. \tag{22.5.10}$$

In the same manner, combining (22.3.8) with (22.1.8), we can write for the effective mass transfer coefficient

$$\frac{D_{\text{eff}}}{D_0 - D_{\text{eff}}} - \frac{1}{\tau_H - 1} = \frac{1}{\varphi}\frac{D_1}{D_0 - D_1} \tag{22.5.11}$$

Taking into account (22.1.9), we can obtain, after some algebra, from (22.5.9) and (22.5.11)

$$D_{\text{eff}} = D_0 \frac{BD_A + \varphi\,(D_0 - D_1)}{\varphi\,(B+1)(D_0 - D_1) + BD_1}, \tag{22.5.12}$$

where

$$B = F - 1 = \frac{\varphi\,(k_0 - k_1)(k_0 - k_{\text{eff}})}{\varphi\,k_{\text{eff}}(k_0 - k_1) - k_1(k_0 - k_{\text{eff}})}. \tag{22.5.13}$$

Figure 22.10 shows behavior of function $B$ for different values of $k_{\text{eff}}$ and $k_1$.

Expressions (22.5.12) and (22.5.13) constitute the cross-property connection between overall mass transport coefficient and electrical conductivity of a two-phase isotropic material where both the phases are electrically conductive and penetrable

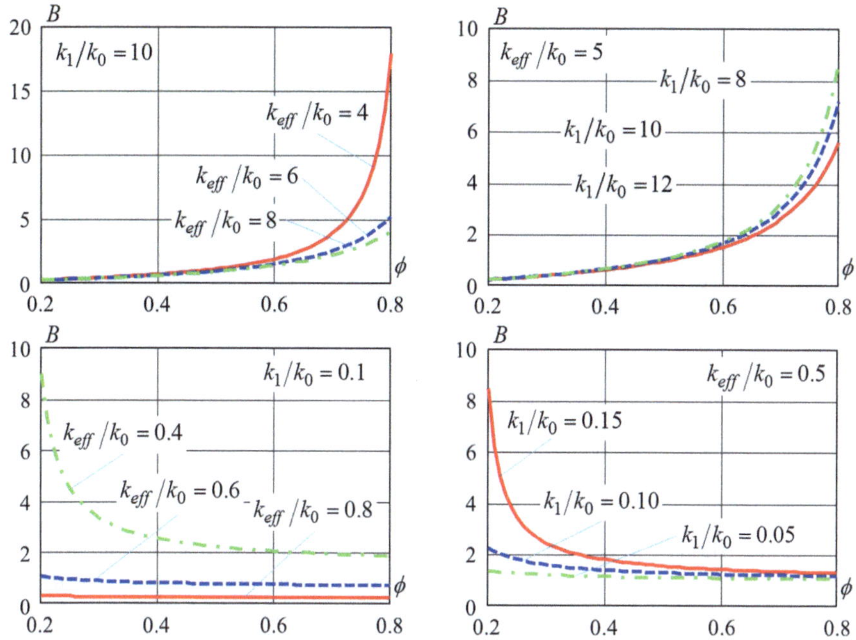


**Fig. 22.10** Dependence of factor $B$ entering cross-property connection (22.5.12) on the volume fraction of phase for different values of $k_{\text{eff}}$ and $k_1$

for a diffusant. Since this cross-property connection is based on approximate replacement relations, their accuracy must be verified. In the next section, we do it using finite element method.

### 22.5.3 *Computer Simulation*

To verify expression (22.5.12), we model the material microstructure using interpenetrating spheres that are randomly located in a reference volume. We start with sphere packing in a unit cell (UC) following procedure described by Drach et al. (2016). The main challenge is the surface re-meshing required to remove overlapping regions and achieve a consistent error-free surface mesh suitable for 3D meshing. It has been done using voxelization method described by Nooruddin and Turk (2003). The idea of the method is to create a regular array (corresponding to a 2D projection of the model) of parallel (voxel-size spaced) rays going through the 3D model and determine the intersection between rays and model polygons. The voxel representation of the model is then constructed based on the determined intersections. Once the voxelization is completed, we can reconstruct the surface mesh. For this goal, we used Marching cubes algorithm (Lorensen and Cline 1987) implemented in MATLAB function called "isosurface.m" that extracts a polygonal mesh of an isosurface from

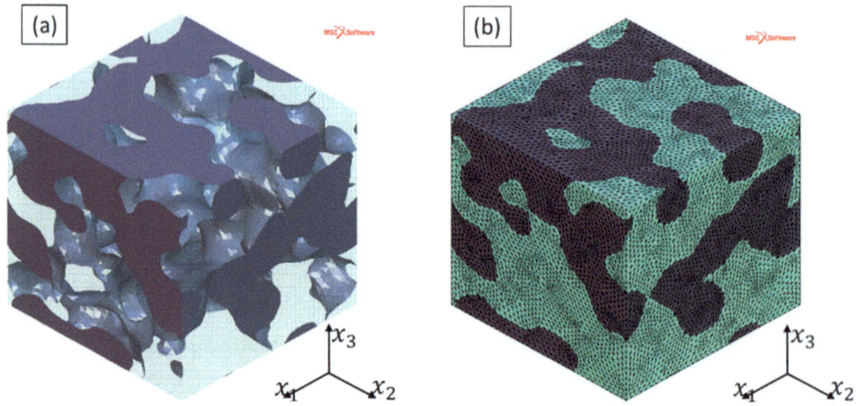


**Fig. 22.11** Example of a two-phase material with equal volume fraction: **a** geometry; **b** 3D meshed structure

voxels (cuberille grid). In order to generate equal element size structure and produce smoothed model, we utilize smoothing method (Taubin 1995). The final structure is presented in Fig. 22.11a. The surface mesh was imported into commercial FEA software MSC Marc/Mentat for preparation of the model and subsequent analysis. The UC is auto-meshed with nonlinear tetrahedral 3D elements (see Fig. 22.11b). In FEM analysis, we considered microstructures with volume fractions of phases varying from 30 to 70%.

After generating the volume mesh material properties have been assigned to the phases. We used the ratio $k_0/k_1 = 1.5$ and varied $D_1/D_0$ from 2 to 10. Two types of boundary conditions are considered to find nonzero components of the tensors of effective properties $D_{\text{eff}}$ and $k_{\text{eff}}$: uniform gradients of diffusant concentration and electric potential on the faces of the reference volume, respectively. The nonzero components of the effective mass transfer tensors are found from the set of three loading cases: concentration gradient along global coordinate axes $x_1$, $x_2$, and $x_3$. In a similar way, effective conductivity tensors are found from the set of the following cases: electric potential gradient along $x_1$, $x_2$, and $x_3$ coordinate axes. Once the boundary conditions are prescribed, FEM simulations are performed, and the result files are processed using a custom Python script to determine $D_{\text{eff}}$ and $k_{\text{eff}}$ tensors utilizing Fick's and Ohm's laws, respectively, and to analyze the possibility of cross-property connection between effective electric conductivity and mass transfer coefficient of a two-phase material.

Figure 22.12 shows comparison of effective diffusion coefficient obtained from FEA calculations and from cross-property connection using FEA calculations for electrical conductivity. For reader's convenience, we also give the results in the form of Table 22.1. It is seen that the even at $D_1/D_0 = 5$ (while $k_0/k_1 = 1.5$) the accuracy of the cross-property connection is better than 8% for the entire range of $\varphi$.

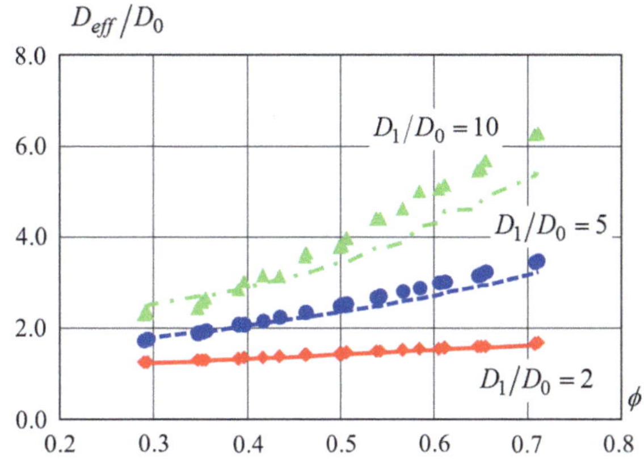


**Fig. 22.12** Effective mass transfer coefficients obtained by direct FEA calculations (symbols) and calculated from effective electric conductivities using cross-property connection (lines). $k_1/k_0 = 2/3$. Solid line and diamonds: $D_1/D_0 = 2$; dashed line and circles $D_1/D_0 = 5$; dot-dashed line and triangles: $D_1/D_0 = 10$

## 22.6 Concluding Remark

The present work focuses on the analysis of the concept of tortuosity of the porous space for evaluation of various physical properties of a porous material. As pointed out by Clennell (1997), "despite its widespread use in petrophysics, tortuosity remains a poorly understood concept. Tortuosity can have various meanings when used by physicists, engineers or geologists to describe different transport processes taking place in a porous material. Values for geometrical, electrical, diffusional and hydraulic tortuosity are in general different from one another." In the present review, we tried to link the concept of tortuosity with solid micromechanical parameters, described, for example, in the book of Kachanov and Sevostianov (2018). In particular, we expressed effective properties of porous materials predicted by various homogenization schemes and variational bounds in terms of tortuosity. It allows several non-trivial cross-property connections which have direct practical applications: connection between electrical conductivity and mass transport coefficient of a conductive porous material filled with electrolyte (Sect. 22.4) and connection between overall electric conductivity and fluid permeability of a random porous media with conducting skeleton (Sect. 22.5).

**Table 22.1** Comparison of $D_{\text{eff}}/D_0$ calculated directly by FEM (**Direct**) with ones obtained using cross-property connection and calculated data for $k_{\text{eff}}$ (CPC). Error is given in %. $k_1/k_0 = 2/3$

| $\varphi$ | $D_1/D_0 = 2$ | | | $D_1/D_0 = 5$ | | | $D_1/D_0 = 10$ | | |
|---|---|---|---|---|---|---|---|---|---|
| | Direct | CPC | Error | Direct | CPC | Error | Direct | CPC | Error |
| 0.290 | 1.243 | 1.232 | 0.875 | 1.737 | 1.765 | −1.621 | 2.296 | 2.499 | −8.843 |
| 0.294 | 1.247 | 1.235 | 0.946 | 1.758 | 1.772 | −0.758 | 2.315 | 2.498 | −7.911 |
| 0.294 | 1.248 | 1.236 | 0.946 | 1.760 | 1.773 | −0.758 | 2.324 | 2.501 | −7.637 |
| 0.346 | 1.293 | 1.281 | 0.916 | 1.899 | 1.911 | −0.611 | 2.420 | 2.708 | −11.890 |
| 0.350 | 1.298 | 1.285 | 1.038 | 1.930 | 1.918 | 0.629 | 2.558 | 2.704 | −5.735 |
| 0.355 | 1.303 | 1.289 | 1.069 | 1.946 | 1.927 | 0.946 | 2.655 | 2.714 | −2.226 |
| 0.389 | 1.335 | 1.319 | 1.220 | 2.079 | 2.018 | 2.897 | 2.844 | 2.853 | −0.313 |
| 0.396 | 1.341 | 1.326 | 1.154 | 2.085 | 2.043 | 2.044 | 3.031 | 2.901 | 4.302 |
| 0.416 | 1.361 | 1.344 | 1.250 | 2.172 | 2.099 | 3.372 | 3.157 | 2.991 | 5.264 |
| 0.434 | 1.378 | 1.360 | 1.284 | 2.239 | 2.153 | 3.835 | 3.135 | 3.082 | 1.678 |
| 0.461 | 1.404 | 1.385 | 1.374 | 2.345 | 2.235 | 4.687 | 3.579 | 3.229 | 9.787 |
| 0.463 | 1.407 | 1.387 | 1.428 | 2.362 | 2.239 | 5.212 | 3.658 | 3.228 | 11.757 |
| 0.498 | 1.441 | 1.421 | 1.378 | 2.487 | 2.363 | 4.986 | 3.777 | 3.466 | 8.226 |
| 0.501 | 1.445 | 1.423 | 1.496 | 2.517 | 2.366 | 6.011 | 3.813 | 3.485 | 8.614 |
| 0.506 | 1.450 | 1.428 | 1.537 | 2.533 | 2.379 | 6.098 | 3.973 | 3.470 | 12.666 |
| 0.537 | 1.483 | 1.458 | 1.668 | 2.684 | 2.485 | 7.382 | 4.408 | 3.787 | 14.080 |
| 0.541 | 1.484 | 1.462 | 1.521 | 2.695 | 2.491 | 7.574 | 4.418 | 3.745 | 15.244 |
| 0.566 | 1.513 | 1.486 | 1.777 | 2.822 | 2.586 | 8.348 | 4.622 | 3.882 | 16.007 |
| 0.584 | 1.532 | 1.505 | 1.772 | 2.902 | 2.661 | 8.292 | 4.990 | 4.243 | 14.972 |
| 0.604 | 1.553 | 1.525 | 1.800 | 2.991 | 2.741 | 8.354 | 5.064 | 4.320 | 14.701 |
| 0.611 | 1.560 | 1.533 | 1.733 | 3.014 | 2.776 | 7.904 | 5.127 | 4.592 | 10.431 |
| 0.645 | 1.596 | 1.568 | 1.758 | 3.171 | 2.919 | 7.942 | 5.467 | 4.609 | 15.691 |
| 0.650 | 1.601 | 1.573 | 1.768 | 3.197 | 2.941 | 8.001 | 5.507 | 4.799 | 12.865 |
| 0.654 | 1.607 | 1.577 | 1.862 | 3.237 | 2.955 | 8.710 | 5.672 | 4.884 | 13.891 |
| 0.706 | 1.662 | 1.635 | 1.617 | 3.455 | 3.225 | 6.652 | 6.246 | 5.364 | 14.134 |
| 0.706 | 1.662 | 1.635 | 1.617 | 3.455 | 3.225 | 6.652 | 6.256 | 5.394 | 13.779 |
| 0.710 | 1.666 | 1.639 | 1.664 | 3.485 | 3.240 | 7.031 | 6.267 | 5.378 | 14.181 |

**Acknowledgements** Financial support from NASA Cooperative Agreement 80NSSC20M0215 is gratefully acknowledged.

## References


Amyx JW, Bass DM Jr, Whiting RL (1960) Petroleum reservoir engineering. McGraw-Hill Book Co., New York

Attia MA (2005) Effects of petrophysical rock properties on tortuosity factor. J Pet Sci Eng 48:185–198

Avellaneda M, Torquato S (1991) Rigorous link between fluid permeability, electrical conductivity, and relaxation times for transport in Porous Media. Physics of Fluids-A 3:2529–2540

Avseth P, Mukerji T, Mavko G (2006) Quantitative seismic interpretation. Cambridge University Press, Cambridge

Beran MJ (1965) Use of the variational approach to determine bounds for the effective permittivity in random media. Nuovo Cimento 38:771–782

Beran MJ (1968) Statistical continuum theories. Wiley, New York

Beran MJ, Molyneux J (1966) Use of classical variational principles to determine bounds for the effective bulk modulus in heterogeneous media. Quart Appl Math 24:107–118

Bergman DJ (1978) The dielectric constant of a composite material—a problem in classical physics. Phys Rep C 43:377–407

Berryman J (1999) Origin of Gassmann's equations. Geophysics 64:1627–1629

Berryman JG, Milton GW (1988) Microgeometry of random composites and porous media. J Phys D 21:87–94

Bruggeman DAG (1935) Berechnung verschiedener physikalisher Konstanten von heterogenen Substanzen. I. Dielectrizitätkonstanten und Leitfähigkeiten der Mischkörper aus isotropen Substanzen. Ann Physik Leipzig 24:636–679

Bruggeman DAG (1937) Berechnung verschiedener physikalisher Konstanten von heterogenen Substanzen. III. Die Elastische Konstanten Der Quasiisotropen Mischkörper Aus Isotropen Substanzen. Ann Physik Leipzig 29:160–178

Cairney J (1924) Tortuosity of the cervical segment of the internal carotid artery. J Anat 59:87–96

Carman PC (1939) Permeability of saturated sands, soils and clays. J Agric Sci 20:262–273

Chen Z, Wang X, Giuliani F, Atkinson A (2013) Analyses of microstructural and elastic properties of porous SOFC cathodes based on focused ion beam tomography. J Power Sources 273:486–494

Chen F, Popov YuA, Sevostianov I, Romushkevich R, Giraud A, Grgic D (2017) Replacement relations for thermal conductivity of a porous rock. Int J Rock Mech Min Sci 97:64–74

Chen F, Sevostianov I, Giraud A, Grgic D (2017) Accuracy of the replacement relations for materials with non-ellipsoidal inhomogeneities. Int J Solids Struct 104–105:73–80

Ciz R, Shapiro A (2007) Generalization of Gassmann equations for porous media saturated with a solid material. Geophysics 72:A75–A79

Clennell MB (1997) Tortuosity: a guide through the maze. Geological Society, London (Special Publications, vol 122, pp 299–344)

Cornell D, Katz DL (1953) Flow of gases through consolidated porous media. Ind Eng Chem 45:2145–2152

Drach B, Tsukrov I, Trofimov A (2016) Comparison of full field and single pore approaches to homogenization of linearly elastic materials with pores of regular and irregular shapes. Int J Solids Struct 96:48–63

Dullien FAL (1975) Prediction of "tortuosity factors" from pore structure data. AZChE J 21:820–822

Edington GH (1901) Tortuosity of both internal carotid arteries. BMJ 2(2134):1526–1527

Feynman RP, Leighton RB, Sands M (1964) The Feynman lectures on physics, vol 2. Addison-Wesley, Reading, MA

Gassmann F (1951) Über die elastizität porpöser medien. Vierteljahrsschrift Der Naturforschenden Gesellscaft in Zurich 96:1–23

Han D-H, Batzle ML (2004) Gassmann's equation and fluid-saturation effects on seismic velocities. Geophysics 69:398–405

Hashin Z, Shtrikman S (1962a) A variational approach to the theory of the effective magnetic permeability of multiphase materials. J Appl Phys 33:3125–3131

Hashin Z, Shtrikman S (1962b) On some variational principles in anisotropic and nonhomogeneous materials. J Mech Phys Solids 10:335–342
Hashin Z, Shtrikman S (1963) A variational approach to the theory of the elastic behaviour of multiphase materials. J Mech Phys Solids 11:127–140
Helmer KG, Dardzinski BJ, Sotak CH (1995) The application of porous-media theory to the investigation of time-dependent diffusion in In-vivo systems. NMR Biomed 8:297–306
Hill R (1963) Elastic properties of reinforced solids: some theoretical principles. J Mech Phys Solids 11:357–372
Kachanov M, Sevostianov I (2018) Micromechanics of materials, with applications. Springer, Cham (Series: Solid mechanics and its applications, vol 249)
Kanaun SK (2003) Dielectric properties of matrix composite materials with high volume concentration of inclusions (effective field approach). Int J Eng Sci 41:1287–1312
Kanaun SK, Levin VM (2008) Self-consistent methods for composites. Springer
Kantor Y, Bergman DJ (1984) Improved rigorous bounds on the effective elastic moduli of a composite material. J Mech Phys Solids 32:41–62
Knyazeva AG, Grabovetskaya GP, Mishin IP, Sevostianov I (2015) On the micromechanical modelling of the effective diffusion coefficient of a polycrystalline material. Philos Mag 95:2046–2066
Koponen A, Kataja M, Timoneny J (1996) Tortuosity flow on porous media. Phys Rev E 54:406–410
Lorensen WE, Cline HE (1987) Marching cubes: a high resolution 3d surface construction algorithm. Comput Gr 21(4):163–169
Lutz MP, Zimmerman RW (2005) Effect of an inhomogeneous interphase zone on the bulk modulus and conductivity of a particulate composite. Int J Solids Struct 42:429–437
Markov K (2000) Elementary micromechanics of heterogeneous media. In: Markov KZ, Preziozi L (eds) Heterogeneous media: micromechanics modeling methods and simulations. Birkhauser, Boston, pp 1–162
McCoy JJ (1970) On the displacement field in an elastic medium with random variation of material properties. In: Eringen AC (ed) Proceedings of the 6-th annual meeting of the society of engineering science. Gordon and Breach, New York, pp 235–254
Michaels AS (1959) Diffusion in a pore of irregular cross section—a simplified treatment. AZChE J 5:270–271
Milton GW, Phan-Thien N (1982) New bounds on effective elastic moduli of two-component materials. Proc Roy Soc L A380:305–331
Mori T, Tanaka K (1973) (1973) Average stress in matrix and average elastic energy of materials with misfitting inclusions. Acta Met 21:571–574
Nooruddin FS, Turk G (2003) Simplification and repair of polygonal models using volumetric techniques. IEEE Trans Visual Comput Graphics 9(2):191–205
Perkins FM, Osoba JS, Ribe KH (1956) Resistivity of sandstones as related to the geometry of their interstitial water. Geophysics XXI:1071–1084
Pride S (1994) Governing equations for the coupled electromagnetics and acoustics of porous media. Phys Rev B50:15678–15696
Revil A, Cathles LM III, Losh S, Nonn JA (1998) Electrical conductivity in shaly sands with geophysical applications. J Geophys Res 103:23925–23936
Sevostianov I, Kachanov M (2002) Explicit cross-property correlations for anisotropic two-phase composite materials. J Mech Phys Solids 50(2002):253–282
Sevostianov I, Kachanov M (2007) Relations between compliances of inhomogeneities having the same shape but different elastic constants. Int J Eng Sci 45:797–806
Sevostianov I, Kachanov M (2008) Connections between elastic and conductive properties of heterogeneous materials. In: van der Giessen E, Aref H (eds) Advances in applied mechanics, vol 42. Academic Press, pp 69–252
Sevostianov I, Kováčik J, Simančík F (2006) Elastic and electric properties of closed-cell aluminum foams. Cross-property connection. Mater Sci Eng A-420:87–99

Sevostianova E, Leinauer B, Sevostianov I (2010) Quantitative characterization of porous material microstructure: evaluation of tortuosity. Int J Eng Sci (in press)

Taubin G (1995) A signal processing approach to fair surface design. In: Proceedings of the 22nd annual conference on computer graphics and interactive techniques, pp 351–358

Thomson W (Lord Kelvin), Tait PG (1879) Treatise on natural philosophy. Cambridge University Press, Cambridge

Weber L, Dorn J, Mortensen A (2003) On the electrical conductivity of metal matrix composites containing high volume fractions of non-conducting inclusions. Acta Mat 51:3199–3211

Weerts AH, Kandhai D, Bouten W, Sloot PMA (2001) Tortuosity of an unsaturated sandy soil estimated using gas iffusion and bulk soil electrical conductivity: comparing analogy-based models and Lattice-Boltzmann simulations. Soil Sci Soc Am J 65:1577–1584

Winsauer WO, Shearin HM, Masson PH, Williams M (1951) Resistivity of brine saturated sands in relation to pore geometry. Bull Am Assoc Pet Geol 36:253–277

Wyllie MRJ, Rose WD (1950) Some theoretical considerations related to the quantitative evaluation of the physical characteristics of reservoir rock from electrical log data. Trans AZME 189:105–118

Wyllie MRJ, de Witte AJ (1956) Comments on paper, “Resistivity of sandstones as related to the geometry of their interstitial water” by F.M. Perkins, J.S. Osoba and K.H. Ribe. Geophysics XXI:1084–1086

Zhao HF, Hu GK, Lu TJ (2004) Correlation between the elastic moduli and conductivity of two-dimensional isotropic two-phase composites. Int J Fract 126:L11–L18

Zimmerman RW (1989) Thermal conductivity of fluid-saturated rocks. J Pet Sci Eng 3:219–227

# Chapter 23
# Effective Viscoelastic Properties of Chiral Structures with Hierarchy

**Yun-Che Wang, Tsai-Wen Ko, and Kai-Wen Tan**

**Abstract** A class of hierarchically chiral metamaterials was analyzed with the finite element method for their effective linear viscoelastic properties under low frequency, uniaxial straining. The viscoelastic property of the solid phase was assumed to be of the standard linear solid. It is found that from Rank-1 to Rank-2 hierarchy effective Young's modulus is enhanced. From Rank-2 to Rank-3, effective modulus may be roughly the same, slightly increased or decreased, depending on the ligament thickness. When the ligament thickness is less than 1.5 mm, increasing hierarchy decreases the overall damping. For larger thickness, overall damping may be slightly enhanced by hierarchy. Gradient lateral deformation is observed in the metamaterials under uniaxial straining, indicating the existence of chiral effects.

**Keywords** Metamaterials · Hierarchical · Chiral · Viscoelastic · Finite element analysis

## 23.1 Introduction

Hierarchical materials are abundant in nature, and their physical properties can be enhanced by the structural hierarchy (Lakes 1993, 2020). Materials with chirality exhibit internally rotational degrees of freedom and can exhibit negative Poisson's ratio, also known as auxeticity (Prall and Lakes 1997). Recently, a class of hierarchical auxetic structure with star-shape geometry has been studied (Attard et al. 2020). Negative Poisson's ratio (NPR) materials have been widely and intensively studied (Lakes 1987; Evans et al. 1991; Lim 2015). In addition to the re-entrant cell shape giving rise to negative Poisson's ratio (Lakes 1987), several other microstructures have been proposed, such as connected stars (Grima 2005), connected rectangles

Y.-C. Wang (✉) · T.-W. Ko · K.-W. Tan
National Cheng Kung University, 1 University Road, Tainan 70101, Taiwan (R.O.C.)
e-mail: yunche@ncku.edu.tw

T.-W. Ko
e-mail: cegel745@gmail.com

 
H. Altenbach et al. (eds.), *Advanced Materials Modelling for Mechanical, Medical and Biological Applications*, Advanced Structured Materials 155,
https://doi.org/10.1007/978-3-030-81705-3_23

(Grima 2011), and rotating units related structures (Mizzi et al. 2015). Isotropic, homogenous, linear elastic materials in three dimensions have a Poisson's ratio ($\nu$) in the range between $-1$ and 0.5. For two-dimensional materials, their allowable Poisson's ratio is in between $-1$ and 1, in order to satisfy pointwise energy stability criterion (Wojciechowski 2003; Wang et al. 2017; Wang 2019). Negative Poisson's ratio is distinct from negative modulus, and it has been shown that negative bulk modulus may simultaneously achieve extreme effective properties and negative Poisson's ratio (Wang and Lakes 2005). Coupled-field properties of the negative-stiffness composites have been analyzed for their extreme effective properties and stability (Wang et al. 2015, 2021).

In addition to auxeticity, chiral materials, containing internally rotational degrees of freedom, are also known for their non-centrosymmetric properties, which lead to the deformation-mode couplings, such as twisting under tension (Lakes and Benedict 1982; Wang et al. 2020a). Such couplings may be utilized to develop novel sensors. The chiral materials can be modeled as the generalized continua in the homogenization sense (Altenbach and Eremeyev 2013). The hierarchical or chiral materials are inhomogeneous. Functionally graded materials, as a class of inhomogeneous materials, are usually analyzed with the consideration of space-dependent material properties in the elasticity equations (Aizikovich et al. 2002, 2011; Krenev et al. 2015).

Viscoelastic properties of chiral materials from the theoretical standpoint have been reported by Lakes (2001). Finite element studies of composite materials have shown damping can be enhanced via viscoelastic inclusions in conventional or auxetic systems (Wang and Ko 2015; Wang et al. 2020b). Rigorous proof of the stability of viscoelastic continua from the view point of center manifold theorem in topology has been reported (Luongo and D'Annibale 2020). Active viscoelastic materials have recently be analyzed for their behavior when both parity and time-reversal symmetries are broken (Banerjee et al. 2021).

In this work, we conduct the finite element analysis (Cook et al. 2001) to study the effective viscoelastic properties of a class of hierarchically chiral metamaterials under the plane strain assumption. Effects of hierarchical rank and ligament thickness on the linear viscoelastic modulus and damping are investigated.

## 23.2 Theoretical Considerations

When an elastic, homogeneous, isotropic solid is under infinitesimal deformation, its displacement fields, described by $u_i$, $i = 1, 2, 3$ for three spatial dimensions, are governed by the Navier equation in terms of the Lamé constants $\lambda$ and shear modulus $\mu$, as follows (Soutas-Little 1973).

$$\mu u_{i,kk} + (\lambda + \mu)u_{k,ik} + \rho b_i = \rho a_i \tag{23.1}$$

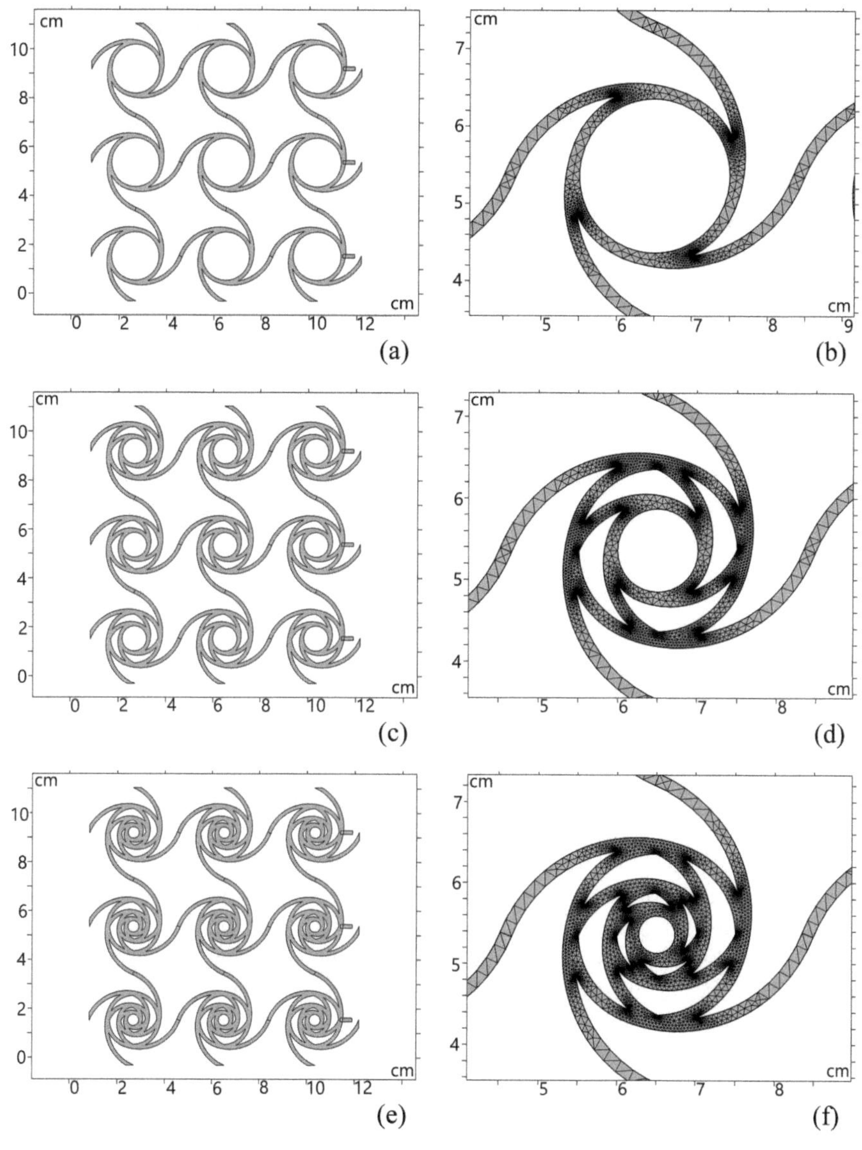


**Fig. 23.1** Numerical specimens of the **a** Rank-1 **c** Rank-2 and **e** Rank-3 model for thickness $t = 1.5$ mm. Corresponding finite element meshes are shown in (**b**) (**d**) and (**f**). For the $t = 1.0$ mm and $t = 2.0$ mm models, please see Fig. 23.7

where $\rho$, $a_i$, $b_i$ are density, acceleration and specific body force, respectively. We adopt the finite element method to numerically solve Eq. (23.1) with suitable displacement boundary conditions to calculate the effective elastic Young's modulus of the metamaterials under in-plane loading. Body force is assumed to be zero, and loading on the boundaries is under displacement control throughout this study. The solid phase of the metamaterial is assumed to have Young's modulus and Poisson's ratio being 100 MPa and 0.3, respectively.

For the viscoelastic calculations, the relaxed bulk modulus and shear modulus of the solid phase are assumed to be 83.33 and 38.46 MPa, respectively. These chosen moduli correspond to a Young's modulus of 100 MPa and a Poisson's ratio of 0.3 in the purely elastic branch of the generalized Maxwell model. The simplest generalized Maxwell model is also known as the standard linear solid, which requires three material parameters, i.e., two time constants $T_1$, $T_2$ and relaxed shear modulus $G$. In our linear viscoelastic calculations, the shear behavior of the solid phase is assumed to be governed by the standard linear solid model, Eq. (23.2), while the bulk behavior is assumed to be purely elastic, Eq. (23.3). This assumption is commonly accepted in viscoelastic analysis.

$$S_{ij} + T_1 \frac{\mathrm{d}s_{ij}}{\mathrm{d}t} = G\left(e_{ij} + T_2 \frac{\mathrm{d}e_{ij}}{\mathrm{d}t}\right), \quad \text{for} \quad i \neq j. \tag{23.2}$$

$$\sigma_{kk} = 3\,K\varepsilon_{kk}. \tag{23.3}$$

Here, the symbols $\sigma$, $\varepsilon$ and $K$ (= 83.33 MPa), respectively, represent stress, strain and three-dimensional bulk modulus, $s_{ij} = \sigma_{ij} - \sigma_{kk}/3$ the deviatoric stress, $e_{ij} = \varepsilon_{ij} - \varepsilon_{kk}/3$ the deviatoric strain, and $G$ (= 38.46 MPa) is the relaxed shear modulus. The deviatoric stress and strain are zero when index $i$ is equal to index $j$. The time derivative is denoted by $\mathrm{d}/\mathrm{d}t$. The time constants are $T_1 = 0.1$ s, and $T_1 = 2T_2$. Physically, $T_1$ is the time constant under constant stress and $T_2$ the time constant under constant strain. The choice of these material parameters is representative to include linear viscoelastic effects in the FEM with suitable frequency dependence (Wang et al. 2020b; Wang and Ko 2013). Figure 23.2a, b show the frequency-dependent viscoelastic modulus and loss tangent of the solid phase of the plates. We remark that the effective properties of the metamaterials are expected to be different from the solid phase material due to the composite effects between the solid phase and empty space. All of our finite element calculations for effective viscoelastic properties are under displacement control. In order to estimate the effective modulus, the volume-averaged stress ($\bar{\sigma}_{ij}$) and strain ($\bar{\varepsilon}_{ij}$) are adopted and defined as follows (Lakes and Drugan 2002).

$$\bar{\sigma}_{ij} = \frac{1}{V}\int_{\Omega} \sigma_{ij}\mathrm{d}\Omega = \frac{1}{V}\int_{\partial\Omega} t_i x_j \mathrm{d}\partial\Omega, \tag{23.4}$$

$$\bar{\varepsilon}_{ij} = \frac{1}{V}\int_{\Omega} \varepsilon_{ij}\mathrm{d}\Omega = \frac{1}{2V}\int_{\partial\Omega} (u_i n_j + u_j n_i)\mathrm{d}\partial\Omega. \tag{23.5}$$

The domain of metamaterial is indicated by $\Omega$, and $\partial\Omega$ denotes the boundary of the domain, and V the area of the two-dimensional domain. The traction, displacement

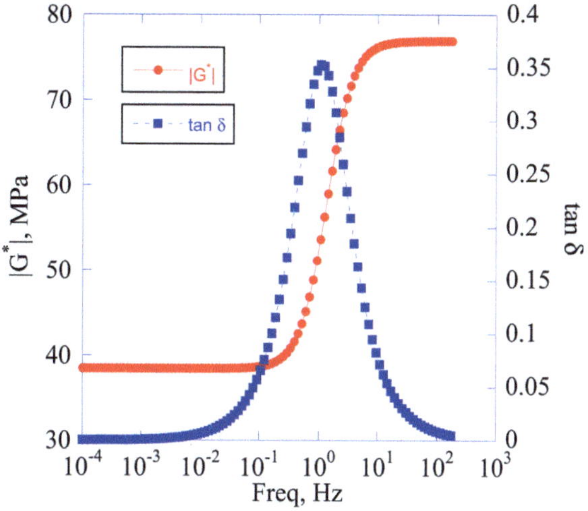


**Fig. 23.2** Frequency response of the **a** absolute value of the viscoelastic shear modulus and **b** tan $\delta$ of the standard linear solid model used for the solid phase of the hierarchically chiral materials

and unit vector are denoted by $t_i$, $u_i$ and $n_i$. With Eqs. (23.4) and (23.5), we calculated the average stress and strain on the boundary of the domain to evaluate the effective Young's modulus of the metamaterials, as the following.

$$E_{\text{eff}} = \frac{\bar{\sigma}_{xx}}{\bar{\varepsilon}_{xx}} \tag{23.6}$$

As for the effective linear viscoelastic properties of the plates, we apply sinusoidal loading with 2 Hz on the right boundary of the metamaterials, and calculate their overall stress-strain Lissajous curves. Effective loss tangent and complex modulus are, respectively, obtained from the width and slope of the Lissajous curves (Lakes 2009). The left and bottom boundaries are the roller boundary condition, hence sliding is allowed but normal displacement is constrained.

## 23.3 Numerical Considerations

Figure 23.1 shows the hierarchically chiral metamaterials and their representative finite element mesh for ligament thickness $t = 1.5$ mm. Triangular elements are adopted to form the mesh. The number of elements was on the order of $10^4$. The ratio of thickness to the unit cell length $t/L_0 = 1.5/38.3 = 0.04$ The geometry of the $t = 1.0$ and $t = 2.0$ mm models can be found in Fig. 23.7. The area fraction, which can be considered as relative density in this work, is listed in Table 23.1. Under uniform, uniaxial, sinusoidal straining on the rightmost edges, the effective Young's modulus and Poisson's ratio can be obtained when the left and bottom edges are on rollers. In the calculations, the meshes were chosen to be sufficiently dense. The shape function was third order Lagrange polynomial. The three small rectangles

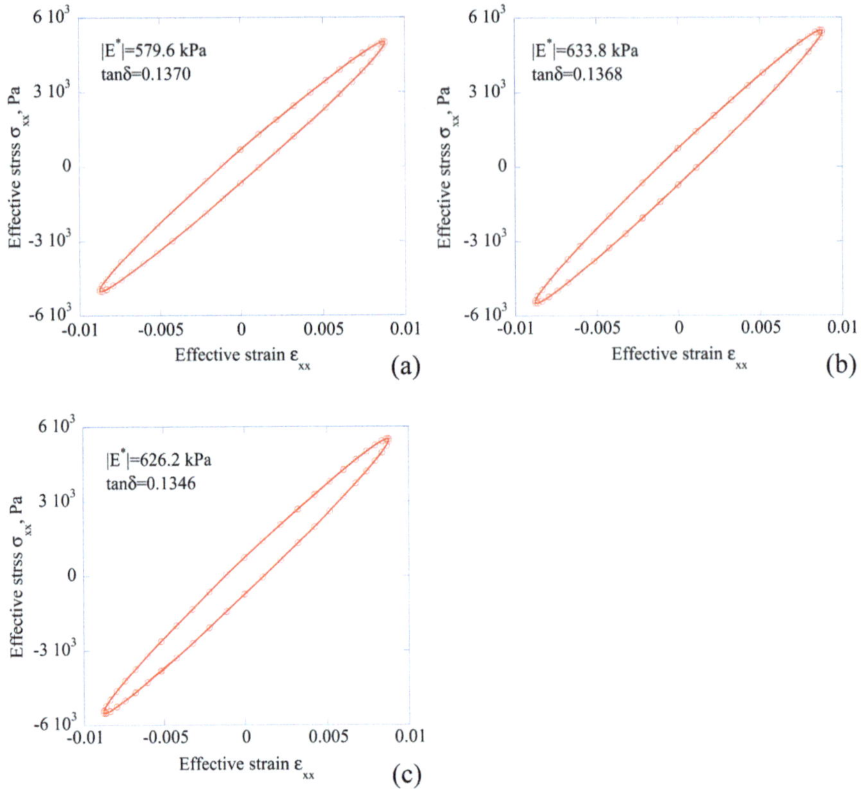


**Fig. 23.3** Lissajous curves of the **a** Rank-1 **b** Rank-2 and **c** Rank-3 hierarchical chiral models with $t = 1.5\,\text{mm}$, driven sinusoidally 2 Hz

**Table 23.1** Area fraction of the metamaterials

| Area fraction | Rank-1 | Rank-2 | Rank-3 |
|---|---|---|---|
| $t = 0.1\,\text{cm}$ | 0.110 | 0.165 | 0.192 |
| $t = 0.15\,\text{cm}$ | 0.161 | 0.239 | 0.271 |
| $t = 0.2\,\text{cm}$ | 0.212 | 0.306 | 0.343 |

attached to the cells on the right are not used in this study. They were constructed for some measurement purposes for other works. The COMSOL (2021) software package was used in this study to carry out the finite element calculations. Under the assumption of affine deformation used in the elasticity and FEM, our calculated results can be scaled down for small size specimens until other physical mechanisms need to be included, such as contact between ribs or weak interaction forces between the surfaces of ribs.

**Table 23.2** Summary of complex modulus and loss tangent of the metamaterials

| Skeleton thickness $t$, mm | $\|E^*\|$, kPa | $\tan\delta$ |
|---|---|---|
| 1.0 | 191.9 | 0.1360 |
| 1.0 | 229.5 | 0.1327 |
| 1.0 | 229.6 | 0.1310 |
| 1.5 | 579.6 | 0.1370 |
| 1.5 | 633.8 | 0.1368 |
| 1.5 | 626.2 | 0.1346 |
| 2.0 | 1200.4 | 0.1361 |
| 2.0 | 1343.3 | 0.1364 |
| 2.0 | 1352.9 | 0.1363 |

## 23.4 Results and Discussion

Figure 23.2 shows the frequency-dependent viscoelastic properties of the ligament modeled by the standard linear solid. The relaxed dynamic shear modulus is about 30 MPa, and instantaneous modulus is about 76 MPa. As shown in Fig. 23.2b, the Debye peak of the ligament is 1 Hz with the tangent delta of 0.35 under shear deformation. The material parameters for the standard linear solid are so chosen to represent a general viscoelastic material in wide frequency range. The frequency responses of the hierarchically chiral materials are studied under a driving frequency 2 Hz to delineate their viscoelastic properties with loss of generality.

Figure 23.3a–c, respectively, show the stress–strain Lissajous curves of the Rank-1, Rank-2 and Rank-3 metamaterials with $t = 0.15$ cm under 2-Hz driving frequency. The transient responses are calculated, but not shown. As can be seen, due to the empty space in the metamaterials, their overall Young's modulus about two orders of magnitude less than that of the solid phase. Furthermore, their effective damping is about one-third of the solid phase. Our calculated effective complex moduli in terms of their absolute value and loss tangent are listed in Table 23.2. Figure 23.4 shows the storage and dissipation strain energy density of the metamaterials with thickness $t = 0.15$ cm. The driving frequency 2 Hz. Dissipation energy curves are shown as the monotonically increasing curves because of accumulation from cycles.

As shown in Figs. 23.5 and 23.6, it can be seen that ligament thickness plays a dominant role in effective modulus and damping. When hierarchical ranks are increased from Rank-1 to Rank-3, effective damping is decreased with thickness, except for the $t = 2.0$ mm case. Hence, overall damping can be controlled by hierarchy and ligament thickness. For effective Young's modulus, regardless of thickness, hierarchy enhances modulus from Rank-1 to Rank-2. However, From Rank-2 to Rank-3, effective modulus is roughly the same for the $t = 1.0$ mm case, slightly decreased for the $t = 1.5$ mm case and slightly increases for the $t = 2.0$ mm case. Hence, ligament thickness may affect elastic hardening or softening in hierarchically chiral materials.

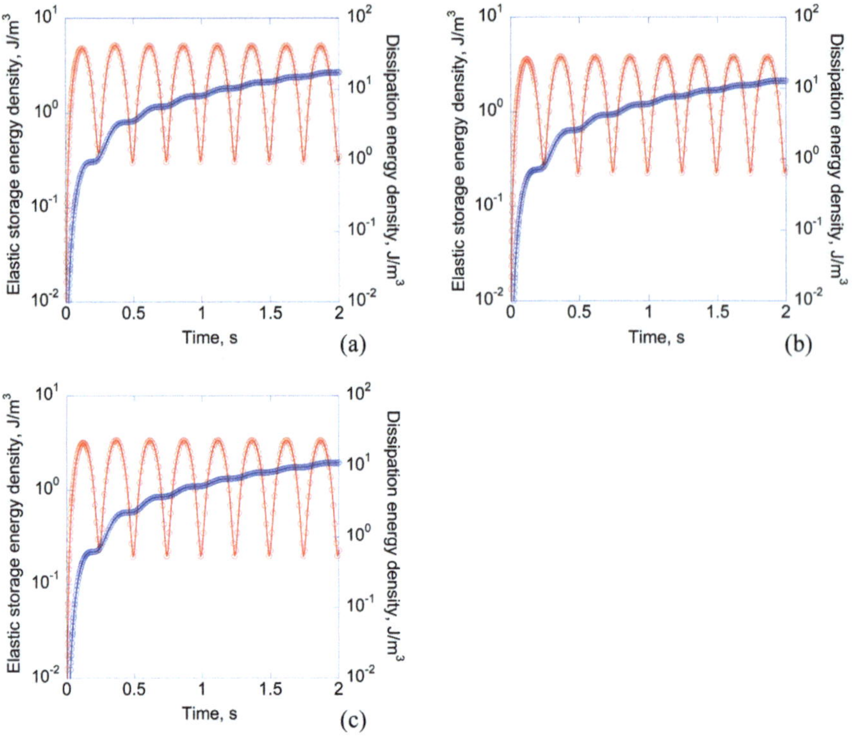


**Fig. 23.4** Storage (red) and dissipation (blue) energy density of the **a** Rank-1 **b** Rank-2 and **c** Rank-3 hierarchical chiral models with $t = 1.5$ cm, driven sinusoidally 2 Hz frequency. Storage energy is oscillatory and dissipation energy is monotonically increasing due to accumulation from cycles

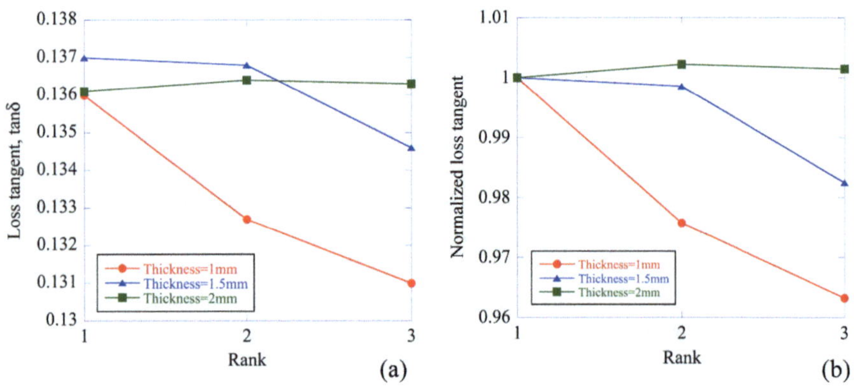


**Fig. 23.5** **a** Loss tangent and **b** normalized loss tangent of the hierarchical chiral models studied in this work

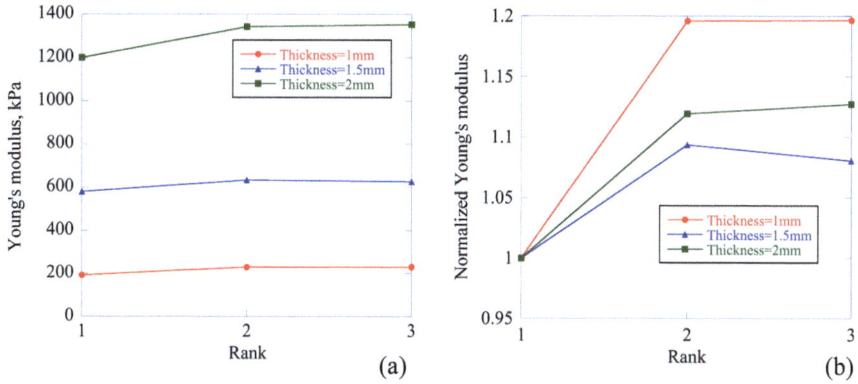


**Fig. 23.6** **a** Young's modulus and **b** normalized Young's modulus of the hierarchical chiral models studied in this work

Figure 23.7 shows the deformed and undeformed shapes of the hierarchically chiral materials under $\varepsilon = 10^{-4}$ straining on the right edge for all the cases studied in this work. The left and bottom edge are under the roller boundary condition. The color indicates the von Mises stress, revealing high stress regions in the heterogeneous material. It is remarked that the lateral deformation is not uniform due to the chiral effects (Wang et al. 2020a). The chiral effects induce deformation-mode couplings, which may be found applications in developing novel sensors. To model the hierarchically chiral materials as generalized continua, our results presented here may be used as comparison data.

## 23.5 Conclusions

The linear viscoelastic properties of the hierarchically chiral materials analyzed by the finite element analysis in this work show that increasing hierarchical ranks may increase the absolute value of effective complex modulus from Rank-1 to Rank-2 hierarchy. From Rank-2 to Rank-3, effective modulus is roughly the same for the $t = 1.0$ mm case, slightly decreased for the $t = 1.5$ mm case, and slightly increases for the $t = 2.0$ mm case. As for effective damping, when ligament thickness is less than 1.5 mm, loss tangent decreases with the hierarchical rank. For larger thickness ($t = 2.0$ mm), overall damping may be slightly enhanced by hierarchy. Under the chiral effects, gradient lateral deformation under uniaxial straining is observed. Further analytical explorations are in need to model the metamaterials in the homogenized sense.

**Acknowledgements** This research was, in part, supported by Taiwan Ministry of Science and Technology, under the contract number MOST 109-2221-E-006-016-MY3, and by the Southern

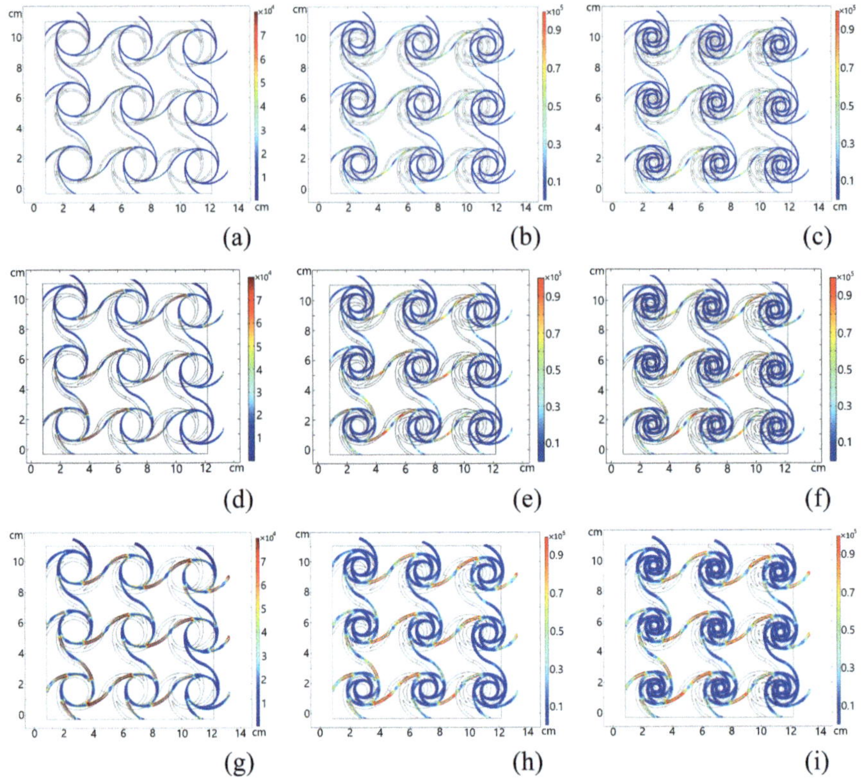

**Fig. 23.7** von Mises stress of the **a**, **d**, **g** Rank-1 **b**, **e**, **h** Rank-2 and **c**, **f**, **i** Rank-3 hierarchical chiral models under uniaxial straining along the horizontal direction. The ligament thickness for (**a**–**c**) is $t = 1$ mm, for (**d**–**f**) $t = 1.5$ mm and for (**g**–**i**) $t = 2$ mm. The samples were under uniaxial straining ($\varepsilon_{xx} = 10^{-4}$) along the horizontal direction

Taiwan Science Park Bureau, Ministry of Science and Technology, Taiwan, R.O.C. under contract 109CP02.

## References

Aizikovich SM, Alexandrov VM, Kalker JJ, Krenev LI, Trubchik IS (2002) Analytical solution of the spherical indentation problem for a half-space with gradients with the depth elastic properties. Int J Solids Struct 39:2745–2772

Aizikovich S, Krenev L, Sevostianov I, Trubchik I, Evich L (2011) Evaluation of the elastic properties of a functionally-graded coating from the indentation measurements. Z Angew Math Mech 91(6):493–515. https://doi.org/10.1002/zamm.201000179

Altenbach H, Eremeyev VA (2013) Generalized continua from the theory to engineering applications. Springer, Wien

Attard D, Farrugia PS, Gatt R, Grima JN (2020) Starchirals–a novel class of auxetic hierarchal structures. Int J Mech Sci 179:105631
Banerjee D, Vitelli V, Jülicher F, Surówka P (2021) Active viscoelasticity of odd materials. Phys Rev Lett 126:138001
COMSOL website (2021). http://www.comsol.com
Cook RD, Malkus DS, Plesha ME, Witt RJ (2001) Concepts and applications of finite element analysis, 4th edn. Wiley, Hoboken
Evans KE, Nkansah MA, Hutchinson IJ, Rogers SC (1991) Molecular network design. Nature 353:124
Grima JN, Gatt R, Alderson A, Evans KE (2005) On the potential of connected stars as auxetic systems. Mol Simul 31:925–935
Grima JN, Manicaro E, Attard D (2011) Auxetic behaviour from connected different-sized squares and rectangles. Proc R Soc A 467:439–458
Krenev LI, Aizikovich SM, Tokovyy YV, Wang Y-C (2015) Axisymmetric problem on the indentation of a hot circular punch into an arbitrarily nonhomogeneous half-space. Int J Solids Struct 59:18–28
Lakes R (1987) Foam structures with a negative Poisson's ratio. Science 235:1038–1040
Lakes R (1993) Materials with structural hierarchy. Nature 361:511–515
Lakes R (2001) Elastic and viscoelastic behavior of chiral materials. Int J Mech Sci 43:1579–1589
Lakes R (2009) Viscoelastic materials. Cambridge University Press, New York
Lakes R (2020) Composites and metamaterials. World Scientific, Singapore
Lakes RS, Benedict RL (1982) Noncentrosymmetry in micropolar elasticity. Int J Eng Sci 20:1161–1167
Lakes R, Drugan W (2002) Dramatically stiffer elastic composite materials due to a negative stiffness phase? J Mech Phys Solids 50:979–1009
Lim T-C (2015) Auxetic materials and structures. Springer, Berlin
Luongo A, D'Annibale F (2020) Modeling the linear dynamics of continuous viscoelastic systems on their infinite-dimensional central subspace. Math Mech Complex Syst 8:127–151
Mizzi L, Azzopardi KM, Attard D, Grima JN, Gatt R (2015) Auxetic metamaterials exhibiting giant negative Poisson's ratios. Phys Status Solidi RRL 9:425–430
Prall D, Lakes RS (1997) Properties of a chiral honeycomb with a Poisson's ratio of -1. Int J Mech Sci 39:305–314
Soutas-Little RW (1973) Elasticity. Dover, New York
Wang YC (2019) Negative Poisson's ratio in plane elasticity. In: Altenbach K, Öchsner A (eds) Encyclopedia of continuum mechanics. Springer, Berlin, p 7
Wang YC, Ko CC (2013) Stability of viscoelastic continuum with negative-stiffness inclusions in low frequency range. Phys Status Solidi B 250:2070–2079
Wang YC, Ko CC (2015) Energy dissipation of steel-polymer composite beam-column connector. Steel Compos Struct 18(5):1161–1176
Wang YC, Lakes RS (2005) Composites with inclusions of negative bulk modulus: extreme damping and negative Poisson's ratio. J Compos Mater 39:1645–1657
Wang YC, Ko CC, Chang KW (2015) Anomalous effective viscoelastic, thermoelastic, dielectric and piezoelectric properties of negative-stiffness composites and their stability. Phys Status Solidi B 252:1640–1655
Wang YC, Shen MW, Liao SM (2017) Microstructural effects on the Poisson's ratio of star-shaped two-dimensional systems. Phys Status Solidi B 254:1700024
Wang YC, Lai HW, Ren XJ (2020a) Enhanced auxetic and viscoelastic properties of filled reentrant honeycomb. Phys Status Solidi B 257:1900184
Wang YC, Ko TW, Ren X (2020b) Effective mechanical responses of a class of 2D chiral materials. Phys Status Solidi B 257:2000277

Wang YC, Ko CC, Chang KW, Ko TW (2021) Negative-stiffness composite systems and their coupled-field properties. Continuum Mech Thermodyn 33(4):857–1872

Wojciechowski KW (2003) Remarks on "Poisson ratio beyond the limits of the elasticity theory." J Phys Soc Jpn 72:1819–1820

# Chapter 24
# Method for Determining the Parameters of the Exponential Shear Modulus of a Functional-Gradient Half-Space

**Vladimir B. Zelentsov, Polina A. Lapina, and Anna D. Zagrebneva**

**Abstract** The basis of the method for determining the parameters of the depth-variable shear modulus of the functional-gradient half-space is the indentation method or the method of shear by the strip indenter of the elastic functional-gradient half-space. Mathematical modeling or description of the indentation process consists in solving the corresponding contact problem of shear by a strip indenter of the surface of an elastic functional-gradient half-space with an exponential shear modulus in depth. The solution of the contact problem is reduced to the solution of an integral equation of the first kind of the Fourier convolution type with a difference kernel. The analytical solution of the integral equation, obtained by asymptotic methods, contains the parameters of the exponential modulus of the half-space shear. To unambiguously determine the parameters of the shear modulus, the theoretical and experimental characteristics of the contact are compared, such as contact stresses, surface displacements outside the contact area and others, and the static condition on the contact is also used. In the space of parameters of the contact problem, methods are developed for studying the boundaries of the regions of existence and uniqueness of the values of the parameters of the exponential shear modulus. Formulas for the effective shear modulus of a functional-gradient half-space are determined. Its dependence on the parameters of the exponential shear modulus is investigated. In conclusion, the integral equations of contact problems for the homogeneous and functional-gradient half-space, as well as their solutions, are compared.

**Keywords** Functional-gradient material · Exponential inhomogeneity · Contact problem · Determination of elastic modulus parameters

## 24.1 Introduction

The development of experimental and theoretical methods for studying the functional-gradient dependence of the elastic moduli of materials and products and

V. B. Zelentsov (✉) · P. A. Lapina · A. D. Zagrebneva
Research and Education Center "Materials", Don State Technical University, Rostov-on-Don 344000, Russia



H. Altenbach et al. (eds.), *Advanced Materials Modelling for Mechanical, Medical and Biological Applications*, Advanced Structured Materials 155,
https://doi.org/10.1007/978-3-030-81705-3_24

their effect on the stress-strain state of an elastic medium has received considerable attention from researchers since the middle of the last century, starting with works of Popov (1959), Aizikovich (1978) and others. The subsequent tightening of requirements for the strength of products, heat resistance, reduction in weight, volume, miniaturization of products and resistance of product materials to chemical influences led to the need to produce first multilayer and then to the creation of functional-gradient materials. Functional-gradient materials are widely used in semiconductor microelectronics when creating modern means of communication (Shiraki and Usami 2011; Miyamoto et al. 1999). The main range of functional-gradient materials is made up of traditional materials such as aluminum, zirconium and their alloys. These materials, coated with appropriate oxides of technological or natural origin, are used in the manufacture of semiconductor products for the benefit of microelectronics. Elastic modules of oxides, nitrides, etc., significantly differ from the elastic moduli of the base material.

When calculating the stress–strain, electro- and magnetostatic state of functional-gradient materials, it is necessary to take into account the change in elastic moduli, electrical and magnetic resistance along the length or depth of the product material.

At the present stage of the evolution of functional-gradient materials, more and more attention is paid to the creation of design schemes for determining the stress–strain state of functional-gradient materials with Young's modulus or shear modulus, varying with the depth of the product material (Giannakopoulos and Suresh 1997a, b; Selvadurai and Katebi 2013; Katebi and Selvadurai 2015; Selvadurai and Katebi 2016; Aizikovich et al. 2015; Kulchytsky-Zhyhailo and Bajkowski 2012; Altenbach and Eremeyev 2009; Awojobi 1976) and design schemes for creating models of functional-gradient materials. Determination of elastic moduli of homogeneous materials and parameters of elastic moduli of inhomogeneous materials is carried out using the nano-indentation method. Modeling of the nano-indentation method is reduced to the mathematical formulation and solution of contact problems on the effect of an indenter on functional-gradient materials. Most often, contact problems are reduced to integral equations (Giannakopoulos and Suresh 1997a, b; Selvadurai and Katebi 2013; Katebi and Selvadurai 2015; Selvadurai and Katebi 2016; Aizikovich et al. 2015), the solution of which is carried out mainly by finite element methods. After comparing the experimental and theoretical data, the dependence of the stress–strain state of the material on the laws of change in elastic moduli with depth is established. Another group of researchers is developing experimental or numerical finite element approaches to solving problems of determining the parameters of elastic moduli, the magnitude of which changes with depth (Nakamuram et al. 2000; Gu et al. 2003; Bocciarelli et al. 2008; Chen et al. 2018).

This work is devoted to an experimental–theoretical method for solving the problem of determining the parameters of the shear modulus of a functionally gradient half-space, which varies with depth according to an exponential law. At the first stage, the problem of determining the parameters of the exponential shear modulus of an elastic half-space is formulated. At the second stage, an auxiliary contact problem on the antiplane shear of the surface of a functionally gradient half-space with an exponential shear modulus of the base of a strip indenter is posed and solved. The solution of the auxiliary problem is reduced to the solution of an integral equation of the first kind with a difference kernel. Analytical solutions of the integral equation were constructed by asymptotic methods for the dimensionless parameter of the problem (Zelentsov et al. 2019; Zelentsov et al. 2020). The dependence of the main mechanical characteristics of the problem on two parameters of the exponential shear modulus of the half-space is studied. In the first formulation of the problem, the shear modulus parameters are determined using two contact characteristics—the coincidence of the theoretical and experimental values of the contact stresses and the coincidence of the theoretical and experimental values of the forces acting on the indenter. In the case of insurmountable difficulties in obtaining experimental data on contact stresses, it is proposed, instead of this formulation, to use the formulation of the problem of equality of experimental and theoretical data on surface displacements outside the contact area in conjunction with the condition of static at the contact. It is proposed to use other characteristics of the contact to determine the parameters of the shear modulus. On the set of parameters of the contact problem, domains of uniqueness of determination of the parameters of the exponential shear modulus are distinguished. The relevance and novelty of the work lie in the fact that the parameters of the exponential shear modulus of the functional-gradient half-space are uniquely determined using the analytical solution of the contact problem.

## 24.2 Statement and General Plan for Solving the Main Task of the Research

An elastic functional-gradient half-space is considered, the shear modulus of which $\mu(y)$ varies with depth $y$ $(0 \leq y < \infty)$ according to an exponential law, written using two parameters in the form

$$\mu(y) = \mu_0 e^{2dy}, 0 \leq y < \infty, \tag{24.2.1}$$

where the parameter $\mu_0 = \mu(0)$ represents the value of the shear modulus on the surface of the half-space, and the parameter $d$ determines the intensity of the change in the shear modulus with depth in the half-space. The task of the study is to determine the parameters $\mu_0$ and $d$ and the depth-variable shear modulus $\mu(y)$ from (24.2.1).

The general plan for solving the task includes experimental, theoretical and comparative parts. In the first experimental part of the solution of the problem posed, the scheme of indentation of a functional-gradient half-space by means of an antiplane shear by a strip indenter of the half-space surface is implemented. During the implementation of the indentation process, the main characteristics of the contact interaction are fixed: the magnitude of the displacement (shear) of the strip indenter, the force of the contact action on the indenter, the distribution of contact stresses based on the indenter, the magnitude of the displacements of the free surface of the half-space outside the contact area, as well as secondary characteristics of the contact, such as,

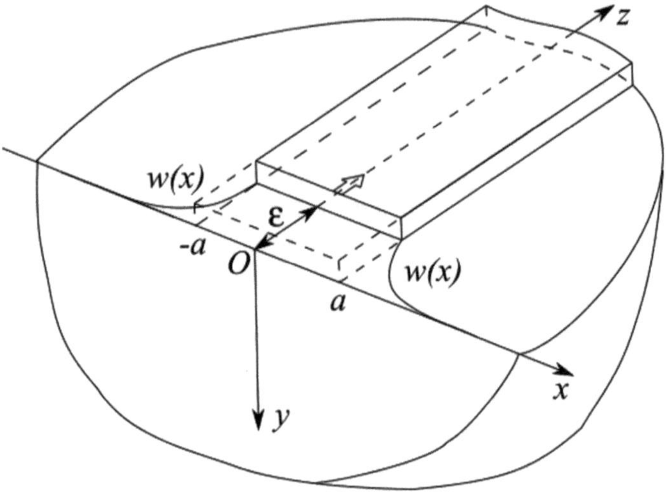


**Fig. 24.1** Statement of the contact problem of the displacement by a strip indenter of the surface of a functional-gradient half-space

for example, the tangent of the angle of inclination or the arc length of the surface displacement curve outside the contact area and others (Fig. 24.1). Following this, to determine the parameters $\mu_0$ and $d$ of the exponential shear modulus, it is necessary to obtain formulas for all experimentally determined characteristics of the contact. For this purpose, in the second part of the solution of the problem of determining the parameters $\mu_0$ and $d$ the mathematical modeling of the process of indentation of the functionally gradient half-space is carried out by solving the contact problem of the antiplane shear of the surface of the functionally gradient half-space with an exponential shear modulus $\mu(y)$ of the form (24.2.1) with two parameters $\mu_0$ and $d$. An analytical solution to such a contact problem will make it possible to obtain formulas for determining the above characteristics: The distribution of contact stresses in the contact area, shear forces, the magnitude of displacements of the free surface outside the contact area and other contact characteristics determined in the experimental part of the solution to the main research problem. In the third part of the solution to the main problem, a comparison is made between the experimental and theoretical results obtained in the first and second parts of the solution. This allows us to reduce the solution of the problem to the solution of nonlinear equations with respect to the parameters $\mu_0$ and $d$ shear modulus $\mu(y)$ from (24.2.1).

**Comment**: In the future, it is believed that the results of the experimental part were obtained for their further use, taking into account special designations.

## 24.3 Statement of the Pure Shear Problem for a Functional-Gradient Half-Space

The implementation of the second part of the solution to the main problem of determining the parameters $\mu_0$ and $d$ the exponential shear modulus $\mu(y)$ from (24.2.1) begins with the formulation of the indentation problem, namely the contact problem of antiplane shear by the base of a rigid strip indenter of width $2a$ ($|x| \leq a, y = 0$) of the surface ($y = 0$) of an elastic functional-gradient half-space ($|x| < \infty, 0 \leq y < \infty$) with the depth-variable shear modulus $\mu(y)$ from (24.2.1). The displacement of the surface of the half-space is carried out by the value $\varepsilon$ ($w(x, 0) = \varepsilon$, $|x| \leq a$, where $w(x, y)$ is the displacement in the half-space). The contact stresses $\varphi(x) = -\sigma_{yz}(x, 0)$, arising on the basis of the indenter, as well as the mechanical characteristics of the contact listed above, are subject to determination. It is assumed that at infinity in the half-space the displacement $w(x, y)$ and stresses $\sigma_{xz}(x, y)$, $\sigma_{yz}(x, y)$ disappear when $\sqrt{x^2 + y^2} \to \infty$.

**Comment**: It is no coincidence that the solution of problems with a shear modulus depending on depth is in demand. So, for example, when studying the Oxford geological deposits (Selvadurai and Katebi 2013), it was found that the shear modulus $\mu(z)$ changes with the depth of the soil according to the exponential law with the parameters $\mu_0 = 3.33$ MPa, $2d = 0.0879\ \text{m}^{-1}$.

In a half-space, when the deformation of pure shear is carried out, the conditions $\sigma_{xx} = \sigma_{xy} = \sigma_{yx} = \sigma_{yy} = \sigma_{zz} = 0$, where $\sigma_{xx}$, $\sigma_{yy}$ and $\sigma_{zz}$ are normal stresses and $\sigma_{xy}$ and $\sigma_{yx}$ are shear stresses. Then the equations of the theory of elasticity in stresses (Alexandrov and Kovalenko 1986) in the case of antiplane deformation have the form

$$\frac{\partial \sigma_{zx}}{\partial x} + \frac{\partial \sigma_{zy}}{\partial y} = 0, \tag{24.3.1}$$

where expressions $\sigma_{zx}$ and $\sigma_{zy}$ in terms of deformations in the case of functional inhomogeneity of the elastic medium are expressed by the formulas

$$\sigma_{zx} = \mu(y)\frac{\partial w}{\partial x}, \sigma_{zy} = \mu(y)\frac{\partial w}{\partial y}. \tag{24.3.2}$$

Substituting (24.3.2) into the differential Eq. (24.3.1) and then dividing it by $\mu(y) \neq 0$, we obtain the differential equation in displacements

$$\frac{\partial^2 w}{\partial y^2} + \frac{\mu'(y)}{\mu(y)}\frac{\partial w}{\partial y} + \frac{\partial^2 w}{\partial x^2} = 0. \tag{24.3.3}$$

Differential Eq. (24.3.3) with a coefficient $\mu'(y)/\mu(y)$, depending on $y$, at $\mu(y) = \mu_0 e^{2dy}$, turns into a differential equation with constant coefficients

$$\frac{\partial^2 w}{\partial y^2} + 2d\frac{\partial w}{\partial y} + \frac{\partial^2 w}{\partial x^2} = 0. \quad (24.3.4)$$

The mixed boundary conditions of the formulated contact problem on the pure shear of the surface ($y = 0$) of a functional-gradient half-space by a strip indenter have the form

$$w(x, 0) = \varepsilon \quad |x| \le a, \quad (24.3.5)$$

$$\sigma_{yz}(x, 0) = \begin{cases} -\tilde{\varphi}(x) & |x| \le a, \\ 0 & a < |x| \le \infty. \end{cases} \quad (24.3.6)$$

where $\tilde{\varphi}(x)$ are the contact stresses based on the indenter to be determined. At infinity for $\sqrt{x^2 + y^2} \to \infty$ for $w$ and $\frac{\partial w}{\partial x}$, $\frac{\partial w}{\partial y}$ the requirements are satisfied

$$\lim_{\sqrt{x^2+y^2} \to \infty} \left\{ w, \frac{\partial w}{\partial x}, \frac{\partial w}{\partial y} \right\} = 0. \quad (24.3.7)$$

## 24.4 Integral Equation of the Problem

To construct a solution to the differential equation of problem (24.3.4) and to satisfy the mixed boundary conditions (24.3.5), (24.3.6) and conditions at infinity, the integral Fourier transform is used

$$w^F(\alpha, y) = \int_{-\infty}^{\infty} w(x, y) \mathrm{e}^{i\alpha x} \mathrm{d}x,$$

$$w(x, y) = \frac{1}{2\pi} \int_{-\infty}^{\infty} w^F(\alpha, y) \mathrm{e}^{-i\alpha x} \mathrm{d}\alpha, \quad (24.4.1)$$

where the superscript $F$ above the function denotes the Fourier transform. Using the integral Fourier transform (24.4.1) and the condition at infinity (24.3.7), differential Eq. (24.3.4) is reduced to solving an ordinary differential equation with constant coefficients with respect to the transform $w^F(\alpha, y)$

$$\frac{\mathrm{d}^2 w^F}{\mathrm{d}y^2} + 2d\frac{\mathrm{d}w^F}{\mathrm{d}y} - \alpha^2 w^F = 0. \quad (24.4.2)$$

The general solution of this equation, which vanishes at infinity at $y \to \infty$, has the form

$$w^F(\alpha, y) = C(\alpha)e^{\lambda_- y}, \lambda_- = -d - \sqrt{d^2 + \alpha^2}, d > 0, \tag{24.4.3}$$

where $C(\alpha)$ is determined from the boundary condition (24.3.6), after applying the integral Fourier transform (24.4.1) to it. The displacements $w(x, y)$ are determined from (24.3.3) using the inverse Fourier transform (24.4.1) and have the form

$$w(x, y) = -(2\pi\mu_0)^{-1} \int_{-\infty}^{\infty} \tilde{\varphi}^F(\alpha)e^{-\lambda_- y} \lambda_-^{-1} e^{-i\alpha x} d\alpha, \tag{24.4.4}$$

where $\tilde{\varphi}^F(\alpha)$ is the Fourier transform of the function $\tilde{\varphi}(x)$.

Satisfying the mixed boundary conditions (24.3.5), (24.3.6) with the help of (24.4.4), we obtain the integral equation of the problem

$$\int_{-a}^{a} \tilde{\varphi}(\xi)k(\xi - x)d\xi = 2\pi\mu_0\varepsilon, |x| \le a, \tag{24.4.5}$$

$$k(t) = \int_{-\infty}^{\infty} K(\alpha)e^{i\alpha t}d\alpha, K(\alpha) = -\lambda_-^{-1}, \tag{24.4.6}$$

where $\lambda_-$ is from (24.4.3).

**Comment**: It should be noted that for $d = 0$ from (24.4.5) and (24.4.6), and we obtain the classical integral equation for a homogeneous half-space, in which $\mu(y) = \mu_0 = const$.

Integral equations (24.4.5), (24.4.6) with the help of replacement $\alpha = d\alpha'$, $d\alpha = d d\alpha'$ in the inner integral, dimensionlization of the inner $\xi = a\xi'$, $d\xi = a d\xi'$ and outer $x = ax'$ variables is reduced to the integral equation in dimensionless form (the primes are removed)

$$\int_{-1}^{1} \varphi(\xi)k\left(\frac{\xi - x}{\lambda}\right)d\xi = 2\pi, |x| \le 1, \lambda = \frac{1}{da}, \tag{24.4.7}$$

$$k(t) = \int_{-\infty}^{\infty} K(\alpha)e^{i\alpha t}d\alpha, K(\alpha) = \left(1 + \sqrt{1 + \alpha^2}\right)^{-1}, \tag{24.4.8}$$

which is an integral equation of the first kind of the Fourier convolution type with a difference kernel with respect to unknown dimensionless contact stresses $\varphi(x) = a\varepsilon^{-1}\mu_0^{-1}\tilde{\varphi}(x)$. The transformant $K(\alpha)$ of the kernel $k(t)$ in (24.4.8) of the integral equation (24.4.7) is an even, multivalued function in the complex plane with branch points of algebraic type $\alpha = \pm i$ and has the following asymptotic properties

$$K(\alpha) = |\alpha|^{-1} + O(\alpha)^{-2} \text{ at } |\alpha| \to \infty, \tag{24.4.9}$$

$$K(\alpha) = K(0) + O(\alpha)^{2} \text{ at } |\alpha| \to 0, \tag{24.4.10}$$

where $K(0) = 0.5$. After calculating the quadrature in (24.4.8), the kernel $k(t)$ takes the form

$$k(t) = 2e^{t}\left(K_0(t) - \int_{t}^{\infty} d\tau \int_{\tau}^{\infty} K_0(\xi)d\xi\right), \tag{24.4.11}$$

where $K_0(t)$ is the Macdonald function. For small $t$ for $k(t)$ the estimate

$$k(t) = -2\ln|t| + O(1) \text{ is fair at } t \to 0. \tag{24.4.12}$$

These properties of $k(t)$ allow us to conclude (Vorovich et al. 1974) that a solution to the integral equation (24.4.7) exists and is unique in the class of functions $\varphi(x) = \omega(x)\left(1 - x^2\right)^{-1/2}$, where $\omega(x) \in C^1[-1, 1]$.

To construct an effective analytical solution to the integral equations (24.4.7) and (24.4.8), asymptotic methods, constructed by a dimensionless parameter $\lambda = (da)^{-1}$ in the entire range of its variation $\lambda \in (0, \infty)$, are used. Parameter $\lambda$-asymptotic solutions of the integral equation (24.4.7) are constructed and presented in Appendix 1, 2.

## 24.5 Analytical Solution of the Contact Problem

Asymptotic solutions of the integral equation (24.4.7), constructed by parameter $\lambda$, for $0 < \lambda < \lambda_0$ in Appendix 1 and for $\lambda_0 < \lambda < \infty$ in Appendix 2 cover the entire interval of parameter $\lambda \in (0, \infty)$ variation. Let us clarify the practical area of joining the obtained asymptotic solutions by determining the value $\lambda_0$.

In Table 24.1, the numerical values of the relative contact stresses $aT^{-1}\varphi(0.00)$ of the integral equations (24.4.7), (24.4.8) in the case $0 < \lambda < \lambda_0$, calculated by the formulas presented in Appendix 1, and in the case $\lambda_0 < \lambda < \infty$, calculated by the formulas presented in Appendix 2, for $\lambda = 1.0, 1.5, 2.0$ at $a = 1$ m are given.

**Table 24.1** Values of the relative contact stresses $aT^{-1}\varphi(0.00)$ obtained by the asymptotic methods

| $aT^{-1}\varphi(0.00)$ | $\lambda = 1.0$ | $\lambda = 1.5$ | $\lambda = 2.0$ |
|---|---|---|---|
| (A.24.16) | 0.4068 | 0.3828 | 0.3668 |
| (A.24.17) | 0.3988 | 0.3803 | 0.3690 |
| (A.24.18) | 0.3965 | 0.3798 | 0.3701 |
| (B.24.4) | 0.4299 | 0.3909 | 0.3749 |

In Table 24.1, the first three rows show the values $aT^{-1}\varphi(0.00)$, obtained by the asymptotic method for $0 < \lambda < \lambda_0$ from Appendix 1. The fourth row of Table 24.1 shows the values $aT^{-1}\varphi(0.00)$, obtained by the method effective at $\lambda_0 < \lambda < \infty$, the formulas of which are contained in Appendix 2.

The analysis of the values $aT^{-1}\varphi(0.00)$ given in Table 24.1 shows that the joining of solutions, that are effective at $0 < \lambda < \lambda_0$ and at $\lambda_0 < \lambda < \infty$, occurs in the interval $1 < \lambda < 2$ with the midpoint $\lambda_0 = 1.5$, since the discrepancy between the values $aT^{-1}\varphi(0.00)$ does not exceed 2%.

The contact stress $\varphi(x)$ formulas presented in Appendix 1 for $0 < \lambda < \lambda_0$ and in Appendix 2 at $\lambda_0 < \lambda < \infty$ allow presenting the solution of a contact problem in analytical form over the entire range of parameter $\lambda \in (0\,,\ \infty)$ variation. Then

$$\begin{aligned}&\varphi(x) = \varphi_1(x)H(\lambda_0 - \lambda) + \varphi_2(x)H(\lambda - \lambda_0),\\&0 < \lambda < \infty,\ |x| < 1,\end{aligned} \tag{24.5.1}$$

where $\varphi_1(x)$ coincides with $\varphi(x)$ from (A.24.1) Appendix 1, $\varphi_2(x)$ coincides with $\varphi(x)$ from (B.24.4) Appendix 2, $H(\lambda)$ is the Heaviside function.

The dimensionless shear force of the indenter $T = \int_{-1}^{1} \varphi(x)\mathrm{d}x$ is calculated by the formula

$$T = T_1 H(\lambda_0 - \lambda) + T_2 H(\lambda - \lambda_0),\ 0 < \lambda < \infty, \tag{24.5.2}$$

where $T_1$ coincides with $T$ from one of Eqs. (A.24.26)–(A.24.28) Appendix 1, $T_2$ coincides with $T$ from (B.24.6) Appendix 2.

Contact stresses $\varphi(x)$, represented by Eq. (24.5.1), allow you to determine other important characteristics of the contact. The dimensionless displacements of the half-space surface outside the contact area $v(x)$ referred to $\varepsilon$ are determined by the formula

$$\begin{aligned}&v(x) = v_1(x)H(\lambda_0 - \lambda) + v_2(x)H(\lambda - \lambda_0),\\&1 < |x| < \infty,\ 0 < \lambda < \infty,\end{aligned} \tag{24.5.3}$$

in which $v_k(x)$ and $k = 1\,,\ 2$ are determined by the formulas

$$\begin{aligned}&v_1(x) = v_+\big((1+x)\,\lambda^{-1}\big)H(-1-x) + v_-\big((1-x)\,\lambda^{-1}\big)H(-1+x),\\&\quad 1 < |x| < \infty,\ 0 < \lambda < \lambda_0,\end{aligned} \tag{24.5.4}$$

$$\begin{aligned}&v_2(x) = -(2\pi)^{-1}\int\limits_{-1}^{1} \varphi_1(\xi)k\big((\xi - x)\,\lambda^{-1}\big)\mathrm{d}\xi,\ 1 < |x| < \infty,\\&\lambda_0 < \lambda < \infty,\end{aligned} \tag{24.5.5}$$

where $\upsilon_{\pm}\left((1 \pm x)\,\lambda^{-1}\right)$ are defined in Apendix A by Eqs. (A.24.19)–(A.24.21) , and the function $\varphi_1(x)$ is defined in (24.5.1).

Other additional characteristics of the contact can be obtained by differentiating and integrating the obtained characteristics, for example, the tangent of the slope of the curve of displacements of the surface of the half-space outside the contact area (Fig. 24.1) or the length of the arc of the curve of displacements of the surface of the half-space outside the contact area.

## 24.6 Determination of $d$ and $\mu_0$ by Shear Contact Stresses

In the introduction, it was noted that researchers are paying more and more attention to the development of methods for determining the parameters of elastic moduli of functional-gradient materials (Nakamuram et al. 2000; Gu et al. 2003; Bocciarelli et al. 2008; Chen et al. 2018), which are based on numerical solutions of contact problems of indentation. The analytical solution of the contact problem obtained in the previous paragraph makes it possible to analyze the relationship between the mechanical characteristics of the contact. Table 24.2 shows the values of the relative contact stresses $aT^{-1}\varphi(x)$, calculated at $x = 0.00,\ 0.25,\ 0.50,\ 0.75$ for various values of the parameters $\lambda(d) = 0.1(10), 0.5(2), 1.0(1.0), 5(0.2), 10(0.1), 1000(0.001)$ at $a = 1$ m. Analysis of the values $aT^{-1}\varphi(0.00)$ shows that for small $d$, for example, for $\lambda(d) = 1000(0.001)$, i.e., $d = 0.001$, the magnitude of the relative contact stresses $aT^{-1}\varphi(0.00)$ practically coincides with the value of the relative contact stresses $aT_0^{-1}\varphi_0(0.00)$ in the classical solution for the shear problem of a homogeneous half-space (Aleksandrov and Belokon' 1967)

$$aT_0^{-1}\varphi_0(x) = \left(\pi\sqrt{1 - x^2}\right)^{-1} \tag{24.6.1}$$

as $aT_0^{-1}\varphi_0(0.00) = \pi^{-1} = 0.3183$. For $\lambda(d) = 0.1(10)$, where $d = 10$, the value $aT^{-1}\varphi(0.00) = 0.4854$, i.e., is 1.52 times the value of $aT^{-1}\varphi(0.00)$ at $\lambda(d) = 1000(0.001)$, where $d = 0.001$, which almost coincides with the classical solution.

**Table 24.2** Values of the relative contact stresses $aT^{-1}\varphi(x)$

| $\lambda(d)$ | $aT^{-1}\varphi(0.00)$ | $aT^{-1}\varphi(0.25)$ | $aT^{-1}\varphi(0.50)$ | $aT^{-1}\varphi(0.75)$ |
|---|---|---|---|---|
| 0.1(10) | 0.4854 | 0.4854 | 0.4854 | 0.4854 |
| 0.5(2) | 0.4349 | 0.4357 | 0.4415 | 0.4759 |
| 1.0(1.0) | 0.3965 | 0.4016 | 0.4217 | 0.4823 |
| 5(0.2) | 0.3451 | 0.3535 | 0.3849 | 0.4794 |
| 10(0.1) | 0.3330 | 0.3423 | 0.3771 | 0.4803 |
| 1000(0.001) | 0.3185 | 0.3289 | 0.3677 | 0.4812 |

This indicates that coefficient $d$ of the shear modulus $\mu(y)$ of the form (24.2.1) has a significant effect on the contact stresses.

The data in Table 24.2 are used to solve the problem of determining the shear modulus $\mu(y)$ parameter $d$ from the values of the relative contact stresses $aT^{-1}\varphi(x)$. If the value $aT_*^{-1}\varphi_*(x)$ obtained experimentally as a result of indentation of the functional-gradient half-space coincides with one of the values $aT^{-1}\varphi(x)$ in Table 24.2, then the parameters $\lambda(d)$ are determined in the first column of the row in which this value is located. If the experimentally obtained value $aT_*^{-1}\varphi_*(x)$ is not found in Table 24.2 and interpolation is difficult, then solving the problem of determining the parameter $d$ is reduced to solving a nonlinear equation

$$aT^{-1}\varphi(x,d) = aT_*^{-1}\varphi_*(x),\ |x| < 1, 0 < d < \infty, \tag{24.6.2}$$

where on the left side, the function $\varphi(x, d)$ coincides with $\varphi(x)$ from (24.5.1), $x$ is fixed, the value $a = 1$ m, and $T_*$ and $\varphi_*(x)$ are determined experimentally.

To determine the parameter $\mu_0$ of the shear modulus $\mu(y)$ from (24.2.1), which coincides with the value of the shear modulus on the surface of the half-space at $y = 0$, the condition of equality of the experimental $T_*$ and theoretical $\tilde{T} = \varepsilon\mu_0 T$ forces is used, where $T$ is the dimensionless force defined in Apendix 1.

$$T = \int_{-1}^{1} \varphi(x)\mathrm{d}x, \tag{24.6.3}$$

where $\varphi(x)$ is from (24.5.1). From the equality of $T_*$ and $\tilde{T}$, it follows that

$$\mu_0 = \frac{T_*}{\varepsilon T}, \tag{24.6.4}$$

where $\varepsilon$ is the magnitude of the indenter shear, $T_*$ is the magnitude of the indenter shear force obtained by experimental means, and the value $T$ is determined by Eq. (24.5.2).

## 24.7 Area of Uniqueness of Determination of the Parameter *d* from Contact Stresses

The area of existence of the solution to the problem of determining the parameter $d$ based on the methodology for comparing the experimental and theoretical values of contact stresses is determined by the set $S$, representing the strip $-1 < x < 1$, $0 < d < \infty$, in which the values $aT^{-1}\varphi(x, d)$ are determined. To determine the region of uniqueness of the solution to the problem of determining the parameter $d$, it is necessary to have an idea of the set of curves $aT^{-1}\varphi(x, d)$ in the strip of existence $-1 < x < 1, 0 < d < \infty$. A more detailed table of values $aT^{-1}\varphi(x, d)$, than Table 24.2, cannot give a visual representation of the region of uniqueness of the solution to the problem by definition $d$ including because the function $\varphi(x, d)$ is unbounded in terms of $-1 < x < 1$. For this purpose, a visual representation of these functions in the form of graphs is used (Fig. 24.2). Figure 24.2 shows graphs $aT^{-1}\varphi(x, d)$ on $-1 < x < 1$ for the following parameter values $\lambda(d) = 0.1(10), 0.5(2), 1.0(1.0), 5(0.2), 10(0.1)$ at a fixed value $a = 1$ m.

Plots $aT^{-1}\varphi(x)$, built for different $\lambda(d)$, intersect. This means that at the point of intersection of two curves $aT^{-1}\varphi(x)$ with different values $\lambda(d)$, the same contact stress value on the contact will be achieved, and from this value, it is impossible to determine the parameter $d$ in a unique way. These points in the strip $\left\{ |x| < 1, \left|aT^{-1}\varphi(x)\right| < \infty \right\}$ should be excluded from consideration, since the problem of determining the parameter $d$ from the value of contact stresses does not have a unique solution.

The area of uniqueness by $x$ will be enclosed in the interval $|x| < x_o$ $(0 < x_o < 1)$. The value $x_o$, which is the abscissa of the intersection point of the two curves $aT^{-1}\varphi(x)$ with the largest value $\lambda_+$ (with the smallest $d_-$) and the smallest value $\lambda_-$ (i.e., with the largest $d_+$), is determined from the solution of the equation (at $a = 1$ m)

$$aT^{-1}\varphi(x, \lambda_+(d_-)) = aT^{-1}\varphi(x, \lambda_-(d_+)), |x| < 1. \tag{24.7.1}$$

The domain of uniqueness $D$ of the definition of the parameter $d$ is described by the inequality

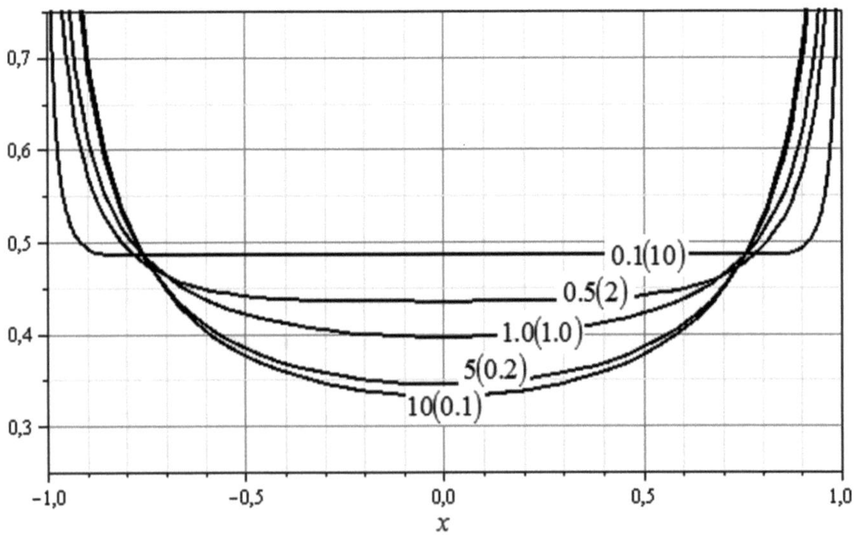


**Fig. 24.2** Graphs of relative contact stresses $aT^{-1}\varphi(x)$ for $\lambda(d) = 0.1(10), 0.5(2), 1.0(1.0), 5(0.2), 10(0.1)$ at $a = 1$ m

$$aT^{-1}\varphi(x, \lambda_+(d_-)) < aT^{-1}\varphi(x, \lambda_-(d_+)), |x| < x_o. \tag{24.7.2}$$

The lower border $D$ is described by the graph of the curve $aT^{-1}\varphi(x, \lambda_+(d_-))$, the points of which do not enter the area $D$, and the upper border is described by $aT^{-1}\varphi(x, \lambda_-(d_+))$, the points of which are included in the area $D$. Thus, the domain of uniqueness $D$ of the shear modulus $\mu(y)$ parameter $d$ from (24.2.1) is open.

## 24.8 Determination of $d$ and $\mu_0$ from Surface Displacements Outside the Contact Area

In Sects. 24.6 and 24.7, a method for determining the parameter $d$ of a depth-variable exponential shear modulus $\mu(y)$ from (24.2.1) of a functional-gradient half-space from the relative contact stresses $aT^{-1}\varphi(x)$ is given. In practical implementation, such a technique causes significant difficulties due to the fact that when trying to measure contact stresses, researchers encounter significant difficulties that do not allow them to be measured directly using sensors. In this regard, contact stresses are determined indirectly (approximately). Indirect methods for determining contact stresses lead to additional errors in determining the parameters of the shear modulus. In this regard, the method proposed in Sects. 24.6 and 24.7 for determining the parameter $d$ by contact stresses should be replaced by another method using a different contact characteristic. Based on the experimental availability of measuring the displacements of the free surface of the half-space outside the contact area during indentation, it is proposed to use the experimentally and theoretically determined value of the relative displacements $v(x)$ of the free surface outside the contact area to determine the parameter $d$.

The columns of Table 24.3 show the values of the relative displacements $v(x)$ $\left(v(x) = \varepsilon^{-1}w(x)\right)$ outside the contact area for fixed values $x = 1.01,\ 1.1,\ 1.2,\ 1.3,\ 1.4,\ 1.5$. The first column of Table 24.3 shows the values

**Table 24.3** Values of the relative displacements $v(x)$ outside the contact area

| $\lambda(d)$ | $v(x)$ | | | | | |
|---|---|---|---|---|---|---|
| | $x = 1.01$ | $x = 1.10$ | $x = 1.20$ | $x = 1.30$ | $x = 1.40$ | $x = 1.50$ |
| 0.1(10) | 0.5271 | 0.0455 | 0.0047 | 0.0005 | 0.0001 | 0.0000 |
| 0.2(5) | 0.6547 | 0.1573 | 0.0455 | 0.0143 | 0.0047 | 0.0016 |
| 0.5(2) | 0.7773 | 0.3711 | 0.2059 | 0.1213 | 0.0736 | 0.0455 |
| 1.0(1.0) | 0.8415 | 0.5271 | 0.3711 | 0.2733 | 0.2059 | 0.1573 |
| 2(0.5) | 0.8921 | 0.6632 | 0.5426 | 0.4609 | 0.3994 | 0.3509 |
| 5(0.2) | 0.9230 | 0.7621 | 0.6715 | 0.6064 | 0.5546 | 0.5115 |
| 10(0.1) | 0.9402 | 0.8142 | 0.7414 | 0.6881 | 0.6450 | 0.6084 |

of the parameters $\lambda(d) = 0.1(10), 0.2(5), 0.5(2), 1.0(1.0), 2(0.5), 5(0.2), 10(0.1)$ at $a = 1$ m.

Thus, in Table 24.3, a one-to-one correspondence is established between the values of the displacements $v(x)$ and the parameters $\lambda(d)$, i.e., each value $v(x)$ is assigned to one value $\lambda(d)$. If the experimentally determined value of the displacement $v_*(x)$ coincides with the theoretical value $v(x)$ from Table 24.3, then the parameters $\lambda(d)$ in the first column correspond to it. If the experimentally determined value $v_*(x)$ does not coincide with the tabulated value $v(x)$, then in this case the parameter $d$ is determined from the coincidence of the theoretical $v(x)$ and experimental $v_*(x)$ at a fixed value $x$. This leads to the need to solve the equation nonlinear with respect to $d$

$$v(x, d) = v_*(x),\ 1 < |x| < \infty, 0 < d < \infty \tag{24.8.1}$$

at a fixed value $x$ and where $v(x, d)$ is determined by the Eq. (24.5.3).

Another parameter of the exponential shear modulus $\mu_0$ is determined, as in Sect. 24.6, by Eq. (24.6.4), which includes the parameter $d$.

## 24.9 Areas of Uniqueness of *d* for Determining the Displacements of the Free Surface

As in Sects. 24.6 and 24.7, Table 24.3 does not make it possible to trace the uniqueness of the one-to-one correspondence of the values $v(x)$ and $d$ established in it in the entire volume of change of the variable $x$ $(1 < |x| < \infty)$ and parameter $d$ $(0 < d < \infty)$. Similar to Sects. 24.6 and 24.7, we will build graphs of changes $v(x)$ in the area of existence $S\{1 < |x| < \infty,\ 0 < d < \infty\}$. Figure 24.3 shows the graphs of the displacements $v(x)$ of the free surface of the functional-gradient half-space outside the contact area at $1 < x < \infty$ to the right of the indenter (due to the symmetry of the problem by variable $x$) for the following values of the parameters $\lambda(d) = 0.1(10), 0.2(5), 0.5(2), 1.0(1.0), 2(0.5), 5(0.2), 10(0.1)$ at $a = 1$ m.

Calculations show that the graphs of the function $v(x)$ do not intersect, i.e., do not have common points, and only one curve $v(x)$ passes through each point of the $D\{1 < x < 2,\ 0 < d < \infty\}$. Hence, we can conclude that the region of existence $S$ of the function $v(x)$ coincides with the region of uniqueness $D$.

## 24.10 Formula for Calculation *d*

Table 24.3, Fig. 24.3 and Eq. (24.8.1) do not create convenience for determining the parameter $d$, which creates the formula for its calculation. To obtain a formula for determining the parameter $d$, the function $v(x)$, the graphs and formulas of which

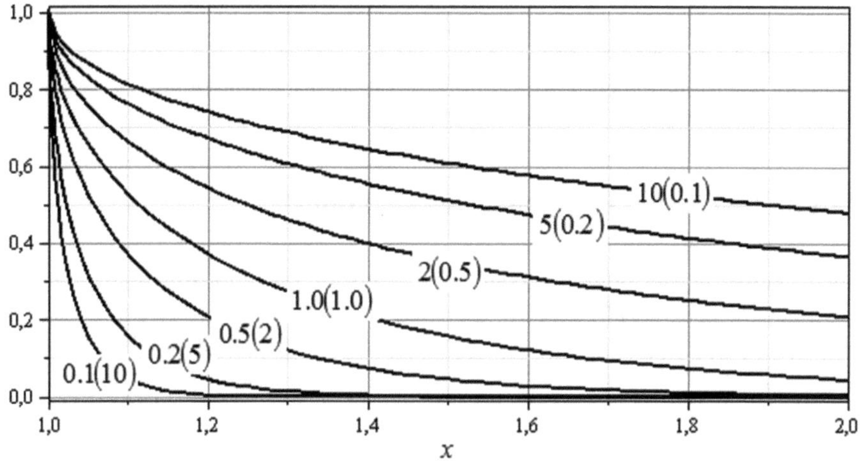


**Fig. 24.3** Graphs of the free surface displacements $v(x)$ for $\lambda(d) = 0.1(10), 0.2(5), 0.5(2), 1.0(1.0), 2(0.5), 5(0.2), 10(0.1)$ at $a = 1$ m

are shown in Fig. 24.3 and in Apendix A, B, respectively, are approximated by the expression

$$v(x,d) = 1 - \operatorname{th}\left(\frac{2}{\sqrt{\pi}}\sqrt{-A(1-x)d}\left(1 + \sum_n B_n x^n\right)\right),\ A = K^{-1}(0) \quad (24.10.1)$$

for all values $0.1 < d < 10$ and $1 \le x \le 2$. Thus, at $B_1 = 0.0544$ the error of such an approximation in the area $S\{1 \le x \le 2\,,\ 0.1 < d < 10\}$ does not exceed 5–7%. Substituting the right-hand side of (24.10.1) into the left-hand side of (24.8.1), we obtain the equation for determining the parameter $d$

$$1 - th\left(\frac{2}{\sqrt{\pi}}\sqrt{-A(1-x)d}\left(1 + \sum_n B_n x^n\right)\right) = v_*(x),$$
$$1 \le x \le 2, 0.1 < d < 10.0. \quad (24.10.2)$$

Solving Eq. (24.10.2) with respect to $d$, we obtain

$$d = \frac{\pi}{16}\frac{K(0)}{x-1}\left(1 + \sum_n B_n x^n\right)^{-2} \ln^2\left(\frac{v_*(x)}{2 - v_*(x)}\right),$$
$$1 \le x \le 2, 0.1 < d < 10.0. \quad (24.10.3)$$

Comparison of the data in Table 24.3 and calculations according to (24.10.3) shows an error that does not exceed the approximation error (24.10.1).

Thus, using the displacements of the surface of the functionally gradient half-space outside the area of contact of the indenter with the surface, it is possible to uniquely determine the parameters of the exponential shear modulus.

## 24.11 Discussion of Some Generalizations

The formulation and solution of contact problems for a functional-gradient medium are a natural generalization of the formulation and solution of the corresponding problems for homogeneous media. The solution of the contact problem posed in Sect. 24.3 on the shear of a functionally gradient half-space with a depth-variable exponential shear modulus $\mu(y)$ of the form (24.2.1) leads to the fact that the relationship between displacements $w(x, y)$ and contact stresses $\varphi(x)$ is uniquely determined by means of the operator

$$w(x, y) = \frac{1}{\pi\mu_0}\int_{-a}^{a} \tilde{\varphi}(\xi)\theta(\xi - x, y)d\xi,\ |x| < \infty, 0 \le y < \infty, \tag{24.11.1}$$

$$\theta(u, v) = \int_{0}^{\infty} \xi^{-1}(\alpha)\exp(-\xi(\alpha)v)\cos u\alpha d\alpha,\ \xi(\alpha) = d + \sqrt{d^2 + \alpha^2}, \tag{24.11.2}$$

which is obtained from (24.4.4). The integral in (24.11.2) is convergent for all $d$ $(0 < d < \infty)$. From Eq. (24.11.1), with the help of substitutions $\alpha d = \alpha'$, $a\xi = \xi'$, $ax = x'$ and constraint $\varphi(x) = a\varepsilon^{-1}\mu_0^{-1}\tilde{\varphi}(x)$ it is quite easy to obtain the integral equation of problem (24.4.7) for $y = 0$.

On the other hand, for $d = 0$ the case of a homogeneous elastic half-space is obtained, since $\mu(y) = \mu_0 = const$ and in which the relationship between displacements $w(x, y)$ and contact stresses $\tilde{\varphi}(x)$ is determined using the irregular integral operator

$$w(x, y) = \frac{1}{\pi\mu_0}\int_{-a}^{a} \tilde{\varphi}(\xi)k(\xi - x, y)\, d\xi,\ |x| < \infty, 0 \le y < \infty, \tag{24.11.3}$$

$$k(t, y) = \int_{0}^{\infty} |\alpha|^{-1}\exp(-y)\cos\alpha t d\alpha, \tag{24.11.4}$$

with diverging integral (24.11.4) for any values of $t$ and $y$. Equations (24.11.3) and (24.11.4) show that when solving the problem of pure shear by a strip indenter of a homogeneous half-space, it is not possible to establish a relationship between the displacements $w(x, y)$ in the half-space and contact stresses $\tilde{\varphi}(x)$. The situation that has arisen that there is no connection between $w(x, y)$ and $\tilde{\varphi}(x)$ in the homogeneous case due to the divergence of the integral in (24.11.4) can be obtained by replacing the unsuitable Eqs. (24.11.3), (24.11.4) for the homogeneous case with Eqs. (24.11.1), (24.11.2) for the case of a functionally gradient half-space with an exponential shear modulus $\mu(y)$ of the form (24.2.1) with a parameter $d$, close to 0.

It should be noted that Eqs. (24.11.3) and (24.11.4) point to one more defect in the solution of the problem of the shear of a homogeneous half-space. In the formulation of the problem of the shear of a homogeneous half-space, as in the formulation of the problem of the shear of a functionally gradient half-space, we accept so-called boundary conditions at infinity at $\sqrt{x^2 + y^2} \to \infty$, which consist in the disappearance of displacements $w(x, y)$ and stresses $\sigma_{xz}(x, y)$, $\sigma_{yz}(x, y)$, i.e., the partial derivatives $\frac{\partial w}{\partial x}$, $\frac{\partial w}{\partial y}$, of the form (24.3.7) indicated in the formulation of the problem of the shear of a functional-gradient half-space. Having obtained a solution to the problem of the shear of a homogeneous half-space in the forms (24.11.3) and (24.11.4), it is clear that for this solution condition (24.3.7) is not satisfied for $0 \leq y < \infty$ due to the diverging integral in (24.11.4). At the same time, using Eq. (24.11.1) for the calculation of $w(x, y)$, it can be shown that $w(x, y)$ with increasing $y$ decreases as $\frac{1}{\sqrt{y\mu(y)}}$ together with its partial derivatives. Thus, in the case of a functionally gradient half-space with an exponentially increasing shear modulus $\mu(y)$ from (24.2.1), the boundary conditions at infinity (24.3.7) are satisfied (Zelentsov et al. 2020).

## 24.12 Conclusions

The method for determining the parameters of the exponential shear modulus along the depth of the functionally gradient half-space, based on the solution of the problem of pure shear by the strip indenter of the surface of the functional-gradient half-space, made it possible to:

- to study the influence of the parameters of the elastic half-space exponential in depth on the main characteristics of the contact—contact stresses, displacements of the half-space surface outside the area of contact with the indenter;
- to obtain formulas and an equation for determining the parameters of the exponential shear modulus from the experimental data of the main characteristics of the contact during indentation;
- to present the solution of the corresponding problem of indentation of a homogeneous half-space as a special case of solving the problem of indentation of a functionally gradient half-space.

**Acknowledgements** This research was supported by Russian Foundation for Basic Research (grant numbers 20-07-00949-a, 20-58-53045-GFEN-a).

## Appendix 1: Asymptotic Solution of the Integral Equation (24.4.7) in the Range $0 < \lambda < \lambda_0$

For the values of the dimensionless parameter $0 < \lambda < \lambda_0$, the solution of the integral equations (24.4.7), (24.4.8) is constructed in the form of a Neumann series, the zero term of which is given by the relation (Vorovich et al. 1974)

$$\phi(x) = \phi_+\big((1+x)\lambda^{-1}\big) + \phi_-\big((1-x)\lambda^{-1}\big) - \phi_\infty(x\lambda^{-1}) \tag{A.24.1}$$

The functions $\varphi_\pm(x)$ are determined from the Wiener–Hopf integral equation corresponding to Eq. (24.4.7) on the semiaxis (Vorovich et al. 1974; Gradshteyn and Ryzhik 1965)

$$\int\limits_0^\infty \varphi_\pm(\xi)k(\xi - x)d\xi = \frac{2\pi}{\lambda}, 0 < x < \infty. \tag{A.24.2}$$

The function $\varphi_\infty(x)$ is determined from the integral equation of convolution type corresponding to Eq. (24.4.7) on an infinite interval

$$\int\limits_{-\infty}^\infty \varphi_\infty(\xi)k(\xi - x)d\xi = \frac{2\pi}{\lambda}, -\infty < x < \infty. \tag{A.24.3}$$

The solution to the integral equation (A.24.3) is constructed using the generalized integral Fourier transform (Brychkov and Prudnikov 1989) and has the form

$$\varphi_\infty(x) = \frac{1}{\lambda K(0)}, -\infty < x < \infty. \tag{A.24.4}$$

The solution to the integral equation on the semiaxis (A.24.2) is constructed by the Wiener–Hopf method (Noble 1958). Integral equation (A.24.2) is extended to the entire axis $-\infty < x < \infty$, so that it was possible to carry out its solution using the Fourier transform. So for the integral equation with respect to $\varphi_+(x)$, redefinition is carried out as follows:

$$\int\limits_0^\infty \varphi_+(\xi)k(\xi - x)d\xi = \begin{cases} \frac{2\pi}{\lambda} & 0 < x < \infty, \\ \frac{2\pi}{\lambda}\upsilon_-(x) & -\infty < x < 0, \end{cases} \tag{A.24.5}$$

where $\upsilon_-(x) = \varepsilon^{-1}w_-(x)$ are the dimensionless displacements of the free surface of the half-space outside the contact area (to the left of the indenter), which are given by the operator

$$v_-(x) = \int_{-\infty}^{0} \varphi_+(\xi)k(\xi - x)d\xi, \ -\infty < x < 0,$$

and $w_-(x)$ are dimensional horizontal displacements of the free surface of the half-space outside the contact area (to the left of the indenter).

Integral equation (A.24.5) using the integral Fourier transform (24.4.1) is reduced to a functional equation with respect to transformants $\varphi_+^F(\alpha)$ and $v_-^F(\alpha)$ of the unknown functions $\varphi_+(x)$ and $v_-(x)$

$$\varphi_+^F(\alpha)K(\alpha) = \frac{1}{\lambda}\left(-\frac{1}{i\alpha}\right) + \frac{1}{\lambda}v_-^F(\alpha), \ \eta_- < \mathrm{Im}(\alpha) < \eta_+, \quad \text{(A.24.6)}$$

in which unknown transformants

$$\varphi_+^F(\alpha) = \int_{0}^{\infty} \varphi_+(\xi)e^{i\alpha\xi}d\xi, \ v_-^F(\alpha) = \int_{-\infty}^{0} v_-(\xi)e^{i\alpha\xi}d\xi$$

are regular in the upper ($\mathrm{Im}(\alpha) > \eta_-$) and lower ($\mathrm{Im}(\alpha) < \eta_+$) half-planes of the complex plane $\alpha$, respectively. Factoring the function $K(\alpha)$ on the left-hand side of (A.24.6) and presenting it as a product

$$K(\alpha) = K_+(\alpha)K_-(\alpha), \quad \text{(A.24.7)}$$

where $K_+(\alpha)$ is regular in the upper ($\mathrm{Im}(\alpha) > \eta_-$), a $K_-(\alpha)$ and regular in the lower ($\mathrm{Im}(\alpha) < \eta_+$) half-plane, and the solution of equation (A.24.6) is carried out by means of the theory of functions of a complex variable (Noble 1958). Defining $\varphi_+^F(\alpha)$ and $v_-^F(\alpha)$ and carrying out the inverse Fourier transform, we obtain a solution to the integral equation (A.24.5)

$$\varphi_+(x) = -\frac{1}{\lambda K_-(0)}\frac{1}{2\pi}\int_{-\infty+ic}^{\infty+ic} \frac{e^{-i\alpha x}}{i\alpha K_+(\alpha)}d\alpha, \ x > 0, \quad \text{(A.24.8)}$$

and function representation $v_-(x)$

$$v_-(x) = \frac{1}{2\pi}\int_{-\infty+id}^{\infty+id} \left(\frac{1}{i\alpha} - \frac{1}{K_-(0)}\frac{K_-(\alpha)}{i\alpha}\right)e^{-i\alpha x}d\alpha, \ x < 0, \quad \text{(A.24.9)}$$

where $\eta_- < c < d < \eta_+$, and as $\eta_\pm$ we take $\pm 1$ taking into account the properties of $K(\alpha)$ from (24.4.8).

Integral representations for $\varphi_-(x)$ и $v_+(x)$ coincide completely with (A.24.8) and (A.24.9), respectively, where $v_+(x) = \varepsilon^{-1} w_+(x)$, a $v_+(x)$ and $w_+(x)$ are dimensionless and dimensional displacements of the free surface of the half-space outside the contact area (to the right of the indenter).

To calculate the quadratures in (A.24.8), (A.24.9) containing the integral $K_\pm(\alpha)$, it is necessary to factorize the kernel transform $K(\alpha)$ (A.24.7). Exact factorization $K(\alpha)$ from (24.4.8) is contained in (Noble 1958), but is not very effective for obtaining an observable solution. To obtain an effective solution, it is necessary to use the approximation $K(\alpha)$, which factorization is carried out by elementary means.

As the simplest approximations of $K(\alpha)$, satisfying properties (24.4.9) and (24.4.10) and not containing zeros and poles in the strip $\eta_- < \mathrm{Im}(\alpha) < \eta_+$, let us take the functions

$$\text{a)}\quad K(\alpha) = \frac{1}{\sqrt{A^2+\alpha^2}},\; A = \frac{1}{K(0)}, \tag{A.24.10}$$

$$\text{b)}\quad K(\alpha) = \frac{1}{\sqrt{A^2+\alpha^2}}\frac{B^2+\alpha^2}{C^2+\alpha^2},\; \frac{1}{A}\frac{B^2}{C^2} = K(0), \tag{A.24.11}$$

$$\text{c)}\quad K(\alpha) = \frac{\sqrt{A^2+\alpha^2}}{D^2+\alpha^2}\frac{B^2+\alpha^2}{C^2+\alpha^2},\; \frac{A}{D^2}\frac{B^2}{C^2} = K(0). \tag{A.24.12}$$

The error of such approximations of $K(\alpha)$ in the strip $|\mathrm{Im}(\alpha)| < \inf(\eta_-, \eta_+)$ along the real axis ($\mathrm{Im}(\alpha) = 0$) does not exceed: (a) 12%, (b) 4% and (c) 2%. The values of the parameters of approximations (A.24.10)–(A.24.12) are given in Table A1.

Factorization (A.24.7) of the reduced approximations (A.24.10)–(A.24.12) is achieved by elementary means

$$\text{a)}\quad K_\pm(\alpha) = \frac{1}{\sqrt{A \mp i\alpha}}, \tag{A.24.13}$$

$$\text{b)}\quad K_\pm(\alpha) = \frac{1}{\sqrt{A \mp i\alpha}}\frac{B \mp i\alpha}{C \mp i\alpha}, \tag{A.24.14}$$

$$\text{c)}\quad K_\pm(\alpha) = \frac{\sqrt{A \mp i\alpha}}{D \mp i\alpha}\frac{B \mp i\alpha}{C \mp i\alpha}, \tag{A.24.15}$$

**Table A1** Values of the parameters of approximations (A.24.10)–(A.24.12)

| $K(\alpha)$ | A | B | C | D |
|---|---|---|---|---|
| (A.24.10) | 2 | – | – | – |
| (A.24.11) | 1.5344 | 6.0332 | 6.8881 | – |
| (A.24.12) | 3.5897 | 3.9419 | 1.9665 | 5.3710 |

where $K_+(\alpha)$ is regular in the upper half-plane ($\mathrm{Im}(\alpha) > \eta_-$), and $K_-(\alpha)$ is regular in the lower half-plane ($\mathrm{Im}(\alpha) < \eta_+$), and $\eta_\pm = \inf(A, B, C, D)$ for each case (a), (b), (c). Calculating the quadrature in (A.24.8), after substitution $K_\pm(\alpha)$ from (A.24.13)–(A.24.15), we obtain the following solutions of the integral equation (A.24.2)

$$\text{a)}\quad \varphi_\pm(x) = \frac{1}{\lambda K_-(0)}\Phi(x, A, 0),\ x > 0, \tag{A.24.16}$$

$$\text{b)}\quad \varphi_\pm(x) = \frac{1}{\lambda K_-(0)}\left(\frac{C}{B}\Phi(x, A, 0) + \frac{B-C}{B}\Phi(x, A, B)\right),\ x > 0, \tag{A.24.17}$$

$$\text{c)}\quad \varphi_\pm(x) = \frac{1}{\lambda K_-(0)}\left(\frac{DC}{AB}\Phi(x, A, 0) + \frac{(D-A)(C-A)}{A(A-B)}\frac{1}{\sqrt{\pi x}}e^{-Ax}\right.$$
$$\left. + \frac{(D-B)(C-B)}{B(B-A)}\Phi(x, A, B)\right),\ x > 0, \tag{A.24.18}$$

where $\Phi(x, A, B) = \frac{1}{\sqrt{\pi x}}e^{-Ax} + \sqrt{A-B}e^{-Bx}\mathrm{erf}\left(\sqrt{(A-B)x}\right)$, $\mathrm{erf}(x)$ is the probability integral (Bateman and Erdelyi 1954).

Calculating the quadrature in (A.24.9), after substitution of $K_-(\alpha)$ from (A.24.13)–(A.24.15), we obtain the displacements outside the indenter

$$\text{a)}\quad v_\mp(x) = \mathrm{erfc}\left(\sqrt{A(-x)}\right),\ x < 0, \tag{A.24.19}$$

$$\text{b)}\quad v_\mp(x) = 1 - \frac{1}{\sqrt{A}}\Phi(-x, A, 0) + \frac{\sqrt{A}C}{B(C-A)}$$
$$\cdot\left(\frac{B-A}{A}\frac{1}{\sqrt{\pi(-x)}}e^{-A(-x)} - \frac{B-C}{C}\Phi(-x, A, C)\right),\ x < 0, \tag{A.24.20}$$

$$\text{c)}\quad v_\mp(x) = 1 - \frac{1}{\sqrt{A}}\Phi(-x, A, 0) + \frac{DC}{\sqrt{A}B(D-C)}$$
$$\cdot\left(\frac{B-D}{D}\Phi(-x, A, D) - \frac{B-C}{C}\Phi(-x, A, C)\right),\ x < 0, \tag{A.24.21}$$

where $\mathrm{erfc}(x) = 1 - \mathrm{erf}(x)$ (Bateman and Erdelyi 1954), $\Phi(x, A, B)$ is given after (A.24.18).

Substituting $\varphi_\pm(x)$ from (A.24.10)–(A.24.12) and $\varphi_\infty(x)$ from (A.24.4) into (A.24.1), we obtain a solution to the integral equation of problem (24.4.7). After returning to the old variable in Eqs. (A.24.19)–(A.24.21), the functions $v_\mp\left((1 \pm x)\lambda^{-1}\right)$ give horizontal displacements of the stress-free surface of the half-space outside the contact area at $1 \pm x < 0$.

Integral characteristic of the problem, the shearing force $\tilde{T}$ of the indenter is determined by integrating the obtained solution over the segment $[-a, a]$

$$\tilde{T} = \int_{-a}^{a} \tilde{\varphi}(x)\mathrm{d}x = a \int_{-1}^{1} \tilde{\varphi}(x)\mathrm{d}x = \varepsilon\mu_0 T, \tag{A.24.22}$$

where

$$T = \int_{-1}^{1} \varphi(x)\mathrm{d}x \tag{A.24.23}$$

represents a dimensionless shear force. Taking $\varphi(x)$ in the form of superposition (A.24.1) and integrating over the interval $[-1,\ 1]$, we obtain the dimensionless shear force in the following form

$$T = T_+ + T_- - T_\infty,\ T_\pm = \int_{-1}^{1} \varphi_\pm\left(\frac{1 \pm x}{\lambda}\right)\mathrm{d}x,\ T_\infty = \int_{-1}^{1} \varphi_\infty\left(\frac{x}{\lambda}\right)\mathrm{d}x. \tag{A.24.24}$$

Using the asymptotic solution of the integral equation (24.4.7) in multiplicative form (Vorovich et al. 1974)

$$\varphi(x) = \varphi_+\big((1+x)\lambda^{-1}\big)\varphi_-\big((1-x)\lambda^{-1}\big)\varphi_\infty^{-1}(x\lambda^{-1}), \tag{A.24.25}$$

where $\varphi_\pm\big((1 \pm x)\lambda^{-1}\big)$ from (A.24.16)–(A.24.18), and $\varphi_\infty(x\lambda^{-1})$ from (A.24.4) allows one to obtain more convenient formulas for $T$ analysis for all cases of approximation (A.24.10)–(A.24.12), of the same order of $\lambda$ accuracy, as in (A.24.22)

$$\text{a)}\quad T = 1 + A\gamma, \tag{A.24.26}$$

$$\begin{aligned}\text{b)}\quad T = {} & \left(\frac{C}{B}\right)^2\left(1 + 2\frac{A}{C}\frac{B-C}{B} + A\gamma\right) \\ & + \left(\frac{C-B}{B}\right)^2\left(1 + 2\frac{C}{B}\frac{A-B}{C-B} + (A-B)\gamma\right)\mathrm{e}^{-\gamma B},\end{aligned} \tag{A.24.27}$$

$$\text{c)}\quad T = \kappa(A) + \kappa_*(0)(z_-(0) + A\gamma) + \kappa(B)(z_+(B) + (A-B)\gamma), \tag{A.24.28}$$

where $\gamma = 2\lambda^{-1} = 2ad$,

$$\kappa(t) = \left(\frac{(D-t)(C-t)}{t\,(A-B)}\right)^2 \mathrm{e}^{-\gamma t},\ \kappa_*(t) = \left(\frac{(D-t)(C-t)}{(A-t)\,(B-t)}\right)^2 \mathrm{e}^{-\gamma t},\ z_\pm(t)$$
$$= -1 + 2\left(\frac{B}{D-t} + \frac{B}{C-t} \pm 1\right)\frac{A-t}{B}.$$

## Appendix 2: Asymptotic Solution of the Integral Equation (24.4.7) in the Range $\lambda_0 < \lambda < \infty$

When constructing a solution to the integral equation of problem (24.4.7), (24.4.8) in the case $\lambda_0 < \lambda < \infty$, it is necessary to take into account the fact that $K(\alpha)$ seems to be near

$$K(\alpha) = \sum_{k=0}^{N} c_k |\alpha|^{-k-1} + O\left(|\alpha|^{-N-2}\right) \text{ at } 0 < m < |\alpha| < \infty, \tag{B.24.1}$$

in which $c_k$ at $k = 0, 1, 2, \ldots, N$ for the case $N = 4$ take the following values:

$$c_0 = 1, c_1 = -1, c_2 = \frac{1}{2}, c_3 = 0, c_4 = -\frac{1}{8}.$$

Using expansion (B.24.1) and relation (24.4.9), it is easy to obtain an expansion $k(t)$ in the form of the following series (Aleksandrov and Belokon' 1967)

$$\frac{1}{2}k(t) = -\ln|t| + a_{20}|t| + a_{11}t^2 \ln|t| + a_{21}|t|^3 + a_{31}t^2 + O\left(t^4 \ln|t|\right) \text{ at } t \to 0 \tag{B.24.2}$$

in which

$$a_{20} = -\frac{\pi}{2}c_1, a_{11} = \frac{1}{2}c_2, a_{21} = \frac{1}{12}\pi c_3, a_{30} = \int_0^\infty \frac{uK(u) - 1 + e^{-u}}{u} du,$$

$$a_{31} = -\frac{3}{4}c_2 + \frac{1}{2}\int_0^\infty \left[\alpha^2 - \alpha^3 K(\alpha) + c_1\alpha + c_2(1 - e^{-\alpha})\right]\alpha^{-1} d\alpha. \tag{B.24.3}$$

After substituting (B.24.2) into (24.4.7) and differentiating with respect to the external variable $x$, we obtain a singular integral equation with respect to $\varphi(x)$. After inverting the singular integral equation, an equivalent integral equation of the second kind is formed (Aleksandrov and Belokon' 1967). The solution of the latter is constructed by the method of successive approximations in the form of a double functional series

$$\varphi(x) = \frac{T}{a\sqrt{1-x^2}} \sum_{m=0}^{\infty}\sum_{n=0}^{\infty} \omega_{mn}(x)\lambda^{-m} \ln^n \lambda, n \le \left[\frac{m}{2}\right] \tag{B.24.4}$$

of the function $\omega_{mn}(x)$ in which they have the foarm

$$\omega_{00}(x) = \pi^{-1}, \omega_{10}(x) = 4\pi^{-3}a_{20}S_1(x), \omega_{11}(x) = 0, \omega_{01}(x) = 0,$$

$$\omega_{20}(x) = \pi^{-1}\left(a_{131}\left(1 - 2x^2\right) + 32\pi^{-4}a_{20}^2(S_2(x) - 0.1508)\right).$$
$$\omega_{21}(x) = -\pi^{-1}a_{11}\left(1 - 2x^2\right), \omega_{31}(x) = -2\pi^{-3}a_{11}a_{20}S_4(x),$$
$$\omega_{30}(x) = \pi^{-3}\left(\frac{8}{3}a_{21} + \sum_{k=0}^{4} b_{1k}(x)S_{1+k}(x)\right), \tag{B.24.5}$$

where

$$b_{12} = \frac{8}{9}a_{11}a_{20}, b_{10}(x) = 6a_{21}\left(1 + 2x^2\right) - 19.3024\pi^{-4}a_{20}^3, b_{11} = 0,$$

$$b_{13} = 9a_{21} + 2a_{131}a_{20}, a_{131} = a_{11}\left(\frac{3}{2} - \ln 2\right) + a_{31}, b_{14}(x) = 64\pi^{-4}a_{20}^3,$$

$$S_1(x) = \left(1 - 2x^2\right) + 2\sqrt{1 - x^2}\sum_{k=1}^{\infty} c_{2k+1}(x), c_n(x) = n^{-2}\sin(n \arccos x),$$

$$S_2(x) = \left(0.4356 + 0.1321x^2\right)\left(1 - x^2\right) + 0.2494xl(x),$$

$$S_3(x) = -\left(1 - 2x^2\right) + 144\sqrt{1 - x^2}\sum_{k=1}^{\infty}(2k - 1)^{-2}(2k + 3)^{-2}c_{2k+1}(x),$$

$$S_4(x) = \frac{1}{3} + \left(1 - 2x^2\right) + xl(x), l(x) = \left(1 - x^2\right)\ln\frac{1 - x}{1 + x},$$

$$S_5(x) = r_1(x) + x\,l(x)\,r_2(x) - 0.04156\,l^2(x) + 0.3026\,S_1(x),$$

$$r_1(x) = 0.3547 - 0.8463\,x^2 + 0.3442\,x^4, r_2(x) = 0.1180 + 0.03305\,x^2.$$

The dimensionless shearing force $T = \int_{-1}^{1} \varphi(x)\mathrm{d}x$ takes the form

$$\begin{aligned} T = {} & \pi\left(\ln(2\lambda)\left(1 - a_{11}\lambda^{-2} + 0.1801a_{11}a_{20}\lambda^{-3}\right)\right) + a_{30} \\ & + 0.8106a_{20}\lambda^{-1} + \left(a_{31} + a_{11} - 0.03287a_{20}^2\right)\lambda^{-2} \\ & + \left(1.442a_{21} - 0.2702a_{11}a_{20} - 0.1807a_{31}a_{20} - 0.02450a_{20}^3\right)\lambda^{-3} + O\left(\lambda^{-4}\right)\Big)^{-1}. \end{aligned} \tag{B.24.6}$$

## References


Aizikovich SM (1978) Die-caused shear of an inhomogeneous elastic half-space of special form. Mech Solids 13(5):66–72

Aizikovich SM, Vasil'ev AS, Volkov SS (2015) The axisymmetric contact problem of the indentation of a conical punch into a half-space with a coating inhomogeneous in depth. J Appl Math Mech 79(5):500–505

Aleksandrov VM, Belokon' AV (1967) Asymptotic solution of a class of integral equations and its application to contact problems for cylindrical elastic bodies. J Appl Math Mech 31(4):718–724. https://doi.org/10.1016/0021-8928(67)90011-1

Alexandrov VM, Kovalenko EV (1986) Zadachi mekhaniki sploshnykh sred so smeshannymi granichnymi usloviyami [Problems in continuum mechanics with mixed boundary conditions]. Nauka, Moscow (in Russ.)

Altenbach H, Eremeyev VA (2009) Eigen-vibrations of plates made of functionally graded material. Comput Mater Con 9(2):153–177

Awojobi AO (1976) On the hyperbolic variation of elastic modulus in a non-homogeneous stratum. Int J Sol Struct 12:739–748

Bateman H, Erdelyi A (1954) Tables of Integral Transforms. McGraw-Hill, New York

Bocciarelli M, Bolzon G, Maier G (2008) A constitutive model of metal-ceramic functionally graded material behavior: Formulation and parameter identification. Comput Mater Sci 43(1):16–26

Brychkov YuA, Prudnikov AP (1989) Integral transformations of generalized functions. Gordon & Breach Science Publishers, CRC-Press, New York

Chen B, Chen W, Wei X (2018) Characterization of elastic parameters for functionally graded material by a meshfree method combined with the NMS approach. Inverse Probl Sci Eng 26(4):601–617

Giannakopoulos AE, Suresh S (1997) Indentation of solids with gradients in elastic properties: Part I. Point force. Int J Sol Struct 34(19):2357–2428

Giannakopoulos AE, Suresh S (1997) Indentation of solids with gradients in elastic properties: Part II. Axisymmetric indentors. Int J Sol Struct 34(19):2393–2428

Gradshteyn IS, Ryzhik IM (1965) Table of integrals, series, and products. Academic Press, New York

Gu Y, Nakamura T, Prchlik L, Sampath S, Wallace J (2003) Micro-indentation and inverse analysis to characterize elastic—plastic graded materials. Mater Sci Eng A 345(1–2):223–233

Katebi A, Selvadurai APS (2015) A frictionless contact problem for a flexible circular plate and an incompressible nonhomogeneous elastic halfspace. Int J Mech Sci 90:239–245

Kulchytsky-Zhyhailo R, Bajkowski A (2012) Analytical and numerical methods of solution of three-dimensional problem of elasticity for functionally graded coated half-space. Int J Mech Sci 54:105–112

Miyamoto Y et al (eds) (1999) Functionally graded materials: design, Processing and applications. Springer, New York

Nakamura T, Wang T, Sampath S (2000) Determination of properties of graded materials. Acta Mater 48:4293–4306

Noble B (1958) Methods based on the Wiener-Hopf technique for the solution of partial differential equations. Pergamon Press, London

Popov GJ (1959) K teorii izgiba plit na uprugom neodnorodnom poluprostranstve [On the theory of bending of plates on an elastic inhomogeneous half-space]. Izvestiya Vuzov. Stroitel'stvo i Arkhitektura 11–12:11–19 (in Russ.)

Selvadurai APS, Katebi A (2013) Mindlin's problem for an incompressible elastic half-space with an exponential variation in the linear elastic shear modulus. Int J Eng Sci 65:9–21

Selvadurai APS, Katebi A (2016) The Boussinesq–Mindlin problem for a non-homogeneous elastic halfspace. Zeitschrift für angewandte Mathematik und Physik, 67(3). https://doi.org/10.1007/s00033-016-0661-z

Shiraki Y, Usami N (eds) (2011) Silicon-Germanium (SiGe) nanostructures: production, properties and applications in electronics. Woodhead Publishing, Cambridge
Vorovich II, Aleksandrov VM, Babeshko VA (1974) Neklassicheskiye smeshannyye zadachi teorii uprugosti [Non-classical mixed problems of the theory of elasticity]. Nauka Publishers, Moscow (In Russ.)
Zelentsov VB, Lapina PA, Mitrin BI, Kudish II (2019) An antiplane deformation of a functionally graded half-space. Continuum Mech Thermodyn https://doi.org/10.1007/s00161-019-00783-1
Zelentsov VB, Lapina PA, Mitrin BI, Eremeyev VA (2020) Characterization of the functionally graded shear modulus of a half-space. Mathematics 8(4). article no. 640. https://doi.org/10.3390/math8040640
Zelentsov VB, Sadyrin EV, Sukiyazov AG, Shubchinskaya NY (2018) On a method for determination of Poisson's ratio and Young modulus of a material. MATEC Web of Conferences, 226, article no. 03027. https://doi.org/10.1051/matecconf/201822603027

MIX
Papier aus verantwortungsvollen Quellen
Paper from responsible sources
FSC® C105338

If you have any concerns about our products,
you can contact us on
**ProductSafety@springernature.com**

In case Publisher is established outside the EU,
the EU authorized representative is:
**Springer Nature Customer Service Center GmbH**
**Europaplatz 3, 69115 Heidelberg, Germany**

Printed by Libri Plureos GmbH
in Hamburg, Germany